THALES

PLATO

PYTHAGORAS

EUCLID

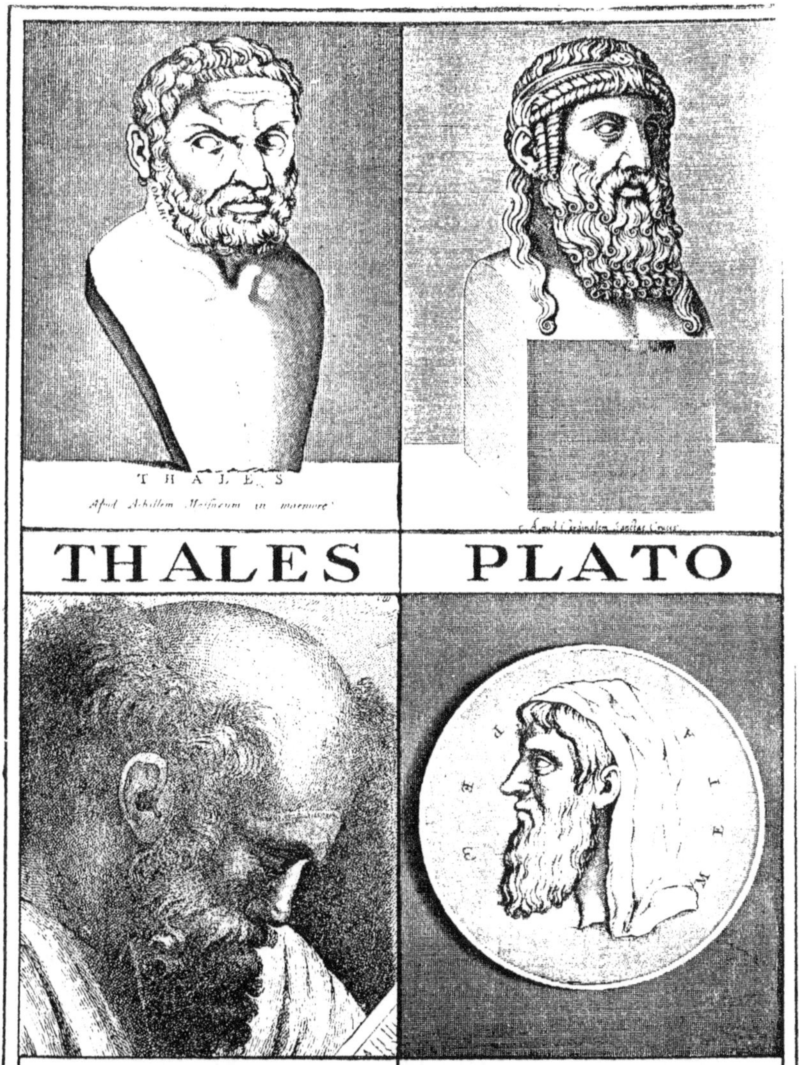

FOUR DISCOVERERS AND FOUNDERS OF GEOMETRY

(See pages 490-498)

PLANE AND SOLID GEOMETRY

BY

FLETCHER DURELL, Ph.D.

HEAD OF THE MATHEMATICAL DEPARTMENT, THE LAWRENCEVILLE SCHOOL

NEW YORK

CHARLES E. MERRILL CO.

44–60 East Twenty-third Street

1911

Durell's Mathematical Series

Plane Geometry
341 pages, 12mo, half leather . . . 75 cents

Solid Geometry
213 pages, 12mo, half leather . . . 75 cents

Plane and Solid Geometry
514 pages, 12mo, half leather . . . $1.25

Plane Trigonometry
184 pages, 8vo, cloth $1.00

Plane Trigonometry and Tables
298 pages, 8vo, cloth $1.25

Plane Trigonometry, with Surveying and Tables
In preparation . . . .

Plane and Spherical Trigonometry, with Tables
351 pages, 8vo, cloth $1.40

Plane and Spherical Trigonometry, with Surveying and Tables
In preparation . . . .

Logarithmic and Trigonometric Tables
114 pages, 8vo, cloth . . . . 75 cents

PREFACE

One of the main purposes in writing this book has been to try to present the subject of Geometry so that the pupil shall understand it not merely as a series of correct deductions, but shall realize the value and meaning of its principles as well. This aspect of the subject has been directly presented in some places, and it is hoped that it pervades and shapes the presentation in all places.

Again, teachers of Geometry generally agree that the most difficult part of their work lies in developing in pupils the power to work original exercises. The second main purpose of the book is to aid in the solution of this difficulty by arranging original exercises in groups, each of the earlier groups to be worked by a distinct method. The pupil is to be kept working at each of these groups till he masters the method involved in it. Later, groups of mixed exercises to be worked by various methods are given.

In the current exercises at the bottom of the page, only such exercises are used as can readily be solved in connection with the daily work. All difficult originals are included in the groups of exercises as indicated above.

Similarly, in the writer's opinion, many of the numerical applications of geometry call for special methods of solution, and the thorough treatment of such exercises should be taken up separately and systematically. [See pp. 304–318, etc.] In the daily extempore work only such numerical problems are included as are needed to make clear and definite the meaning and value of the geometric principles considered.

Every attempt has been made to create and cultivate the heuristic attitude on the part of the pupil. This has been done by the method of initiating the pupil into original work described above, by queries in the course of proofs, and also at the bottom of different pages, and also by occasional queries in the course of the text where definitions and discussions are presented. In the writer's opinion, the time has not yet come for the purely heuristic study of Geometry in most schools, but it is all-important to use every means to arouse in the pupil the attitude and energy of original investigation in the study of the subject.

In other respects, the aim has been to depart as little as possible from the methods most generally used at present in teaching geometry.

The Practical Applications (Groups 88-91) have been drawn from many sources, but the author wishes to express his especial indebtedness to the Committee which has collected the Real Applied Problems published from time to time in *School Science and Mathematics*, and of which Professor J. F. Millis of the Francis W. Parker School of Chicago is the chairman. Page 360 is due almost entirely to Professor William Betz of the East High School of Rochester, N. Y.

FLETCHER DURELL.

LAWRENCEVILLE, N. J., Sept. 1, 1904.

TO THE TEACHER

1. In working original exercises, one of the chief difficulties of pupils lies in their inability to construct the figure required and to make the particular enunciation from it. Many pupils, who are quite unable to do this preliminary work, after it is done can readily discover a proof or a solution. In many exercises in this book the figure is drawn and the particular enunciation made. It is left to the discretion of the teacher to determine for what other exercises it is best to do this for pupils.

2. It is frequently important to give partial aid to the pupil by eliciting the outline of a proof by questions such as the following: "On this figure (or, in these two triangles) what angles are equal, and why?" "What lines are equal, and why?" etc.

3. In many cases it is also helpful to mark in colored crayon pairs of equal lines, or of equal angles. Thus, in the figure on p. 37 lines AB and DE may be drawn with red crayon, AC and DF with blue, and the angles A and D marked by small arcs drawn with green crayon. If colored crayons are not at hand, the homologous equal parts may be denoted by like symbols placed on them, thus:

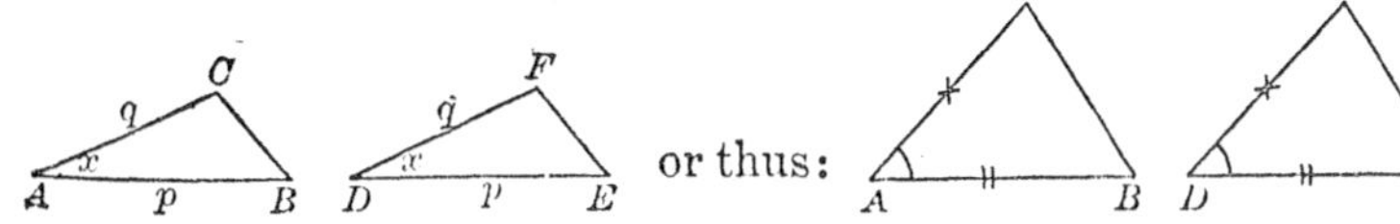

In solving theorems concerning proportional lines, it is occasionally helpful to denote the lines in a proportion (either given or to be proved) by figures denoting the order in which the lines are to be taken. Thus, if $OA : OC = OD : OB$, the relation may be indicated as on the figure.

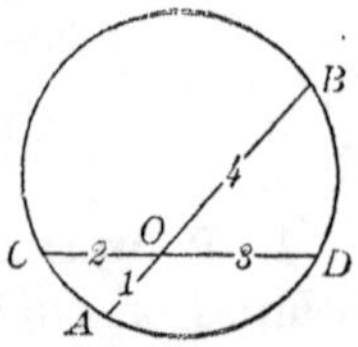

4. It is sometimes helpful to vary the symbolism of the book. Thus, in dealing with inequalities a convenient symbol for "angle" is $\measuredangle$,

$$\text{as } \measuredangle A > \measuredangle B.$$

5. Each pupil need be required to work only so many originals in each group as will give him a mastery of the particular method involved. A large number of exercises is given in order that the teacher may have many to select from and may vary the work with successive classes.

6. It is important to insist that the solutions of exercises for the first few weeks be carefully written out; later, for many pupils, oral demonstration will be sufficient and ground can be covered more rapidly by its use.

7. In leading pupils to appreciate the meaning of theorems, it is helpful at times to point out that not every theorem has for its object the demonstration of a new and unexpected truth (i. e., not all are "synthetic"), but that some theorems are analytic, it being their purpose to reduce an obvious truth to the certain few principles with which we start in Geometry. Their function is, therefore, to simplify and clarify the subject rather than to extend its content.

TABLE OF CONTENTS

SYMBOLS AND ABBREVIATIONS

Symbol	Meaning	Abbreviation	Meaning
$+$	*plus*, or *increased by.*	Adj.	*adjacent.*
$-$	*minus*, or *diminished by.*	Alt.	*alternate.*
$\times$	*multiplied by.*	Art.	*article.*
$\div$	*divided by.*	Ax.	*axiom.*
$=$	*equals; is* (or *are*) *equal to.*	Constr.	*construction.*
$\doteq$	*approaches* (*as a limit*).	Cor.	*corollary.*
$\approx$	*is* (or *are*) *equivalent to.*	Def.	*definition.*
$>$	*is* (or *are*) *greater than.*	Ex.	*exercise.*
$<$	*is* (or *are*) *less than.*	Ext.	*exterior.*
$\therefore$	*therefore.*	Fig.	*figure.*
$\perp$	*perpendicular*, *perpendicular to*, or, *is perpendicular to.*	Hyp.	*hypothesis.*
		Ident.	*identity.*
$\perp$s	*perpendiculars.*	Int.	*interior.*
$\parallel$	*parallel*, or, *is parallel to.*	Post.	*postulate.*
$\parallel$s	*parallels.*	Prop.	*proposition.*
$\angle$, $\angle$s	*angle, angles*	Rt.	*right.*
$\triangle$, $\triangle$s	*triangle, triangles.*	Sug.	*suggestion.*
▱, ▱s	*parallelogram, parallelograms.*	Sup.	*supplementary.*
$\odot$, $\odot$s	*circle, circles.*	St.	*straight.*

Q. E. D. quod erat demonstrandum; that is, which was to be proved.

Q. E. F. quod erat faciendum; that is, which was to be made.

A few other abbreviations and symbols will be introduced and their meaning indicated later on.

DEFINITIONS AND FIRST PRINCIPLES

INTRODUCTORY ILLUSTRATIONS. DEFINITIONS

1. Computation of an area. Practical experience has taught men that certain ways of dealing with objects in the world about us are more advantageous than others; thus, if it be desired to find the number of square yards in the area of a floor, we do not mark off the floor into actual square yards and count the number of square yards thus made, but pursue the much easier course of measuring two lines, the length and breadth of the floor, obtaining 7 yards, say, as the length, and 5 yards as the width, and multiplying the length by the breadth. The area is thus found to be 35 square yards.

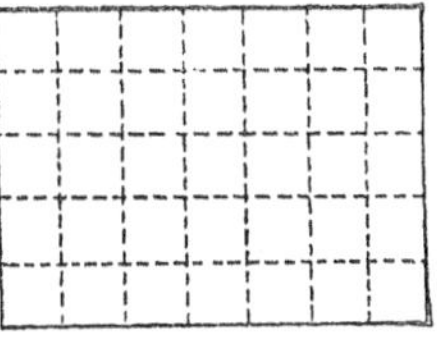

Let the pupil determine the area of some convenient floor in each of these two ways, and compare the labor of the two processes.

2. Computation of a volume. Similarly, for example, in order to determine the number of cubic feet which a box contains, instead of filling the box with blocks of wood, each of the size of a cubic foot, and counting the number of blocks, we pursue the much easier course of measuring the number of linear feet in each inside edge of the box and multiplying together the three dimensions obtained; thus, if

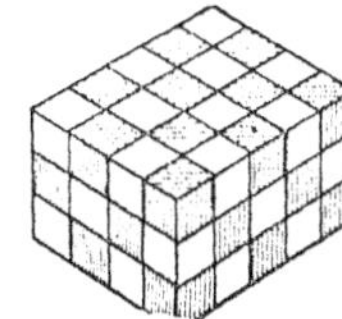

the inside dimensions are 5, 4 and 3 feet, the volume is $5 \times 4 \times 3$, or 60 cubic feet.

Similarly, the direct method of measuring the number of bushels of wheat in a bin is to fill a bushel measure with wheat from the bin, time after time, till the bin is exhausted, and count the number of times the bushel measure is used. But a much less laborious method is to measure the three dimensions of the bin in inches and divide their product by the number of cubic inches in a bushel.

Let the pupil in like manner compare the labor of finding the number of feet of lumber in a given block of wood by actually sawing the block up into lumber feet, with the labor of measuring the dimensions of the block and computing the number of lumber feet by taking the product of the linear dimensions obtained.

3. Unknown line determined from known lines. The student is also probably familiar with the fact that, by computations based on the relations of certain lines whose lengths are known, the lengths of other unknown lines may be determined without the labor of measuring these unknown lines.

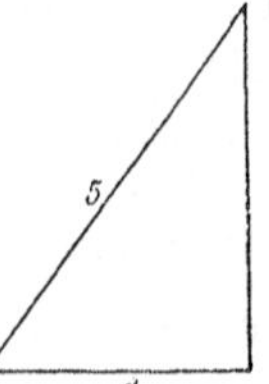

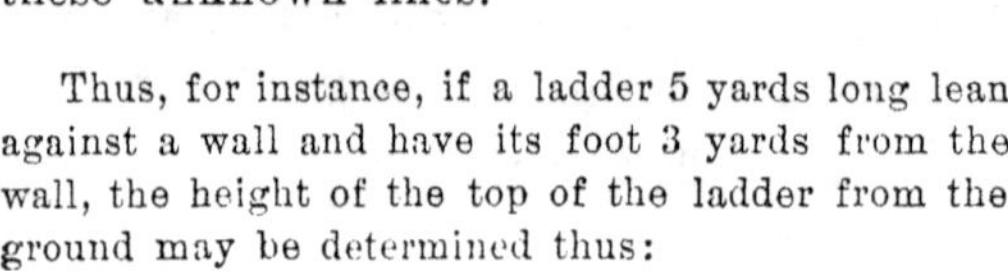

Thus, for instance, if a ladder 5 yards long lean against a wall and have its foot 3 yards from the wall, the height of the top of the ladder from the ground may be determined thus:

$$(\text{Height})^2 = 5^2 - 3^2 = 25 - 9 = 16.$$
$$\therefore \text{height} = 4 \text{ yards}.$$

4. Economies in representing surfaces, lines, etc. Other principles of advantage of an even more general character occur in dealing with geometric objects. Thus, since only one straight line can be passed through two given points, the two points may be taken as a highly economized symbol or representative of the line, by the use of which much labor is saved in dealing with lines, and new results are made attainable.

Similarly, since only one flat, or plane surface can be passed through three given points (not in the same straight line), these three points may be taken as a symbol or representative of the flat surface. This gives the advantage not only of reducing an unlimited surface to three points, but also of giving for the plane a symbol made up of three parts. By varying one of these parts and not the others, the plane may be varied in one respect and not in others; also planes having certain properties in common may be grouped together, and dealt with in the groups formed.

5. Geometry as a science. Definition. The above illustrations serve to show how advantageous it often is to deal with geometric magnitudes by certain methods rather than others.

In the study of Geometry as a science we proceed to make a systematic examination of these methods.

Geometry is the science which treats of the properties of continuous magnitudes and of space.

GEOMETRIC MAGNITUDES

6. Solids. A **physical solid** is a portion of matter, as a block of wood, an iron weight, or a piece of marble.

The portion of space occupied by a physical solid may be considered apart from the physical solid itself, hence

A **geometric solid** is the portion of space occupied by a physical solid, or definitely determined in any way. Hence, also, a geometric solid is a limited portion of space.

One advantage in using geometric solids lies in this. If we dealt with physical solids only, as blocks of wood, marble, iron, etc., we should need to determine the properties of each kind of physical solid, separately; but, by determining the properties of a geometric solid, we determine once for all the properties of every physical solid, no matter what its material, that will exactly fill the space occupied by the given geometric solid.

Hereafter in this book the term "solid" is understood to mean geometric solid, unless it be otherwise specified.

7. Other geometric magnitudes defined as boundaries.

A **surface** is the boundary of a solid.

A **line** is the boundary of a surface.

A **point** is the boundary of a line.

The solid, surface, line, and point are the fundamental geometric magnitudes.

8. Geometric magnitudes defined by their dimensions.

A **solid** has three dimensions; viz., length, breadth and thickness.

A **surface** has two dimensions, length and breadth.

A **line** has one dimension, length.

A **point** has no dimension. Hence a **point** is that which has position, but no magnitude.

9. Geometric magnitudes defined as generated by motion.

A **line** is that which is generated by the motion of a point.

A **surface** is that which is generated by the motion of a line (not moving along itself).

A **solid** is that which is generated by the motion of a surface (not moving along itself).

The three independent motions by which a solid is generated illustrate the fact that a solid has three dimensions.

10. Geometric magnitudes as intersections. The intersection of two surfaces is a *line;* of two lines is a *point.*

It is sometimes more advantageous to regard geometric magnitudes from one of the above points of view, sometimes from another.

11. A **geometric figure** is any combination of points, lines, surfaces, or solids.

12. The **form** or **shape** of a figure is determined by the relative position of its parts.

13. **Similar geometric figures** are those which have the same *shape*.

Equivalent figures are those which have the same *size*.

Equal or **congruent figures** are those which have the same *shape* and *size*, and can, therefore, be made to coincide.

14. A **point** is represented to the eye by a dot and is named by a letter affixed to the dot, as the point A, ◦A.

LINES

15. A **straight line** is a line such that, if any two points in it be fixed and the line rotated, every point in the line will retain its original position.

A straight line is also sometimes described as a line which has the same direction throughout its whole extent; or, as the shortest line connecting two points.

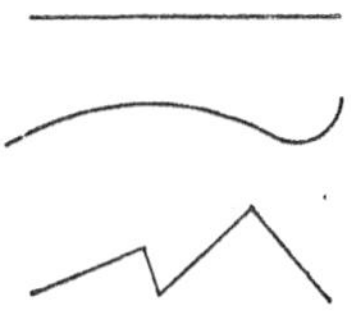

The word "line" may be used for "straight line," if no ambiguity results.

16. A **curved line** is a line no portion of which is straight. The word "curve" is often used for "curved line."

17. A **broken line** is a line made up of different straight lines.

18. A **rectilinear figure** is a figure composed only of straight lines; a **curvilinear figure** is a figure composed only of curved lines; a **mixtilinear figure** is a figure containing both straight and curved lines.

19. Kinds of straight line. A straight line may be definite or indefinite in length.

The line of definite length is sometimes termed a *segment* or *sect*.

Other kinds of straight line are defined in Arts. 26 and 41.

20. Naming a straight line. A straight line is named by naming two of its points, as the line AB (a sect); or the line CD (indefinite in length). A segment or sect may also be denoted by a single (small) letter, as the line a.

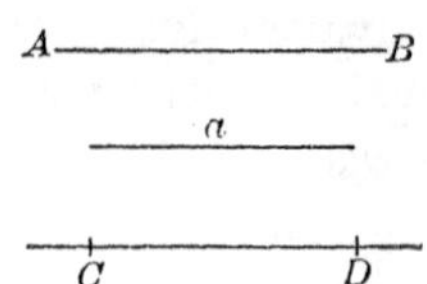

21. The **circumference** of a circle is a line every point of which is equally distant from a fixed point within, called the center.

ANGLES

22. An **angle** is the amount of opening between two straight lines which meet at a point.

The **sides** of an angle are the lines whose intersection forms the angle; the **vertex** of an angle is the point in which the sides intersect.

23. Naming angles. (1) The most precise way of naming an angle is to use three letters, one for a point on each side of the angle with the letter at the vertex between these two, as the angle ABC.

Since the size of an angle is independent of the length of its sides, the points named on its sides may be taken at any place on its sides. Thus, the angles AOD, BOD, BOE, AOF are all the same angle.

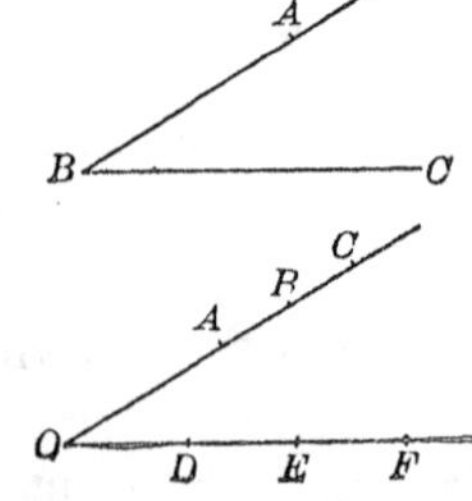

(2) In case there is but one angle at a given vertex, the letter at the vertex alone is sufficient to denote the angle, as the angle O in the last figure.

(3) Sometimes a letter or figure placed inside the angle and near the vertex is a convenient symbol, as the angle a.

24. A **straight angle** is an angle whose sides lie in the same straight line, but which extend in opposite directions from the vertex, as the angle BOA.

25. A **right angle** is one of two equal angles made by one straight line meeting another straight line. Thus, if the line PQ meets line AB so as to make angle PQA equal to angle PQB, each of these angles is a right angle. A right angle is also half of a straight angle.

26. A **perpendicular** is a line that makes a right angle with a given line; thus PQ in the last figure is perpendicular to BA. The *foot* of a perpendicular is the point in which the perpendicular meets the line to which it is drawn, thus Q is the foot of the perpendicular PQ.

27. An **acute angle** is an angle less than a right angle, as the angle AOC.

28. An **obtuse angle** is an angle greater than a right angle but less than a straight angle, as angle AOD.

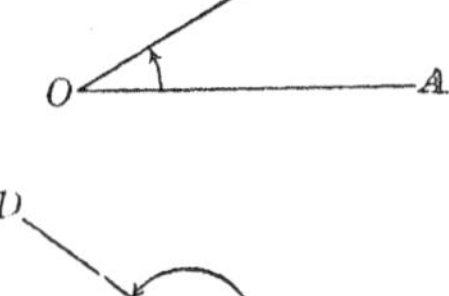

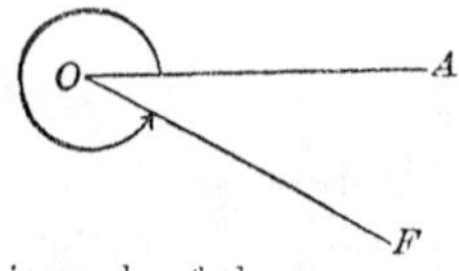

29. A **reflex angle** is an angle greater than a straight angle, but less than two straight angles, as angle AOF.

In this book, angles larger than a straight angle are not considered unless special mention is made of them.

30. An **oblique angle** is an angle that is neither a right nor a straight angle. Hence, oblique angle is a general term for acute, obtuse, and reflex angles.

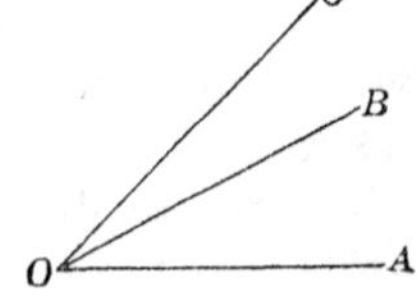

31. Adjacent angles are angles which have a common vertex and a common side between them, as angles AOB and BOC.

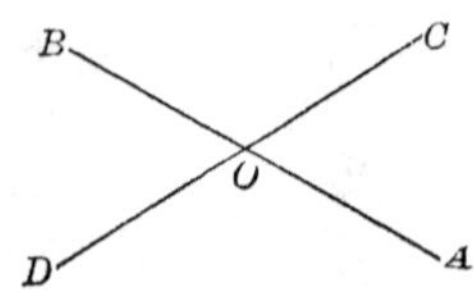

32. Vertical angles are angles which have a common vertex and the sides of one angle the prolongations of the sides of the other angle, as the angles AOC and BOD.

33. Complementary angles are two angles which together equal a right angle, as the angles AOP and POQ.

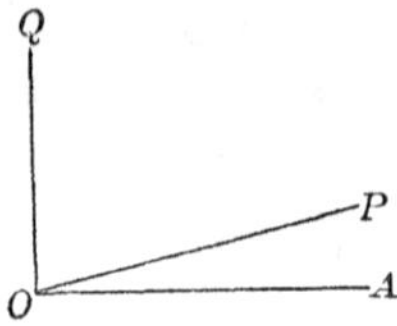

Hence, the **complement of an angle** is the difference between that angle and one right angle.

34. Supplementary angles are two angles which together equal two right angles (or a straight angle), as the angles AOP and POR.

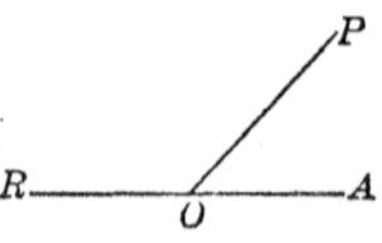

Hence, the **supplement of an angle** is the difference between that angle and two right angles.

35. Angles as formed by a rotating straight line. If the line OB start in the position OA and rotate to the position OB, it is said to generate the $\angle AOB$.

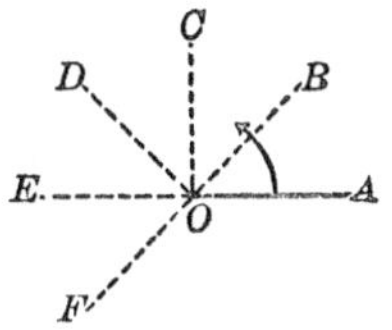

The size of an angle may, therefore, be considered as the amount of rotation of a line about a point, from the original position of the line.

If the rotation is continued far enough, a right angle ($\angle AOC$) is formed; afterward an obtuse angle ($\angle AOD$); then a straight angle ($\angle AOE$), and a reflex angle ($\angle AOF$), etc.

An advantage of this method of forming or conceiving angles is that by continuing the rotation of the moving line an angle of indefinite size may be formed.

36. Units of angle. A **right angle** is a unit of angle useful for many purposes. Sometimes a smaller unit of angle is needed.

A **degree** is one-ninetieth of a right angle; a **minute** is one-sixtieth of a degree; a **second** is one-sixtieth of a minute.

Besides these, other units of angle are used for certain purposes.

SURFACES. DIVISIONS OF GEOMETRY. PARALLEL LINES

37. A **plane** is a surface such that, if any two points in the surface be joined by a straight line, the straight line lies wholly in the surface.

Hence, a **plane figure** is a figure such that all its points lie in the same plane.

38. A **curved surface** is a surface no part of which is plane.

39. Plane Geometry is that branch of Geometry which treats of plane figures.

40. Solid Geometry is that branch of Geometry which treats of figures all points of which are not in the same plane.

41. Parallel lines are straight lines in the same plane which do not meet, however far they be produced.

EXERCISES. GROUP 1

Ex. 1. Draw a straight line. A curved line. A broken line.

Ex. 2. Which of the capital letters of the alphabet are straight, which curved, which broken, which curved and straight lines combined?

Ex. 3. Draw an acute angle. An obtuse angle. A reflex angle.

Ex. 4. Draw two adjacent angles. Two vertical angles.

Ex. 5. What is the complement of an angle of 43°? What is its supplement?

Ex. 6. What is the complement of 57° 19′? of 62° 23′ 43″? What is the supplement of each of these?

Ex. 7. At two o'clock, what is the angle made by the hour and minute hands of a clock? at three o'clock? at five o'clock?

Ex. 8. At 1:30 o'clock, what angle do the hands of a clock make? at 2:15? at 8:45?

Ex. 9. What kind of an angle is the supplement of an obtuse angle? of an acute angle? of a right angle?

Ex. 10. What kind of a surface is the floor of a room? the surface of a baseball? the surface of an egg? the surface of a hemisphere?

Ex. 11. If the surface $ABCD$ move to the right, what solid is generated by it? What surfaces are generated by its bounding lines? What lines by the vertices of its angles?

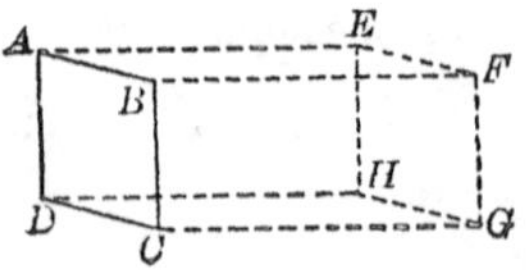

Ex. 12. Draw two supplementary adjacent angles. Also two supplementary angles that are not adjacent. Also two adjacent angles that are not supplementary.

Ex. 13. The sum of a right angle and an acute angle is what kind of an angle? Their difference is what kind of an angle?

Ex. 14. The sum of an obtuse angle and a right angle is what kind of an angle? Their difference is what kind?

Ex. 15. If three straight lines meet (but do not intersect) at a point in a plane, how many angles have this point as their common vertex? Draw a figure illustrating this and name the angles.

Ex. 16. How many angles are formed if four lines meet (but do not intersect) at a point in a plane?

Ex. 17. How many degrees are in an angle which equals twice its complement?

Ex. 18. How many degrees in an angle which equals one-third its supplement?

Ex. 19. What kind of an angle is greater than its supplement? What kind is less?

PRIMARY RELATIONS OF GEOMETRIC MAGNITUDES

42. Certain **primary relations of geometric objects** have already been given in the definitions used for geometric objects. We now proceed to investigate the relations of geometric magnitudes more generally and systematically.

43. An **axiom** is a truth accepted as requiring no demonstration.

44. **Two kinds of axioms** are used in geometry:

1. **General axioms,** or axioms which apply to other kinds of quantity as well as to geometric magnitudes; for instance, to numbers, forces, masses, etc.

2. **Geometric axioms,** or axioms which apply to geometric magnitudes alone.

45. The **general axioms** may be stated as follows:

GENERAL AXIOMS

1. *Things which are equal to the same thing, or to equal things, are equal to each other.*

2. *If equals be added to equals, the sums are equal.*

3. *If equals be subtracted from equals, the remainders are equal.*

4. *Doubles of equals are equal; or, in general, if equals be multiplied by equals the products are equal.*

5. *Halves of equals are equal; or, in general, if equals be divided by equals the quotients are equal.*

6. *The whole is equal to the sum of its parts.*

7. *The whole is greater than any of its parts.*

8. *A quantity may be substituted for its equal in any process.*

9. *If equals be added to, or subtracted from, unequals, the results are unequal in the same order; if unequals be added to unequals in the same order, the results are unequal in that order.*

10. *Doubles, or halves, of unequals are unequal in the same order.*

11. *If unequals be subtracted from equals, the remainders are unequal in the reverse order.*

12. *If, of three quantities, the first is greater than the second, and the second is greater than the third, then the first is greater than the third.*

46. The value of the general axioms. The axioms given above seem so obvious that the student at first is not likely to realize their value. This value may be illustrated as follows:

If the distance from Washington to Philadelphia be known, and also the distance from Philadelphia to New York, the distance from Washington to New York may be obtained by adding together the two distances named; for, by axiom 6, the whole is equal to the sum of its parts. Thus the labor of actually measuring the distance from Washington to New York is saved.

Again, if the height of a schoolboy of a given age in Paris be measured, and the height of a like schoolboy in New York be measured, and the result of the measurement in each case is the same, we know that the boys are of the same height, without the labor and cost of bringing the boys together and comparing their heights directly; for, by axiom 1, things which are equal to the same thing are equal to each other.

Thus the general axioms are to be looked at not merely as fundamental equivalences, but also as fundamental economies. For many purposes the latter point of view is more important than the former.

47. The **geometric axioms** may be stated as follows:

GEOMETRIC AXIOMS

1. *Through two given points only one straight line can be passed.*

2. *A geometric figure may be freely moved in space without any change in form or size.*

This axiom is equivalent to regarding space as *uniform*, or *homaloidal;* that is, as having the same properties in all its parts. It has already been assumed in some of the definitions given. See Arts. 15 and 35.

3. *Through a given point one straight line and only one can be drawn parallel to another given straight line.*

By Art. 13, *geometric figures which coincide are equal.*

48. Utility or uses of the geometric axioms. By means of the first geometric axiom we are able to shrink or condense any straight line into two points. Later this advantage gives rise to many other advantages. By the second geometric axiom, the knowledge which we have of one geometric object may be transferred to another like object, however widely separated in space. The utility of the third geometric axiom can be made more evident when we come to use it in proving new geometric truths.

49. A **postulate** in geometry is a construction of a geometric figure admitted as possible.

50. The **postulates of geometry** may be stated as follows:

1. *Through any two points a straight line may be drawn.*

2. *A straight line may be extended indefinitely, or it may be limited at any point.*

3. *A circumference may be described about any given point as center, and with any given radius.*

These postulates limit the pupil to the use of the straight-edged ruler and the compasses in constructing figures in geometry. One of the objects of the study of geometry is to discover what geometric figures can be constructed by a combination of the elementary constructions allowed in the postulates; that is, by the use of the two simplest drawing instruments.

51. Logical postulates. Besides the postulates which are used in the actual construction of figures, there are certain other postulates which are used only in the processes of reasoning. Thus, for purposes of reasoning, a given angle may be regarded as divided into any convenient number of equal parts. Whether it is possible actually thus to divide this angle on paper by use of the ruler and compasses, is another question.

DEMONSTRATION OF GEOMETRIC RELATIONS

52. A **geometric proof,** or *demonstration*, is a course of reasoning by which a relation between geometric objects is established.

53. A **geometric theorem** is a statement of a truth concerning geometric objects which requires demonstration.

Ex. The sum of the angles of a triangle equals two right angles.

54. A **geometric problem** is a statement of the construction of a geometric figure which is required to be made.

Ex. On a given line to construct a triangle containing three equal angles.

55. A **proposition** is a general term for either a theorem or a problem.

Thus, propositions are subdivided into two classes: 1, Theorems; 2, Problems.

56. **Immediate inference** is of two kinds:

1. Changing the point of view in a given statement. Thus the statement, "two straight lines drawn through two given points must coincide," may be changed to "two straight lines cannot inclose a space."

2. Reasoning which involves but a single step.

Ex. "All straight angles are equal;"
∴ "All right angles are equal." (Ax. 5.)

57. A **corollary** is a truth obtained by immediate inference from another truth just stated or proved.

58. A **scholium** is a remark made upon some particular feature of a proposition, or upon two or more propositions which are compared.

59. Hypothesis and conclusion. A proposition consists of two parts:

1. The **hypothesis**, or that which is known or granted.

2. The **conclusion**, or that which is to be proved or constructed.

Thus, in the proposition, "if two straight lines are perpendicular to the same line, they are parallel," the hypothesis is, that two given lines are perpendicular to another given line; the conclusion is, that the two given lines are parallel.

60. The **converse** of a proposition is another proposition formed by interchanging the hypothesis and the conclusion of the original proposition.

Thus, *theorem*, "every point in the perpendicular bisector of a line is equidistant from the extremities of the line;"

Converse, "every point equidistant from the extremities of a line lies in the perpendicular bisector of the line."

Or, in general, *theorem*, "if A is B, then X is Y;"
converse, "if X is Y, then A is B;"

Frequently a converse is formed by interchanging part only of an hypothesis with part or all of the conclusion, or vice versa.

Thus, *theorem*, "if A is B and C is D, then X is Y;"
converse, "if A is B and X is Y, then C is D."

The converse of a theorem is not necessarily true. Thus, it is true that all right angles are equal, but it is not true that all equal angles are right angles.

61. The **opposite** of a theorem is a theorem formed by making both the hypothesis and the conclusion of the original theorem negative.

Ex. *theorem*, "if A is B, then X is Y;"
opposite, "if A is not B, then X is not Y."

If a direct theorem and its opposite are both true, the converse is known to be true without proof.

Also, if a theorem and its converse are both true, the opposite is known to be true without proof.

Thus certain economies arise in the demonstration of theorems.

62. Methods of geometric proof. Several principal methods of proving theorems are used in geometry.

1. **Direct demonstration.**

2. **Proof by superposition,** in which two figures are proved equal by making one of them coincide with the other.

3. **Indirect demonstration,** which consists essentially in showing that a given statement is true by showing that its negative cannot be true.

Other special methods of proof will be pointed out as they occur in the course of the work.

63. Form of a proof. The statement of a theorem and its proof consist of certain distinct parts which it is important to keep clearly in mind. These parts are:

1. The **general enunciation,** which is the statement of the theorem in general terms.

2. The **particular enunciation,** or statement of the theorem as applied to a particular figure used to aid the mind in carrying forward the proof.

3. The **construction** of supplementary parts of the figure (not necessary in all proofs).

4. The **proof.** This must include a reason for every statement.

5. The **conclusion.**

The letters Q. E. D., standing for "quod erat demonstrandum," and meaning "which was to be proved," are usually annexed at the end of a completed demonstration.

EXERCISES. GROUP 2

Ex. 1. In the figure measure AB; then measure BC. Now find AC without measuring it. What axiom have you used?

A B C

Ex. 2. If $\angle AOB = 60°$, $\angle BOC = 80°$ and $\angle COD = 130°$; find without measuring them ∡ AOC and BOD (reflex). What axiom have you used?

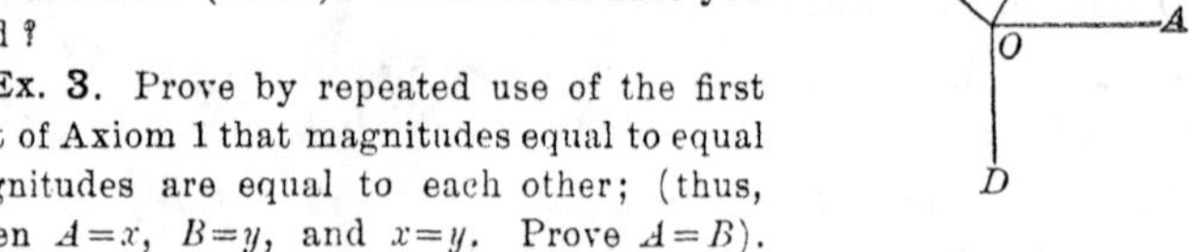

Ex. 3. Prove by repeated use of the first part of Axiom 1 that magnitudes equal to equal magnitudes are equal to each other; (thus, given $A = x$, $B = y$, and $x = y$. Prove $A = B$).

Ex. 4. Give a numerical illustration of Axiom 11.

Ex. 5. Show by the axioms that a part is equal to a whole diminished by the remaining part.

Ex. 6. Show that Axiom 1 is a special case of Axiom 8.

Ex. 7. If $a = x + y$ and $x = y$, show by use of the axioms that $x = \frac{1}{2}a$.

Ex. 8. Draw a line and produce it so that the produced part shall equal another given line.

Ex. 9. By use of the compasses, mark off on a given line a part twice as long as another given line.

Ex. 10. On a given line mark off, by fewest uses of the compasses, a part four times as long as another given line.

Ex. 11. Draw three straight lines and denote them by l, m, and n. Then draw a line $l + m - n$, and also a line $l - 2m + 3n$.

PROPERTIES OF LINES INFERRED IMMEDIATELY

64. *If two straight lines have two points in common, the lines coincide throughout their whole extent* (Art. 47, Geom. Ax. 1).

Hence, *two straight lines can intersect in but one point.*

65. *If two straight lines coincide in part, they coincide throughout.*

66. *Only one straight line can be drawn connecting two given points.*

67. *Two straight lines cannot enclose a surface.*

68. *A given straight line (sect) can be divided into two equal parts at but one point.*

For (by Ax. 5) halves of equals (or of the same thing) are equal.

PROPERTIES OF ANGLES INFERRED IMMEDIATELY

69. *All straight angles are equal.*

70. *A straight angle can be divided into two equal angles by but one line at a given point in the given straight line.*

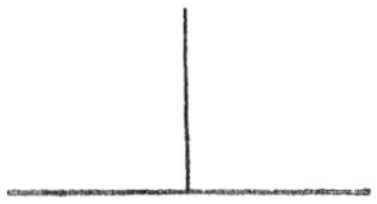

For (Ax. 5) halves of the same magnitude are equal.

71. Hence, *at a given point in a straight line but one perpendicular can be erected to the line.*

72. *All right angles are equal.*

For all straight angles are equal (Art. 69) and halves of equals are equal (Ax. 5).

73. *The sum of the two adjacent angles formed by one straight line meeting another straight line equals two right angles.*

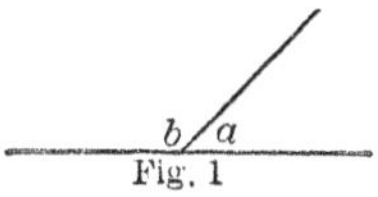

b a
Fig. 1

For the angles formed are supplementary adjacent angles (Arts. 31, 34).

74. *If two adjacent angles are together equal to a straight angle (or two right angles), their exterior sides form one and the same straight line.*

For their exterior sides form a straight angle, and hence must lie in a straight line (Art. 24).

75. *The complements of two equal angles are equal* (Art. 72 and Ax. 3); *the supplements of two equal angles are equal* (Art. 69, Ax. 3).

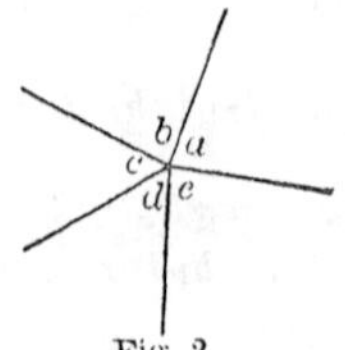

Fig. 2

76. *The sum of all the angles about a point equals four right angles.*

Thus, $\angle a + \angle b + \angle c + \angle d + \angle e =$ 4 rt. $\measuredangle$.

77. *The sum of all the angles about a point on the same side of a straight line passing through the point equals two right angles.*

Thus, $\angle p + \angle q + \angle r = 2$ rt. $\measuredangle$.

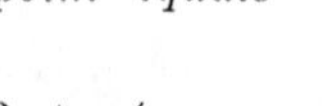

Fig. 3

EXERCISES. GROUP 3

Ex. 1. How many different straight lines are determined by three points not in the same straight line?

Ex. 2. How many straight lines are determined by four points in a plane, no three of them being in the same straight line?

Ex. 3. If, in Fig. 2 above, $\measuredangle\ a, b, c, d = 40°, 50°, 60°, 70°$ respectively, find $\angle e$.

Ex. 4. If, in Fig. 3 above, the lines forming the angle q are perpendicular to each other and $\angle p = 47°$, find the other angles of the figure.

Ex. 5. Measure $\angle a$ of Fig. 1 on preceding page. Find $\angle b$ without measuring it. Now measure $\angle b$ and compare the two results.

Ex. 6. Given $QB \perp AB$, $PB \perp BC$, and $\angle ABC = 130°$; find the other angles of the figure.

Ex. 7. Arrange five points in a plane so that the fewest number of straight lines may pass through them, no line to pass through more than three points.

PLANE GEOMETRY

BOOK I

RECTILINEAR FIGURES

PROPOSITION I. THEOREM

78. *If one straight line intersects another straight line, the opposite or vertical angles are equal.*

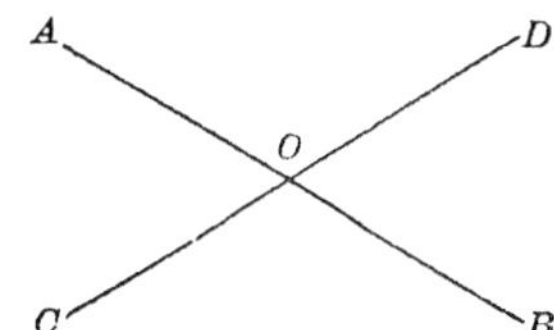

Given the straight lines AB and CD intersecting at the point O.

To prove $\angle AOC = \angle DOB$ and $\angle AOD = \angle COB$.

Proof. $\angle AOC + \angle AOD = 2$ rt. $\angle$s, Art. 73.

(*the sum of two adjacent angles formed by one straight line meeting another straight line equals two right angles*).

Also $\angle BOD + \angle AOD = 2$ rt. $\angle$s,

(*same reason*).

$\therefore \angle AOC + \angle AOD = \angle BOD + \angle AOD$, Ax. 1.

(*things equal to the same thing are equal to each other*).

Subtracting $\angle AOD$ from the two equals,

$\angle AOC = \angle BOD$, Ax. 3.

(*if equals be subtracted from equals, the remainders are equal*).

In like manner it may be proved that

$\angle AOD = \angle COB$. Q. E. D.

Ex. If in the above figure $\angle DOB = 70°$, find the other angles without measuring them.

PROPOSITION II. THEOREM

79. *If, from a point in a perpendicular to a given line, two oblique lines be drawn cutting off on the given line equal segments from the foot of the perpendicular, the oblique lines are equal and make equal angles with the perpendicular.*

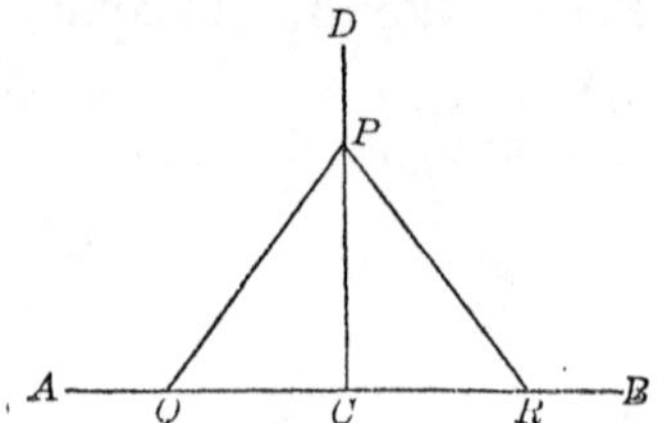

Given a line AB with $CD \perp$ to it at the point C, and PR and PQ drawn from any point as P in CD, cutting off $CR = QC$ on AB.

To prove $PR = PQ$ and $\angle CPR = \angle CPQ$.

Proof. Fold over the figure DCB about DC as an axis till it comes in the plane DCA. Geom. Ax. 2.

Then $\angle DCB = \angle DCA$ (*all right* ∡ *are* =). Art. 72.

∴ line CB will take the direction of CA.

But $CR = CQ$. Hyp.

∴ point R will fall on point Q.

Hence line PR will coincide with line PQ, Art. 66.
(*only one straight line can be drawn connecting two given points*).

And $\angle CPR$ will coincide with $\angle CPQ$.

∴ $PR = PQ$, and $\angle CPR = \angle CPQ$, Art. 47.
(*geometric figures which coincide are equal*).

Q. E. D.

Ex. 1. Point out the hypothesis and the conclusion in the general enunciation of Prop. I. Also point them out in the particular enunciation. Do the same for Prop. II.

Ex. 2. If three straight lines intersect at a point, how many of the angles formed is it necessary to measure, in order to determine all the angles?

PROPOSITION III. THEOREM

80. *From a given point without a straight line but one perpendicular can be drawn to the line.*

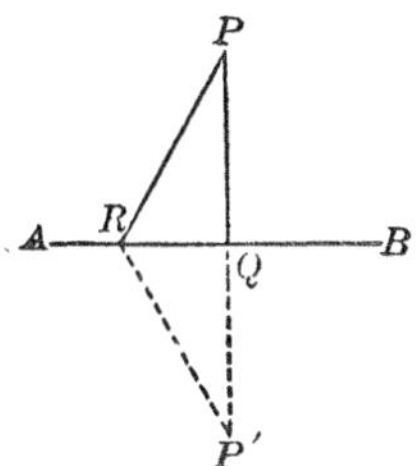

Given the straight line AB, P any point without AB, $PQ \perp AB$, and PR any other line drawn from P to AB.

To prove that PR is not $\perp$ AB.

Proof. Produce PQ to P' making $QP' = PQ$. Draw RP'.

Then $RQ \perp PP'$. Hyp.

$P'Q = PQ$. Constr.

$\therefore$ $RP = RP'$ and $\angle PRQ = \angle P'RQ$, Art. 79.

(*if, from a point in a $\perp$ to a given line, two oblique lines be drawn cutting off on the given line equal segments from the foot of the $\perp$, the oblique lines are equal and make equal $\measuredangle$ with the $\perp$*).

But PRP' is not a straight line, Art. 66.

(*only one straight line can be drawn connecting two given points*).

$\therefore$ PRP' is not a straight $\angle$.

$\therefore$ $\angle PRQ$, the half of $\angle PRP'$, is not a right $\angle$. Ax. 10.

$\therefore$ PR is not $\perp$ AB.

$\therefore$ only one perpendicular can be drawn from P to AB.

Q. E. D.

Ex. Three straight lines intersect at a point. Two of the adjacent angles formed at the point are 30° and 40°. Find all the other angles at the point.

TRIANGLES

81. A **triangle** is a portion of a plane bounded by three straight lines, as the triangle ABC.

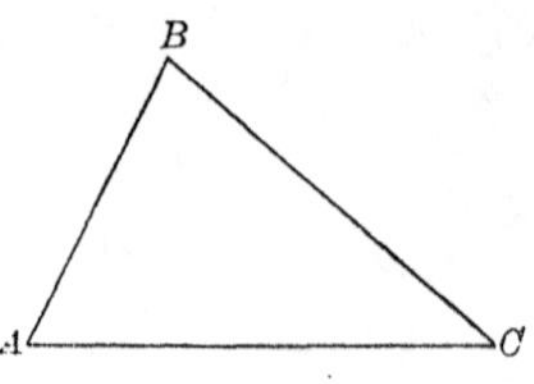

82. The **sides** of a triangle are the lines which bound it; the **perimeter** of a triangle is the sum of the sides; the **angles** of a triangle are the angles formed by the sides, as the angles A, B and C; the **vertices** of a triangle are the vertices of the angles of the triangle.

83. An **exterior angle** of a triangle is an angle formed by one side and by another side produced, as the angle BCD. With reference to the angle BCD, the angles A and B are termed the **opposite interior angles.**

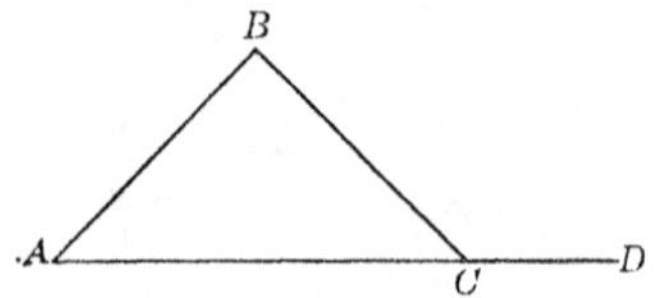

84. Classification of triangles according to relative length of the sides. A **scalene triangle** is a triangle in which no two sides are equal. An **isosceles triangle** is one in which two sides are equal. An **equilateral triangle** is one in which **all three sides** are equal.

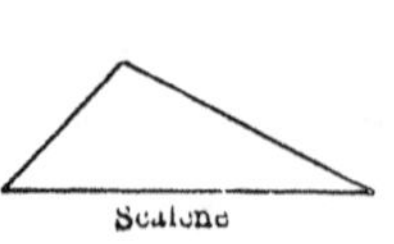
Scalene

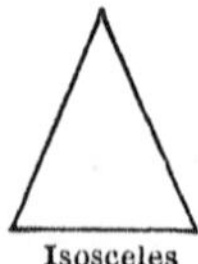
Isosceles

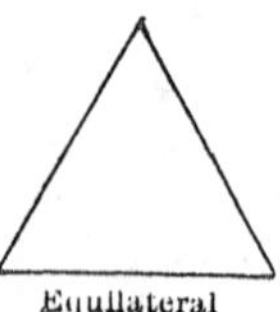
Equilateral

85. Classification of triangles with reference to character of their angles. A **right triangle** is a triangle *one* of whose angles is a right angle. An **obtuse triangle** is a triangle *one* of whose angles is an obtuse angle. An **acute triangle** is a triangle *all* of whose angles are acute angles. An **equiangular triangle** is one in which all the angles are equal.

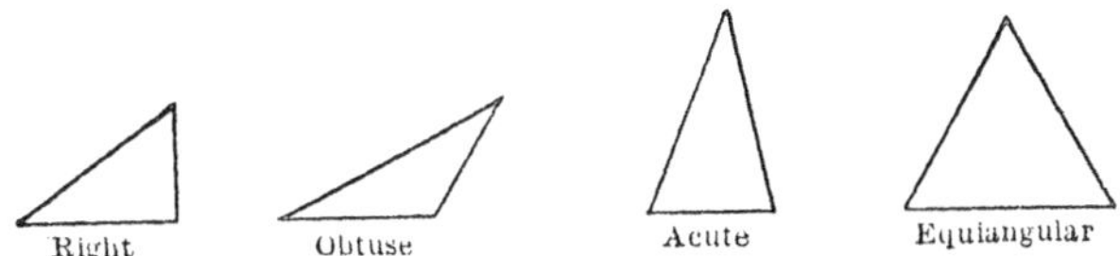

86. The **base** of a triangle is the side upon which the triangle is supposed to stand, as AB. The angle opposite the base is called the **vertex angle,** as angle ACB; the **vertex** of a triangle is the vertex of the vertex angle of the triangle.

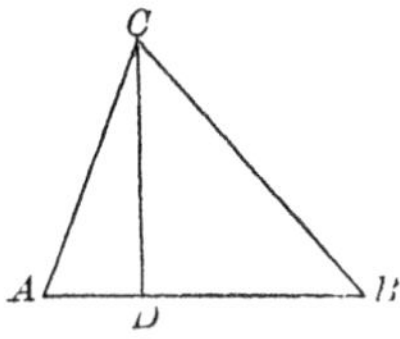

The **altitude of a triangle** is the perpendicular from the vertex to the base or base extended, as CD.

87. In an **isosceles** triangle, the **legs** are the equal sides, and the **base** is the remaining side.

88. In a **right** triangle, the **hypotenuse** is the side opposite the right angle, and the **legs** are the sides adjacent to the right angle.

89. Altitudes, bisectors, medians. In any triangle, any side may be taken as the base; hence the **altitudes** of a triangle are the three perpendiculars drawn one from each vertex to the side opposite.

A **bisector** of an angle of a triangle is a line which divides this angle into two equal parts. This bisector is usually produced to meet the side opposite the given angle.

A **median** of a triangle is a line drawn from a vertex of the triangle to the middle point of the opposite side. How many medians has a triangle?

90. Two mutually equiangular triangles are triangles having their corresponding angles equal.

91. Homologous angles of two mutually equiangular triangles are corresponding angles in those triangles.

Homologous sides of two mutually equiangular triangles are sides opposite homologous angles in those triangles.

We shall now proceed to determine first, the properties of a single triangle, as far as possible, then those of two triangles.

92. Property of a triangle immediately inferred. *The sum of any two sides of a triangle is greater than the third side.* For a straight line is the shortest line between two points (Art. 15.)

Ex. 1. Point out the hypothesis and conclusion in the general enunciation of Prop. III; also point them out in the particular enunciation.

(As each of the next fifteen Props. is studied, let the pupil do the same for it.)

Ex. 2. Find the angle whose complement is 18°; whose supplement is 76°

Ex. 3. If the complement of an angle is known, what is the shortest way of finding the supplement of the angle? If the supplement is known, what is the shortest way of finding the complement?

Ex. 4. In 25 minutes, how many degrees does the minute-hand of a clock travel? How many does the hour-hand?

Ex. 5. Draw three straight lines so that they shall intersect in three points; in two points; in one point.

Proposition IV. Theorem

93. *Any side of a triangle is greater than the difference between the other two sides.*

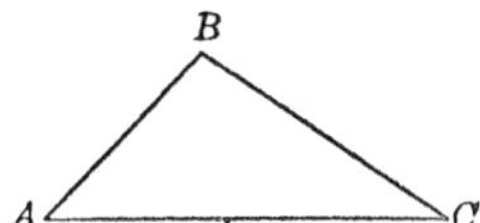

Given AB any side of the $\triangle ABC$, and $AC > BC$.

To prove $AB > AC - BC$.

Proof. $AB + BC > AC$, Art. 92.

(*the sum of any two sides of a triangle is greater than the third side*).

Subtracting BC from each member of the inequality,

$AB > AC - BC$, Ax. 9.

(*if equals be subtracted from unequals, the remainders are unequal in the same order*). Q. E. D.

94. Cor. *The perpendicular is the shortest line that can be drawn from a given point to a given line.*

For, in the Fig. page 31,

$PP' < PR + RP'$, Art 92.

Or, $2\,PQ < 2\,PR$. Ax. 8.

$\therefore PQ < PR$. Ax. 10.

Hence, Def. The **distance** from a point to a line is the perpendicular drawn from the point to the line.

Ex. 1. If one side of an equilateral triangle is 4 inches, what is its perimeter?

Ex. 2. Is it possible to form a triangle whose sides are 6, 9 and 17 inches? Try to do this with the compasses and ruler.

Ex. 3. Is it possible to form a triangle in which one side is 10 inches and the difference of the other two sides is 12 inches?

Ex. 4. On a given line as base, by exact use of ruler and compasses, construct an equilateral triangle.

Proposition V. Theorem

95. *If, from a point within a triangle, two lines are drawn to the extremities of one side of the triangle, the sum of the other two sides of the triangle is greater than the sum of the two lines so drawn.*

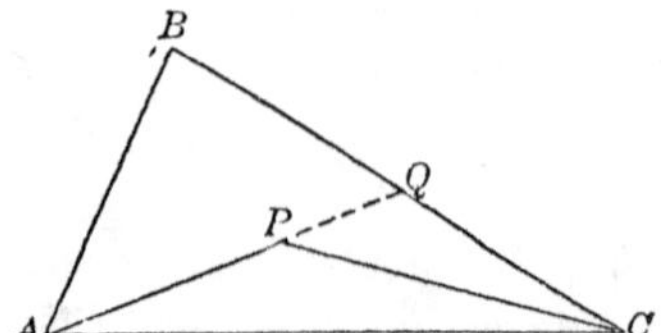

Given P any point within the triangle ABC, and PA and PC lines drawn from P to the extremities of the side AC.

To prove $AB + BC > AP + PC$.

Proof. Produce the line AP to meet BC at Q.

Then $AB + BQ > AP + PQ$, Art. 15.

(*a straight line is the shortest line connecting two points*).

Also $PQ + QC > PC$,

(*same reason*).

Adding these inequalities,

$$AB + BQ + PQ + QC > AP + PQ + PC. \quad \text{Ax. 9.}$$

Substituting BC for its equal $BQ + QC$,

$$AB + BC + PQ > AP + PQ + PC. \quad \text{Ax. 8.}$$

Subtracting PQ from each side,

$$AB + BC > AP + PC. \quad \text{Ax. 9.}$$

Q. E. D.

Ex. On a given line as base, by exact use of ruler and compasses, construct an isosceles triangle each of whose legs is double the base.

PROPOSITION VI. THEOREM

96. *Two triangles are equal if two sides and the included angle of one are equal, respectively, to two sides and the included angle of the other.*

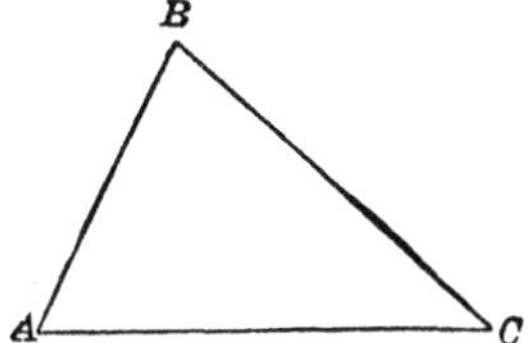

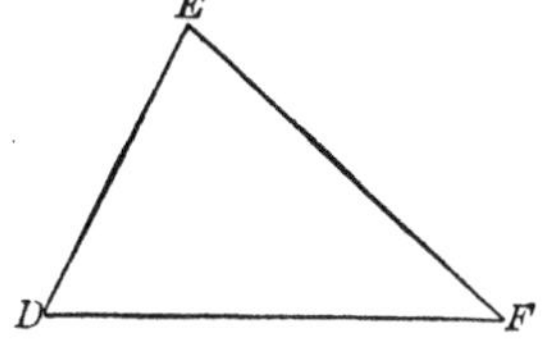

Given the triangles ABC and DEF in which $AB = DE$, $AC = DF$, and $\angle A = \angle D$.

To prove $\triangle ABC = \triangle DEF$.

Proof. Place the $\triangle ABC$ upon the $\triangle DEF$ so that the line AC concides with its equal DF. Geom. Ax. 2.

Then the line AB will take the direction of DE,
(*for* $\angle A = \angle D$ *by hyp.*).

Also the point B will fall on E,
(*for line* $AB =$ *line* DE *by hyp.*).

Hence the line BC will coincide with the line EF, Art. 66.
(*only one straight line can be drawn connecting two points*).

$\therefore$ $\triangle$s ABC and DEF coincide.

$\therefore$ $\triangle ABC = \triangle DEF$, Art. 47.
(*geometric figures which coincide are equal*).

Q. E. D.

Ex. 1. What kind of proof is used in Prop. VI? (See Art. 62).

Ex. 2. If $\angle$s A, B and $C = 60°, 70°, 50°$, $AB = 16$, $AC = 19$, $BC = 18$; also $\angle D = 60°$, $DE = 16$, $DF = 19$: find $\angle$s E and F and side EF without measuring them.

Proposition VII. Theorem

97. *Two triangles are equal if two angles and the included side of one are equal, respectively, to two angles and the included side of the other.*

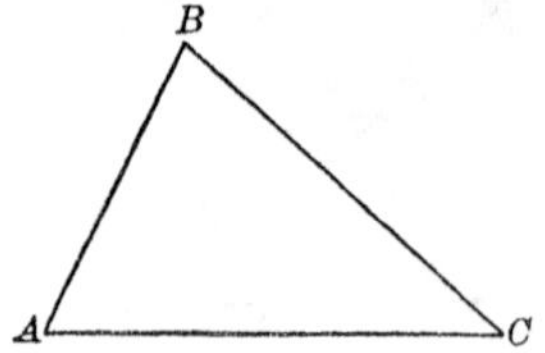

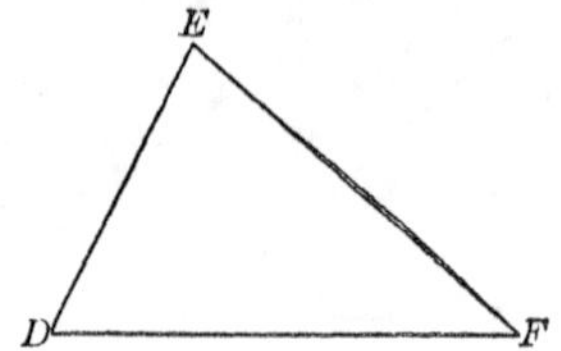

Given the $\triangle s$ ABC and DEF in which $\angle A = \angle D$, $\angle C = \angle F$, and $AC = DF$.

To prove $\triangle ABC = \triangle DEF$.

Proof. Place the $\triangle ABC$ upon the $\triangle DEF$ so that AC shall coincide with its equal DF. Geom. Ax. 2.

Then AB will take the direction of DE,
(*for* $\angle A = \angle D$ *by hyp.*),

and the point B will fall somewhere on the line DE or DE produced.

Also the line CB will take the direction of FE,
(*for* $\angle C = \angle F$ *by hyp.*),

and the point B will fall on FE or FE produced.

$\therefore$ point B falls on point E, Art. 64.
(*two straight lines can intersect in but one point*).

$\therefore$ $\triangle s$ ABC and DEF coincide.

$\therefore$ $\triangle ABC = \triangle DEF$, Art. 47.
(*geometric figures which coincide are equal*). Q. E. D.

Ex. 1. What kind of proof is used in Prop. VII. ?

Ex. 2. If $\angle s$ A, B, $C = 65°$, $55°$, $60°$, $AB = 24$, $AC = 18$, $BC = 27$; also $\angle s$ D, $F = 65°$, $60°$, and $DF = 18$; find DE, EF, and $\angle E$.

Ex. 3. Construct by exact use of ruler and compasses a scalene triangle whose sides are 2, 3 and 4 times a given line.

Proposition VIII. Theorem

98. *Two right triangles are equal if the hypotenuse and an acute angle of one are equal to the hypotenuse and an acute angle of the other.*

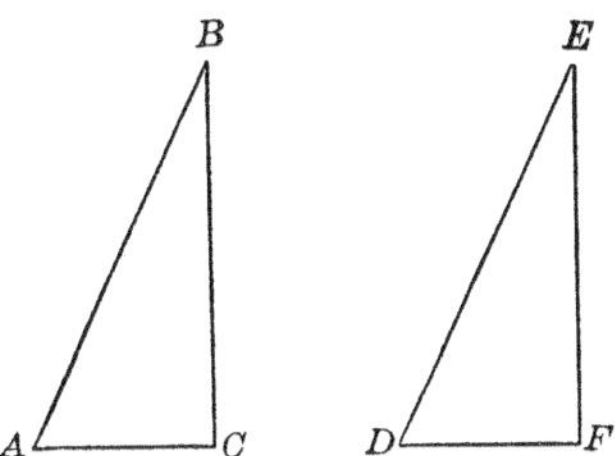

Given the right ⊿s ABC and DEF in which hypotenuse AB = hypotenuse DE, and $\angle A = \angle D$.

To prove $\triangle ABC = \triangle DEF$.

Proof. Place the $\triangle ABC$ upon the $\triangle DEF$ so that the side AB shall coincide with its equal, the side DE, the point A coinciding with the point D.

Then the line AC will take the direction of DF,
(*for* $\angle A = \angle D$ *by hyp.*)

Also the side BC will coincide with the side EF, Art. 80.
(*from a given point, E, without a straight line, DF, but one* ⊥ *can be drawn to the line*).

∴ ⊿s ABC and DEF coincide.

∴ $\triangle ABC = \triangle DEF$, Art. 47.
(*geometric figures which coincide are equal*).

Q. E. D.

Ex. Construct exactly an equilateral triangle, each of whose sides shall be double a given line.

Proposition IX. Theorem

99. *In an isosceles triangle the angles opposite the equal sides are equal.*

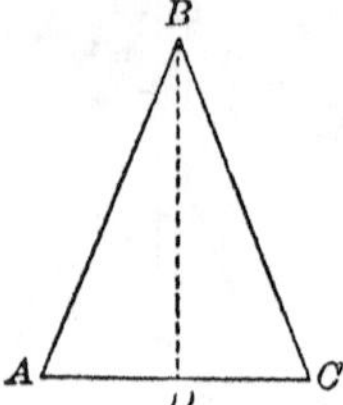

Given the isosceles $\triangle ABC$ in which $AB = BC$.

To prove $\angle A = \angle C$.

Proof. Let BD be drawn so as to bisect $\angle ABC$.

Then, in the $\triangle ABD$ and DBC,

$AB = BC$. Hyp.

Also $BD = BD$, Ident.

And $\angle ABD = \angle CBD$. Constr.

$\therefore \triangle ABD = \triangle CBD$, Art. 96.

(*two* △ *are equal if two sides and the included* ∠ *of one are equal, respectively, to two sides and the included* ∠ *of the other.*)

$\therefore \angle A = \angle C$,

(*homologous* ∠ *of equal* △).

Q. E. D.

Ex. 1. On a given line as base, construct exactly an equilateral triangle above the line and another below it.

Ex. 2. On a given line as base, construct exactly an isosceles triangle whose leg shall be equal to a given line; make the same construction below the given line and join the vertices of the two isosceles triangles.

PROPOSITION X. THEOREM (CONVERSE OF PROP. IX)

100. *If two angles of a triangle are equal, the sides opposite are equal, and the triangle is isosceles.*

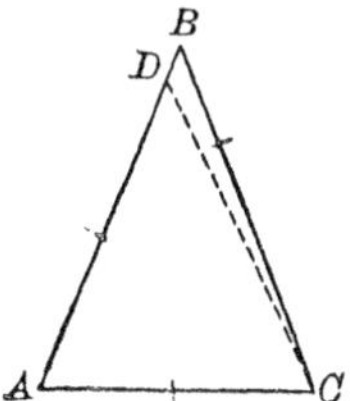

Given the triangle ABC in which $\angle A = \angle BCA$.

To prove $AB = BC$.

Proof. If the sides AB and BC are not equal, one of them must be longer than the other. Let AB be longer than BC.

On AB mark off $AD = BC$ Draw DC.

Then, in the ▲ ABC and ADC,

$AD = BC$, Constr.

$AC = AC$, Ident.

$\angle DAC = \angle BCA$. Hyp.

$\therefore \triangle DAC = \triangle BCA$, Art. 96.

(*two* ▲ *are equal if two sides and the included* ∠ *of one are equal, respectively, to two sides and the included* ∠ *of the other*).

Or a part is equal to the whole, which is impossible. Ax. 7.

Hence AB cannot be greater than BC.

In like manner it may be shown that AB is not less than BC.

Hence $AB = BC$.

Q. E. D.

Ex. 1. What method of proof is used in Prop. X?

Ex. 2. If in a triangle DEF, $\angle D = 36°$, $\angle E = 36°$ and $DF = 12$, find EF. Draw a figure and on it mark the value of the parts named.

PROPOSITION XI. THEOREM

101. *Two triangles are equal if the three sides of one are equal, respectively, to the three sides of the other.*

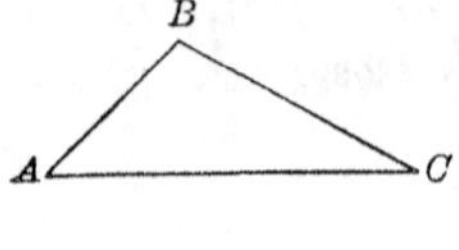

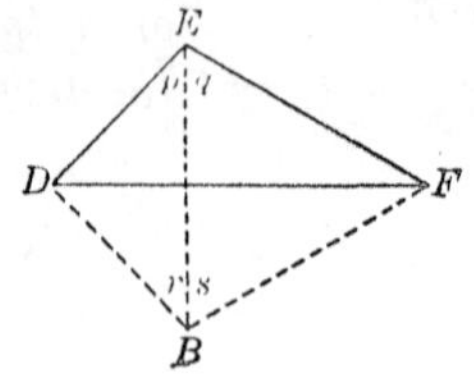

Given the ⊿s ABC and DEF in which $AB = DE$, $BC = EF$, and $AC = DF$.

To prove $\triangle ABC = \triangle DEF$.

Proof. Place the $\triangle ABC$ so that its longest side AC shall coincide with its equal DF in the $\triangle DEF$ and the vertex B shall fall on the opposite side of DF from E. Geom. Ax. 2.

Draw the line EB.

Then $DE = DB$ (Hyp.) $\therefore \triangle DEB$ is isosceles. Def.

$\therefore \angle p = \angle r$, Art. 99.

(*in an isosceles* △ *the* ∡ *opposite the equal sides are equal*).

In like manner, in the $\triangle BEF$, $\angle q = \angle s$.

Adding,

$\angle p + \angle q = \angle r + \angle s$, Ax. 2.

Or $\angle DEF = \angle DBF$. Ax. 6.

$\therefore \triangle DEF = \triangle DBF$, Art. 96.

(*two* ⊿ *are equal if two sides and the included* ∠ *of one are equal, respectively, to two sides and the included* ∠ *of the other*).

$\therefore \triangle ABC = \triangle DEF$. Ax. 1.

Q. E. D.

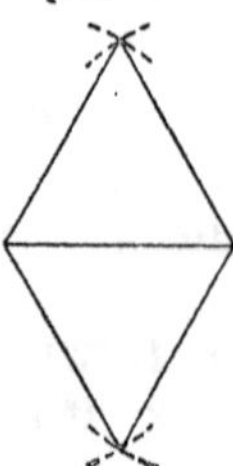

Ex. 1. Construct two equilateral triangles on the same base, one above and the other below, and join the two vertices. Prove that the line joining the vertices bisects the vertex angles, and also bisects the base at right angles.

Ex. 2. Hence, at any point in a given straight line, construct exactly by use of ruler and compasses a perpendicular to that line.

Proposition XII. Theorem

102. *Two right triangles are equal if the hypotenuse and a leg of one are equal to the hypotenuse and a leg of the other.*

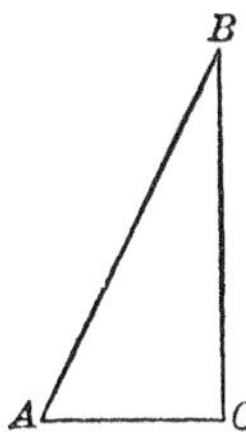

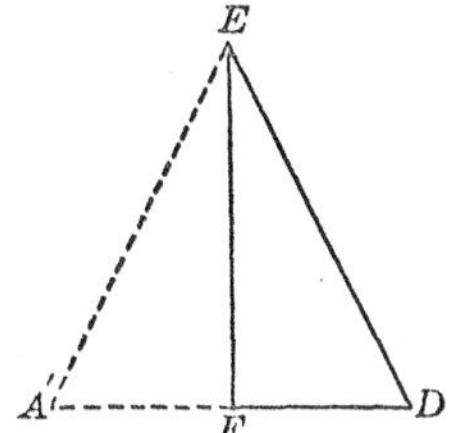

Given two right ⊿s ABC and DEF having the hypotenuse AB = hypotenuse DE, and $BC = EF$.

To prove $\triangle ABC = \triangle DEF$.

Proof. Place the $\triangle ABC$ so that BC shall coincide with its equal, EF, and A fall on the opposite side of EF from D, at A'. Geom. Ax. 2.

Then $A'F$ and FD will form a straight line, $A'FD$, Art. 74.
(*if two adj. ∠s are together equal to two rt. ∠s, their ext. sides form one and the same straight line*).

But $A'E = ED$. Hyp.

$\therefore \triangle A'ED$ is isosceles. Def.

$\therefore \angle A' = \angle D$, Art. 99.
(*in an isosceles △ the ∠s opposite the equal sides are equal.*)

$\therefore \triangle A'EF = \triangle DEF$, Art. 98.
(*two right ⊿s are equal if the hypotenuse and an acute ∠ of one are equal to the hypotenuse and an acute ∠ of the other*).

Or $\triangle ABC = \triangle DEF$. Ax. 1.

Q. E. D.

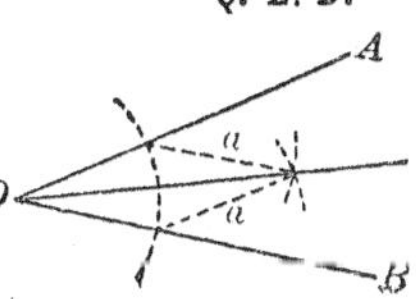

Ex. 1. By making the same construction as in Ex. 1, p. 42, bisect any given straight line.

Ex. 2. Bisect any given angle AOB.

Proposition XIII. Theorem

103. *An exterior angle of a triangle is greater than either opposite interior angle.*

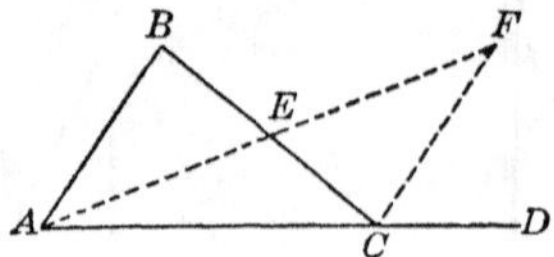

Given $\angle BCD$ an exterior $\angle$ of the $\triangle ABC$.

To prove $\angle BCD$ greater than $\angle ABC$ or $\angle BAC$.

Proof. Let E be the middle point of the line BC.

Draw AE and produce it to F, making $FE = AE$. Draw FC.

Then, in the $\triangle$s AEB and FEC,

$AE = FE$, and $BE = CE$, Constr.

$\angle BEA = \angle FEC$ (*being vertical* $\angle$s). Art. 78.

$\therefore \triangle AEB = \triangle FEC$, Art. 96.

(*two* $\triangle$s *are equal if two sides and the included* $\angle$ *of one are equal, respectively, to two sides and the included* $\angle$ *of the other*).

$\therefore \angle ABE = \angle FCE$,

(*being homologous* $\angle$s *of equal* $\triangle$s).

But $\angle BCD$ is greater than $\angle FCE$, Ax. 7.

(*the whole is greater than any of its parts*).

Substituting $\angle ABE$ for its equal $\angle FCE$, Ax. 8.

$\angle BCD$ is greater than $\angle ABE$, that is, than $\angle ABC$.

Similarly, by drawing a line from B through the midpoint of AC and by producing BC through C to a point H, it may be shown that $\angle ACH$ ($= \angle BCD$) is greater than $\angle BAC$.

Q. E. D.

Ex. On a given line, as base, construct exactly an isosceles triangle each of whose legs equals half a given line.

PROPOSITION XIV. THEOREM

104. ***If two sides of a triangle are unequal, the angles opposite are unequal, and the greater angle is opposite the greater side.***

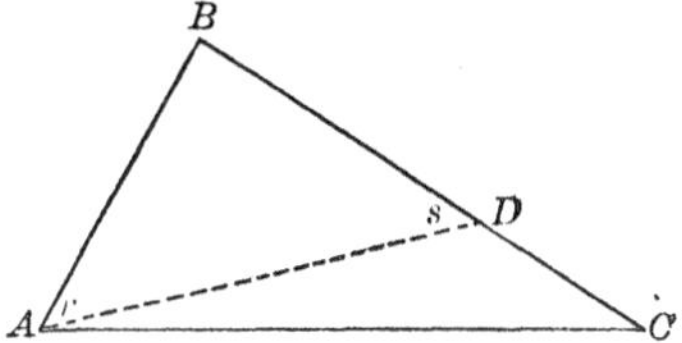

Given the side $BC >$ side AB in the $\triangle ABC$.

To prove $\angle BAC$ greater than $\angle C$.

Proof. On the side BC take BD equal to AB and draw AD.

Then, in the isosceles $\triangle ABD$, $\angle r = \angle s$, Art. 99.
(*in an isosceles $\triangle$ the $\angle s$ opposite the equal sides are equal*).

$\angle BAC$ is greater than $\angle r$, Ax. 7.
(*the whole is greater than any of its parts*).

$\therefore \angle BAC$ is greater than $\angle s$. Ax. 8.

But $\angle s$ is an exterior $\angle$ of the $\triangle ADC$.

$\therefore \angle s$ is greater than $\angle C$, Art. 103.
(*an ext. $\angle$ of a $\triangle$ is greater than either opposite int. $\angle$*).

Much more, then, is $\angle BAC$ (*which is greater than $\angle s$*) greater than $\angle C$, Ax. 12.
(*if, of three quantities, the first is greater than the second, and the second is greater than the third, then the first is greater than the third*).

Q. E. D.

105. NOTE. The essential steps of the above proof may be arranged in a single statement, thus:

$\angle BAC > \angle r = \angle s > \angle C$ $\therefore$ $\angle BAC$ is greater than $\angle C$.

Ex. 1. Which is the longest side of a right triangle? of an obtuse triangle?

Ex. 2. Construct exactly an equilateral triangle, each of whose sides is half a given line.

PROPOSITION XV. THEOREM (CONVERSE OF PROP. XIV)

106. *If two angles of a triangle are unequal, the sides opposite are unequal, and the greater side is opposite the greater angle.*

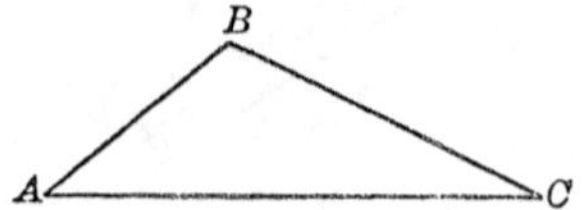

Given $\angle A$ greater than $\angle C$ in the $\triangle ABC$.

To prove $BC > AB$.

Proof. BC either equals AB, or is less than AB, or is greater than AB.

But BC cannot equal AB,

for, if it did, $\angle A$ would equal $\angle C$, Art. 99.
(being opposite equal sides in an isosceles $\triangle$).

But this is contrary to the hypothesis.

Also BC cannot be less than AB,

for, if it were, $\angle A$ would be less than $\angle C$, Art. 104.
(if two sides of a $\triangle$ are unequal, the $\measuredangle$ opposite are unequal, and the greater $\angle$ is opposite the greater side).

This is also contrary to the hypothesis.

$\therefore BC > AB$,

(for it neither equals AB, nor is less than AB).

Q. E. D.

Ex. 1. Draw a triangle the altitude of which falls on the base produced. What kind of a triangle is this?

Ex. 2. Draw a triangle the altitude of which coincides with one side. What kind of a triangle is this?

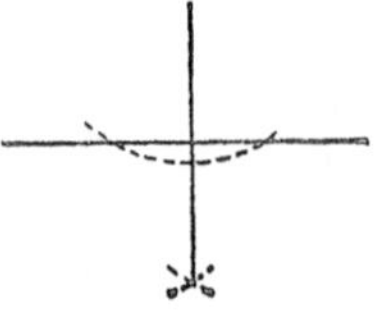

Ex. 3. By exact use of the ruler and compasses, draw a perpendicular to a given line from a given point without the line.

PROPOSITION XVI. THEOREM

107. ***If two triangles have two sides of one equal, respectively, to two sides of the other, but the included angle of the first greater than the included angle of the second, then the third side of the first is greater than the third side of the second.***

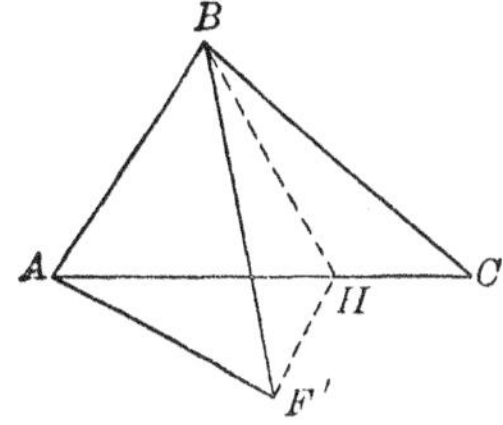

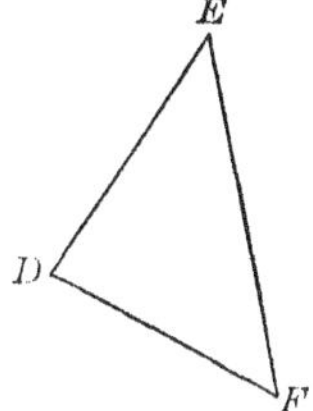

Given the ▲ ABC and DEF in which $AB=DE$, $BC=EF$, and $\angle ABC$ is greater than $\angle E$.

To prove $AC > DF$.

Proof. Place the $\triangle DEF$ so that the side DE coincides with its equal, the side AB, and F takes the position F'. Geom. Ax. 2.

Let the line BH bisect the $\angle F'BC$ and meet the line AC at H. Draw $F'H$.

Then, in the ▲ $F'BH$ and BHC, $F'B=BC$, Hyp.

$BH=BH$, Ident.

$\angle F'BH = \angle CBH$. Constr.

$\therefore \triangle F'BH = \triangle BHC$. (Why?)

$\therefore F'H=CH$, (*homologous sides of equal* ▲).

But $AH + HF' > AF'$, Art. 92.

(*the sum of any two sides of a* $\triangle$ *is greater than the third side*).

Substituting for HF' its equal HC,

$AH + HC$, or $AC > AF'$. Ax. 8.

$\therefore AC > DF$. Ax. 8.

Q. E. D.

Ex. 1. Draw a figure for Prop. XVI in which the sides and angles are of such a size that F' falls within the triangle ABC.

Ex. 2. Draw another figure in which F' falls on the side AC.

PROPOSITION XVII. THEOREM (CONVERSE OF PROP. XVI)

108. *If two sides of a triangle are equal, respectively, to two sides of another triangle, but the third side of the first is greater than the third side of the second, then the angle opposite the third side of the first triangle is greater than the angle opposite the third side of the second.*

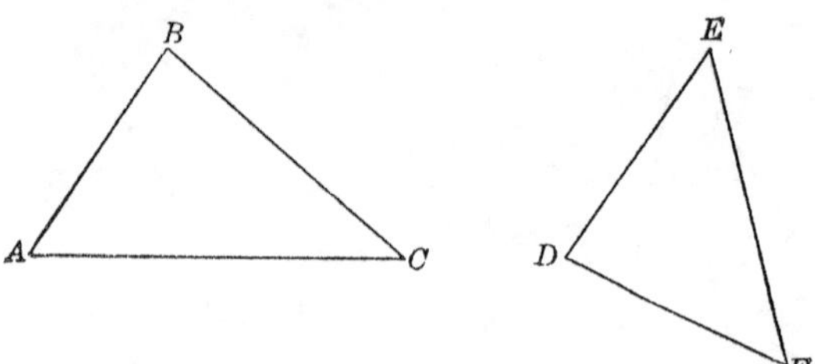

Given the ⧍ ABC and DEF having $AB = DE$, $BC = EF$, but $AC > DF$.

To prove $\angle B$ greater than $\angle E$.

Proof. The $\angle B$ either equals $\angle E$, or is less than $\angle E$, or is greater than $\angle E$.

But $\angle B$ does not equal $\angle E$,

for, if it did, $\triangle ABC$ would $= \triangle DEF$, Art. 96.

(*two ⧍ are equal if two sides and the included ∠ of one are equal, respectively, to two sides and the included ∠ of the other*),

and AC would equal DF (*homologous sides of equal* ⧍), which is contrary to the hypothesis.

Also if $\angle B$ were less than $\angle E$,

side AC would be less than side DF, Art. 107.

(*if two ⧍ have two sides of one equal, respectively, to two sides of the other, but the included ∠ of the first greater than the included ∠ of the second, then the third side of the first is greater than the third side of the second*).

But this is also contrary to the hypothesis.

Hence $\angle B$ is greater than $\angle E$,

(*for it neither equals* $\angle E$, *nor is less than* $\angle E$).

Q. E. D.

Ex. On a given line (l) construct a triangle whose other two sides are equal to two given lines (m and n).

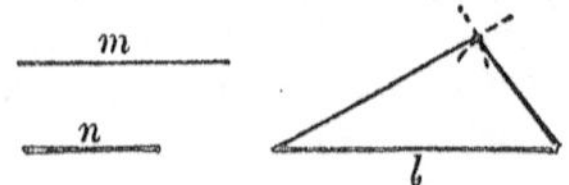

PROPERTIES OF LINES PROVED BY USE OF TRIANGLES

Proposition XVIII. Theorem

109. *Of lines drawn from the same point in a perpendicular and cutting off unequal segments from the foot of the perpendicular, the more remote is the greater.*

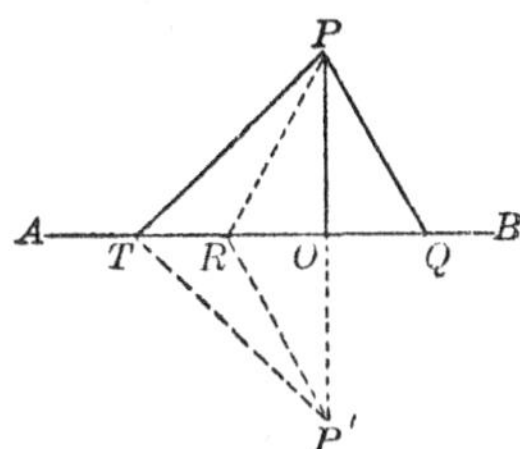

Given $PO \perp AB$, PT and PQ oblique to AB, and $OT > OQ$.

To prove $PT > PQ$.

Proof. Produce PO to the point P' making $OP' = OP$.

On AB take $OR = OQ$. Draw PR, $P'Q$, $P'R$, $P'T$.

Then $PQ = PR$, Art. 79.

(*if from a point in a* $\perp$ *a given line, two oblique lines be drawn, cutting off on the given line equal segments from the foot of the* $\perp$, *the oblique lines are equal*).

In $\triangle PTP'$, $PT + TP' > PR + RP'$, Art. 95.

(*if, from a point within a* $\triangle$, *two lines be drawn to the extremities of a side of the* $\triangle$, *the sum of the other two sides of the* $\triangle$ *is greater than the sum of the two lines so drawn*).

But OT is $\perp PP'$ and PT and $P'T$ cut off equal segments, PO and $P'O$, from the foot of the $\perp AO$.

Hence $PT = P'T$. In like manner $PR = P'R$. Art. 79.

Hence, by substitution, $2\ PT > 2\ PR$. Ax. 8.

$\therefore PT > PR$. Ax. 10.

$\therefore PT > PQ$. Ax. 8.

Q. E. D.

PROPOSITION XIX. THEOREM (CONVERSE OF PROP. II)

110. *Equal oblique lines drawn from a point in a perpendicular cut off equal segments from the foot of the perpendicular.*

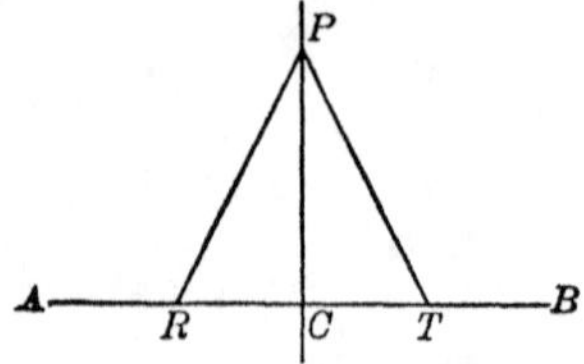

Given $PC \perp AB$, PR and PT oblique to AB, and $PR = PT$.

To prove $CR = CT$.

Proof. In the right ▵ RPC and CPT,

$PC = PC$. Ident.

Also $PR = PT$. Hyp.

$\therefore \triangle RPC = \triangle CPT$, Art. 102

(*two right ▵ are equal if the hypotenuse and a leg of one are equal to the hypotenuse and a leg of the other*).

$\therefore RC = CT$,

(*homologous sides of equal ▵*).

Q. E. D.

111. COR. *Of two unequal lines drawn from a point in a perpendicular, the greater line cuts off the greater segment from the foot of the perpendicular.*

Thus, if $PT > PQ$ (Fig. of Prop. XVIII), OT cannot $= OQ$ (Art. 110); nor is $OT < OQ$ (Art. 109) $\therefore OT > OQ$.

Hence, also, *from a given point only two equal straight lines can be drawn to a given line.*

PROPOSITION XX. THEOREM.

112. I. *Every point in the perpendicular bisector of a line is equally distant from the extremities of the line; and*

II. *Every point not in the perpendicular bisector is unequally distant from the extremities of the line.*

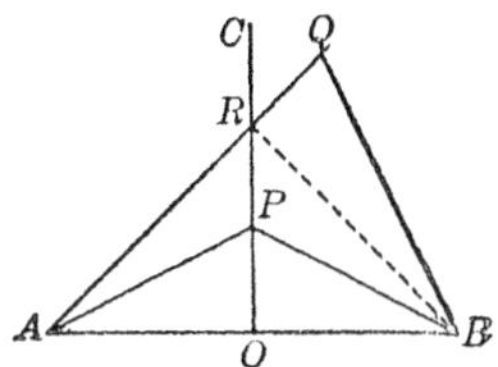

Given O the middle point of the line AB, $OC \perp AB$, P any point in OC, and Q any point not in OC.

To prove $AP=PB$, but QA and QB unequal.

Proof. I. $AP=PB$, Art. 79.

(*if, from a point in a $\perp$ to a given line, two oblique lines be drawn cutting off on the given line equal segments, etc.*).

II. Since Q is not in the line OC, either AQ or QB must cut the line OC.

Let AQ intersect OC in the point R and join RB.

Then $AR=RB$, (*by first part of this theorem*).

To each of these equals add RQ.

Then $AR+RQ=RB+RQ$. Ax. 2.

But $RB+RQ>QB$. (Why?)

$\therefore$ by substitution, $AR+RQ$, or $AQ>QB$. Ax. 8.

Q. E. D.

113. COR. *Two points each equidistant from the extremities of a line determine the perpendicular bisector of the line.*

This corollary gives a useful method of determining the perpendicular bisector of a given straight line, by determining two points only of the perpendicular bisector.

LOCI

114. Def. The **locus of a point** is the path of a point moving according to a given geometric law.

Thus, if a point move in a plane so as to be always two inches distant from a given point, its locus is the circumference of a circle whose center is the given point, and whose radius is a line two inches in length.

Thus, also, the locus of a point moving so as to be equidistant from two given parallel lines is a straight line lying midway between the two given lines.

The locus of a point may consist of two or more separate lines or parts.

Thus, the locus of a point moving so as to be always at a given distance from a given line is two lines, one on either side of the given line, at the given distance from it.

115. Demonstration of loci. In order to prove that a given line is the locus of a given point moving according to a given geometric law, it is necessary:

1. *To prove that every point* in *the given line satisfies the given law or condition.*

2. *To prove that every point* not in *the given line does* not *satisfy the given law or condition.*

Instead of 2, it may be proved that every point which satisfies the given condition lies in the given line.

Hence, in Prop. XX it has been proved that *the perpendicular bisector of a line is the locus of all points equidistant from the extremities of the line.*

116. Use of loci. Loci are useful in determining a point (or points) which shall satisfy two or more geometrical conditions. For, by finding the locus of all points which satisfy one of the given conditions, and also finding the locus of all points which satisfy a second condition, and then finding the intersection of these two loci, we obtain the point (or points) which satisfy both conditions at the same time

Thus, if it be required to find the points which are two inches from one given point and three inches from another given point, the two given points being four inches apart, the required points are the intersections of the circumferences of two circles.

Let the pupil make a construction and obtain the required points.

Ex. 1. Draw the locus of a point moving at the distance of one inch from a given point.

Ex. 2. Draw exactly the locus of a point moving at a distance of one inch from a given line.

Ex. 3. Draw exactly the locus of a point moving so as to be equidistant from the extremities of a given line one inch long.

Ex 4. How many points in a plane are necessary to determine two parallel lines? Three parallel lines? (See Art. 47.)

Ex. 5. Are two triangles equal if three angles of one equal the corresponding three angles of the other? Illustrate by drawing a figure.

Ex. 6. Draw three isosceles triangles on the same base and connect their vertices. What truth is illustrated by this figure?

Ex. 7. Draw a straight line and locate a point 2 inches from it. By the use of loci, locate the points which are $1\frac{1}{2}$ inches from the given line and at the same distance from the given point.

PROPOSITION XXI. THEOREM

117. I. *Every point in the bisector of an angle is equidistant from the sides of the angle;* and

II. CONVERSELY, *every point equidistant from the sides of an angle lies in the bisector of the angle.*

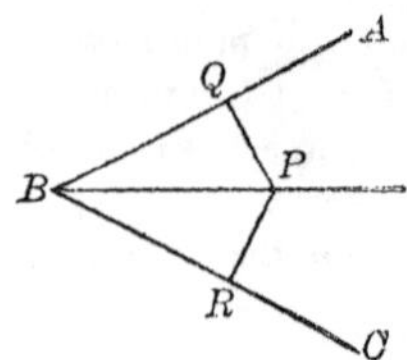

I. **Given** PB the bisector of the angle ABC, P any point in PB, $PQ \perp BA$, and $PR \perp BC$.

To prove $PQ = PR$.

Proof. In the rt. $\triangle$ PBQ and PBR,

$PB = PB$. Ident.

Also $\angle PBQ = \angle PBR$. Hyp.

$\therefore \triangle PBQ = \triangle PBR$, Art. 98.

(*two rt.* $\triangle$ *are equal if the hypotenuse and an acute* $\angle$ *of one, etc.*).

Hence $PQ = PR$,

(*homologous sides of equal* $\triangle$).

II. **Given** $\angle ABC$, $PQ \perp AB$, $PR \perp BC$, and $PQ = PR$.

To prove that PB is the bisector of $\angle ABC$.

Proof. In the right $\triangle$ PBQ and PBR,

$PB = PB$. Ident.

Also $PQ = PR$. Hyp.

$\therefore \triangle PBQ = \triangle PBR$, Art. 102.

(*two rt.* ⊿ *are equal if the hypotenuse and a leg of one, etc.*).

$\therefore \angle ABP = \angle CBP$ (*homologous* ∠ *of equal* ⊿).

Or $\angle ABC$ is bisected by BP.

Q. E. D.

118. Cor. In Prop. XXI it has been proved that *the bisector of an angle is the locus of all points equidistant from the sides of the angle*, for it has been proved that every point in the given line satisfies the given law or condition, and that every point which satisfies the given condition lies in the given line (see Art. 115).

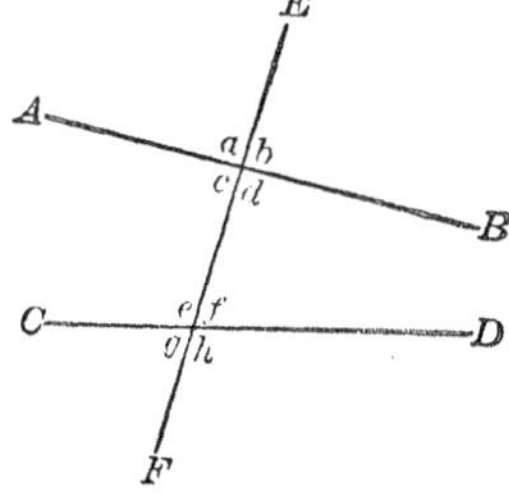

119. Def. A **transversal** is a line that intersects two or more other lines. Thus, EF is a transversal of the l'nes AB and CD.

If two lines are cut by a transversal, it is convenient to give the eight angles of intersection special names.

a, b, g, h, are called **exterior** angles.

c, d, e, f, are called **interior** angles.

c, f, form a pair of **alternate-interior** angles.

b, f, form a pair of **exterior-interior** angles on the same side of the transversal.

Let the pupil name another pair of alternate-interior angles; also name another pair of exterior-interior angles on the same side of the transversal; also name a pair of interior angles on the same side of the transversal.

PARALLEL LINES

120. DEF. **Parallel lines** have already been defined (Art. 41) as straight lines which lie in the same plane and do not meet, however far they be produced.

What is the fundamental axiom concerning parallel lines? (see Art. 47.)

PROPOSITION XXII. THEOREM

121. *Two straight lines in the same plane, perpendicular to the same straight line, are parallel.*

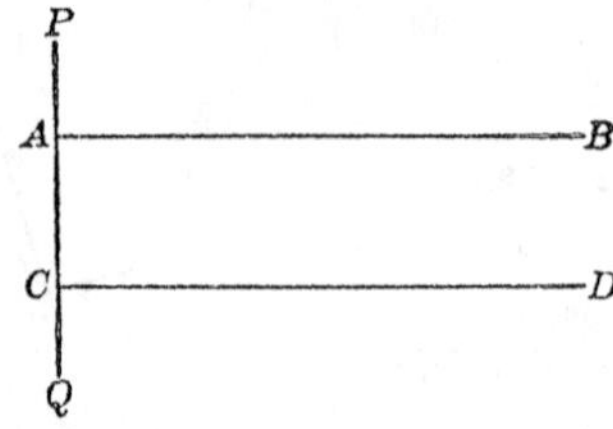

Given the lines AB and $CD \perp$ line PQ and in the same plane.

To prove $AB \parallel CD$.

Proof. If AB and CD are not parallel, they will meet if sufficiently produced. Art. 120.

We shall then have two $\perp$s from the same point to the line PQ.

But this is impossible, Art. 80.

(*from a given point without a straight line but one* $\perp$ *can be drawn to the line*).

Hence AB and CD never meet.

$\therefore$ AB and CD are parallel. Def.

Q. E. D.

122. COR. *Two straight lines parallel to a third straight line are parallel to each other;*

Lines parallel to parallel lines are parallel;

Lines perpendicular to parallel lines are parallel;

Lines perpendicular to non-parallel lines are not parallel.

PROPOSITION XXIII. THEOREM

123. *If a straight line is perpendicular to one of two given parallel lines it is perpendicular to the other also.*

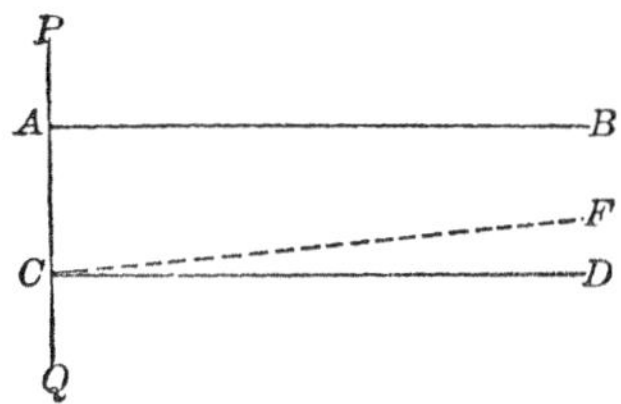

Given $AB \parallel CD$, and $PQ \perp AB$.

To prove $PQ \perp CD$.

Proof. Let CF be drawn $\perp PQ$ at C.

Then $CF \parallel AB$, Art. 121.

(*two straight lines in the same plane* $\perp$ *same straight line are* $\parallel$).

But $CD \parallel AB$. Hyp.

$\therefore$ CF coincides with CD, Geom. Ax. 3.

(*through a given point one straight line, and only one, can be drawn* $\parallel$ *another given straight line*).

But $PQ \perp CF$. Constr.

Hence $PQ \perp CD$,

(*for CD coincides with CF, to which PQ is* $\perp$). Q. E. D.

PROPOSITION XXIV. THEOREM

124. *If two parallel straight lines are cut by a transversal, the alternate interior angles are equal.*

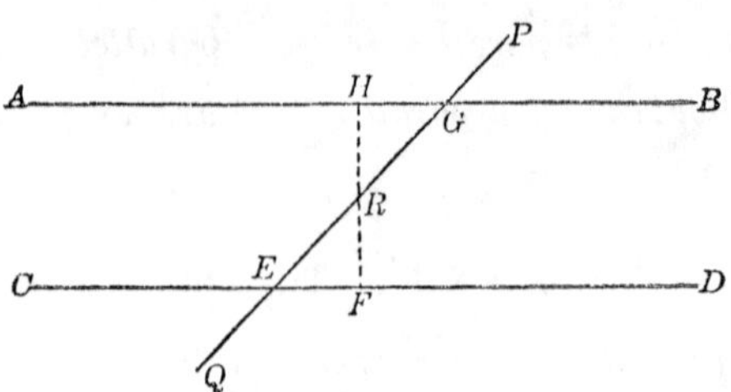

Given the $\parallel$ lines AB and CD cut by the transversal PQ at the points G and E respectively.

To prove $\angle AGE = \angle GED$.

Proof. Through R, the middle point of EG, let the line HF be drawn $\perp AB$.

Then $HF \perp CD$, Art. 123.
(*if a straight line is* $\perp$ *one of two* $\parallel$ *lines, it is* $\perp$ *the other also*).

In the right $\triangle$ GRH and ERF,

$GR = ER$, Constr.

$\angle GRH = \angle ERF$. (Why?)

$\therefore \triangle GRH = \triangle ERF$, Art. 98.
(*two right* $\triangle$ *are equal if the hypotenuse and an acute* $\angle$ *of one = the hypot. and an acute* $\angle$ *of the other*).

$\therefore \angle HGR = \angle REF$, or, $\angle AGE = \angle GED$,
(*homologous* $\angle$ *of equal* $\triangle$).

Q. E. D.

Ex. In the above figure let the pupil show that $\angle BGE = \angle GEC$.

PROP. XXV. THEOREM (CONVERSE OF PROP. XXIV)

125. *If two straight lines are cut by a transversal, making the alternate interior angles equal, the two straight lines are parallel.*

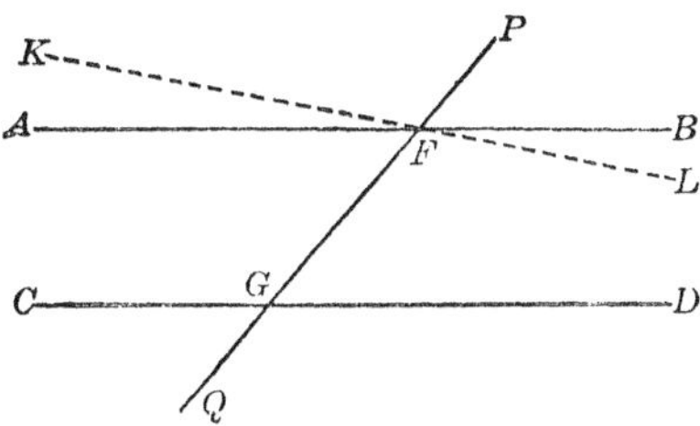

Given the two lines AB and CD cut by the transversal PQ at the points F and G, making $\angle AFG = \angle FGD$.

To prove $AB \parallel CD$.

Proof. Through F let the line KL be drawn $\parallel CD$.

Then $\angle KFG = \angle FGD$, Art. 124.

(*if two $\parallel$ st. lines are cut by a transversal, the alt. int. ∠s are equal*).

But $\angle AFG = \angle FGD$. Hyp.

Hence $\angle KFG = \angle AFG$. Ax. 1.

$\therefore$ KL coincides with AB.

But $KL \parallel CD$. Constr.

Hence $AB \parallel CD$,

(*for AB coincides with KL, which is $\parallel CD$*).

Q. E. D.

Ex. 1. If $\angle BFG = \angle FGC$, prove that AB and CD are parallel.

Ex. 2. By exact use of ruler and compasses, at a given point (P) in a given straight line (OA) construct an angle equal to a given $\angle(B)$.

Ex. 3. By exact methods, through a given point draw a line parallel to a given line.

PROPOSITION XXVI. THEOREM

126. *If two parallel lines are cut by a transversal, the exterior interior angles are equal.*

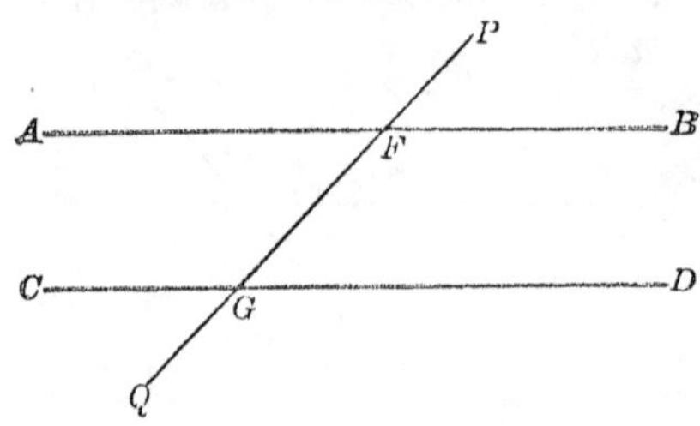

Given the ∥ lines AB and CD cut by the transversal PQ at the points F and G respectively.

To prove $\angle PFB = \angle FGD$.

Proof. $\angle PFB = \angle AFG$. (Why?)

$\angle FGD = \angle AFG$, Art. 124.

(*being alt. int. ∠s of parallel lines*).

$\therefore \angle PFB = \angle FGD$. Ax. 1.

In like manner it may be shown that $\angle PFA = \angle FGC$.

Q. E. D.

PROP. XXVII. THEOREM (CONVERSE OF PROP. XXVI)

127. *If two straight lines are cut by a transversal, making the exterior interior angles equal, the two straight lines are parallel.*

Given, on Fig. of Prop. XXV, $\angle PFB = \angle FGD$.

To prove $AB \parallel CD$.

Let the pupil supply the proof.

Ex. If, in the Fig. to Prop. XXVI, $\angle PFB$ equals 67°, find the other seven angles in the figure without measuring them.

PROPOSITION XXVIII. THEOREM

128. *If two parallel lines are cut by a transversal, the sum of the interior angles on the same side of the transversal is equal to two right angles.*

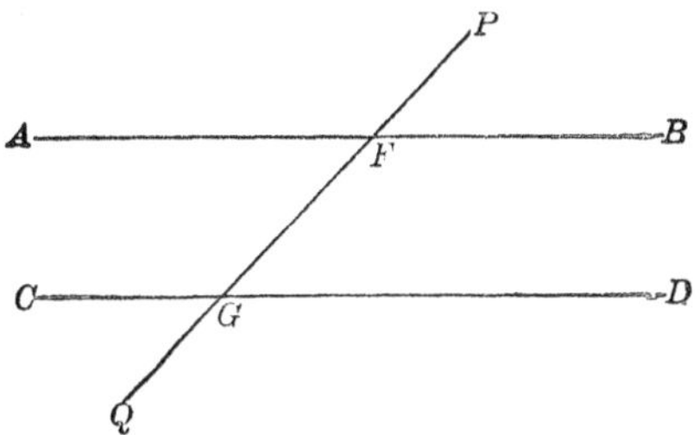

Given the || lines AB and CD cut by the transversal PQ at the points F and G respectively.

To prove $\angle BFG + \angle FGD = 2$ rt. $\angle s$.

Proof. $\angle FGD = \angle PFB$. Art. 126.

To each of these equals add $\angle BFG$.

Then $\angle BFG + \angle FGD = \angle PFB + \angle BFG$. Ax. 2.

But $\angle PFB + \angle BFG = 2$ rt. $\angle s$. Art. 73.

$\therefore \angle BFG + \angle FGD = 2$ rt. $\angle s$. Ax. 1.

Q. E. D.

PROP. XXIX. THEOREM (CONVERSE OF PROP. XXVIII)

129. *If two straight lines are cut by a transversal, making the sum of the interior angles on the same side of the transversal equal to two right angles, the two lines are parallel.*

Given, on Fig. of Prop. XXV, $\angle BFG + \angle FGD = 2$ rt. $\angle s$.

To prove $AB \parallel CD$.

Let the pupil supply the proof.

Ex. If, on Fig. of Prop. XXVIII, $\angle PFB + \angle QGD = 180°$, are AB and CD || ?

Proposition XXX. Theorem

130. *Two angles whose sides are parallel, each to each, are either equal or supplementary.*

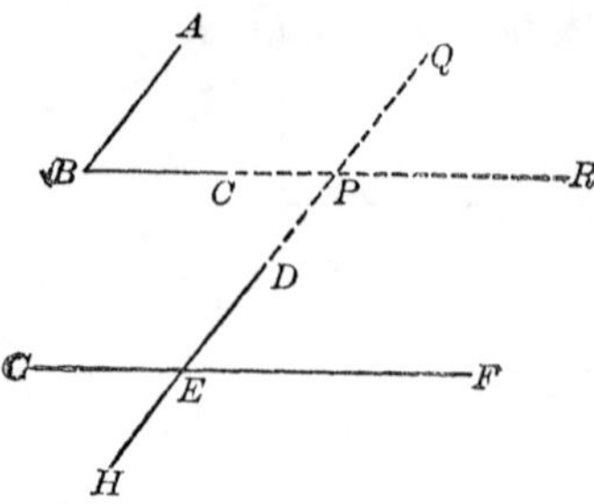

Given $AB \parallel DH$, and $BC \parallel GF$.

To prove $\measuredangle s\ ABC$, DEF and GEH equal, and $\measuredangle s\ ABC$ and DEG supplementary.

Proof. Produce the lines BC and HD to intersect in P.

Then $\angle ABC = \angle QPR$, and $\angle QPR = \angle DEF$, Art. 126.
(being ext. int. $\measuredangle s$ *of* $\parallel$ *lines).*

$\therefore \angle ABC = \angle DEF.$ (Why?)

Also $\angle DEF = \angle GEH.$ (Why?)

$\therefore \angle ABC = \angle GEH.$ (Why?)

Again $\measuredangle s\ DEF$ and DEG are supplementary. Art. 73.

$\therefore \measuredangle s\ ABC$ and DEG are supplementary. Ax. 8.

Q. E. D.

131. Note. It is to be observed that in the above theorem the two angles are *equal* if, in the pairs of parallel sides, both pairs extend in the *same* direction from the vertices ($\measuredangle s\ B$ and DEF), or both pairs in *opposite* directions ($\measuredangle s\ B$ and GEH); and that they are *supplementary* if *one pair* extends in the *same* direction, and the *other pair* in *opposite* directions ($\measuredangle s\ B$ and DEG). The directions of the sides are determined by connecting the vertices of the angles and observing whether the lines considered lie on the same side or on opposite sides of the line drawn.

PROPOSITION XXXI. THEOREM

132. *Two angles whose sides are perpendicular each to each are either equal or supplementary.*

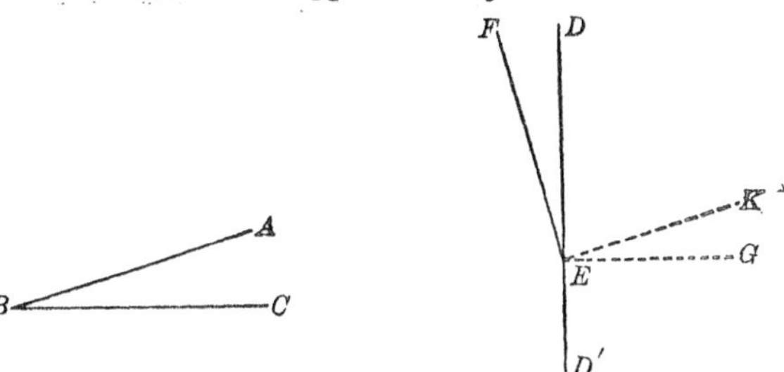

Given $BA \perp EF$, and $BC \perp DD'$.

To prove $\angle ABC = \angle FED$, and $\angle s\ ABC$ and FED' supplementary.

Proof. At E let line EK be drawn $\perp EF$ and in the same direction with BA; also $EG \perp DD'$ and in the same direction with BC.

Then $BA \parallel EK$, and $BC \parallel EG$, Art. 121.

(*two straight lines in the same plane, $\perp$ the same straight line, are $\parallel$*).

$\therefore \angle ABC = \angle KEG$, Arts. 130, 131.

(*two $\angle s$ whose sides are $\parallel$, each to each, and extend in the same direction from the vertices are $=$*).

But $\angle KEG$ is complement of $\angle DEK$, Art. 33.

(*for $\angle DEG$ is a rt. $\angle$ by constr.*).

Also $\angle FED$ is complement of $\angle DEK$, Art. 33.

(*for $\angle FEK$ is a rt. $\angle$ by constr.*).

$\therefore \angle FED = \angle KEG$, Art. 75.

(*complements of the same $\angle$ are $=$*).

$\therefore \angle ABC = \angle FED$. Ax. 1.

But $\angle FED'$ is supplement of $\angle FED$. Art. 34.

$\therefore \angle FED'$ is supplement of $\angle ABC$. Ax. 8.

Q. E. D.

133. NOTE. In the above theorem the two angles are equal if the sides, considered as rotating about the vertices, are taken in the same order (thus BC is to the right of BA, and ED to the right of EF $\therefore \angle ABC = \angle FED$); but the angles are supplementary if the corresponding sides are taken in the opposite order (thus, BC is to the right of BA but ED' is to the left of EF $\therefore \angle ABC =$ supplement of $\angle FED'$).

PROPOSITION XXXII. THEOREM

134. *The sum of the angles of a triangle is equal to two right angles.*

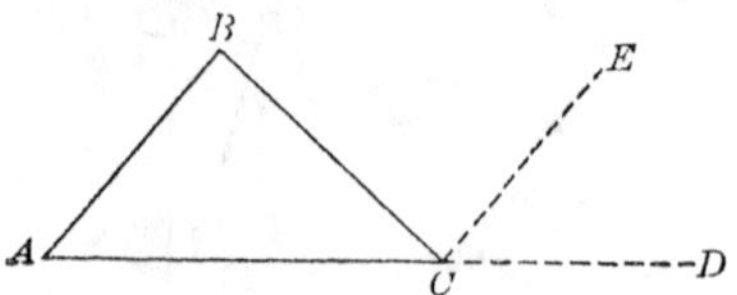

Given the △ ABC.

To prove $\angle A + \angle B + \angle BCA = 2$ rt. ∠s.

Proof. Produce the side AC to the point D and through C let the line CE be drawn ∥ AB.

Then $\angle ECD + \angle BCE + \angle BCA = 2$ rt. ∠s, Art. 77.
(*the sum of all the ∠s about a point on the same side of a straight line passing through the point = 2 rt. ∠s*).

But $\angle ECD = \angle A$, Art. 126.
(*being ext. int. ∠s of parallel lines*).

Also $\angle BCE = \angle B$, Art. 124.
(*being alt. int. ∠s of parallel lines*).

Substituting for $\angle ECD$ its equal, $\angle A$, and for $\angle BCE$ its equal, $\angle B$, Ax. 8.

$\angle A + \angle B + \angle BCA = 2$ rt. ∠s.

Q. E. D.

135. COR. 1. *An exterior angle of a triangle is equal to the sum of the two opposite interior angles.*

136. COR. 2. *The sum of any two angles of a triangle is less than two right angles.*

137. COR. 3. *In a right triangle the sum of the two acute angles equals one right angle.*

138. Cor. 4. *A triangle can have but one right, or one obtuse angle.*

139. Cor. 5. *If two angles of one triangle equal two angles of another triangle, the third angle of the first triangle equals the third angle of the second.*

140. Cor. 6. *If an acute angle of one right triangle equals an acute angle of another right triangle, the remaining acute angles of the triangles are equal.*

141. Cor. 7. *Two triangles are equal if two angles and a side of one are equal to two angles and the homologous side of the second.*

142. Cor. 8. *Two right triangles are equal if a leg and an acute angle of one are equal to a leg and the homologous acute angle of the other.*

Ex. 1. If two angles of a triangle are 56° and 62°, find the remaining angle.

Ex. 2. If one acute angle of a right triangle is 36° 15′, find the other acute angle.

Ex. 3. How many degrees in each angle of an equilateral triangle?

Ex. 4. How many degrees in each acute angle of an isosceles right triangle?

Ex. 5. Is it possible to have a triangle whose angles are 45°, 62°, 72°?

Ex. 6. If one angle of a triangle is 42°, find the sum of the other two angles.

Ex. 7. If two angles of a triangle are 38° and 65°, find all the exterior angles of the triangle.

Ex. 8. If the vertex angle of an isosceles triangle is 38°, find each angle at the base.

Ex. 9. If an angle at the base of an isosceles triangle is 50°, find the vertex angle.

Ex. 10. An exterior angle at the base of an isosceles triangle is 102°, find all the angles of the triangle.

QUADRILATERALS

143. A **quadrilateral** is a portion of a plane bounded by four straight lines.

The **sides** of a quadrilateral are the bounding lines; the **angles** are the angles made by the bounding lines; the **vertices** are the vertices of the angles of the quadrilateral.

The **perimeter** of a quadrilateral is the sum of the sides.

144. A **diagonal** is a straight line joining two vertices that are not adjacent.

145. A **trapezium** is a quadrilateral no two of whose sides are parallel.

146. A **trapezoid** is a quadrilateral which has two, and only two, of its sides parallel.

147. A **parallelogram** is a quadrilateral whose opposite sides are parallel.

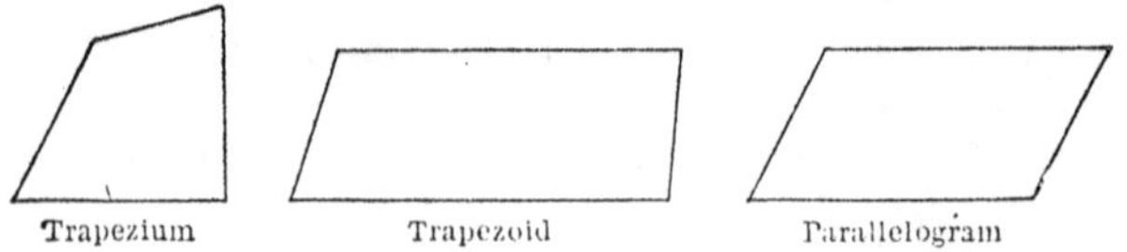

148. A **rhomboid** is a parallelogram whose angles are oblique angles.

149. A **rhombus** is a rhomboid whose sides are equal.

150. A **rectangle** is a parallelogram whose angles are right angles.

151. A **square** is a rectangle whose sides are equal.

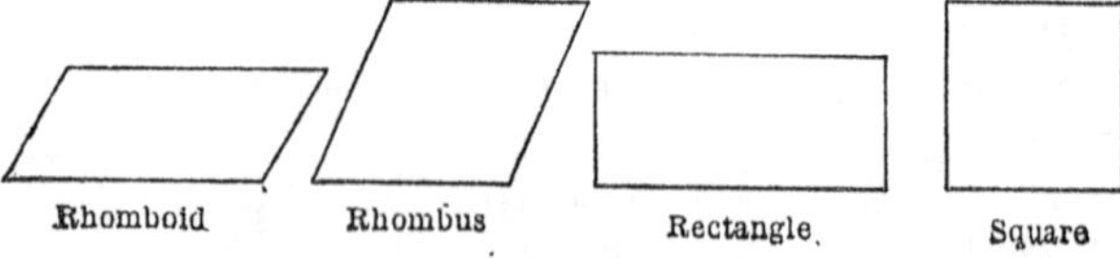

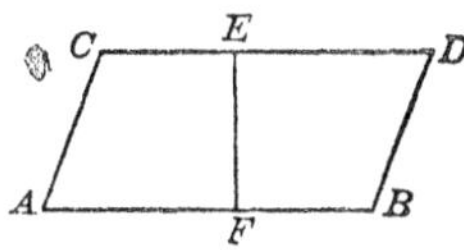

152. The **base** of a parallelogram is the side upon which it is supposed to stand, as AB. The opposite side is called the upper base (CD).

The **altitude** of a parallelogram is the perpendicular distance between the bases, as EF.

153. The **bases of a trapezoid** are its two parallel sides. The **legs** of a trapezoid are the sides which are not parallel. The **altitude** of a trapezoid is the perpendicular distance between the bases. The **median** of a trapezoid is the line joining the midpoints of the legs.

154. An **isosceles trapezoid** is a trapezoid whose legs are equal.

Ex. 1. Draw a quadrilateral with three acute angles and one obtuse angle.

Ex. 2. Is every rhombus a rhomboid? Is every rhomboid a rhombus?

Ex. 3. What is the difference between a square and a rhombus? What properties do they have in common?

Ex. 4. Find the perimeter of a square foot in inches.

By aid of the following classification:

<table>
<tr><td rowspan="4">Quadrilateral .</td><td colspan="2">Trapezium.</td><td></td></tr>
<tr><td>Trapezoid . . . . .</td><td colspan="2">Isosceles trapezoid.</td></tr>
<tr><td rowspan="2">Parallelogram . .</td><td>Rectangle</td><td>. . . square.</td></tr>
<tr><td>Rhomboid</td><td>. . . rhombus.</td></tr>
</table>

Ex. 5. Determine what four names the rhombus is entitled to.

Ex. 6. Determine what properties the rhombus, square and rectangle have in common.

Ex. 7. A diagonal of a rhombus divides the rhombus into how many triangles? What kind of triangles are these?

PROPOSITION XXXIII. THEOREM

155. *The opposite sides of a parallelogram are equal, and its opposite angles are also equal.*

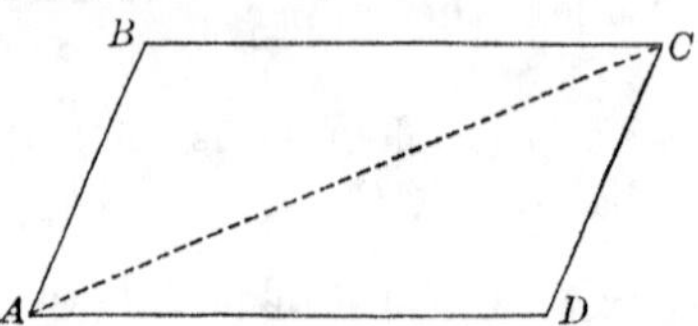

Given the parallelogram $ABCD$.

To prove $AD = BC$, $AB = DC$, $\angle B = \angle D$, and $\angle BAD = \angle BCD$.

Proof. Draw the diagonal AC.

Then, in the $\triangle$ ABC and ADC,

$AC = AC$, (Why?)

$\angle BCA = \angle CAD$, Art. 124.

(*being alt. int.* $\angle$ *of parallel lines*).

$\angle BAC = \angle ACD$,

(*same reason*).

$\therefore \triangle ABC = \triangle ACD$, Art. 97.

(*two* $\triangle$ *are equal if two* $\angle$ *and the included side of one are equal respectively to two* $\angle$ *and the included side of the other*).

$\therefore AD = BC$, $AB = DC$ and $\angle B = \angle D$,

(*homologous parts of equal* $\triangle$).

In like manner, by drawing the diagonal BD, it may be proved that $\angle BAD = \angle BCD$.

Q. E. D.

156. COR. 1. *A diagonal divides a parallelogram into two equal triangles.*

157. COR. 2. *Parallel lines comprehended between parallel lines are equal.*

158. COR. 3. *Two parallel lines are everywhere equidistant.*

Ex. 1. In the above figure, prove $\angle BAD = \angle BCD$ by use of Ax. 2.

Ex. 2. Prove the opposite angles of a parallelogram equal, by use of Art. 130.

PROPOSITION XXXIV. THEOREM

159. *If the opposite sides of a quadrilateral are equal, the figure is a parallelogram.*

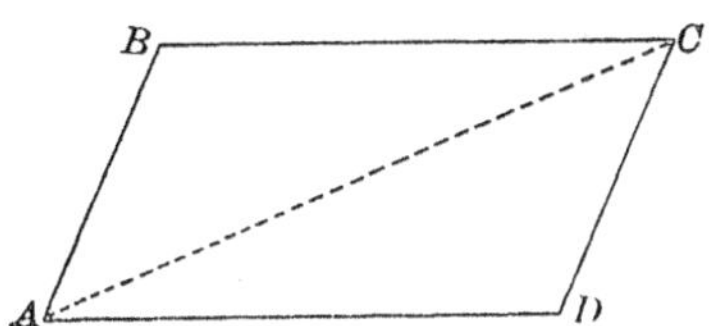

Given the quadrilateral $ABCD$ in which $AB = CD$ and $BC = AD$.

To prove $ABCD$ a ▱.

Proof. Draw the diagonal AC.

Then, in the △s ABC and ADC,

$AC = AC$. (Why?)

$BC = AD$. (Why?)

$AB = CD$. (Why?)

$\therefore \triangle ABC = \triangle ADC$. (Why?)

$\therefore \angle BAC = \angle ACD$. (Why?)

$\therefore AB \parallel CD$, Art. 125.

(*if two lines are cut by a transversal, making the alt. int.* ∠s *equal, the lines are* ∥).

Also $\angle BCA = \angle CAD$. (Why?)

$\therefore BC \parallel AD$. Art. 125.

$\therefore ABCD$ is a ▱, Art. 147.

(*a* ▱ *is a quadrilateral whose opposite sides are* ∥).

Q. E. D.

PROPOSITION XXXV. THEOREM

160. *If two sides of a quadrilateral are equal and parallel, the other two sides are equal and parallel and the figure is a parallelogram.*

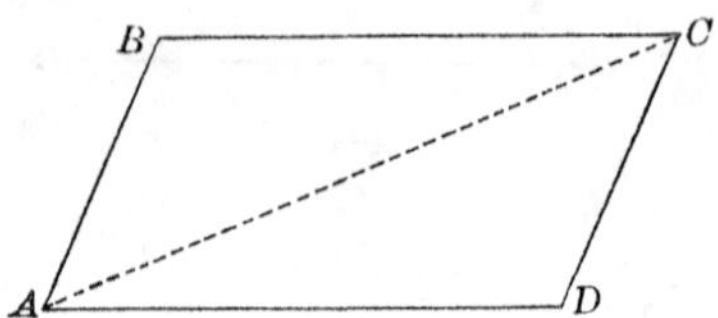

Given the quadrilateral $ABCD$ in which $BC =$ and $\parallel AD$.

To prove $ABCD$ a ▱.

Proof. Draw the diagonal AC.

Then, in the △ ABC and ADC,

$AC = AC$. (Why?)

$BC = AD$. (Why?)

$\angle BCA = \angle CAD$, Art. 124.

(*being alt. int.* ∠s *of parallel lines*),

$\therefore \triangle ABC = \triangle ADC$. (Why?)

$\therefore \angle BAC = \angle ACD$. (Why?)

$\therefore AB \parallel CD$, Art. 125.

(*if two lines are cut by a transversal, making the alt. int.* ∠s *equal, the lines are* ∥).

$\therefore ABCD$ is a ▱. Art. 147.

Q. E. D.

Ex. 1. Show that in a ▱ each pair of adjacent angles is supplementary.

Ex. 2. One angle of a parallelogram is 43°; find the other angles.

Ex. 3. If, in the triangle ABC, $\angle A = 60°$, $\angle B = 70°$, which is the longest side in the triangle? Which the shortest?

Proposition XXXVI. Theorem

161. *The diagonals of a parallelogram bisect each other.*

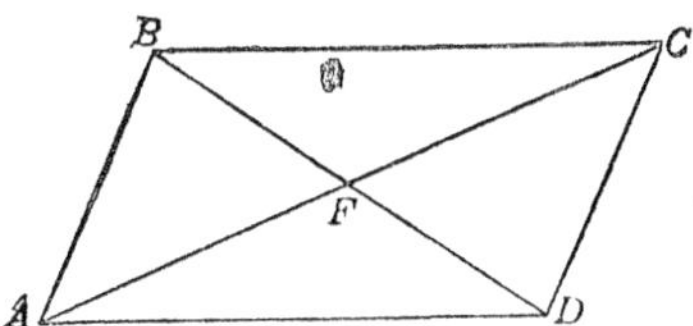

Given the diagonals AC and BD of the ▱ $ABCD$, intersecting at F.

To prove $AF = FC$, and $BF = FD$.

Proof. Let the pupil supply the proof.

[Sug. In the △ BFC and AFD what sides are equal, and why? What ∠ are equal, and why? etc.]

Q. E. D.

Ex. 1. How many pairs of equal triangles are there in the above figure?

Ex. 2. If one angle of a parallelogram is three times another angle, find all the angles of the parallelogram.

Ex. 3. If two angles of a triangle are p° and q°, find the third angle.

Ex. 4. If two angles of a triangle are x° and $90^\circ + x^\circ$, find the third angle.

Ex. 5. If one angle of a parallelogram is a°, find the other angles.

Ex. 6. Construct exactly an angle of 60°.

Ex. 7. How large may the double of an acute angle be? how small?

Ex. 8. How large may the double of an obtuse angle be? how small?

PROPOSITION XXXVII. THEOREM

162. *Two parallelograms are equal if two adjacent sides and the included angle of one are equal, respectively, to two adjacent sides and the included angle of the other.*

Given the ▱s $ABCD$ and $A'B'C'D'$ in which $AB = A'B'$, $AD = A'D'$, and $\angle A = \angle A'$.

To prove ▱ $ABCD$ = ▱ $A'B'C'D'$.

Proof. Apply the ▱ $A'B'C'D'$ to the ▱ $ABCD$ so that $A'D'$ shall coincide with its equal AD.

Then $A'B'$ will take the direction of AB (*for* $\angle A' = \angle A$); and point B' will fall on B (*for* $A'B' = AB$).

Then $B'C'$ and BC will both be ∥ AD and will both pass through the point B.

∴ $B'C'$ will take the direction of BC, Geom. Ax. 3.
(*through a given point one straight line, and only one, can be drawn* ∥ *another given straight line*).

In like manner, $D'C'$ must take the direction of DC.

∴ C' must fall on C, Art. 64.
(*two straight lines can intersect in but one point*).

∴ ▱ $ABCD$ = ▱ $A'B'C'D'$, Art. 47.
(*geometric figures which coincide are equal*).

Q. E. D.

163. COR. *Two rectangles which have equal bases and equal altitudes are equal.*

Ex. Construct exactly an angle of 30°.

POLYGONS

164. A **polygon** is a portion of a plane bounded by straight lines, as *ABCDE*.

The **sides** of a polygon are its bounding lines; the **perimeter** of a polygon is the sum of its sides; the **angles** of a polygon are the angles formed by its sides; the **vertices** of a polygon are the vertices of its angles.

A **diagonal** of a polygon is a straight line joining two vertices which are not adjacent, as *BD* in Fig. 1.

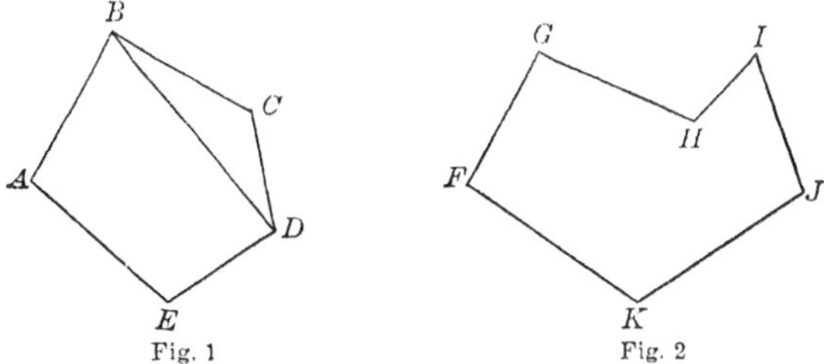

Fig. 1 Fig. 2

165. An **equilateral polygon** is a polygon all of whose sides are equal.

166. An **equiangular polygon** is a polygon all of whose angles are equal.

What four-sided polygon is equilateral but not equiangular? Also, what four-sided polygon is both equilateral and equiangular?

167. A **convex polygon** is a polygon in which no side, if produced, will enter the polygon, as *ABCDE* (Fig. 1).

Each angle of a convex polygon is less than two right angles and is called a **salient angle.**

168. A **concave polygon** is a polygon in which two or more sides, if produced, will enter the polygon, as *FGHIJK* (Fig. 2),

Some angle of a concave polygon must be greater than two right angles, as angle *GHI* of Fig. 2. Such an angle is termed a **re-entrant angle.**

If the kind of polygon is not specified in this respect, a convex polygon is meant.

169. Two mutually equiangular polygons are polygons whose corresponding angles are equal, as Figs. 3 and 4.

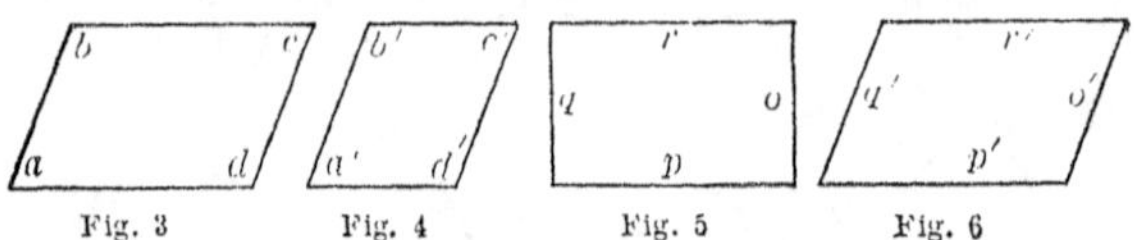

Fig. 3 Fig. 4 Fig. 5 Fig. 6

170. Two mutually equilateral polygons are polygons whose corresponding sides are equal, as Figs. 5 and 6.

From Figs. 3 and 4 it is seen that two polygons may be mutually equiangular without being mutually equilateral.

What similar truth may be inferred from Figs. 5 and 6?

171. Names of particular polygons. Some polygons are used so frequently that special names have been given to them. A polygon of three sides is called a **triangle**; one of four sides, a **quadrilateral**; one of five sides, a **pentagon**; of six sides, a **hexagon**; of seven sides, a **heptagon**; of eight sides, an **octagon**; of ten sides, a **decagon**; of twelve sides, a **dodecagon**; of fifteen sides, a **pentedecagon**; of n sides, an **n-gon.**

Ex. 1. Let the pupil illustrate Arts. 169 and 170 by drawing two pentagons that are mutually equilateral without being mutually equiangular, and another pair of which the reverse is true.

Ex. 2. Can two triangles be mutually equilateral without being mutually equiangular? What polygons can?

Ex. 3. How does the number of vertices in a polygon compare with the number of sides?

Proposition XXXVIII. Theorem

172. *The sum of the angles of any polygon is equal to two right angles taken as many times, less two, as the polygon has sides.*

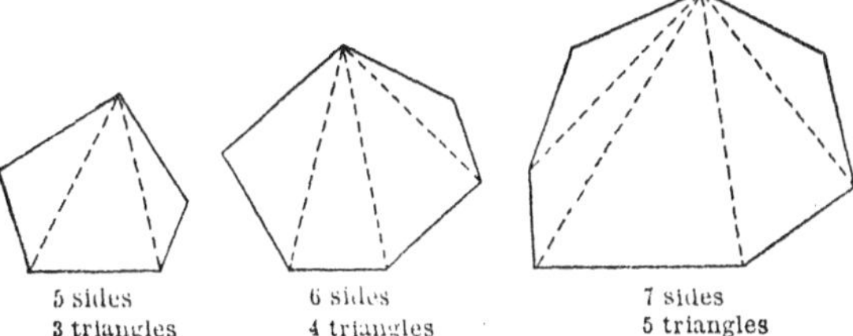

Given a polygon of n sides (the above polygons of 5, 6, 7 sides being used merely as particular illustrations, to aid in carrying forward the proof).

To prove the sum of its ∠s $= (n-2)$ 2 rt. ∠s.

Proof. By drawing diagonals from one of its vertices the polygon is divided into $(n-2)$ triangles.

Then the sum of the ∠s of each triangle = 2 rt. ∠s. Art. 134.
(*the sum of the ∠s of a △ is equal to 2 rt. ∠s*).

Hence the sum of the ∠s of the $(n-2)$ △s $= (n-2)$ 2 rt. ∠s. Ax. 4.

But the sum of the ∠s of the polygon is equal to the sum of the ∠s of the $(n-2)$ △s. Ax. 8.

Hence the sum of the ∠s of the polygon $= (n-2)$ 2 rt. ∠s. Ax. 1.

Q. E. D.

173. Cor. 1. *The sum of the angles of a polygon equals* $(2n-4)$ *rt. ∠s.*

174. Cor. 2. *In an equiangular polygon of n sides each angle equals* $\frac{(n-2)\ 2 \text{ rt. ∠s}}{n}$, *or* $\frac{2n-4}{n}$ *rt. ∠s.*

PROPOSITION XXXIX. THEOREM

175. *The sum of the exterior angles of a polygon formed by producing its sides in succession at one extremity equals four right angles.*

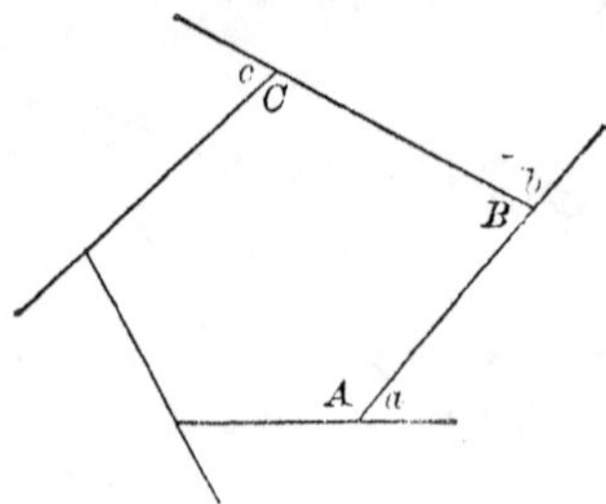

Given a polygon of n sides having its sides produced in succession.

To prove the sum of the exterior ∠s = 4 rt. ∠s.

Proof. If the interior ∠s of the polygon be denoted by A, B, C, and the corresponding exterior ∠s by a, b, c,

$$\angle A + \angle a = 2 \text{ rt. } \angle s, \qquad \text{(Why?)}$$

$$\angle B + \angle b = 2 \text{ rt. } \angle s, \qquad \text{(Why?)}$$

etc.

Adding, int. ∠s + ext. ∠s = n times 2 rt. ∠s = $2n$ rt. ∠s. Ax. 2.

But int. ∠s = $(n-2)$ 2 rt. ∠s = $2n$ rt. ∠s − 4 rt. ∠s. Art. 173.

∴ Ext. ∠s = 4 rt. ∠s.

Q. E. D.

Ex. 1. What does the sum of the interior angles of a hexagon equal? of a heptagon? of a decagon?

Ex. 2. Each angle of an equiangular pentagon contains how many degrees? of an equiangular hexagon? octagon? decagon?

Ex. 3. Would a quadrilateral constructed of rods hinged at the ends (i. e., at the vertices of the quadrilateral) be rigid? Would a triangle so constructed be rigid? Would a pentagon?

MISCELLANEOUS THEOREMS

Proposition XL. Theorem

176. *If three or more parallels intercept equal parts on one transversal, they intercept equal parts on every transversal.*

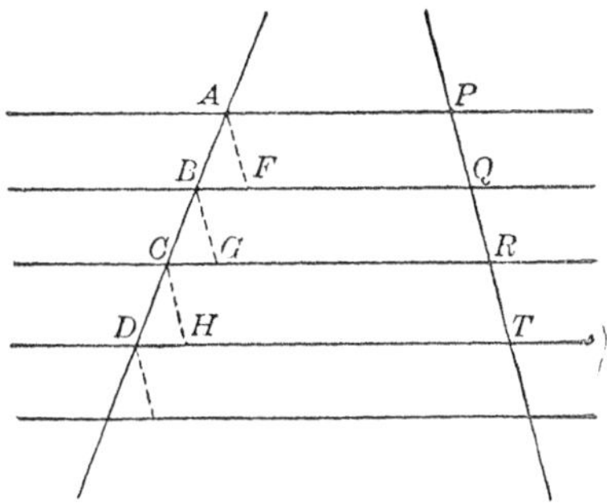

Given AP, BQ, CR and DT parallel lines intercepting equal parts AB, BC and CD on the transversal AD.

To prove that they intercept equal parts PQ, QR and RT on the transversal PT.

Proof. Through A, B and C let AF, BG and CH be drawn parallel to PT and meeting the lines BQ, CR and DT in the points F, G and H respectively.

Then the lines AF, BG and CH are $\parallel$, Art. 122.

(*two straight lines $\parallel$ a third straight line are $\parallel$ each other*).

In the $\triangle$s ABF, BCG and CDH, $\angle ABF = \angle BCG = \angle CDH$, Art. 126.

(*being ext. int.* $\angle$s *of* $\parallel$ *lines*).

Also $\angle BAF = \angle CBG = \angle DCH$, (*same reason*).

And $AB = BC = CD$. Hyp.

$\therefore \triangle ABF = \triangle BCG = \triangle CDH$. Art. 97.

$\therefore AF = BG = CH$. (Why ?)

But $AF = PQ$, $BG = QR$, $CH = RT$, Art. 157.

(*parallel lines comprehended between parallel lines are equal*).

$\therefore PQ = QR = RT$. Ax. 1.

Q. E. D.

PROPOSITION XLI. THEOREM

177. *The line which joins the midpoints of two sides of a triangle is parallel to the third side, and is equal to one-half the third side.*

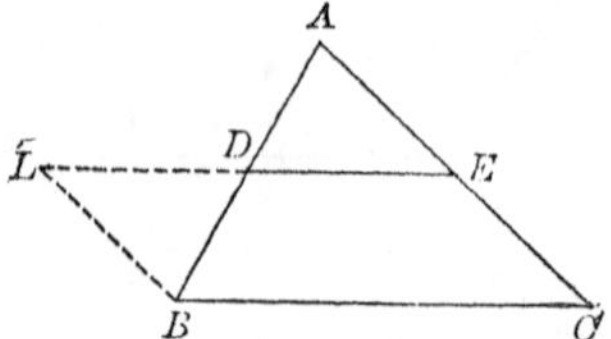

Given D the midpoint of AB, and E the midpoint of AC in the triangle ABC.

To prove $DE \parallel BC$ and $=\frac{1}{2}BC$.

Proof. Through B let BL be drawn $\parallel AC$, and meeting DE produced at L.

Then, in the $\triangle$ ADE and BDL,

$AD=BD$, (Why?)

$\angle ADE=\angle BDL$. (Why?)

$\angle DAE=\angle DBL$. (Why?)

$\therefore \triangle ADE \doteq \triangle BDL$. (Why?)

$\therefore DE=DL$, or $DE=\frac{1}{2}LE$, and $AE=BL$. (Why?)

But $AE=EC$ (Hyp.) $\therefore EC=BL$. Ax. 1.

Also $EC \parallel BL$. Constr.

$\therefore BLEC$ is a $\square$, Art. 160.

(*if two sides of a quadrilateral are equal and parallel the figure is a* $\square$).

$\therefore DE \parallel BC$. Art. 147.

Also $LE=BC$, (*opp. sides of a* $\square$ *are* $=$). Art. 155.

$\therefore \frac{1}{2}LE$, or $DE=\frac{1}{2}BC$. Ax. 5.

Q. E. D.

178. COR. *The line which bisects one side of a triangle and is parallel to another side bisects the third side.*

Thus, given $AD=DB$ and $DE \parallel BC$, then will $AE=EC$.

For, suppose a line FG drawn through $A \parallel BC$,

Then the three parallels FG, DE, BC will intercept equal parts on AB. Hyp.

$\therefore$ they intercept equal parts on AC. Art. 176.

$\therefore AE = EC$.

Proposition XLII. Theorem

179. *The line which joins the midpoints of the legs of a trapezoid is parallel to the bases and equal to one-half their sum.*

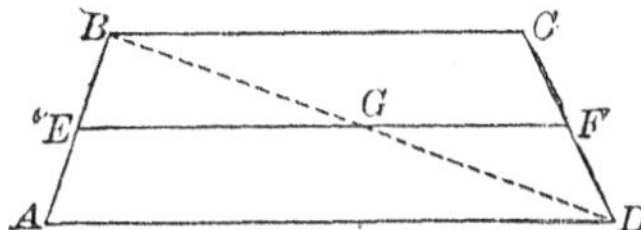

Given the trapezoid $ABCD$, E the midpoint of the leg AB, and F the midpoint of the leg CD.

To prove that a line joining E and F is $\parallel AD$ and BC and $= \frac{1}{2}(AD + BC)$.

Proof. Draw the diagonal BD and take G its midpoint. Draw EG and GF.

Then, in the $\triangle ABD$, $EG \parallel AD$ and $= \frac{1}{2} AD$, Art. 177.

(*the line which joins the midpoints of two sides of a $\triangle$ is $\parallel$ the third side and $=$ one-half the third side*).

Also, in the $\triangle BDC$, $GF \parallel BC$ and $= \frac{1}{2} BC$, (*same reason*).

$\therefore GF$ and AD both $\parallel BC$; $\therefore GF \parallel AD$, Art. 122.

(*two straight lines $\parallel$ a third straight are $\parallel$ each other*):

$\therefore EG$ and GF are both $\parallel AD$.

$\therefore EG$ and GF form one and the same straight line EF, Geom. Ax. 3.

(*through a given point one line, and only one, can be drawn $\parallel$ to another given line*).

$\therefore EF \parallel AD$ and BC.

Also $EG = \frac{1}{2} AD$, and $GF = \frac{1}{2} BC$.

Adding, $EG + GF$, or $EF = \frac{1}{2}(AD + BC)$. Ax. 2.

Q. E. D.

180. Cor. *A line drawn bisecting one leg of a trapezoid and parallel to the base bisects the other leg also.*

Proposition XLIII. Theorem

181. *The bisectors of the angles of a triangle intersect at a common point (called the* **in-center**).

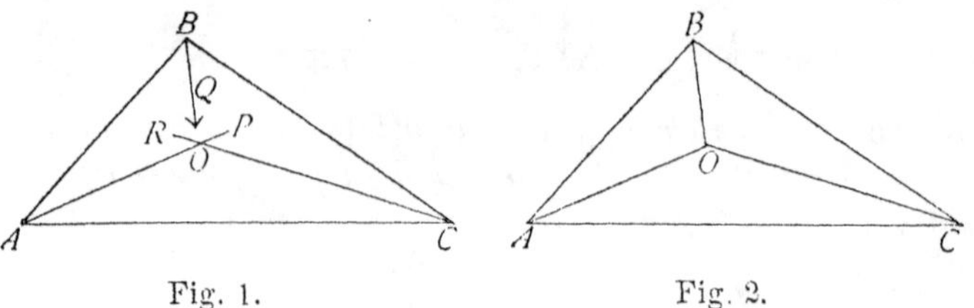

Fig. 1. Fig. 2.

Given the △ ABC with the lines AP, BQ, CR (Fig. 1) bisecting the ∠s BAC, ABC, ACB, respectively.

To prove that AP, BQ, CR intersect in a common point.

Proof. Let AP, the bisector of ∠ BAC, and CR, the bisector of ∠ BCA, intersect at the point O.

Then O, being in bisector AP, is equidistant from AB and AC, Art. 117.

(*every point in the bisector of an ∠ is equidistant from the sides of the ∠*).

Also O, being in bisector CR, is equidistant from AC and BC, (*same reason*).

Hence O is equidistant from the sides AB and BC. Ax. 1.

∴ O is in the bisector of ∠ ABC, Art. 117.

(*every point equidistant from the sides of an ∠ lies in the bisector of the ∠*).

Hence BQ, or BQ produced, passes through O.

Hence the bisectors, AP, BQ, CR, of the three ∠s of the △ ABC intersect at O. Q. E. D.

182. Cor. *The point in which the three bisectors of the angles of a triangle intersect is equidistant from the three sides of the triangle.*

183. Def. **Concurrent lines** are lines which pass through the same point.

Ex. Find other ∠ of above figure if ∠ $BAC = 72°$ and ∠ $BCA = 44°$.

PROPOSITION XLIV. THEOREM

184. *The perpendicular bisectors of the sides of a triangle intersect at a common point (called the* **circumcenter***).*

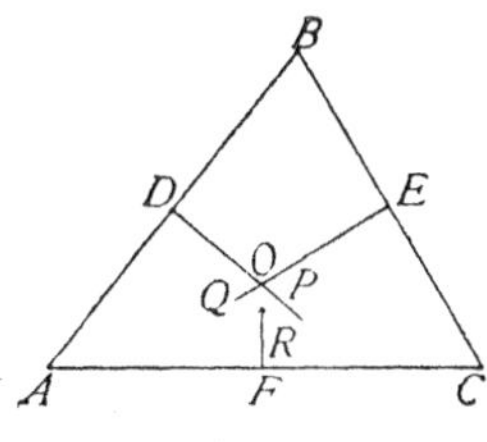

Fig. 1.

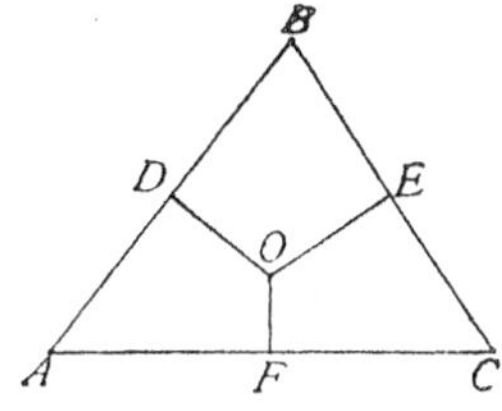

Fig. 2.

Given the $\triangle$ ABC with DP, EQ, FR (Fig. 1), the $\perp$ bisectors of the sides AB, BC, CA, respectively.

To prove that DP, EQ, FR intersect at a common point.

Proof. Let DP and EQ intersect at the point O.

Then O, being in $\perp$ DP, is equidistant from the points A and B, Art. 112.

(*every point in the perpendicular bisector of a line is equally distant from the extremities of the line*).

Also O, being in $\perp$ EQ, is equidistant from the points B and C,

(*same reason*).

Hence O is equidistant from A and C. Ax. 1.

$\therefore$ O is in the $\perp$ bisector of AC, Art. 115.

(*the $\perp$ bisector of a line is the locus of all points equidistant from the extremities of the line*).

Hence FR, or FR produced, passes through O.

Hence the perpendicular bisectors, DP, EQ, FR, of the three sides of the $\triangle$ ABC meet in the point O. Q. E. D.

185. COR. *The point in which the perpendicular bisectors of the sides of a triangle meet is equidistant from the vertices of the triangle.*

PROPOSITION XLV. THEOREM

186. *The perpendiculars from the vertices of a triangle to the opposite sides meet in a point* (*called the* **ortho-center**).

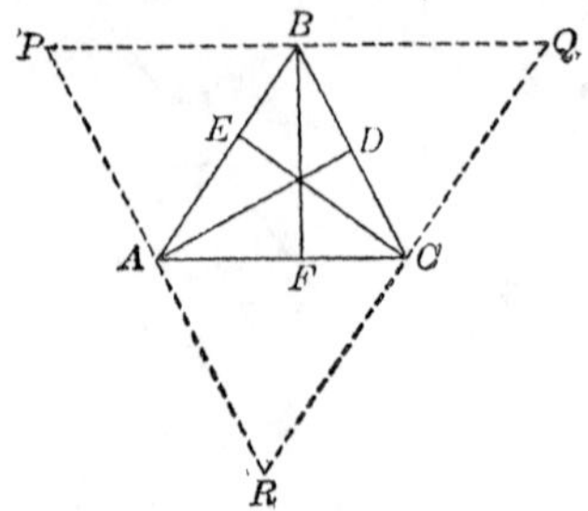

Given AD, BF, and CE the perpendiculars from the vertices A, B, and C of the $\triangle ABC$ to the opposite sides.

To prove that AD, BF, and CE intersect in a common point.

Proof. Through A, B, and C let the lines PR, PQ and QR be drawn $\parallel$ BC, AC, and AB, respectively, and forming the $\triangle PQR$.

Then $AD \perp PR$, Art. 123.

(*for* $AD \perp BC$, *and a line* $\perp$ *one of two* $\parallel$ *lines is* $\perp$ *the other also*).

Also $APBC$ and $ABCR$ are ▱s. Constr.

$\therefore AP = BC$, and $AR = BC$, Art. 155.

(*the opposite sides of a* ▱ *are* $=$).

$\therefore AP = AR$. Ax. 1.

Hence, in the $\triangle PQR$, AD is the perpendicular bisector of side PR.

In like manner it may be shown that BF is the perpendicular bisector of PQ, and that CE is the perpendicular bisector of QR.

Hence AD, BF, and CE are the perpendicular bisectors of the sides of the $\triangle PQR$.

$\therefore AD$, BF, and CE meet in a common point, Art. 184.

(*the perpendicular bisectors of the sides of a* $\triangle$ *are concurrent*).

Q. E. D.

PROPOSITION XLVI. THEOREM

187. *The medians of a triangle intersect (or are concurrent) in a point (called the* **centroid**) *which cuts off two-thirds of each median from its vertex.*

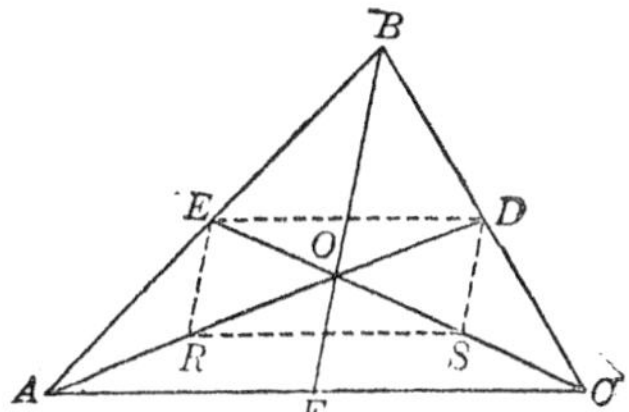

Given AD, BF and CE the three medians of $\triangle ABC$.

To prove that AD, BF and CE intersect at a point which cuts off two-thirds of each median from its vertex.

Proof. Let the medians AD and CE intersect at O.

Take R the midpoint of AO, and S the midpoint of OC, and draw RE, ED, DS, and SR.

Then $ED \parallel AC$ and $= \frac{1}{2} AC$. Art. 177.

Also, in the $\triangle AOC$,

$RS \parallel AC$ and $= \frac{1}{2} AC$. Art. 177.

Hence $ED \parallel RS$ and $= RS$. Art. 122 and Ax. 1.

$\therefore$ $REDS$ is a ▱. (Why?)

Hence ES and RD bisect each other. (Why?)

But $AR = RO$, and $CS = SO$. Constr.

$\therefore$ $AR = RO = OD$, and $CS = SO = OE$. Ax. 1.

Hence CE crosses AD at a point O, such that $AO = \frac{2}{3} AD$.

In like manner it may be shown that BF crosses AD at the point O.

Hence the medians AD, BF, and CE intersect at point O, which cuts off two-thirds of each median from its vertex.

Q. E. D.

188. Properties of rectilinear figures to be proved by the pupil. Proof of equality of triangles. Other properties of rectilinear figures will now be given which are to be demonstrated by the pupil. These theorems will be arranged in groups, according to the method of proof to be used, followed by a group of general or mixed exercises.

Let the pupil form a list of the conditions that make two triangles equal.

(See Arts. 96, 97, 98, 101, 102, 141, 142.)

EXERCISES. GROUP 4

EQUALITY OF TRIANGLES

Ex. 1. Given ABC any triangle, BO the bisector of $\angle ABC$, and $AD \perp BO$; prove $\triangle ABO = \triangle BOD$.

[SUG. In the $\triangle$ ABO and DBO what lines are equal? What $\angle$ are equal? etc.]

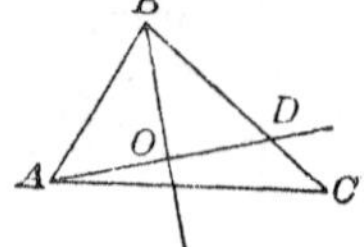

Ex. 2. If, at any point in the bisector of an angle, a $\perp$ be erected and produced to meet the sides of the angle, how many triangles are formed? Are these triangles equal? Prove this.

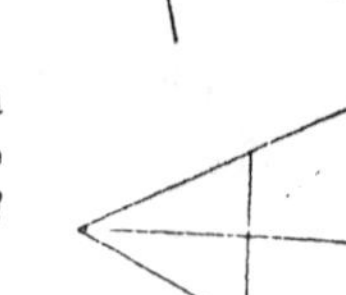

Ex. 3. If, through the midpoint of a given straight line, another line be drawn, and produced to meet the perpendiculars erected at the ends of the given line, the triangles so formed are equal.

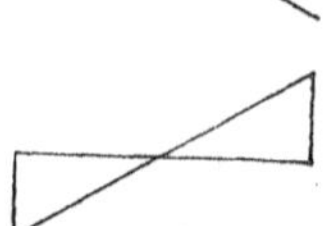

Ex. 4. If two straight lines bisect each other and their extremities be joined, how many pairs of equal triangles are formed? Prove this.

Ex. 5. If equal segments from the base be laid off on the sides of an isosceles triangle, and lines be drawn from the extremities of the segments to the opposite vertices, prove that two pairs of equal triangles are formed.

Ex. 6. If, upon the sides of an angle, equal segments be laid off from the vertex, and lines be drawn from the ends of these segments to any point in the bisector of the angle, prove that the triangles formed are equal.

Ex. 7. If two sides of a triangle be produced, each its own length, through the vertex in which they meet, and the extremities of the produced parts be joined, prove that a new triangle is formed which equals the original triangle.

Ex. 8. Given $AB=DC$, and $BC=DA$; prove $\triangle BAC=\triangle DAC$. What other pair of equal triangles is there in the figure?

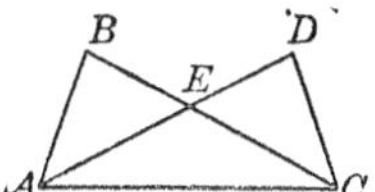

Ex. 9. Two right triangles are equal if their corresponding legs are equal.

Ex. 10. The altitudes from the extremities of the base of an isosceles triangle upon the legs of the triangle divide the figure into how many pairs of equal triangles? Prove this.

Ex. 11. In a given quadrilateral two adjacent sides are equal and a diagonal bisects the angle between these sides. Prove that the diagonal bisects the quadrilateral.

Ex. 12. If, from the ends of the shorter base of an isosceles trapezoid, lines be drawn parallel to the legs and produced to meet the other base, prove that a pair of equal triangles is formed.

189. Proof of the equality of lines. There are several methods of proving that two lines (segments) are equal. *One of the principal methods of proving that two lines are equal is by proving that two triangles, in which the given lines form homologous parts, are equal.*

EXERCISES. GROUP 8

EQUALITY OF LINES

Ex. 1. Given ABC any triangle, BO the bisector of $\angle ABC$, and $AD \perp BO$; prove $AO = OD$.

Ex. 2. If, at any point in the bisector of an angle, a perpendicular be erected to the bisector and produced to meet the sides of the angle, the perpendicular is divided into two equal parts at the given point.

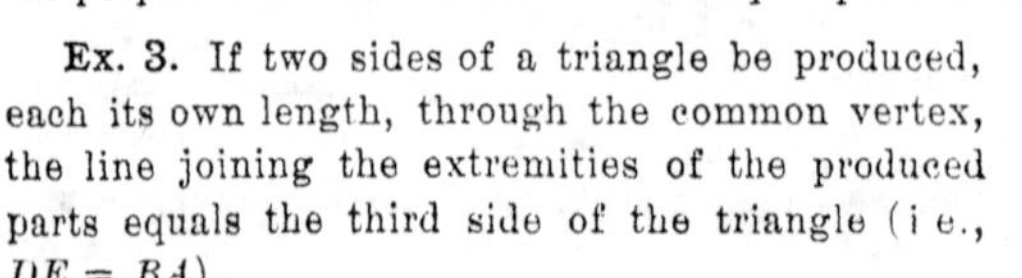

Ex. 3. If two sides of a triangle be produced, each its own length, through the common vertex, the line joining the extremities of the produced parts equals the third side of the triangle (i e., $DE = BA$).

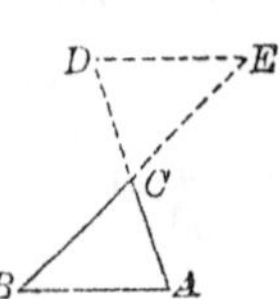

Ex. 4. If equal segments from the base be laid off on the legs of an isosceles triangle, lines drawn from the ends of these segments to the opposite vertices are equal.

Ex. 5. Given $AR \parallel QB$, and $AP = PB$; prove $RP = PQ$.

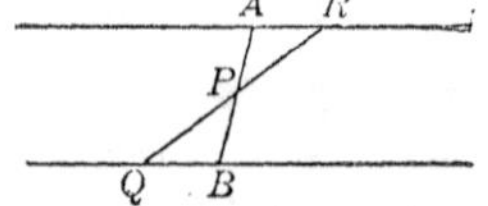

Ex. 6. The bisector of the vertical angle of an isosceles triangle bisects the base.

Ex. 7. The altitudes of an isosceles triangle upon the legs are equal.

Ex. 8. The diagonals of a rectangle are equal.

Ex. 9. The medians of an isosceles triangle to the legs are equal.

Ex. 10. If two altitudes of a triangle are equal, the triangle is isosceles.

Ex. 11. The perpendiculars to a diagonal of a parallelogram from a pair of opposite vertices are equal.

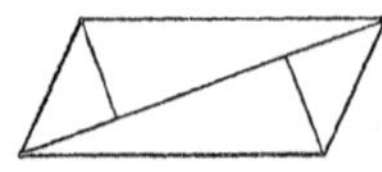

Ex. 12. If the equal sides of an isosceles triangle be produced through the vertex so that the produced parts are equal, the lines joining the extremities of the produced parts to the extremities of the base are equal.

Ex. 13. If the base of an isosceles triangle be trisected, lines drawn from the vertex to the points of trisection are equal.

Lines may also be proved equal by showing that they are:
opposite equal angles in a triangle;
or *opposite sides of a parallelogram;*
or *parallel lines comprehended between parallel lines, etc.*
(See Arts. 100, 155, 157, etc.).

Ex. 14. If the exterior angles at the base of a triangle are equal, the triangle is isosceles.

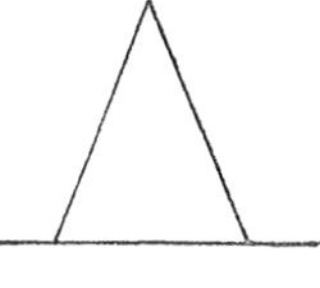

Ex. 15. In the bisectors of the equal angles of an isosceles triangle, the segments next to the base are equal ($AO = OP$).

Ex 16. In $\triangle ABC$, given $AB = AC$, $DE \parallel BC$; prove $AD = AE$.

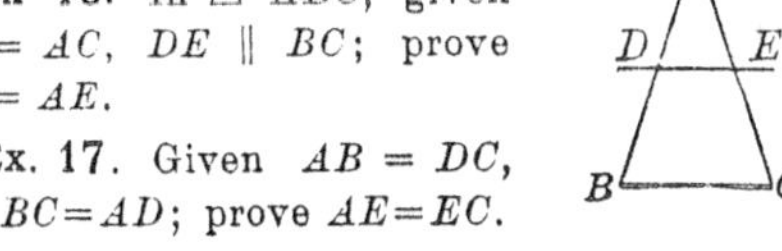

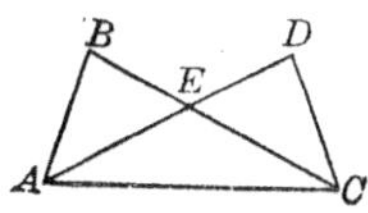

Ex. 17. Given $AB = DC$, and $BC = AD$; prove $AE = EC$.

190. Proof of the equality of angles may be obtained in several different ways.

One of the principal methods of proving that two angles are equal is by proving that two triangles, in which the given angles form homologous parts, are equal.

EXERCISES. GROUP 6

EQUALITY OF ANGLES

Ex. 1. Given ABC any $\triangle$, BO the bisector of the $\angle ABC$, and $AD \perp BO$; prove $\angle BAO = \angle BDO$.

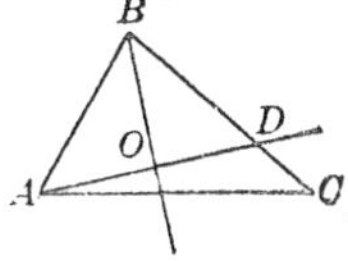

Ex. 2. If, at any point in the bisector of an angle, a perpendicular be erected and produced to meet the sides of the angle, the perpendicular makes equal angles with the sides of the angle.

Ex. 3. Given $AB = DC$, and $AD = BC$; prove $\angle B = \angle D$.

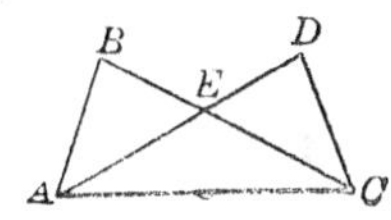

Ex. 4. If equal segments from the base be laid off on the legs of an isosceles triangle, the lines drawn from the extremities of the segments to the opposite vertices make equal angles with the base.

Ex. 5. The median to the base of an isosceles triangle is perpendicular to the base.

Ex. 6. The altitudes upon the legs of an isosceles triangle make equal angles with the base.

Ex. 7. The diagonals of a rhombus bisect its angles.

Angles may also be proved equal by proving that they:

are opposite equal sides in an isosceles triangle;
or *are vertical angles;*
or *are complements (or supplements) of equal angles;*
or *by the use of the properties of parallel lines;*
or *that their sides are parallel, or perpendicular.*

(See Arts. 75, 78, 99, 124, 126, 130, 132.)

Ex. 8. In an isosceles triangle the exterior angles made by producing the base are equal.

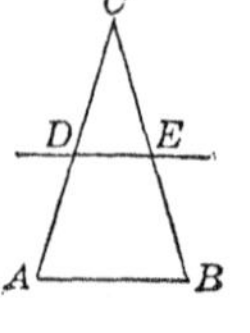

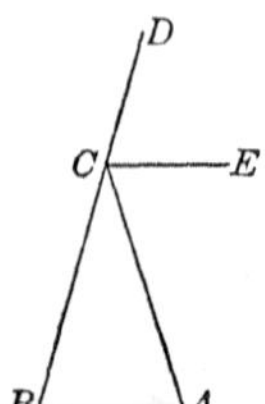

Ex. 9. Given $AC = CB$, and $DE \parallel AB$; prove $\angle CDE = \angle CED$.

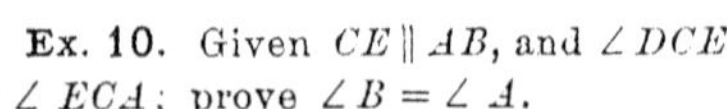

Ex. 10. Given $CE \parallel AB$, and $\angle DCE = \angle ECA$; prove $\angle B = \angle A$.

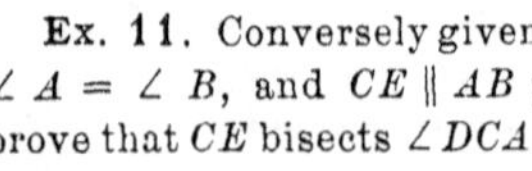

Ex. 11. Conversely given $\angle A = \angle B$, and $CE \parallel AB$; prove that CE bisects $\angle DCA$.

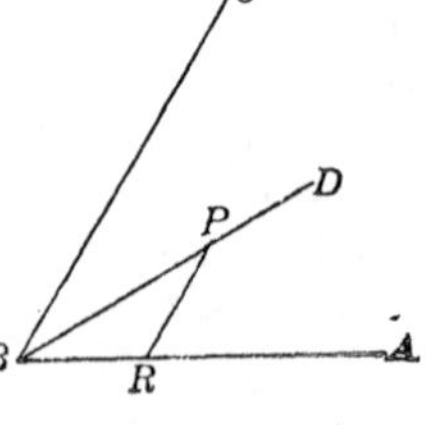

Ex. 12. Given BD the bisector of the angle ABC, and $PR \parallel CB$; prove PBR an isosceles $\triangle$. Let the pupil state this theorem in general language.

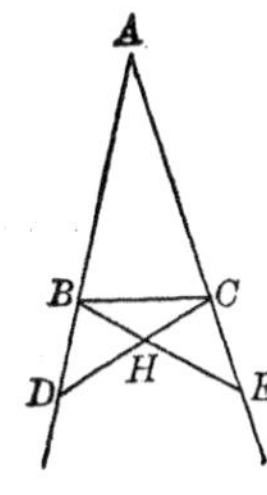

Ex. 13. Given $AB = AC$, and $BD = CE$; prove $\angle BCD = \angle CBE$. How many pairs of equal $\triangle$ in the figure? of equal lines? of equal $\angle$?

Ex. 14. A line drawn through the vertex of an angle, perpendicular to the bisector of the angle, makes equal angles with the sides of the given angle.

Ex. 15. If a straight line which bisects one of two vertical angles be produced, it bisects the other vertical angle also.

191. Proof that two lines are parallel may be obtained by showing that:

the lines are cut by a transversal, making the alternate interior angles equal;

or *making the exterior interior angles equal;*

or *making the interior angles on the same side of the transversal supplementary;*

or *that the lines are opposite sides of a parallelogram;*

or *that one of the lines joins the midpoints of two sides of a triangle, and the other line is the third side of the triangle.*

(See Arts. 125, 127, 129, 147, 177, etc.)

EXERCISES. GROUP 7

PARALLEL LINES

Ex. 1. If two sides of a triangle be produced, each its own length, through the common vertex, the line joining their extremities is parallel to the third side of the triangle.

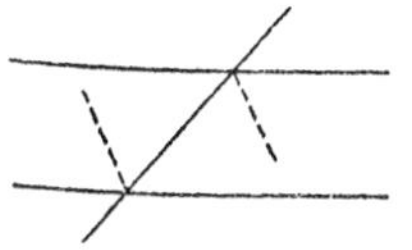

Ex. 2. The bisectors of two alternate interior angles of parallel lines are parallel.

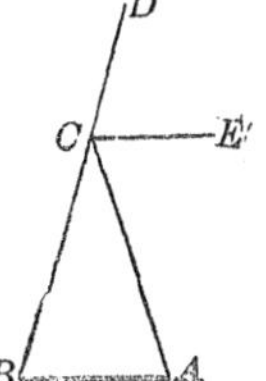

Ex. 3. Given $\angle A = \angle B$, and $\angle DCE = \angle ECA$; prove $CE \parallel BA$.

Ex. 4. The bisectors of the opposite angles of a parallelogram are parallel.

Ex. 5. Lines perpendicular to parallel lines are parallel (or coincide).

Ex. 6. Two straight lines are parallel if two points on one line are equidistant from the other line.

192. The **proof of a numerical property of the rectilinear figures** (of Book I) usually depends on one of the following:

The sum of the angles about a given point on the same side of a straight line passing through the point is 180°;

or *the sum of the angles about a point is* 360°;

or *the sum of the angles of a triangle is* 180°;

or *the sum of the interior angles of parallel lines on the same side of a transversal is* 180°;

or *the sum of the interior angles of a polygon of* n *sides is* $(n-2)$ 180°;

or *the sum of the exterior angles of a polygon is* 360°.

(See Arts. 76, 77, 128, 134, 172, 175.)

EXERCISES. GROUP 8

NUMERICAL PROPERTIES

Ex. 1. If an exterior angle of a triangle is 123° and an opposite interior angle is 38°, find the other two angles of the triangle.

Ex. 2. Find the angle formed by the bisectors of the two acute angles of a right triangle.

Ex. 3. If two angles of a triangle are 50° and 60°, find the angle formed by their bisectors. Find the same if the two angles contain $p°$ and $q°$.

Ex. 4. If the vertex angle of an isosceles triangle is 40° and a perpendicular is drawn from an extremity of the base to the opposite side, find the angles of the figure.

Ex. 5. If the vertex angle of an isosceles triangle is 40°, find the angle included between the altitudes drawn from the extremities of the base to the opposite sides.

Ex. 6. How many degrees in each angle of an equiangular dodecagon? of an equiangular n-gon?

Ex. 7. How many diagonals are there in a pentagon? in a hexagon? in a decagon? in an n-gon?

The **methods of proving that a given angle is a right angle** (or **that a given line is perpendicular to another given line**), or **that one angle is the supplement of another,** are closely related to the *above methods of obtaining the numerical values of given angles.*

Ex. 8. Any pair of adjacent angles of a parallelogram is supplementary.

Ex. 9. If one angle of a parallelogram is a right angle the figure is a rectangle.

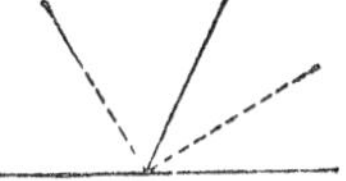

Ex. 10. The bisectors of two supplementary adjacent angles form a right angle (are perpendicular).

Ex. 11. The bisectors of two interior angles on the same side of a transversal to two parallel lines form a right angle.

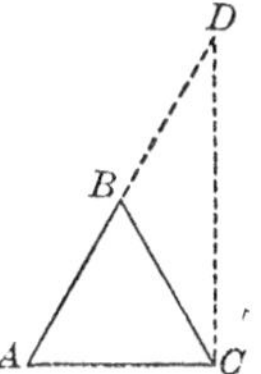

Ex. 12. If one of the legs (AB) of an isosceles triangle be produced its own length (BD) and its extremity (D) be joined to the other end of the base (C), the line last drawn (DC) is perpendicular to the base.

193. Algebraic method of proving theorems. The proof of certain properties of a geometric figure is often facilitated by the *use of an algebraic symbol for an unknown angle or an unknown line of the figure, and the use of an equation or other algebraic method of solution.*

EXERCISES. GROUP 9

ALGEBRAIC METHOD

Ex. 1. Find the number of degrees in an angle which equals twice its complement?

[SUG. Let x = the complement, etc.]

Ex. 2. Find the number of degrees in an angle which equals its supplement? in one which equals one-third its supplement?

Ex. 3. The angular space about a point is divided into four angles which are in the ratio 1, 2, 3, 4. Find the number of degrees in each angle.

[SUG. $x + 2x + 3x + 4x = 360°$, etc.].

Ex. 4. The angles of a triangle are in the ratio 1, 2, 3; find the angles.

Ex. 5. Two angles are supplementary and the greater exceeds the less by 30°; find the angles.

Ex. 6. Find all the angles of a parallelogram if one of them is double another angle.

Ex. 7. One of the base angles of a triangle is double the other, and the exterior angle at the vertex is 105°. Find the angles of the triangle.

Ex. 8. How many sides has a polygon the sum of whose angles is fourteen right angles?

[SUG. $2(n-2)=14$; find n.]

Ex. 9. How many sides has a polygon the sum of whose angles is ten right angles? twenty right angles? 720°?

Ex. 10. How many sides has an equiangular polygon one of whose angles is seven-fourths of a right angle?

Ex. 11. How many sides has a polygon the sum of whose interior angles equals the sum of the exterior angles?

Ex. 12. How many sides has a polygon the sum of whose interior angles equals three times the sum of the exterior angles?

Ex. 13. If the base of any triangle be produced in both directions, the sum of the exterior angles thus formed, diminished by the vertex angle, is equal to two right angles.

[SUG. $180° - a + 180° - b - (180° - a - b) =$, etc.]

a b

Ex. 14. The bisectors of the base angles of an isosceles triangle include an angle which is equal to the exterior angle at the base.

[SUG. To prove $a = b$, denote one of the base ∠s by $2x$, etc.]

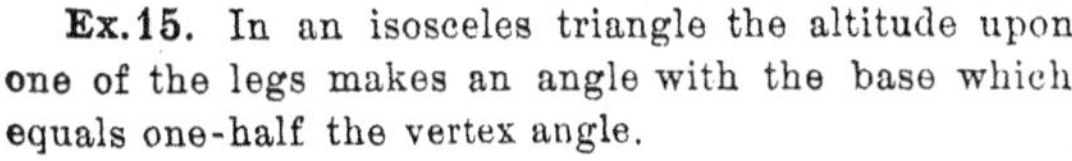

Ex. 15. In an isosceles triangle the altitude upon one of the legs makes an angle with the base which equals one-half the vertex angle.

[SUG. To prove $a = \frac{1}{2}b$, show that $a = 90° - x$, $b = 180° - 2x$, etc.]

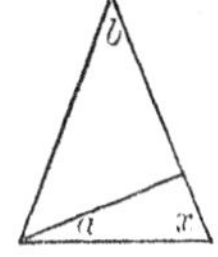

Ex. 16. If the opposite angles of a quadrilateral are equal, the figure is a parallelogram.

[SUG. $2x + 2y = 360°$, etc.]

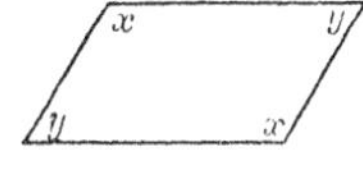

194. Use of auxiliary lines. Inequalities. The demonstration of a property of a geometrical figure is frequently facilitated by drawing one or more auxiliary lines on the figure. For examples of the use of such lines, see Props. III, V, IX, etc., of Book I.

Some of the principal auxiliary lines used on rectilinear figures are:

a line connecting two given points;
a line through a given point parallel to a given line;
a line through a given point perpendicular to a given line;
a line making a given angle with a given line;
a line produced its own length, etc.

EXERCISES. GROUP 10

AUXILIARY LINES

Ex. 1. In the quadrilateral $ABCD$ given $AB = AD$, and $BC = CD$; prove $\angle B = \angle D$.

[SUG. Draw AC, etc.]

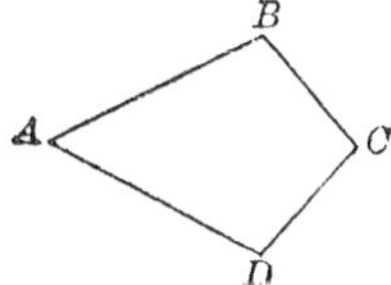

Ex. 2. Prove that the angles at the base of an isosceles trapezoid are equal.

Ex. 3. State and prove the converse of Ex. 2.

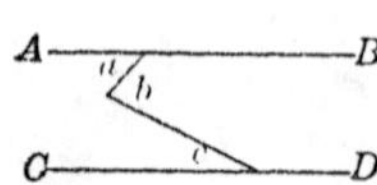

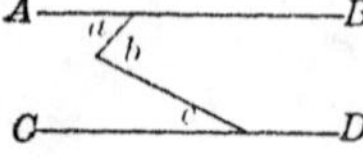

Ex. 4. Given $AB \parallel CD$; prove $\angle b = \angle a + \angle c$.

Ex. 5. Conversely, given $\angle b = \angle a + \angle c$; prove $AB \parallel CD$.

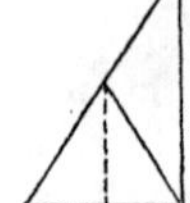

Ex. 6. The median to the hypotenuse of a right triangle is one-half the hypotenuse.

Ex. 7. If one acute angle of a right triangle is double the other, the hypotenuse is double the shorter leg.

[SUG. Draw the median to the hypotenuse, etc.]

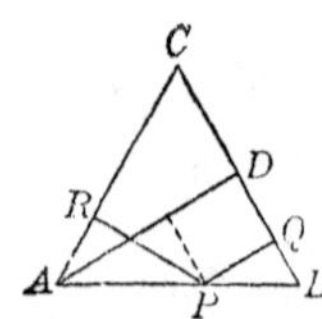

Ex. 8. In an isosceles triangle, the sum of the perpendiculars drawn from any point in the base to the legs is equal to the altitude upon one of the legs.

In some cases it is useful to draw *two or more auxiliary lines.*

Ex. 9. Show that the median of a trapezoid equals one-half the sum of the two bases by drawing a line through the midpoint of one leg of the trapezoid, parallel to the other leg and meeting one base and the other base produced.

Ex. 10. Lines joining the midpoints of the sides of a quadrilateral taken in order form a parallelogram.

[SUG. Draw the diagonals of the quadrilateral and use Art. 177.]

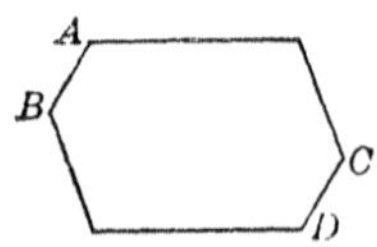

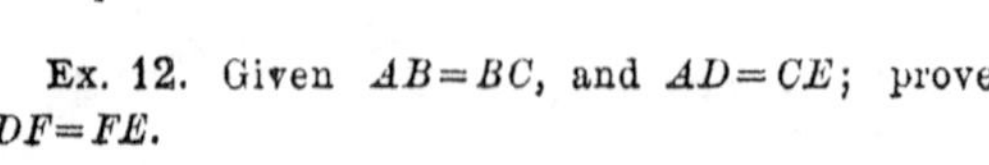

Ex. 11. If the opposite sides of a hexagon are equal and one pair of sides (AB and CD) are parallel, the opposite angles of the hexagon are equal.

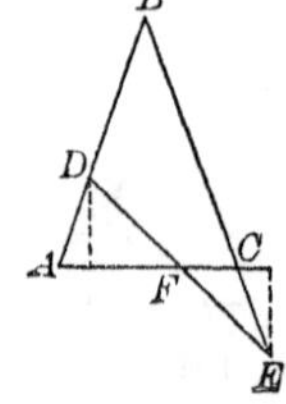

Ex. 12. Given $AB = BC$, and $AD = CE$; prove $DF = FE$.

Let the student form a list of the principles proved in Book I concerning unequal lines and unequal angles.
(See Arts. 92, 93, 95, etc.).

EXERCISES. GROUP 11

INEQUALITIES

Ex. 1. In the triangle ABC let P be any point in the side BC; prove $AB + BC > AP + PC$.

Ex. 2. In the quadrilateral $ABCD$ let F be any point in the side BC; prove perimeter of $ABCD >$ perimeter of AFD.

Ex 3. In the triangle ABC let D be any point in AB and F any point in BC. Prove $AB + BC > AD + DF + FC$.

Ex. 4. The sum of the four sides of a quadrilateral is greater than the sum of the diagonals.

Ex. 5. If, from any point within a triangle, lines be drawn to the vertices, the sum of the lines drawn is greater than one-half the sum of the sides of the triangle.

[Sug. Use Art. 92 three times, etc.]

Ex. 6. If, from any point, within a triangle, lines be drawn to the vertices, the sum of the lines drawn is less than the sum of the sides of the triangle.

[Sug. Use Art. 95.]

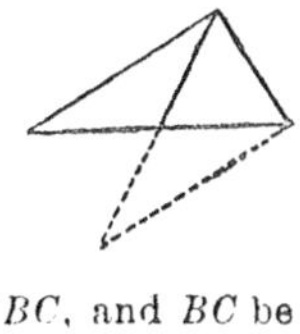

Ex. 7. The median to any side of a triangle is less than half the sum of the other two sides.

[Sug. Produce the median its own length.]

Ex. 8. If B is the vertex of an isosceles triangle ABC, and BC be produced to the point D, $\angle BAD$ is greater than $\angle BDA$.

Ex. 9 In the figure p. 71, $BC > AB$; prove $\angle BFC$ greater than $\angle BFA$. [Sug. Use Art. 108.]

Ex. 10 In the same figure, show that $FC < BC$.

Ex. 11. In the quadrilateral $ABCD$, AD is the longest side and BC the shortest; prove $\angle ABC$ greater than $\angle ADC$, and $\angle BCD$ greater than $\angle BAD$.

Ex. 12 Lines are drawn from A, B, C, D, four points in a straight line, to the point F outside of the line. Which angles on the figure are less than angle ACF? Which angles are greater?

195. Indirect demonstrations. Loci. In Book I three methods of indirect proof have been used.

1. The **reduction to an absurdity** (reductio ad absurdum), that is, the *proof that the negative of a given theorem leads to an absurdity* (see Prop. X).

2. The **method of exclusion,** that is, *showing that any other statement than the given theorem cannot be true* (see Props. XV, XVII). This method is a special case of the preceding, the negative of a given theorem being divided in it into two parts which are separately shown to be impossible.

3. The **method of coincidence,** that is, *proof that a given line coincides with another line, which fulfils certain required conditions* (see Props. XVII, XXIII, XXV, etc.).

EXERCISES. GROUP 12

INDIRECT, OR NEGATIVE DEMONSTRATIONS

Prove the following by an indirect method:

Ex. 1. Every point within an angle and not in the bisector of the angle is unequally distant from the sides of the angle.

[SUG. In the given angle take P any point not in the bisector of the angle. Then, if P is not unequally distant from AC and OB, it must be equally distant from them, etc.]

Ex. 2. If two straight lines are cut by a transversal, making the alternate interior angles unequal, the lines are not parallel.

Ex. 3. The line joining the midpoints of two sides of a triangle is parallel to the third side.

[SUG. Through one of the midpoints draw a line ∥ to the third side, show that it bisects the second side and that the line joining the midpoints coincides with it.]

Ex. 4. If, from a point P in a line AB, lines PC and PD be drawn on opposite sides of AB making the angle APC equal to the angle BPD, PC and PD are in the same straight line.

[SUG. From P draw PQ in the same straight line with PC and show that PD coincides with it.]

Ex. 5. The bisectors of two vertical angles are in the same straight line.

Ex. 6. In the triangle ABC, D is any point in the side AB, and E is any point in the side AC. Prove that BE and DC cannot bisect each other.

EXERCISES. GROUP 13

LOCI

Ex. 1. What is the locus of all points at a given distance, a, from a given line? Prove this.

Ex. 2. What is the locus of all points equidistant from two given parallel lines? Prove this.

By use of known loci (see Arts. 112–118), prove the following:

Ex. 3. The diagonals of a rhombus are perpendicular to each other. [SUG. See Art. 113.]

Ex. 4. The median of an isosceles triangle is perpendicular to the base.

Ex. 5. The line that joins the vertices of two isosceles triangles on the same base is perpendicular to the base.

196. General method of obtaining a demonstration of a theorem. Analysis.

A clue to the solution of some of the more difficult theorems is often obtained by proceeding thus:

Assume the proposed theorem as true; observe what other relation among the parts of the figure must then be true; proceed backward thus, step by step, till the required theorem is found to depend on some known truth; then, starting with this known truth, reverse the steps taken, and thus build up a direct proof of the required theorem.

This method is called **solution by analysis.**

The following is a simple example of the use of this method:

Ex. Given AB and AC the legs of an isosceles triangle and D any point on AB; prove DC greater than DB.

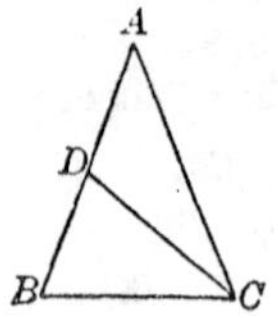

ANALYSIS. If $DC > DB$,

$\angle B$ is greater than $\angle DCB$. Art. 104.

Hence substituting for $\angle B$ its equal, $\angle ACB$, we have $\angle ACB$ is greater than $\angle DCB$. Ax. 8.

But we know that $\angle ACB > \angle DCB$. Ax. 7.

Hence, DIRECT PROOF (or SYNTHESIS)

$\angle ACB$ is greater than $\angle DCB$, Ax. 7.

$\therefore \angle B$ is greater than $\angle DCB$. Ax. 8.

$\therefore DC > DB$. Art. 106.

Q. E. D.

The first part (analysis) of the above process is to be purely mental work on the part of the pupil, in investigating a given theorem; the second part (the direct proof, or synthesis) is to be written out as the required solution.

In working the following exercises, this method will be found to be necessary in the solution of only a few of the more difficult theorems.

EXERCISES. GROUP 14

THEOREMS PROVED BY VARIOUS METHODS

Ex. 1. If two opposite angles of a quadrilateral are bisected by the diagonal connecting their vertices, the quadrilateral is bisected by this diagonal.

Ex. 2. Perpendiculars drawn from the extremities of the base of a triangle to the median to the base, are equal.

Ex. 3. If the perpendiculars from the extremities of the base of a triangle to the other two sides are equal, (1) these perpendiculars make equal angles with the base, (2) the triangle is isosceles.

Ex. 4. If the lines AB and CD intersect, then $AB+CD > AC+DB$.

Ex. 5. The vertex angle of an isosceles triangle is 44°, and one of the base angles is bisected by a line produced to meet the opposite side. Find all the angles of the figure.

Ex. 6. In the figure of Prop. V prove that $\angle APC$ is greater than $\angle ABC$. Also prove the same in another way by means of an auxiliary line drawn through B and P.

Ex. 7. Perpendiculars drawn from the midpoint of the base of an isosceles triangle to the legs are equal.

Ex. 8 State and prove the converse of Ex. 7.

Ex. 9. In an isosceles triangle an exterior angle at the base equals a right angle increased by one-half the vertical angle.

Ex 10. In a re-entrant quadrilateral the exterior angle at the re-entrant vertex equals the sum of the three opposite interior angles ($\angle d = \angle a + \angle b + \angle c$).

[SUG Draw an auxiliary line.]

Ex 11 In the equilateral triangle ABC, BC is produced to D; prove $BD > AD > AB$.

Ex. 12. If from a point in the bisector of an oblique angle lines are drawn parallel to the sides of the angle, the quadrilateral formed is a rhombus.

Ex. 13. If two angles of a quadrilateral are supplementary, the other two angles are supplementary.

Ex 14. If the median of a triangle is perpendicular to the base, the triangle is isosceles.

Ex. 15. Given $AB = AC$,

$\angle BAC = 4 \angle B$,

and $DF \perp BC$,

prove $\triangle EFA$ equilateral.

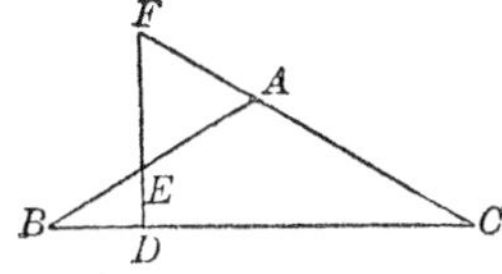

[ANALYSIS. If $\triangle FAE$ is equilateral, $\angle FEA = 60°$; $\therefore \angle DEB = 60°$; $\therefore \angle B = 30°$. Hence, to get a direct proof, show that $\angle C = 30°$ by using $\angle BAC = 4 \angle C$.]

Ex. 16. If the diagonals of a quadrilateral bisect each other at right angles, what kind of a figure is the quadrilateral? Prove this.

Ex. 17. From the point in which the altitudes drawn to the legs of an isosceles triangle intersect, a line is drawn to the vertex. Prove that this line bisects the angle at the vertex.

Ex. 18. If from a point within an acute angle perpendiculars are drawn to the sides of the angle, the angle formed by these perpendiculars is the supplement of the given angle.

Ex. 19. Lines joining the midpoints of the sides of a triangle divide the triangle into four equal triangles.

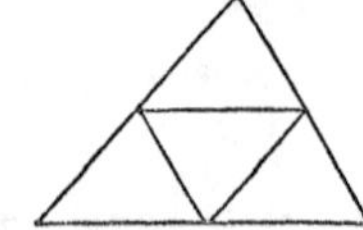

Ex. 20. If, in the parallelogram $ABCD$, $BP = DQ$, then $AQCP$ is a parallelogram.

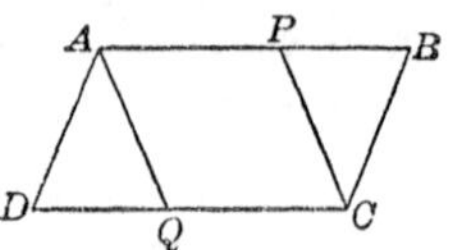

[ANALYSIS. If $AQCP$ is a ▱, AP = and ∥ QC. ∴ begin the direct proof by showing that AP = and is ∥ QC.]

Ex. 21. If the diagonals of a parallelogram are equal, what kind of a figure is the parallelogram? Prove this.

Ex. 22. If the angle A of the triangle ABC is 50° and the exterior angle BCD is 120°, which is the largest side in the triangle?

Ex. 23. Two triangles are equal if two sides and the median to one of these sides in one triangle are equal, respectively, to two homologous sides and a median in the other.

Ex. 24. Two isosceles triangles are equal if the base and an angle of one are equal to the base and the homologous angle of the other.

Ex. 25. Two equilateral triangles are equal if an altitude of one is equal to an altitude of the other.

Ex. 26. If two medians of a triangle are equal, the triangle is isosceles.

[SUG. On Fig. to Prop. XLVI, taking $AD = EC$, prove △ AOC isosceles, △AEC = △ADC, etc. How could this theorem be investigated by analysis?]

Ex. 27. Prove the sum of the angles of a triangle equal to two right angles by drawing a line through the vertex of the triangle parallel to the base.

Ex. 28. The homologous medians of two equal triangles are equal.

Ex. 29. The bisectors of an angle of a triangle and of the two exterior angles at the other vertices are concurrent.

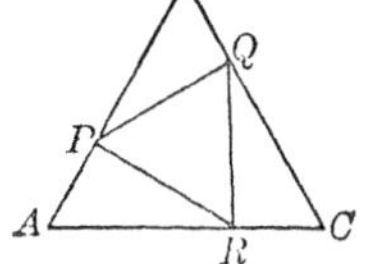

Ex. 30. If ABC is an equilateral triangle and $AP = BQ = CR$, then PQR is an equilateral triangle.

Ex. 31. If the two base angles of a triangle be bisected, and through the point of intersection of the two bisectors a line be drawn parallel to the base, the part of this line intercepted between the two sides equals the sum of the segments of the sides included between the parallel and the base (i. e., prove $PQ = AP + QC$).

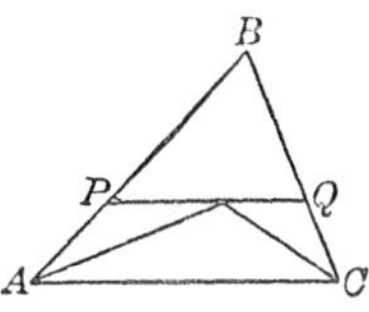

Ex. 32. Two quadrilaterals are equal, if three sides and the two included angles of one are equal to three sides and the two included angles of the other, respectively.

Ex. 33. If the diagonals of a quadrilateral bisect each other, the figure is a parallelogram.

Ex. 34. Lines joining the midpoints of the sides of a rectangle in order form a rhombus.

Ex. 35. Lines joining the midpoints of the sides of a rhombus form a rectangle.

Ex. 36. The bisectors of the angles of a parallelogram form a rectangle.

Ex. 37. The bisectors of the angles of a rectangle form a square.

Ex. 38. If lines be drawn through the vertices of a quadrilateral parallel to the diagonals, a parallelogram is formed which is twice as large as the original quadrilateral.

Ex. 39. Lines drawn from two opposite vertices of a parallelogram to the midpoints of a pair of opposite sides trisect a diagonal of the parallelogram.

Ex. 40. On the diagonal AC of a parallelogram $ABCD$ equal parts, AP and CQ, are marked off. Prove $BPDQ$ a parallelogram. How many pairs of equal triangles does the figure contain?

Ex. 41. The opposite angles of an isosceles trapezoid are supplementary.

Ex. 42. In an isosceles trapezoid, the diagonals are equal.

Ex. 43. If the upper base of an isosceles trapezoid equals the sum of the legs, and lines be drawn from the midpoint of the upper base to the extremities of the lower base, how many isosceles triangles are formed? Prove this.

Ex. 44. The lines joining the midpoints of the sides of an isosceles trapezoid, taken in order, form a rhombus or a square.

Ex. 45. The bisectors of the angles of a trapezoid form a quadrilateral whose opposite angles are supplementary.

Book II

THE CIRCLE

197. A **circle** is a portion of a plane bounded by a curved line, all points of which are equally distant from a point within called the **center.**

The **circumference** of a circle is the curved line bounding the circle. The term circle may also be used for the bounding line, if no ambiguity results.

A circle is named by naming its center, as the circle O; or by naming two or more points on its circumference, as the circle ACD.

198. A **radius** of a circle is a straight line drawn from the center to any point on the circumference, as AO. A **diameter** of a circle is a straight line drawn through the center and terminated by the circumference, as BC.

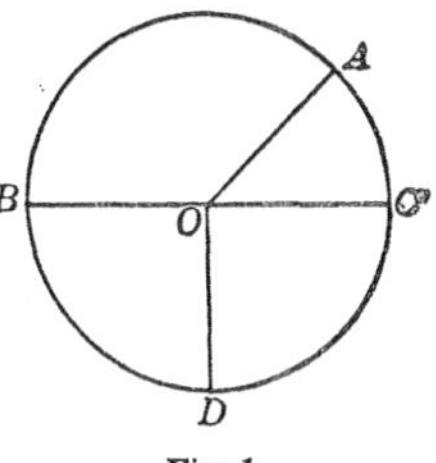

Fig. 1

199. An **arc** is any portion of a circumference, as AC. A **semicircumference** is an arc equal to one-half the circumference, as BAC. A **quadrant** is an arc equal to one-fourth of a circumference, as BD.

200. A **chord** is a straight line joining the extremities of an arc, as EF.

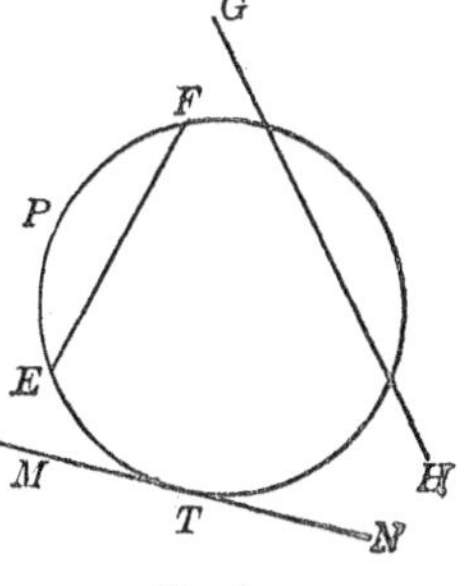

Fig. 2

Every chord subtends two arcs. A **minor arc** is the smaller of two arcs subtended by a chord. A **major arc** is the larger of two arcs subtended by a chord. Thus, for the chord *EF* the minor arc is *EPF*, and the major arc is *ETF*.

Conjugate arcs is a general term for a pair of minor and major arcs.

If the arc subtended by a given chord is mentioned, unless it is otherwise specified, the minor arc is meant.

201. A **tangent** to a circle is a straight line which, if produced, has but one point in common with the circle, as *MN*. Hence, a tangent touches the circumference in one point only.

A **secant** is a straight line which, if produced, intersects the circumference in two points, as *GH*.

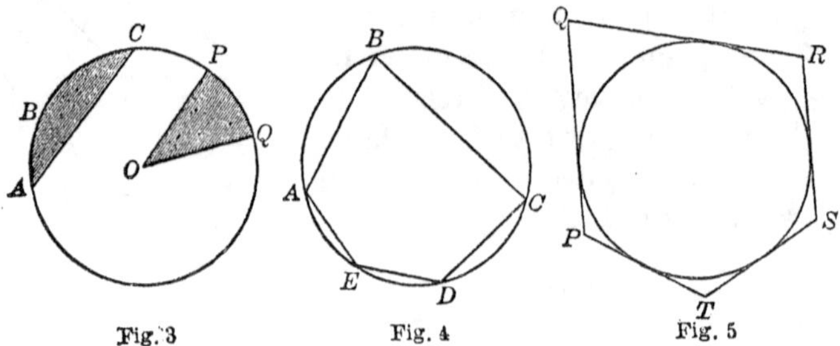

Fig. 3 Fig. 4 Fig. 5

202. A **segment** of a circle is a portion of the circle bounded by an arc and its chord, as *ABC* (Fig. 3).

Into how many segments does each chord divide a circle?

A semicircle is a segment bounded by a semicircumference and its diameter.

203. A **sector** of a circle is a portion of a circle bounded by two radii and the arc included by them, as *POQ* (Fig. 3).

204. A **central angle** is an angle whose vertex is at the center and whose sides are radii, as the angle POQ (Fig. 3).

An **inscribed angle** is an angle whose vertex is in the circumference and whose sides are chords, as the angle ABC (Fig. 4).

An **angle inscribed in a segment** is an angle whose vertex is in the arc of the segment and whose sides are chords drawn from the vertex to the extremities of the arc. Let the pupil draw a circle, a segment in it, and an angle inscribed in the segment.

205. Two circles tangent to each other are circles which are tangent to the same straight line at the same point. They are *tangent internally* or *externally* according as one circle lies entirely within or entirely without the other. See the figures, page 122.

Concentric circles are circles which have the same center. Let the pupil draw a pair of concentric circles.

206. A **polygon inscribed in a circle** is a polygon all of whose vertices lie in the circumference of the circle, as $ABCDE$ (Fig. 4).

A **circle circumscribed about a polygon** is a circle whose circumference passes through every vertex of the polygon.

207. A **polygon circumscribed about a circle** is a polygon all of whose sides are tangent to the circle, as $PQRST$ (Fig. 5).

A **circle inscribed in a polygon** is a circle to which all the sides of the polygon are tangent.

Concyclic points are points lying on the same circumference.

PROPERTIES OF THE CIRCLE INFERRED IMMEDIATELY

208. *Radii of the same circle, or of equal circles, are equal.*

209. *The diameter of a circle equals twice its radius.*

210. *Diameters of the same circle, or of equal circles, are equal.*

211. *If two circles are equal* (i. e., *may be made to coincide*, Art. 13), *their radii are equal, and conversely.*

212. *A diameter of a circle bisects the circle.* For, by placing the two parts of the circle so that the diameters coincide and their arcs fall on the same side of the diameter, these arcs will coincide (Art. 197).

213. *A straight line cannot intersect a circle in more than two points.* For, if a straight line can intersect a circle in three (or more) points, three or more equal lines (radii) can be drawn from the same point (the center) to the straight line. But this is impossible (Art. 111).

Proposition I. Theorem

214. *A diameter of a circle is longer than any other chord.*

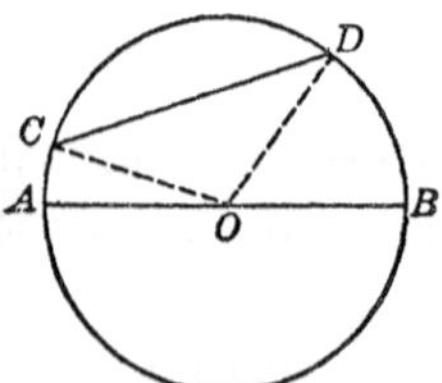

Given AB a diameter, and CD any other chord in the circle O.

To prove $AB > CD$.

Proof. Draw the radii OC and OD.

Then, in the $\triangle OCD$, $OC + OD > CD$. (Why?)

Substituting for OC its equal OA, and for OD its equal OB, Ax. 8.

$$OA + OB, \text{ or } AB > CD.$$

Q. E. D.

PROPOSITION II. THEOREM

215. *In the same circle, or in equal circles, equal central angles intercept equal arcs on the circumference.*

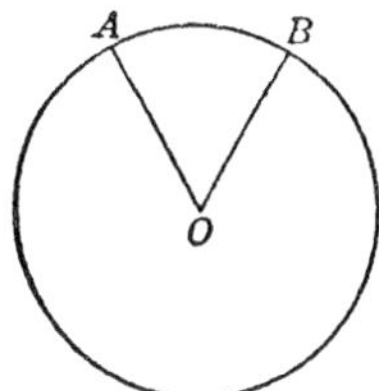

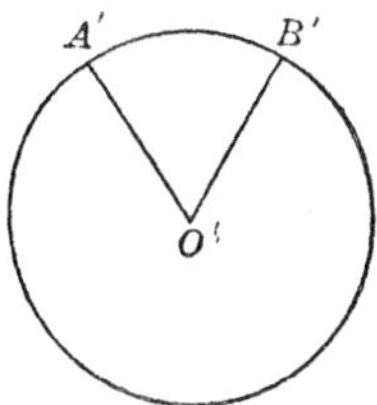

Given the equal circles O and O', and the equal central $\angle s$ AOB and $A'O'B'$.

To prove the arc AB = arc $A'B'$.

Proof. Apply the circle O' to the circle O so that the center O' coincides with the center O, and the radius $O'A'$ with the radius OA.

Then the radius $O'B'$ will fall on the radius OB,
(*for by hyp.* $\angle AOB = \angle A'O'B'$).

And B' will fall on B, Art. 208.
(*for* $OB = O'B'$, *being radii of equal* ⊙).

Hence arc $A'B'$ will coincide with AB, Art. 197.
(*for all points of each arc are equidistant from the center*).

$\therefore$ arc $A'B'$ = arc AB. Art. 47.

Q. E. D.

PROPOSITION III. THEOREM (CONVERSE OF PROP II)

216. *In the same circle, or in equal circles, equal arcs subtend equal angles at the center.*

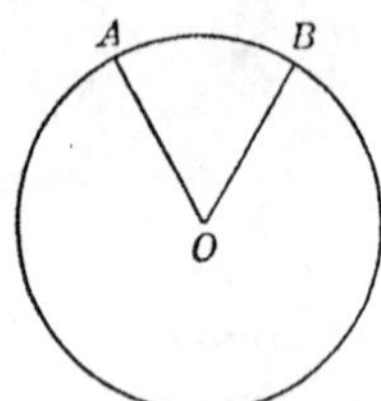

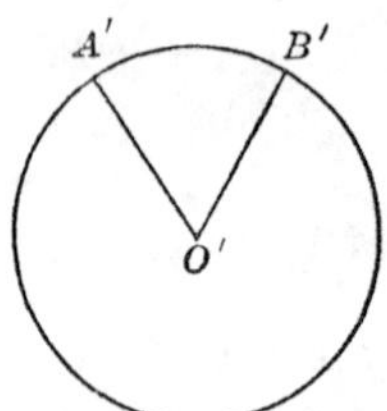

Given the equal circles O and O', and the equal arcs AB and $A'B'$ subtending the central ∠s O and O'.

To prove $\angle O = \angle O'$.

Proof. Apply the circle O' to the equal circle O so that the center O' coincides with the center O and the point A' with the point A.

Then the point B' will fall on B,

(*for arc $A'B'$=arc AB by hyp.*).

Hence the radius $O'A'$ will coincide with OA, and radius $O'B'$ with OB, Art. 66.

(*between two points only one straight line can be drawn*).

$\therefore \angle O$ and $\angle O'$ coincide.

$\therefore \angle O = \angle O'$. Art. 47.

Q. E. D.

217. COR. *In the same circle, or in equal circles, of two unequal central angles the greater angle intercepts the greater arc, and, conversely, of two unequal arcs, the greater arc subtends the greater angle at the center.*

Ex. Draw a circle and in it a segment which is less than the sector having the same arc. Also one that is greater. Also one that is equal.

PROPOSITION IV. THEOREM

218. *In the same circle, or in equal circles, equal chords subtend equal arcs.*

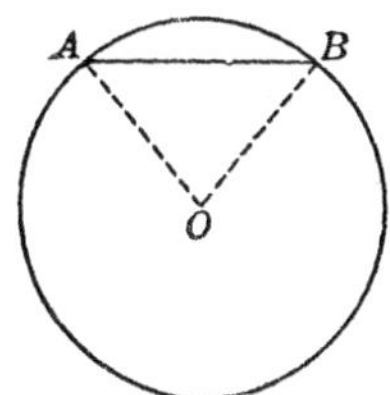

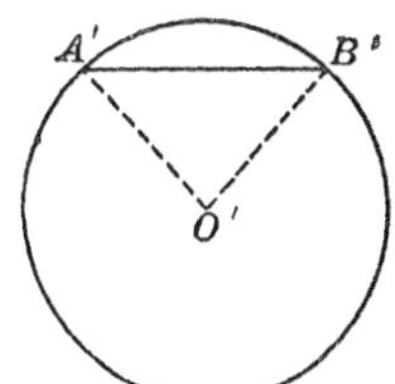

Given the equal circles O and O', and the chord $AB=$ chord $A'B'$.

To prove arc $AB=$ arc $A'B'$.

Proof. Draw the radii OA, OB, $O'A'$, $O'B'$.

Then, in the $\triangle$s AOB and $A'O'B'$,

$AB=A'B'$. (Why?)

$AO=A'O'$, and $BO=B'O'$. (Why?)

$\therefore \triangle AOB=\triangle A'O'B'$. (Why?)

$\therefore \angle O=\angle O'$. (Why?)

$\therefore$ arc $AB=$ arc $A'B'$. Art. 215.

(*in the same* ⊙, *or in equal* ⊙s, *equal central* ∠s *intercept equal arcs on the circumference*).

Q. E. D.

Ex. 1. In the above figure, if chord $AB=1$ in., chord $A'B'=1$ in., and arc $AB=1\frac{1}{4}$ in., find the length of arc $A'B'$.

Ex. 2. Draw a circle and mark off a part of it that is both a segment and a sector.

Ex. 3. If the distance from the center of a circle to a line is greater than the radius, will the line intersect the circumference?

Proposition V. Theorem

219. *In the same circle, or in equal circles, equal arcs are subtended by equal chords.*

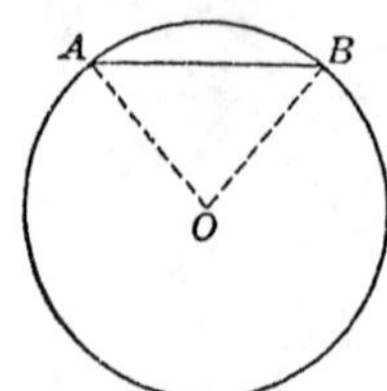

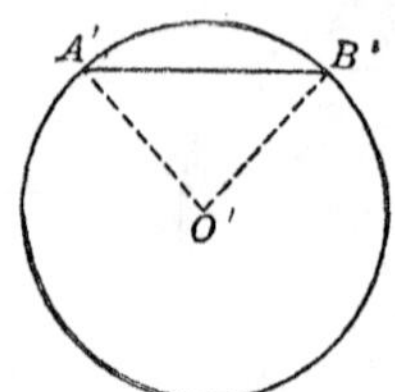

Given the equal circles O and O', and arc AB = arc $A'B'$.

To prove chord AB = chord $A'B'$.

Proof. Draw the radii AO, BO, $A'O'$, $B'O'$.

Then, in the $\triangle$ AOB and $A'O'B'$,

$\angle O = \angle O'$, Art. 216.

(*for arc $AB = A'B'$, and, in the same ⊙, or in equal ⊙, equal arcs subtend equal ∠ at the center*).

Also $OA = O'A'$, and $OB = O'B'$. (Why?)

$\therefore \triangle AOB = \triangle A'O'B'$. (Why?)

$\therefore AB = A'B'$. (Why?)

Q. E. D.

Ex. 1. In the above figure if arc $AB = 1\frac{1}{4}$ in., arc $A'B' = 1\frac{1}{4}$ in., and chord $AB = 1$ in., find chord $A'B'$ without measuring it.

Ex. 2. Draw two circles so that the radius of one is the diameter of the other.

Ex. 3. To which of the classes of figures mentioned in Art. 18 does a sector belong? a segment?

PROPOSITION VI. THEOREM

220. *In the same circle, or in equal circles, the greater of two (minor) arcs is subtended by the greater chord; and,* CONVERSELY, *the greater of two chords subtends the greater (minor) arc.*

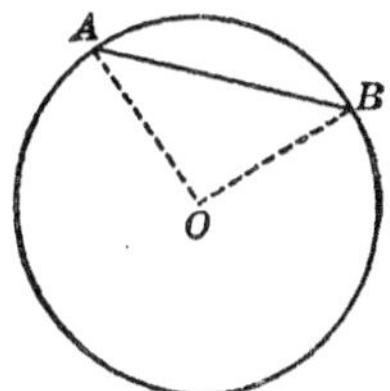

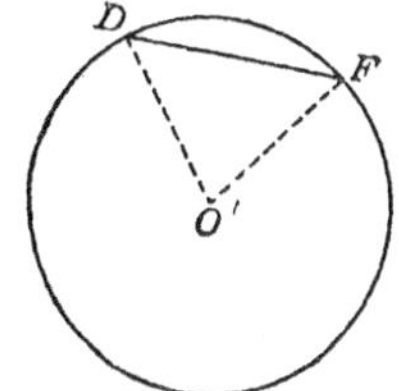

Given the equal circles O and O', and arc $AB >$ arc DF.

To prove chord $AB >$ chord DF.

Proof. Draw the radii OA, OB, $O'D$, $O'F$.

Then, in the △ AOB and $DO'F$,

$OA = O'D$, and $OB = O'F$. (Why?)

$\angle O$ is greater than $\angle O'$, Art. 217.

(*for arc $AB >$ arc DF, and, in the same ⊙, or in equal ⊙, of two unequal arcs the greater arc subtends the greater angle at the center*).

∴ chord $AB >$ chord DF, Art. 107.

(*if two △ have two sides of one equal to two sides of the other, but the included angle of the first greater, etc.*).

CONVERSELY. **Given** the equal circles O and O', and chord $AB >$ chord DF.

To prove arc $AB >$ arc DF.

Proof. In the △ AOB and $DO'F$,

$OA = O'D$, and $OB = O'F$. (Why?)

$AB > DF$. (Why?)

∴ $\angle O$ is greater than $\angle O'$. (Why?)

∴ arc $AB >$ arc DF, Art. 217.

(*in the same ⊙, or in equal ⊙, of two unequal central ∠ the greater angle intercepts the greater arc*).

Q. E. D.

Proposition VII. Theorem

221. *A diameter perpendicular to a chord bisects the chord and the arcs subtended by the chord.*

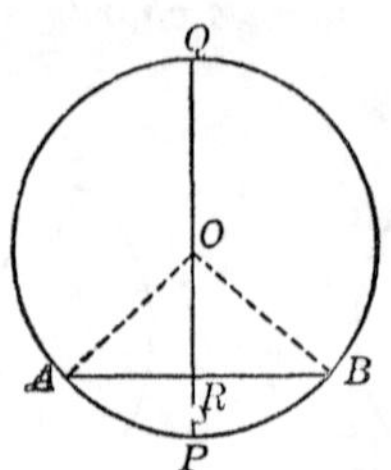

Given the circle O, and the diameter $PQ \perp$ chord AB.

To prove that PQ bisects the chord AB and the arcs APB and AQB.

Proof. Draw the radii OA and OB.

Then, in the rt. $\triangle$ OAR and OBR,

$$OA = OB. \qquad \text{(Why?)}$$

$$OR = OR. \qquad \text{(Why?)}$$

$$\therefore \triangle OAR = \triangle OBR. \qquad \text{(Why?)}$$

$$\therefore AR = BR, \text{ and } \angle AOR = \angle BOR. \qquad \text{(Why?)}$$

$$\therefore \text{arc } AP = \text{arc } BP, \qquad \text{Art. 215.}$$

(*in the same* ⊙, *or in* = ⊙, = *central* ∠ *intercept* = *arcs on the circf.*)

$\angle AOQ = \angle BOQ$ (Art. 75). $\therefore$ arc AQ = arc BQ. (Why?)

Q. E. D.

222. Cor. 1. *A diameter which bisects a chord (shorter than a diameter) is perpendicular to the chord.*

223. Cor. 2. *The perpendicular bisector of a chord passes through the center of the circle, and bisects the arcs subtended by the chord.*

224. Cor. 3. *A line from the center perpendicular to a chord bisects the chord.*

225. Cor. 4. *A line passing through the midpoints of a chord and its arc passes through the center, and is a diameter perpendicular to the chord.*

PROPOSITION VIII. THEOREM

226. *In the same circle, or in equal circles, equal chords are equidistant from the center; and,* CONVERSELY, *chords which are equidistant from the center are equal.*

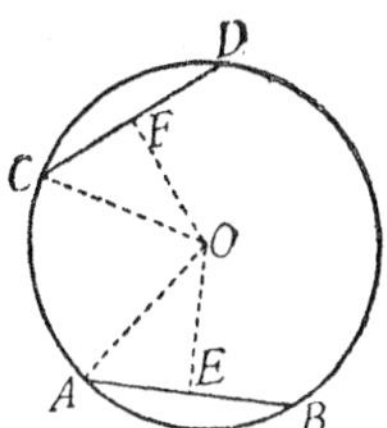

Given the circle O in which the chords AB and CD are equal.

To prove that AB and CD are equidistant from the center.

Proof. Let OE be drawn $\perp$ AB, and $OF \perp CD$.

Draw the radii OA and OC.

Then OE bisects AB, and OF bisects CD, Art. 224.
(*a line from the center* $\perp$ *a chord bisects the chord*).

Hence, in the rt. $\triangle$ OAE and OCF,

$OA = OC$. (Why?)

$AE = CF$. Ax. 5.

$\therefore \triangle OAE = \triangle OCF$. (Why?)

$\therefore OE = OF$. (Why?)

CONVERSELY. **Given** circle O, and AB and CD equidistant from the center.

To prove $AB = CD$.

Proof. Let the pupil supply the proof.

Proposition IX. Theorem

227. *In the same circle, or in equal circles, if two chords are unequal, they are unequally distant from the center, and the less chord is at the greater distance from the center.*

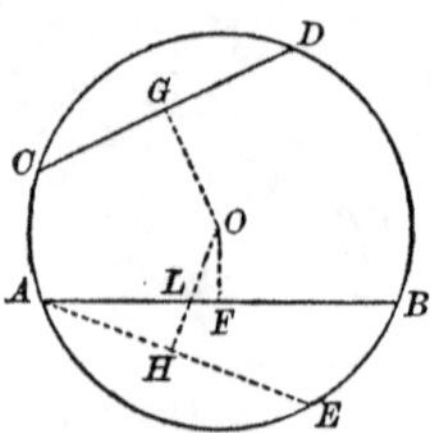

Given in the circle O the chord $CD <$ chord AB.

To prove that chord CD is at a greater distance from the center than chord AB.

Proof. Let OG be drawn $\perp$ CD, and $OF \perp AB$.

Then chord $AB >$ chord CD. Hyp.

$\therefore$ arc $AB >$ arc CD, Art. 220.

(*in the same* ⊙, *or in equal* ⊙, *the greater of two minor arcs is subtended by the greater chord, and conversely*).

Mark off on the arc AB the arc AE=arc CD, and draw the chord AE.

Chord AE = chord CD, Art. 219.

(*in the same* ⊙, *or in equal* ⊙, *equal arcs are subtended by equal chords*).

Let OH be drawn $\perp$ AE, and intersecting AB at L.

$\therefore OH = OG$, Art. 226.

(*in the same* ⊙, *or in equal* ⊙, *equal chords are equidistant from the center*).

But $OH > OL$. Ax. 7.

Also $OL > OF$. (Why?)

Much more then OH, or its equal $OG > OF$. Ax. 12.

Q. E. D.

PROPOSITION X. THEOREM (CONVERSE OF PROP. IX)

228. *In the same circle, or in equal circles, if two chords are unequally distant from the center, the more remote is the less.*

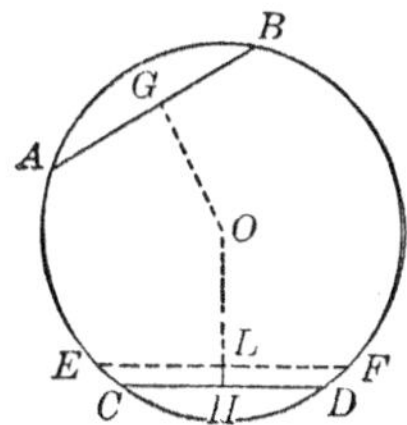

Given in the circle O the chord CD farther from the center than the chord AB.

To prove chord $CD <$ chord AB.

Proof. Let OH be drawn $\perp$ CD, and $OG \perp AB$.

$$OH > OG. \qquad \text{(Why ?)}$$

On OH mark off $OL = OG$.

Through L let the chord ELF be drawn $\perp$ OH.

Then chord $EF =$ chord AB, Art. 226.

(*in the same ⊙, or in equal ⊙, chords which are equidistant from the center are equal*).

But the arc $CD <$ arc EF. Ax. 7.

$\therefore$ chord $CD <$ chord EF, or its equal AB, Art. 220.

(*in the same ⊙, or in equal ⊙, the greater of two minor arcs is subtended by the greater chord, and conversely*).

Q. E. D.

PROPOSITION XI. THEOREM

229. *A straight line perpendicular to a radius at its extremity is tangent to the circle.*

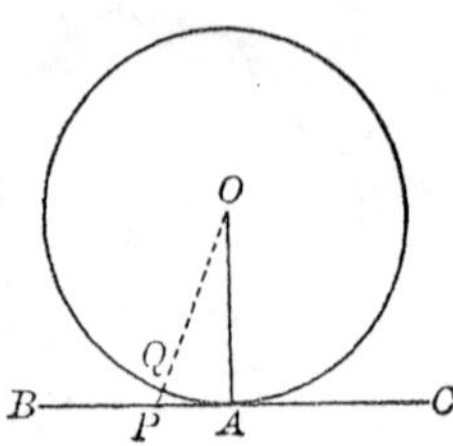

Given the circle O, the radius OA, and the line $BC \perp OA$ at its extremity A.

To prove that BC is tangent to the $\odot$.

Proof. Take P, any point on the line BC except A, and draw OP.

Then $OP > OA$. (Why?)

Hence the point P lies without the circle.

$\therefore$ every point in the line BC, except A, lies outside the $\odot$.

$\therefore$ BC is tangent to the circle, Art. 201.
(*a tangent to a* $\odot$ *is a straight line which, etc.*).

Q. E. D.

230. COR. 1. *The radius drawn to the point of contact is perpendicular to a tangent to a circle.*

231. COR. 2. *A perpendicular to a tangent at the point of contact passes through the center of the circle.*

232. COR. 3. *The perpendicular drawn from the center of a circle to a tangent passes through the point of contact.*

PROPOSITION XII. THEOREM

233. *Two parallel lines intercept equal arcs on a circumference.*

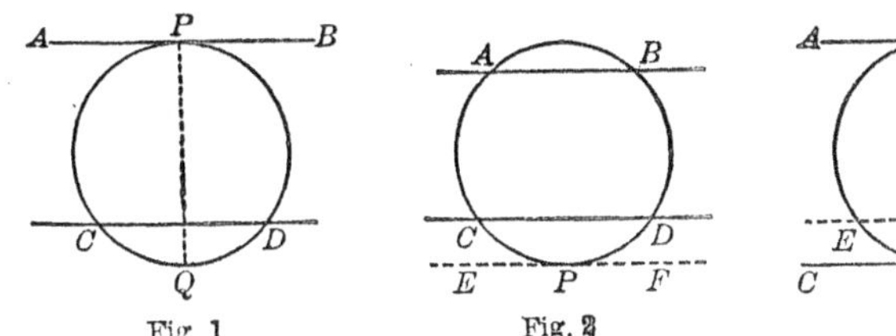

Fig. 1 Fig. 2 Fig. 3

CASE I. **Given** AB (Fig. 1) tangent to the ⊙ PCD at P, CD a secant ∥ AB and intersecting the circumference in C and D.

To prove arc PC = arc PD.

Proof. Draw the diameter PQ.

Then $PQ \perp AB$. Art. 230.

$PQ \perp CD$. Art. 123.

∴ arc PC = arc PD, Art. 221.

(*a diameter ⊥ chord bisects the chord and the arcs subtended by the chord*).

CASE II. **Given** AB and CD (Fig. 2) ∥ secants intersecting the circumference in A, B and C, D respectively.

To prove arc AC = arc BD.

Proof. Let a tangent EF be drawn ∥ AB and touching the circle at P.

Then $EF \parallel CD$. (Why?)

Then arc AP = arc BP. Case I.

Also arc CP = arc DP. (Why?)

Hence arc AC = arc BD. Ax. 3.

CASE III. **Given** AB and CD (Fig. 3) ∥ tangents touching the ⊙ at P and Q respectively.

To prove arc $PEQ =$ arc PFQ.

Proof. Let the pupil supply the proof.

234. COR. *The straight line which joins the points of contact of two parallel tangents is a diameter.*

PROPOSITION XIII. THEOREM

235. *Through three points, not in the same straight line, one circumference, and only one, can be drawn.*

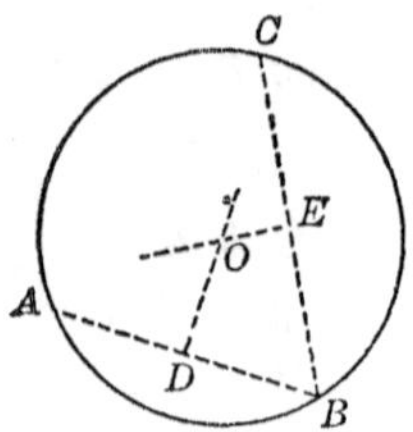

Given A, B and C any three points not in the same straight line.

To prove that one circumference, and only one, can be drawn through A, B and C.

Proof. Draw the straight lines AB and BC, and let ⊥s be erected at the midpoints, D and E, of AB and BC respectively.

These ⊥s will intersect at some point O, Art. 122.
(*lines ⊥ non-parallel lines are not* ∥).

But O is in the ⊥ bisector of AB. Const.

∴ O is equidistant from the points A and B. Art. 112.

In like manner, O is in the ⊥ bisector of BC, and is equidistant from the points B and C.

Hence O is equidistant from the three points A, B and C. Ax. 1.

Hence if a circumference be described with O as a center and OA as a radius, it will pass through A, B and C.

Also DO and EO intersect in but one point. Art. 64.

Hence there is but *one center*.

Again, O is equally distant from the points A, B and C; hence there is but *one radius*.

With only one center and only one radius, but one circumference can be described.

Hence one circumference, and only one, can be drawn through the points A, B and C.

Q. E. D.

236. Note. The theorem of Art. 235 enables us to shrink or economize a circle into three points; or to expand any three points into a circle.

Ex. 1. How many circumferences can be passed through four given points in a plane, each circumference passing through three, and only three, of the given points?

Ex. 2. Draw two circles so that they can have a common chord.

Ex. 3. Can two circles which are tangent to each other have a common chord?

Ex. 4. Can two circles which are tangent to each other have a common secant?

Ex. 5. Draw two circles which can have neither a common chord nor a common tangent.

Ex. 6. Is it possible to draw two circles which cannot have a common secant?

Proposition XIV. Theorem

237. *The two tangents drawn to a circle from a point outside the circle are equal, and make equal angles with a line drawn from the point to the center.*

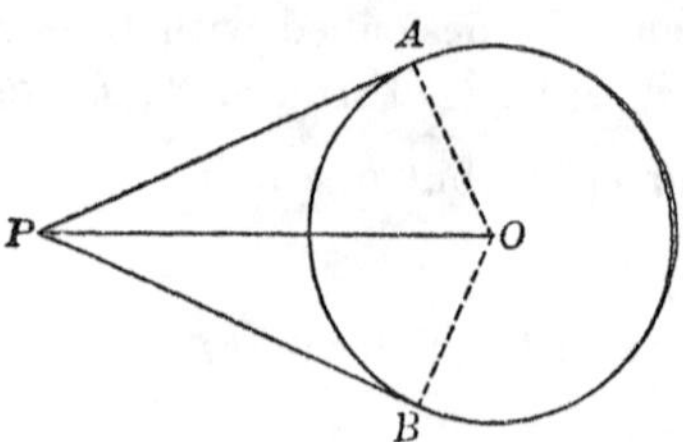

Given PA and PB two tangents drawn from the point P to the circle O.

To prove $PA = PB$, and $\angle APO = \angle BPO$.

Proof. Let the pupil supply the proof.

238. Def. The **line of centers** of two circles is the line joining their centers.

239. Def. Two circles which do not meet may have four common tangents.

A **common internal tangent** of two circles is a tangent which cuts their line of centers.

240. Def. A **common external tangent** of two circles is a tangent which does not cut their line of centers.

Ex. 1. In the above figure, prove that the line drawn to the center from the point in which the two tangents meet makes equal angles with the radii to the points of contact.

Ex. 2. Draw a circle with a radius of 3 in. and another with a radius of 2 in., with their centers 4 in. apart. Will these circles intersect? If their centers were 6 in. apart, would they intersect?

Proposition XV. Theorem

241. *If two circles intersect, their line of centers is perpendicular to their common chord at its middle point.*

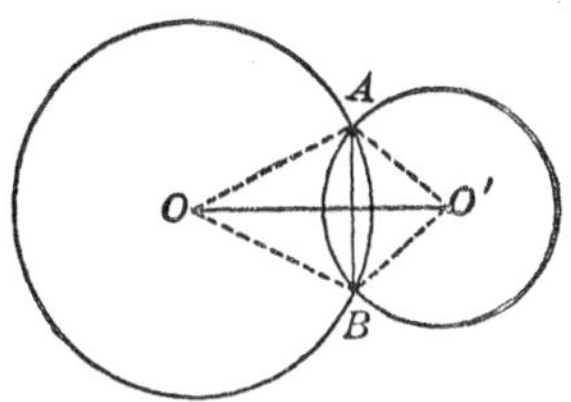

Given the circles O and O', intersecting at the points A and B.

To prove $OO' \perp AB$ at its middle point.

Proof. Draw the radii OA, OB, $O'A$, $O'B$.

Then $OA = OB$, and $O'A = O'B$. (Why?)

Hence O and O' are two points each equidistant from A and B.

$\therefore OO'$ is the $\perp$ bisector of AB. Art. 113.

Q. E. D.

Ex. 1. Draw two intersecting circles and show that the line of centers is less than the sum of the radii.

Draw two circles in which the line of centers

Ex. 2. Equals the sum of the radii.

Ex. 3. Is greater than the sum of the radii.

A and B.

$\therefore OO'$ is the $\perp$ bisector of AB. Art. 113.

Q. E. D.

Ex. 1. Draw two intersecting circles and show that the line of centers is less than the sum of the radii.

Draw two circles in which the line of centers

PROPOSITION XVI. THEOREM

242. *If two circles are tangent to each other, the line of centers passes through the point of contact.*

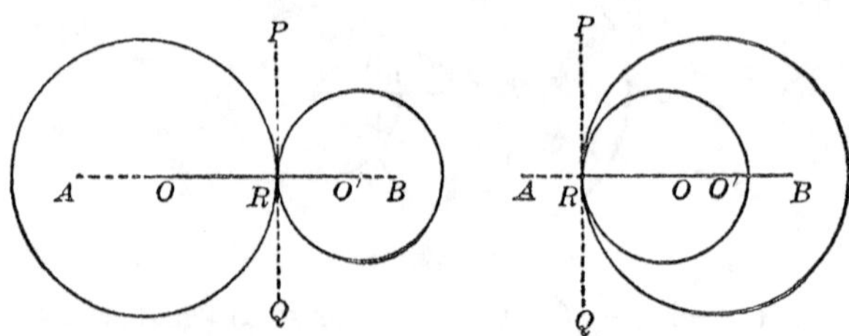

Given the circles O and O' tangent to each other at the point R.

To prove that the line OO' passes through R.

Proof. At the point R let PQ, a tangent to the given ⊙s, be drawn.

Also let AB be drawn $\perp$ PQ at R.

Then AB passes through O and also through O', Art. 231.
(*a* $\perp$ *to a tangent at the point of contact passes through the center*).

$\therefore$ line AB coincides with the line OO'. Art. 64.

$\therefore$ OO' passes through the point R,
(*for it coincides with AB which passes through R*).

Q. E. D.

How many common internal, and how many common external tangents have two circles

Ex. 1. If they touch externally?

Ex. 2. If they touch internally?

Ex. 3. If they intersect?

Ex. 4. If one circle lies wholly within the other?

Ex. 5. If one circle lies wholly without the other?

EXERCISES. GROUP 15

Ex. 1. The line joining the center of a circle to the midpoint of a chord is perpendicular to the chord.

Ex. 2. A, B, C and D are four points taken in succession on the circumference of a circle, and arc AB=arc CD. Prove that chord AC=chord BD.

Ex. 3. Tangents drawn at the extremities of a diameter are parallel.

Ex. 4. PA and PB are tangents to a circle drawn from the point P. O is the center of the circle. Prove that PO is the perpendicular bisector of the chord AB.

Ex. 5. If the perpendiculars from the center upon two chords are equal, the arcs subtended by these chords are equal.

Ex. 6. A, B, C and D are points taken in succession on a semi-circumference, and arc AC is greater than arc BD; prove that chord $AB >$ chord CD.

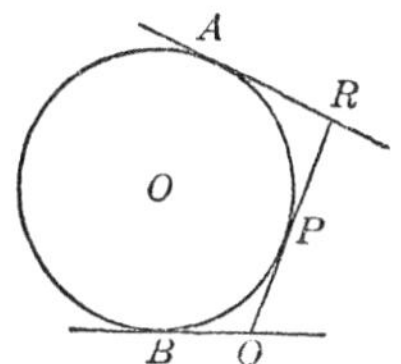

Ex. 7. State the converse of the preceding theorem and prove it.

Ex. 8. Given RA, RQ and QB tangents of the circle O; prove $RQ=RA+QB$.

Ex. 9. If a quadrilateral be circumscribed about a circle, show that the sum of one pair of opposite sides equals the sum of the other pair.

Ex. 10. If a hexagon be circumscribed about a circle, show that the sum of three alternate sides equals the sum of the other three sides.

Ex. 11. If a polygon of $2n$ sides be circumscribed about a circle, the sum of n alternate sides equals the sum of the other n sides.

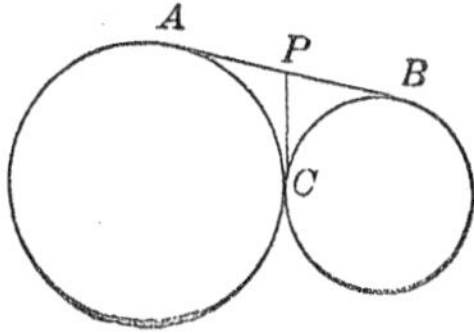

Ex. 12. A circumscribed parallelogram is equilateral.

Ex. 13. If two circles are tangent externally, the common internal tangent bisects the common external tangent (i. e., prove $PA=PB$).

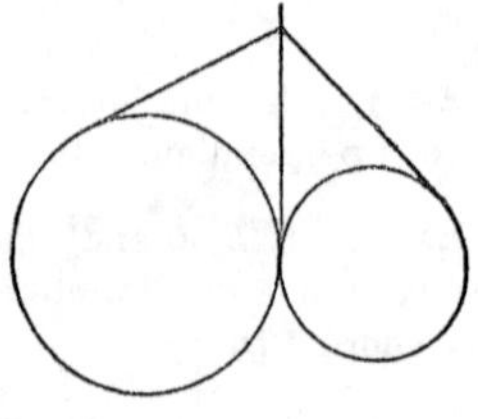

Ex. 14. If two circles are tangent (either externally or internally) tangents drawn to them from any point in the common tangent are equal.

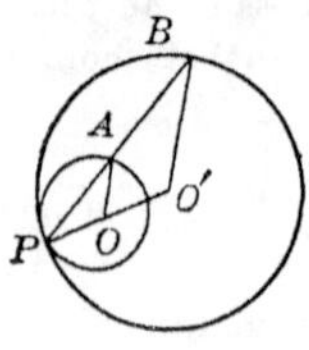

Ex. 15. Two circles whose centers are O and O' are tangent internally at P. The line PAB is drawn intersecting the circumferences at A and B. Prove that OA and $O'B$ are parallel.

MEASUREMENT. RATIO.

243. Measurement. For many purposes, the most advantageous way of dealing with a given magnitude is to take a certain definite part of the magnitude as a unit, and to determine the number of times this unit must be taken in order to make up the given magnitude. Ease and precision in dealing with magnitudes are thus obtained.

Geometric magnitudes thus far have been treated as wholes, the object being simply to determine whether two given magnitudes are equal, or unequal, or to determine some similar general relation. Hereafter geometric magnitudes will frequently be treated as if composed of units.

To measure a given magnitude is to find how many times the given magnitude contains another magnitude of the same kind taken as a unit.

244. The **numerical measure** of a magnitude is the number which expresses how many times the unit of measure is contained in the given magnitude.

Thus, when a boy says that he is five feet tall, he means that, if a foot rule be applied to his height, the foot rule will be contained five times.

A quantity is often measured to best advantage by measuring a related but more accessible quantity, which has the same numerical measure as the original quantity. This process is **indirect measurement.** Thus, the temperature of the air is measured indirectly by measuring the height of a column of mercury in a thermometer tube. So the number of times a unit of angle is contained in a given angle is often ascertained most readily by determining the number of times a unit of arc is contained in a given arc.

245. The **ratio** of two magnitudes of the same kind is their relative magnitude as determined by the number of times one is contained in the other. Hence, it is the quotient, or indicated quotient, of the two magnitudes.

Thus, the ratio of 3 ft. to 1 ft. 7 in. is $\frac{36 \text{ in.}}{19 \text{ in.}}$, or $\frac{36}{19}$.

The *use of ratio* is illustrated by the fact that several indicated quotients when taken together may be simplified by cancellation before a final determination of their value is made.

Two magnitudes of the same kind have the same ratio as their numerical measures.

246. Measurement as a ratio. An important particular instance of ratio is that ratio in which one of the two magnitudes compared is a unit of measurement. Hence, the numerical measure of a magnitude is the ratio of the magnitude to the unit of measure.

Thus, the numerical measure of the height of a boy is the ratio of his height (5 ft.) to the unit of measure (1 ft.), or 5.

247. Commensurable magnitudes are magnitudes of the same kind which have a common unit of measure.

Thus, 12 ft. and 25 ft. have the common unit of measure, 1 ft., and hence are commensurable magnitudes; also $13\frac{1}{2}$ bus. and $7\frac{3}{4}$ bus. have a common unit, 1 peck, and are commensurable.

248. Incommensurable magnitudes are magnitudes of the same kind which have no common unit of measure.

Ills. \$5 and $\$\sqrt{3}$; 7 yrs. and $\sqrt[3]{15}$ yrs.; so the side and the diagonal of a square may be proved to have no common unit of measure; likewise the diameter and the circumference of a circle.

In general, a ratio which is expressed by a surd number, as $\sqrt{2}$ or $\sqrt{3}$, is a ratio between incommensurable magnitudes (called an **incommensurable ratio**).

METHOD OF LIMITS

249. A **variable quantity**, or a **variable**, is a quantity which may have an indefinite number of different values under the conditions of a theorem or problem.

Thus, the distance a railroad train goes varies with the number of hours which the train travels.

250. A **constant** is a quantity which remains unchanged in value in a given problem or discussion.

Thus, if a polygon be inscribed in a circle and the number of sides of the polygon be doubled, quadrupled, etc., the perimeter of the polygon will be variable, but the circumference of the circle will remain unchanged, and hence be a constant.

251. The **limit** of a variable quantity is a constant quantity which, under the conditions of a given discussion, the given variable may approach as nearly as we please in value, but which the variable can never equal.

Thus, in the illustration of Art. 250, the circumference of the circle is the limit of the perimeter of the inscribed polygon; also, the area of the circle is the limit of the area of the inscribed polygon.

From the definition of limit, it follows that the difference between a variable and its limit may be made as small as we please but can not become zero.

As another illustration of a variable and its limit, we may take the case of a point P travelling along a given line AB, in such a way that in the first second it passes over AP_1, one-half the line AB; in the second second over half the remaining part of the line, and arrives at P_2; in the third second over one-half the remainder of the line, and arrives at P_3, etc. It is evident that the point P can never arrive at B; for, in order to do this, in some one second the point would need to pass over the whole of the remaining distance.

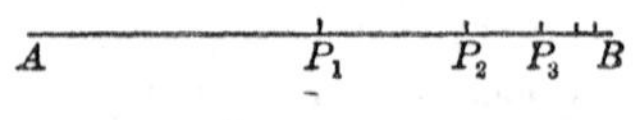

In this illustration, AP, the distance traveled by the moving point, is a variable (depending on the number of seconds), and AB is its limit.

If the distance AB be denoted by 2, the distance traveled will be denoted by $1 + \frac{1}{2} + \frac{1}{4} + \frac{1}{8} + \ldots$ and will vary according to the number of terms of the series taken.

252. Use of variables and limits. Many of the properties of limits are the same as the properties of the variables approaching them. Hence a demonstration of a difficult theorem may often be obtained by first finding the properties of relatively simple variables and then transferring these properties to their more complex limits.

253. Properties of variables and limits.

1. *The limit of the sum of a number of variables equals the sum of the limits of these variables.* For, since the difference between each variable and its limit may be made as small as we please, the sum of all these differences may be made as small as we please (since it is a finite number of differences, with each difference approaching zero).

2. *The limit of* a *times a variable equals* a *times the limit of the variable,* a *being a constant.* For, if the difference between a variable and its limit may be made as small as we please, a times this difference may be made as small as we please.

3. *The limit of* $\frac{1}{a}$*th part of a variable is* $\frac{1}{a}$*th part of the limit of the variable,* a *being a constant.* For, if the difference between a variable and its limit may be made as small as we please, $\frac{1}{a}$th part of this difference may be made as small as we please.

4. *If a variable* $\doteq 0$, a *times* (a *being finite*) *or* $\frac{1}{a}$ *part of the variable* $\doteq 0$; *and if* a *diminish in the one, or increase in the other process, the limit is still zero.*

Ex. 1. Are $2\frac{1}{8}$ gal. and $3\frac{1}{2}$ qt. commensurable? If so, what is the common unit of measure?

Ex. 2. If c denote a constant and v and v' variables, is the value of each of the following a variable or a constant: $\frac{v}{c}$, $\frac{c}{v}$, $v \times v'$, $\frac{v'}{v}$? Illustrate.

PROPOSITION XVII. THEOREM

254. *If two variables are always equal, and each approaches a limit, their limits are equal.*

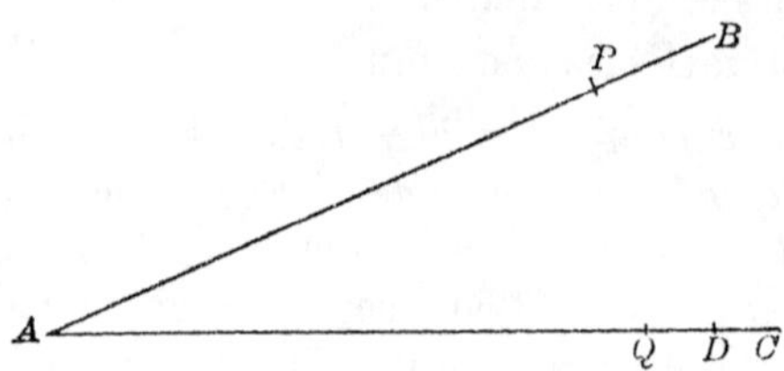

Given AB the limit of the variable AP, AC the limit of the variable AQ, and $AP = AQ$ always.

To prove $AB = AC$.

Proof. If the limit AB does not equal the limit AC, one of these limits, as AC, must be larger than the other.

Then, on AC, take AD equal to AB.

But AQ may have a value greater than AD. Art. 251.

Hence AQ would be greater than AB,
(*for* $AQ > AD$ *which* $= AB$).

$\therefore AQ > AP$, Ax. 12.
(*for* $AB > AP$).

But this is contrary to the hypothesis that AQ and AP are always equal.

$\therefore AC$ cannot be greater than AB.

In like manner it may be shown that AB is not greater than AC.

$\therefore AB = AC$.

Q. E. D.

Ex. What method of proof is used in Prop. XVII?

PROPOSITION XVIII. THEOREM

255. *In the same circle, or in equal circles, two central angles have the same ratio as their intercepted arcs.*

CASE I. *When the intercepted arcs are commensurable.*

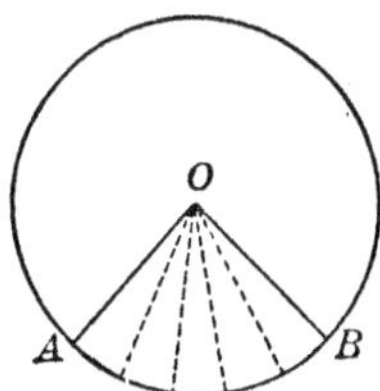

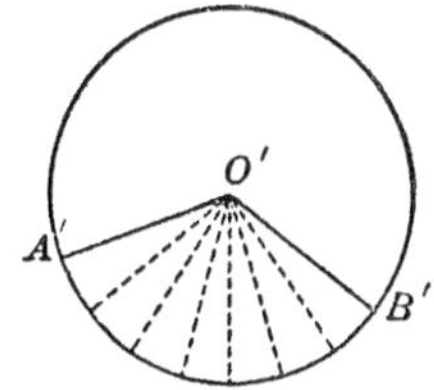

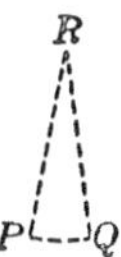

Given the equal ⊙ O' and O, with the central ∡ $A'O'B'$ and AOB intercepting the commensurable arcs $A'B'$ and AB.

To prove $$\frac{\angle A'O'B'}{\angle AOB}=\frac{A'B'}{AB}.$$

Proof. Let arc PQ (from a circle $=O$ and O') be a common measure of the arcs $A'B'$ and AB.

PQ will be contained in arc $A'B'$ an exact number of times, as 7 times, and in AB an exact number of times, as 5 times.

Then $$\frac{\text{arc } A'B'}{\text{arc } AB}=\frac{7}{5}.$$ Art. 24[illegible]

From O' and O draw radii to the several points of division of the arcs $A'B'$ and AB.

Then the $\angle A'O'B'$ will be divided into 7, and the $\angle AOB$ into 5 small angles, all equal, Art. 216.
(*in the same ⊙, or in = ⊙ equal arcs subtend equal ∡ at the center*).

$$\therefore \frac{\angle A'O'B'}{\angle AOB}=\frac{7}{5}.$$ Art. 245.

Hence $$\frac{\angle A'O'B'}{\angle AOB}=\frac{A'B'}{AB}.$$ Ax. 1.

CASE II. *When the intercepted arcs are incommensurable.*

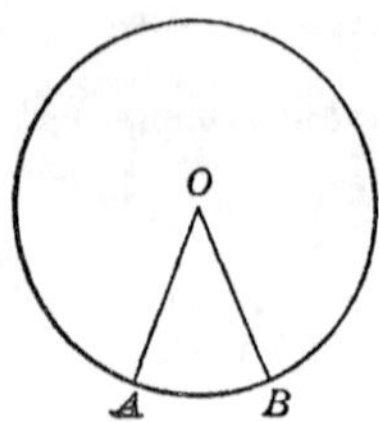

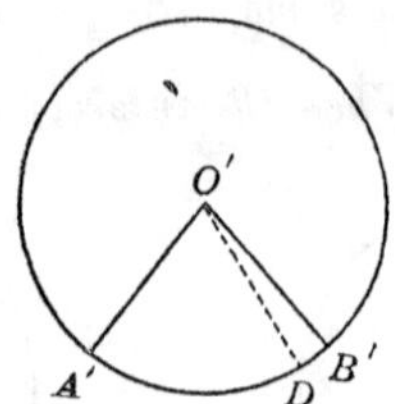

Given the equal ⊙ O' and O, with the central ∡ $A'O'B'$ and AOB intercepting the incommensurable arcs $A'B'$ and AB.

To prove $$\frac{\angle A'O'B'}{\angle AOB}=\frac{A'B'}{AB}.$$

Proof. Let the arc AB be divided into any number of equal parts, and let one of these parts be applied to the arc $A'B'$. It will be contained in $A'B'$ a certain number of times, with an arc DB' as a remainder.

Hence the arcs $A'D$ and AB have a common unit of measure. Cons.r.

$$\therefore \frac{\angle A'O'D}{\angle AOB}=\frac{A'D}{AB}.$$ Case I.

If now we let the unit of measure be indefinitely diminished, the arc DB', which is less than the unit of measure, will be indefinitely diminished.

Hence arc $A'D \doteq$ arc $A'B'$ as a limit, and $\angle A'O'D \doteq \angle A'O'B'$ as a limit. Art 251.

Hence $\frac{\angle A'O'D}{\angle AOB}$ becomes a variable approaching $\frac{\angle A'O'B'}{\angle AOB}$ as its limit; Art. 253, 3.

Also $\frac{A'D}{AB}$ becomes a variable approaching $\frac{A'B'}{AB}$ as its limit. Art. 253, 3.

But the variable $\frac{\angle A'O'D}{\angle AOB}$ = the variable $\frac{A'D}{AB}$ always. Case I.

$\therefore$ the limit $\frac{\angle A'O'B'}{\angle AOB}$ = the limit $\frac{A'B'}{AB}$, Art. 254.

(*if two variables are always equal, and each approaches a limit, their limits are equal*).

256. Def. A **degree of arc** is one three hundred and sixtieth part of the circumference of a circle.

257. Cor. *The number of degrees in a central angle equals the number of degrees in the intercepted arc;* that is, *a central angle is measured by its intercepted arc.*

Ex. 1. What is the ratio of a quadrant to a semi-circumference?

Ex. 2. What is the ratio of an angle of an equilateral triangle to one of the acute angles of an isosceles right triangle?

Ex. 3. Draw two circles so that the center of each circle is on the circumference of the other.

Ex. 4. Draw three circles so that the center of each is on the circumference of the other two.

[Sug. First draw an equilateral triangle.]

Ex. 5. Draw three circles each of which shall be tangent to the other two.

Ex. 6. Draw two concentric circles and a line which is a tangent to one of these circles and a chord of the other.

Ex. 7. Draw two concentric circles and a line which is a secant of one and a chord of the other.

PROPOSITION XIX. THEOREM

258. *An inscribed angle is measured by one-half its intercepted arc.*

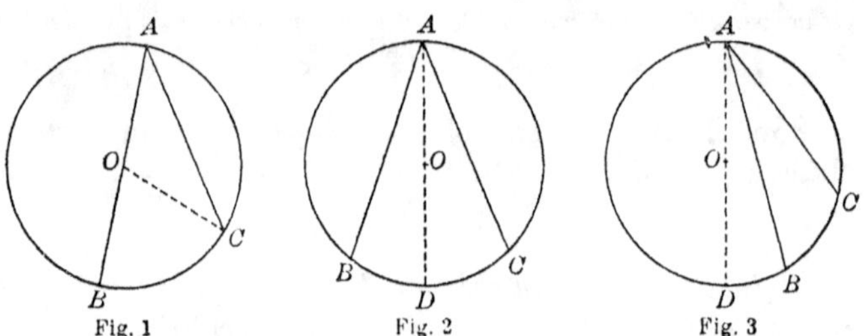

Fig. 1 Fig. 2 Fig. 3

CASE I. *When the center of the circle lies in one side of the inscribed angle.*

Given $\angle BAC$ (Fig. 1) inscribed in the circle O, and AB passing through the center O.

To prove that $\angle BAC$ is measured by $\frac{1}{2}$ arc BC.

Proof. Draw the radius OC.

Then, in the $\triangle OAC$, $OA = OC$. (Why?)

$\angle A = \angle C$. (Why?)

But $\angle BOC = \angle A + \angle C$. Art. 135.

$\therefore \angle BOC = 2\angle A$. Ax. 8.

But $\angle BOC$ is measured by arc BC, Art. 257.

(*a central $\angle$ is measured by its intercepted arc*).

$\therefore$ No. of angular degrees in $\angle BOC \doteq$ No. of arc degrees in BC.

Hence $\angle A$ is measured by $\frac{1}{2}$ arc BC. Ax. 5.

CASE II. *When the center of the circle lies within the inscribed angle.*

Given the inscribed $\angle BAC$ (Fig. 2), with the center of the circle O lying within the angle.

To prove that $\angle BAC$ is measured by $\frac{1}{2}$ arc BC.

Proof. Draw the diameter AD.

Then $\angle BAD$ is measured by $\frac{1}{2}$ arc BD. **Case I.**

Let the pupil complete the proof.

CASE III. *When the center of the circle is without the inscribed angle.*

Given the inscribed $\angle BAC$ (Fig. 3) with the center O outside the angle.

To prove that $\angle BAC$ is measured by $\frac{1}{2}$ arc BC.

Proof. Let the pupil supply the proof.

259. NOTE. By use of the above theorem, if the number of degrees in the intercepted arc be known, the number of degrees in the inscribed angle can be determined immediately. Thus, if the arc BC (Fig. 3) contains 48°, the angle BAC contains 24°; also, if it be known that the angle BAC contains, say 27°, the arc must contain 54°.

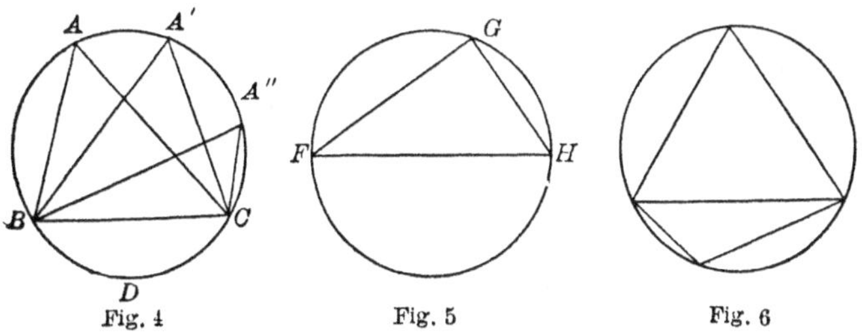

Fig. 4 Fig. 5 Fig. 6

260. COR. 1. *All angles inscribed in the same segment, or in equal segments, are equal.*

Thus ∡ A, A', A'' (Fig. 4) are all equal, for each of them is measured by one-half the arc BDC.

261. COR. 2. *An angle inscribed in a semicircle is a right angle*, for it is measured by one-half a semicircumference.

Thus FH (Fig. 5) is a diameter ∴ $\angle FGH$ is a rt. ∠.

262. COR. 3. *An angle inscribed in a segment greater than a semicircle is an acute angle; an angle inscribed in a segment less than a semicircle is an obtuse angle.*

Ex. If, in Fig. 1, p. 132, $\angle A$ contains 23°, how many degrees are there in arc BC? in arc AC?

Proposition XX. Theorem

263. *An angle formed by two chords intersecting within a circumference is measured by half the sum of the intercepted arcs.*

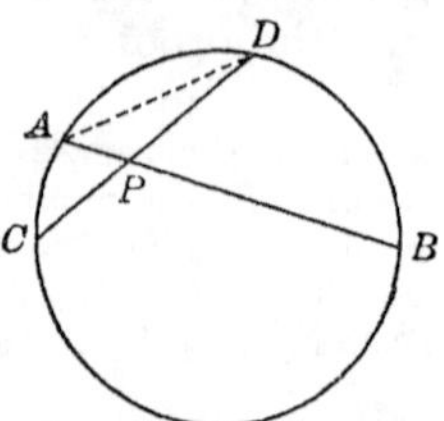

Given the chords AB and CD in the $\odot$ $ADBC$, intersecting within the circumference at the point P.

To prove that $\angle DPB$ is measured by $\frac{1}{2}$ (arc DB + arc AC).

Proof. Draw the chord AD.

Then, in $\triangle$ ADP,

$$\angle DPB = \angle A + \angle D,$$ Art. 135.

(*an ext. $\angle$ of a $\triangle$ is equal to the sum of the two opp. int. $\angle$s*).

But $\angle A$ is measured by $\frac{1}{2}$ arc DB, Art. 258.

(*an inscribed $\angle$ is measured by one-half its intercepted arc*).

Also $\angle D$ is measured by $\frac{1}{2}$ arc AC. (Why?)

Hence $\angle DPB$ is measured by $\frac{1}{2}$ (arc DB + arc AC). Ax. 2.

Q. E. D.

Ex. 1. In the above figure, if arc DB contains 64° and arc AC contains 38°, how many degrees in $\angle APC$? in $\angle APD$?

Ex. 2. If arc $AD = 52°$ and $\angle APD = 124°$, find arc CB.

Ex. 3. If, in Fig. 1, p. 132, arc AC contains 112°, how many degrees are there in the $\angle A$?

Proposition XXI. Theorem

264. ***An angle formed by a tangent and a chord drawn from the point of contact is measured by half the intercepted arc.***

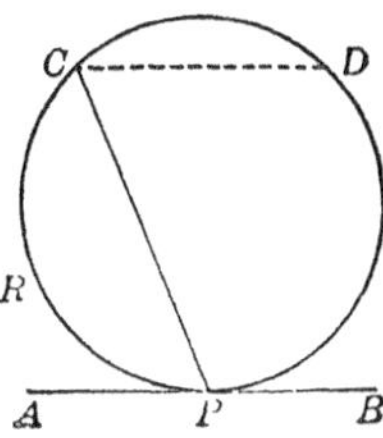

Given the ⊙ PCD, and $\angle APC$ formed by the tangent APB and the chord PC.

To prove that $\angle APC$ is measured by $\frac{1}{2}$ arc PRC.

Proof. Let the chord CD be drawn $\parallel AB$.

Then $\angle APC = \angle PCD$. (Why?)

But $\angle PCD$ is measured by $\frac{1}{2}$ arc PD. (Why?)

$\therefore$ $\angle APC$ is measured by $\frac{1}{2}$ arc PD. Ax. 8.

But arc PRC = arc DP, Art. 233.

(*two $\parallel$ lines intercept equal arcs on a circumference*).

$\therefore$ $\angle APC$ is measured by $\frac{1}{2}$ arc PRC. Ax. 8.

Also $\angle BPC$, the supplement of $\angle APC$, is measured by $\frac{1}{2}$ arc PDC, Ax. 3,

(*for arc PDC is the conjugate of arc PRC*).

Q. E. D.

Ex. 1. In the above figure, if arc PRC contains 124°, how many degrees are in the angle APC?

Ex. 2. If arc CD = 96°, find the angles on the figure.

PROPOSITION XXII. THEOREM

265. *An angle formed by two secants, or by two tangents, or by a secant and a tangent meeting without the circumference, is measured by half the difference of the intercepted arcs.*

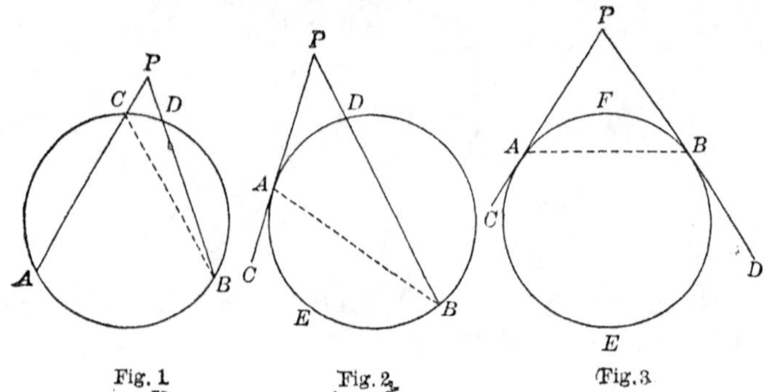

Fig. 1 Fig. 2 Fig. 3

I. **Given** the ⊙ $ACDB$ (Fig. 1), and the $\angle APB$ formed by the two secants PA and PB, meeting at the point P without the circumference.

To prove that $\angle P$ is measured by $\frac{1}{2}$ (arc AB — arc CD).

Proof. Draw the chord CB.

Then $\angle ACB = \angle P + \angle B$. (Why?)

$\therefore \angle P = \angle ACB - \angle B$. Ax. 3.

But $\angle ACB$ is measured by $\frac{1}{2}$ arc AB. Art. 258.

Also $\angle B$ is measured by $\frac{1}{2}$ arc CD. (Why?)

$\therefore \angle P$ is measured by $\frac{1}{2}$ (arc AB — arc CD). Ax. 3.

II. **Given** the ⊙ ADB (Fig. 2), and $\angle CPB$ formed by the tangent PC and the secant PB.

To prove that $\angle P$ is measured by $\frac{1}{2}$ (arc AEB — arc AD).

Proof. Let the pupil supply the proof.

III. **Given** $\angle APB$ (Fig. 3) formed by the tangents PC and PD.

To prove that $\angle P$ is measured by $\frac{1}{2}$ (arc AEB — arc AFB).

Proof. Let the pupil supply the proof.

266. Note. By means of Props. XVIII–XXII, angles formed by chords, secants, or tangents, or combinations of these, are all reduced to central angles and hence may readily be compared.

Ex. 1. If, in Fig. 1, p. 136, arc $CD = 34°$ and arc $AB = 108°$, draw AD and find all the angles of the figure.

Ex. 2. If, in Fig. 3, p. 136, angle $P = 80°$, find arcs AFB and AEB.

EXERCISES. GROUP 16

Ex. 1. What new methods of proving two lines equal are afforded by Book II ?

Ex. 2. What new methods of proving two angles equal ? of proving two angles supplementary ? of proving an angle a right angle ?

Ex. 3. What methods of proving two arcs equal ?

Ex. 4. In a quadrilateral inscribed in a circle, each pair of opposite angles is supplementary.

Ex. 5. A chord forms equal angles with the tangents at its extremities.

Ex. 6. If an isosceles triangle be inscribed in a circle, the tangent at its vertex makes equal angles with two of its sides and is parallel to the third side. (Is the converse of this theorem true ?)

Ex. 7. If two chords in a circle intersect within the circle at right angles, the sum of a pair of alternate arcs equals a semicircumference.

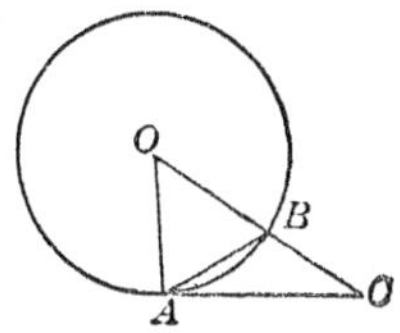

Ex. 8. Given O the center of a circle and AC a tangent; prove $\angle BAC = \frac{1}{2} \angle O$.

Ex. 9. If one side of an inscribed quadrilateral be produced, the exterior angle so formed equals the opposite interior angle of the quadrilateral.

Ex. 10. If two tangents to a circle include an angle of 60° at their point of intersection, the chord joining the points of contact forms with the tangents an equilateral triangle.

Ex. 11. The chord AB equals the chord CD in a given circle, and the chords, if produced, intersect at the point P. Prove secant $PA =$ secant PC.

Ex. 12. A circle is circumscribed about the triangle ABC, and P is the midpoint of the arc AB. Prove that the angle ABP equals one-half the angle C.

Ex. 13. If A, B, C, D and E be points taken in succession on the circumference of a circle, and the arcs AB, BC, CD and DE be equal, prove that the angles ABC, BCD and CDE are equal.

Ex. 14. An inscribed angle formed by a diameter and a chord has its intercepted arc bisected by a radius which is parallel to the chord.

Ex. 15. Two secants, PAB and PCD, intersect the circle $ABDC$. Prove that the triangles PBC and PAD are mutually equiangular.

Ex. 16. Given AC a tangent and $AB \parallel CE$; prove △ ACD and ABE mutually equiangular.

Ex. 17. Given ABC an inscribed triangle, $AE \perp BC$, and $CD \perp AB$; prove arc $BD =$ arc BE.

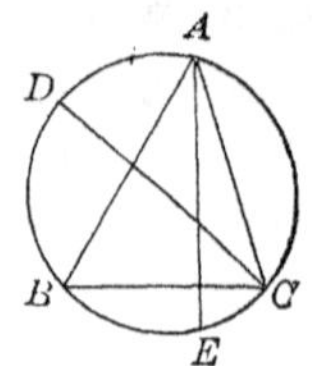

Ex. 18. If the diagonals of an inscribed quadrilateral are diameters, the quadrilateral is a rectangle.

Ex. 19. Tangents through the vertices of an inscribed rectangle form a rhombus.

Ex. 20. Two circles intersect at P and Q. PA and PB are diameters. Prove that QA and QB form a straight line.

Ex. 21. Two equal circles intersect at P and Q, through P a line is drawn terminated by the circumferences at A and B. Show that QA equals QB.

[SUG. ∠s A and B are measured by what arcs?]

267. Use of auxiliary lines. In demonstrating theorems relating to the circle, it is often helpful to draw one or more of the following auxiliary lines: *A radius, a diameter, a chord, a perpendicular from the center upon a chord, an arc, a circumference, etc.*

EXERCISES. GROUP 17

AUXILIARY LINES

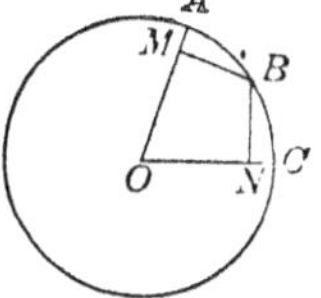

Ex. 1. Given circle O, arc AB=arc BC, $BM \perp OA$, and $BN \perp OC$; prove $BM=BN$.

Ex. 2. If from any point in the circumference of a circle a chord and a tangent be drawn, the perpendiculars drawn to them from the midpoint of the arc subtended by the chord are equal.

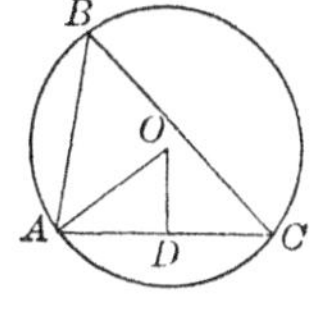

Ex. 3. Given O the center of a circle, and OD $\perp$ chord AC; prove $\angle AOD = \angle B$.

Ex. 4. From the extremity of a diameter chords are drawn making equal angles with the diameter. Prove that these chords are equal.

[SUG. Draw ⊥s from the center to the chords, etc.]
Is the converse of this theorem true ?

Ex. 5. If through any point within a circle equal chords be drawn, show that the line drawn from the center to the point of intersection of the chords bisects their angle of intersection.

Ex. 6. Tangents PA and PB are drawn to a circle whose center is O. Prove that angle P equals twice angle OAB.

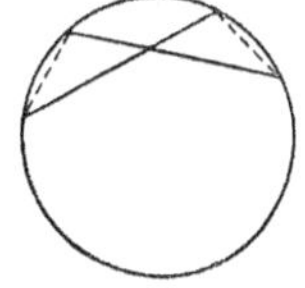

Ex. 7. If in a circle two equal chords intersect, the segments of one chord equal the segments of the other chord.

Ex. 8. A parallelogram inscribed in a circle is a rectangle.

[SUG. Draw the diagonals of the ▱, use Art. 218, and Ex. 21, p. 100.]

Ex. 9. If a quadrilateral be circumscribed about a circle, the angles subtended at the center by a pair of opposite sides are supplementary.

[SUG. Draw radii to the points of contact and show that there are four pairs of equal ∠s at the center.]

Ex. 10. If an equilateral triangle ABC be inscribed in a circle and any point P be taken in the arc AB, show that $PC = PA + PB$.

[SUG. On PC take PM equal to PA, draw AM and prove △s PAB and MAC equal.]

Ex. 11. Two radii perpendicular to each other are produced to intersect a tangent, and from the points of intersection other tangents are drawn to the circle. Prove that the tangents last drawn are parallel.

[SUG. Draw radii to the three points of contact, etc.]

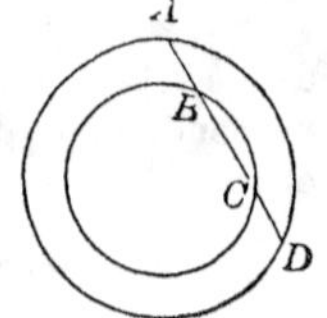

Ex. 12. A straight line intersects two concentric circles. Show that the segments of the line intercepted between the circumferences are equal (prove $AB = CD$).

Ex. 13. A common tangent is drawn to two circles which are exterior to each other. Show that the chords drawn from the points of tangency to the points where the line of centers cuts the circumferences are parallel.

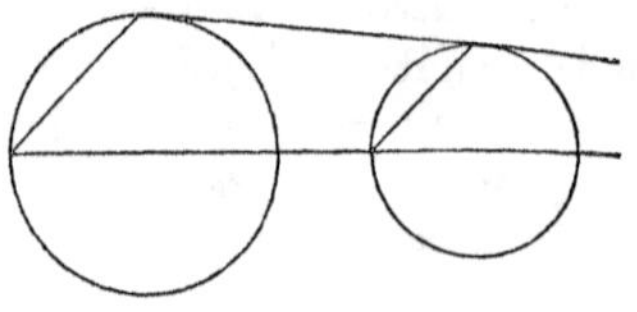

Ex. 14. A circle is described on the radius of another circle as a diameter, and a chord of the larger circle is drawn from the point of contact of the two circles. Prove that this chord is bisected by the circumference of the smaller circle.

[SUG. If the chord is bisected, a ⊥ from the center of the larger to the chord will also bisect the chord, etc.]

Ex. 15. Two circles are tangent externally at the point P. Through P any two lines APB and CPD are drawn terminated by the circumferences. Show that the chords AC and BD are parallel.

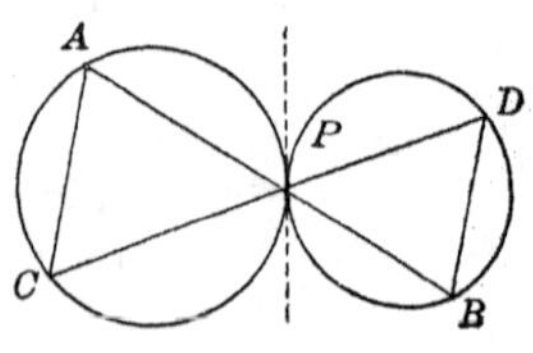

[SUG. Draw the common tangent at P. If AC and BD are ∥, what ∠s must be equal?]

Ex. 16. Two circles intersect at the points P and Q. Lines APB and CQD are drawn, terminated by the circumferences. Show that AC and BD are parallel.

Ex. 17. If a square be described on the hypotenuse of a right triangle, a line drawn from the center of the square to the vertex of the right angle bisects the right angle.

[SUG. Describe a circumference on the hypotenuse of the right triangle as a diameter.]

Ex. 18. If two circles are tangent externally at P, and a common tangent touches them at A and B, respectively, the angle APB is a right angle.

Ex. 19. If in the triangle ABC the two altitudes BD and AE are drawn, the angle ABD equals the angle AED.

[SUG. Describe a semicircumference on AB as a diameter.]

268. DEF. A **maximum** is the greatest of a class of magnitudes satisfying certain given conditions, and a **minimum** is the least (see Arts. 470, 471). For instance, of the chords in a given circle, which is the maximum?

EXERCISES. GROUP 18

MAXIMA AND MINIMA

Ex. 1. Of the chords drawn through a given point within a circle, determine which is the greatest, and also which is the least.

Ex. 2. Find the shortest line, and also the longest line, that can be drawn from a given external point to the circumference of a circle.

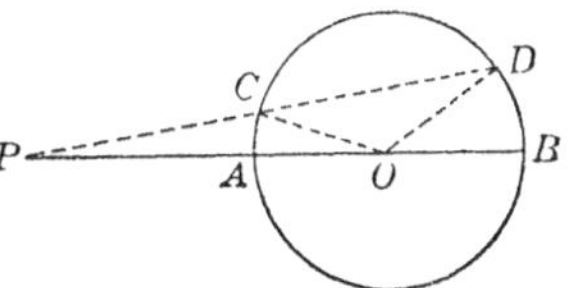

[SUG. O being the center, prove in $\triangle PCO$, $PA < PC$; by $\triangle PDO$, $PB > PD$.]

Ex. 3. Find the shortest line, and also the longest line, that can be drawn to the circumference of a circle from a point within the circle.

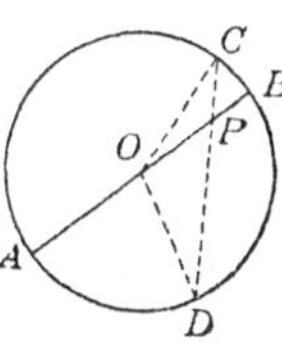

[SUG. O being the center, prove, by use of $\triangle OPC$, $PB < PC$, etc.]

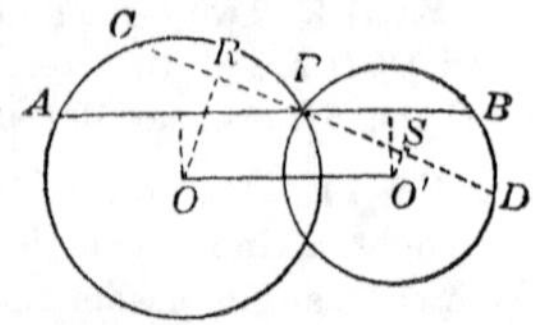

Ex. 4. If two circles intersect, show that, of lines drawn through a point of intersection and terminated by the circumferences, that line is a maximum which is parallel to the line of centers.

[Sug. Prove $RS < OO'$ $\therefore$ $CPD < APB$.]

Ex. 5. Given $AB \perp OB$ in circle O; prove $\angle OAB$ the maximum of all $\angle$s having their vertices on the circumference and their sides passing through O and B respectively.

[Sug. Draw a circle on OA as a diameter.]

EXERCISES. GROUP 19

DEMONSTRATIONS BY INDIRECT METHODS

Prove the following by an indirect method (see Art. 195):

Ex. 1. A segment of a circle which contains a right angle is a semicircle.

Ex. 2. If a rectangle be inscribed in a circle, its diagonals are diameters.

Ex. 3. Prove the second part of Prop. VI by an indirect method.

Ex. 4. Prove Prop. X by an indirect method.

Ex. 5. A straight line connecting the midpoint of a chord and the midpoint of the arc subtended by the chord is perpendicular to the chord (use the method of coincidence).

Ex. 6. A line joining the midpoints of two parallel chords passes through the center.

[Sug. Draw a $\perp$ to each chord from the center and show that these $\perp$s are in the same line, etc.]

Ex. 7. If the opposite angles of a quadrilateral are supplementary, a circle can be circumscribed about the quadrilateral.

[Sug. Pass a circumference through three vertices of the quadrilateral; if it does not pass through the remaining vertex, etc.]

Ex. 8. Prove the converse of Prop. XII.

269. A **geometrical constant** is a geometrical magnitude which varies in some respect, as in position, but remains constant in size. Thus the angles inscribed in a given semicircle vary in position but are all of the same size; viz., a right angle (see Art. 261).

EXERCISES. GROUP 20

DETERMINATION OF CONSTANTS AND LOCI

Ex. 1. AB and AC are tangents to a circle. P is any point on the circumference outside the triangle ABC. As P moves, prove that the sum of the $\angle A$ and $\angle BPC$ is constant.

Ex. 2. In Ex. 8, p. 123, if AR and BQ be produced to meet at T, show that the perimeter of the triangle TRQ equals the sum of TA and TB, and hence that the perimeter of triangle TRQ is constant, no matter how P may vary in position between A and B.

Ex. 3. Show on the same figure that, if O is the center, $\angle ROQ$ is constant as P varies in position.

Ex. 4. Two circles intersect in the points A and B. From any point P on one circumference lines PAC and PBD are drawn, terminated by the other circumference. Show that the chord CD is constant.

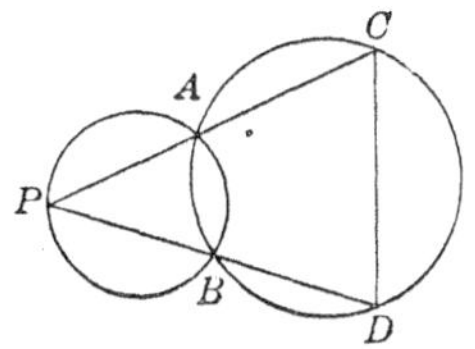

[SUG. Draw BC and prove $\angle CBD$ a constant.]

Ex. 5. Given $OA \perp OB$, and CD a line of given length moving so that D is always in OA and C in OB and P the midpoint of CD; prove that OP is constant in length (see Ex. 6, p. 94).

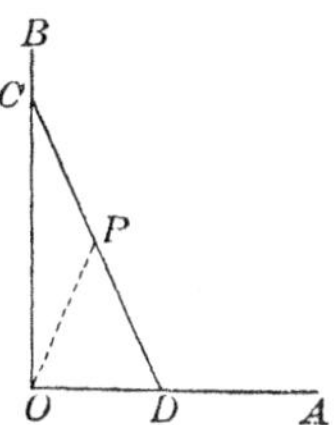

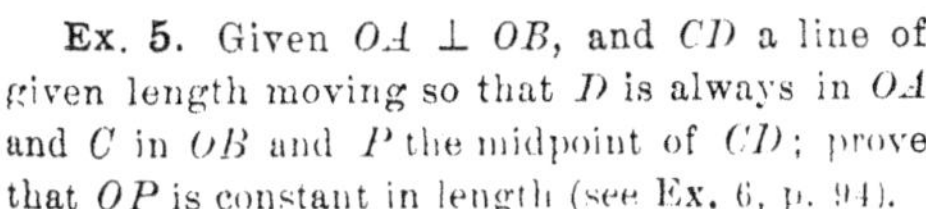
The determination of loci is often facilitated by showing that some given magnitude is a constant.

Ex. 6. Find the locus of a point moving so that it is at a given distance a from a given circumference whose radius is r.

Ex. 7. Find the locus of the midpoints of the radii of a given circle.

Ex. 8. Find the locus of the midpoints of all chords of a given length drawn in a given circle.

Ex. 9. Find the locus of the vertices of all right triangles having a given hypotenuse.

Ex. 10. Find the locus of the midpoints of all the chords drawn from a given point on a given circumference.

[SUG. Draw a line from the center to the given point and perpendiculars from the center upon the chords.]

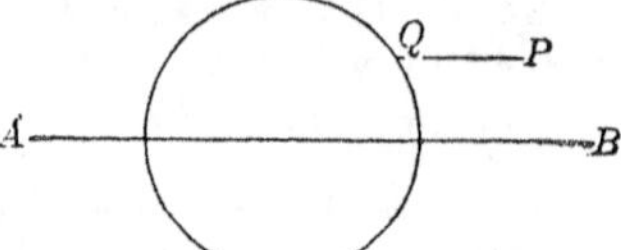

Ex. 11. In Ex. 5, p. 143, find the locus of P.

Ex. 12. QP is a line of given length and moves so that Q is always in a given circumference, and QP is always parallel to a fixed line. Find the locus of P.

EXERCISES. GROUP 21

THEOREMS PROVED BY VARIOUS METHODS

Ex. 1. The line which bisects the angle formed by a tangent and a chord bisects the intercepted arc also.

Ex. 2. An inscribed trapezoid is isosceles.

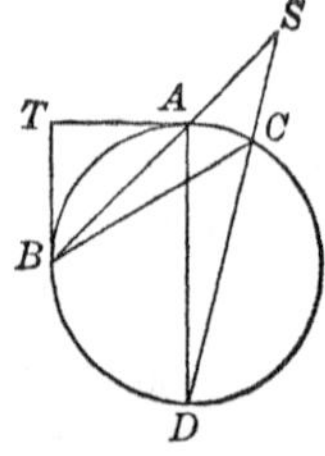

Ex. 3. Given TA and TB tangents, arc $AB = 80°$, arc $BD = 95°$, and arc $DC = 150°$; find all the angles of the figure.

Ex. 4. If two tangents to a circle are parallel, the line joining their points of contact is a diameter.

[SUG. Draw radii to the points of contact and use an indirect method of proof.]

Ex. 5. A rectangle circumscribed about a circle is a square.

[SUG. Use the preceding theorem.]

Ex. 6. Given O the center of a circle, and $BP =$ the radius; prove $\angle AOC = 3 \angle P$.

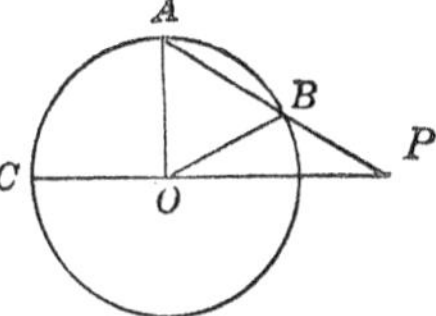

Ex. 7. Find the angle formed by the side of an inscribed square and the tangent through the vertex of the square.

Ex. 8. From an external point a secant is drawn through the center of a circle, and also two other secants making equal angles with the first secant. Show that the secants last drawn are equal.

Ex. 9. ABC is a triangle and on the side AB the point P is taken and on BC the point Q, so that angle BPQ equals angle C. Show that a circle may be circumscribed about the quadrilateral $APQC$.

Ex. 10. Two circles are tangent to each other externally, and a line is drawn through the point of contact terminated by the circumferences. Show that the radii from the extremities of this line are parallel.

Ex. 11. If a circumference be described on the leg of an isosceles triangle as diameter, the circumference will bisect the base of the triangle.

Ex. 12. The chord of an arc is parallel to the tangent at the midpoint of the arc.

Ex. 13. If a triangle be inscribed in a circle, the sum of the angles inscribed in the segments exterior to the triangle is four right angles.

Ex. 14. Find the corresponding theorem for an inscribed quadrilateral.

Ex. 15. Find the locus of the centers of all circles passing through two given points.

Ex. 16. If two unequal chords intersect in a circle, the greater chord makes the less angle with the diameter through the point of intersection of the chords.

State also the converse of this theorem. Is the converse true?

Ex. 17. The sum of the legs of a right triangle equals the sum of the hypotenuse and the diameter of the inscribed circle.

J

Ex. 18. The sides AB, BC and AC of a triangle touch the inscribed circle at the points P, Q and R. Show that angle PQR and one-half angle A are complementary.

[SUG. Draw radii from the center O to P and R. Then $\angle PQR = \angle POA$, etc.]

Ex. 19. From the point in which the bisector of an inscribed angle meets the circumference, a chord is drawn parallel to one side of the angle. Show that this chord equals the other side of the angle.

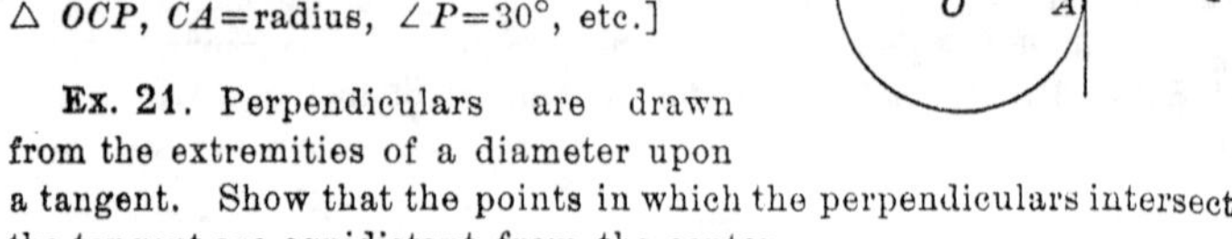

Ex. 20. Given AB a diameter, $AP=$ the radius, AD and PC tangents; prove $\triangle CED$ equilateral.

[SUG. Draw OC and CA; then in rt. $\triangle OCP$, $CA=$radius, $\angle P=30°$, etc.]

Ex. 21. Perpendiculars are drawn from the extremities of a diameter upon a tangent. Show that the points in which the perpendiculars intersect the tangent are equidistant from the center.

CONSTRUCTION PROBLEMS

270. Postulates. As stated in Art. 49, a postulate in geometry is a construction of a geometric figure admitted as possible.

The postulates used in geometry, are as follows (see Art. 50):

1. *Through any two points a straight line may be drawn.*

2. *A straight line may be extended indefinitely, or it may be limited at any given point.*

3. *A circumference may be described about any given point as a center and with any given radius.*

271. The **meaning of the postulates** is that only two drawing instruments are to be used in making geometrical constructions; viz., the *straight-edge ruler* and the *compasses*.

With these two simplest drawing instruments it is desirable to be able to construct as many geometrical figures as possible.

272. Form of solution of a problem. The statement of a problem and of its solution consists of certain distinct parts which it is important to keep in mind. These are

1. The **general enunciation.**

2. The **particular enunciation.**
   (1) **Given,** etc.
   (2) **To construct,** etc. (or some other construction phrase, as "to draw," "to bisect," etc.).

3. The **construction.**

4. The **assertion.**

5. The **proof of the assertion.**

6. The **conclusion** (indicated by Q. E. F., quod erat faciendum, "which was to be done").

7. The **discussion** of special or limiting cases, if such cases occur in the given problem.

In the figures drawn in connection with problems, the *given lines* are drawn as *heavy lines*, the *lines required* as *light lines*, and the *auxiliary lines* as *dotted lines*.

PROPOSITION XXIII. PROBLEM

273. *From a given point without a given line to draw a perpendicular to the line.*

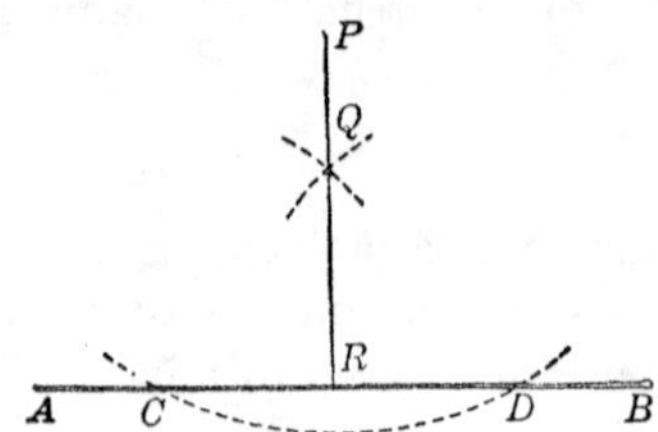

Given the line AB and the point P outside AB.

To construct a perpendicular from the point P to the line AB.

Construction. With P as a center and with any convenient radius, describe an arc intersecting AB in two points as at C and D. Post. 3.

From C and D as centers and with convenient equal radii, greater than $\frac{1}{2}$ CD, describe arcs intersecting at Q. Post. 3.

Draw the line PQ and produce it to meet AB at R. Posts. 1, 2.

[**Assertion**]. Then PR is the $\perp$ required.

Proof. P is equidistant from the points C and D. Constr.

Also Q is equidistant from the points C and D. Constr.

$\therefore PR \perp CD$. Art. 113.

(*two points, each equidistant from the extremities of a line, determine the $\perp$ bisector of the line*).

Q. E. F.

Proposition XXIV. Problem

274. *At a given point in a given line to erect a perpendicular to that line.*

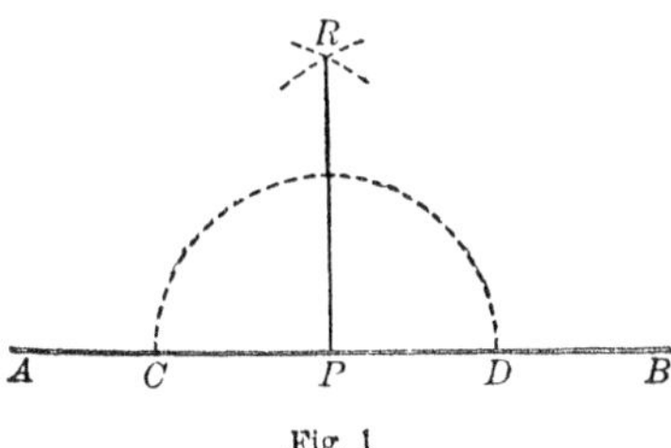

Fig. 1

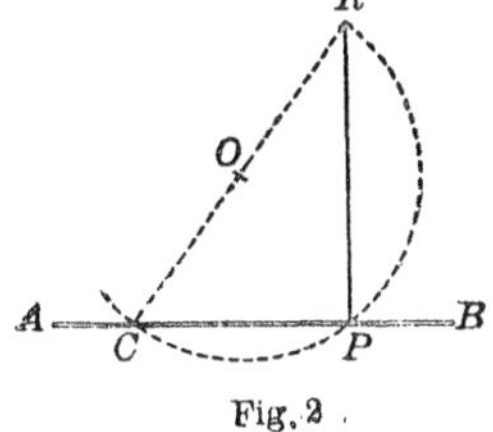

Fig. 2

Given the point P in the line AB

To construct a perpendicular to the line AB at the point P.

Method I. **Construction.** From P (Fig. 1) as a center with a convenient radius, describe an arc cutting off the equal segments PC and PD on the line AB. Post. 3.

From C and D as centers and with equal radii, greater than PD, describe arcs intersecting at R. Post. 3.

Draw PR. Post. 1.

[**Assertion**]. Then PR is the $\perp$ required.

Proof. Let the pupil supply the proof.

Method II. **Construction.** Take any point O (Fig. 2) without the line AB, and with OP as a radius, describe a circumference intersecting the line AB at C. Post. 3.

Draw CO, and produce CO to meet the circumference at R. Draw RP. Posts. 1 and 2.

[**Assertion**]. Then RP is the $\perp$ required.

Proof. Let the pupil supply the proof.

Discussion. When is Method II preferable?

PROPOSITION XXV. PROBLEM

275. *To bisect a given line.*

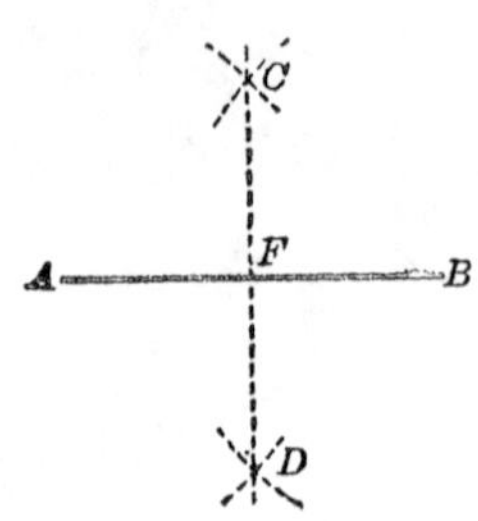

Given the line AB.

To bisect line AB.

Construction. With A and B as centers and with equal radii, greater than $\frac{1}{2}$ AB, describe arcs intersecting in C and D. Post. 3.

Draw the line CD intersecting AB in F. Post. 1.

Then AB is bisected at the point F.

Proof. Let the pupil supply the proof.

Q. E. F.

PROPOSITION XXVI. PROBLEM

276. *To bisect a given arc.*

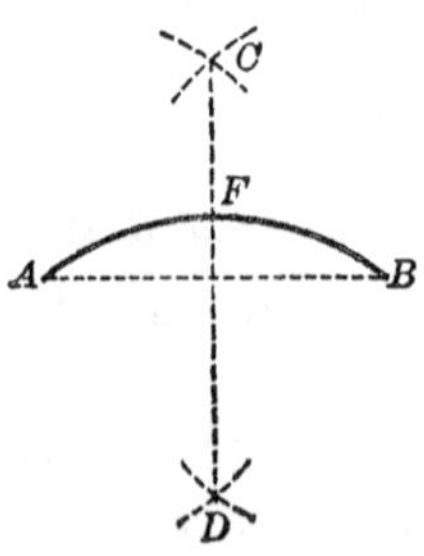

Given the arc AB.

To bisect arc AB.

Construction. With A and B as centers, and with convenient equal radii, describe arcs intersecting at C and D. Post. 3.

Draw the line CD intersecting the arc AB at F. Post. 1.

Then arc AB is bisected at the point F.

Proof. Draw the chord AB.

Then $CD \perp$ chord AB at its middle point. (Why?)

$\therefore$ CD bisects the chord AB, Art. 223.

(*the* $\perp$ *bisector of a chord passes through the center and bisects the arcs subtended by the chord*).

Q. E. F.

PROPOSITION XXVII. PROBLEM

277. *To bisect a given angle.*

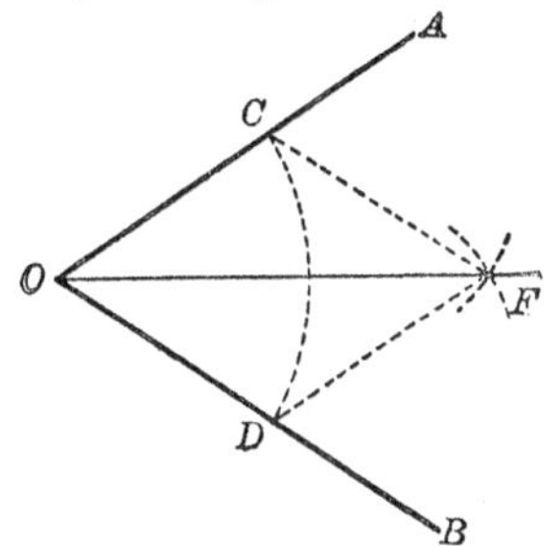

Given angle AOB.

To bisect angle AOB.

Construction. With O as a center and with any convenient radius, describe an arc intersecting OA at C and OB at D. Post. 3.

With C and D as centers and with convenient equal radii, describe arcs intersecting at F. Post. 3.

Draw OF. Post. 1.

Then $\angle AOB$ is bisected by line OF.

Proof. Let the pupil supply the proof.

Q. E. F.

Ex. Construct an $\angle$ of $45°$.

PROPOSITION XXVIII. PROBLEM

278. *At a given point in a given straight line to construct an angle equal to a given angle.*

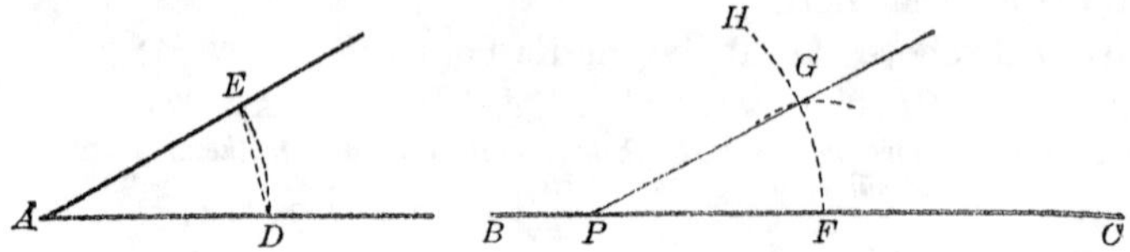

Given the $\angle A$, and the point P in the line BC.

To construct at the point P an angle equal to $\angle A$, and having PC for one of its sides.

Construction. With A as a center and with any convenient radius, describe an arc meeting the sides of $\angle A$ at D and E. Post. 3.

Draw the chord DE. Post. 1.

With P as a center and with a radius equal to AD, describe an arc cutting PC at F. Post. 3.

From F as a center and with a radius equal to the chord DE, describe an arc intersecting the arc HF at G. Post. 3.

Draw PG. Post. 1.

Then $\angle GPF$ is the angle required.

Proof. Let the pupil draw the chord FG and complete the proof.

Q. E. F.

Ex. 1. On a given line construct a square.

Ex. 2. Construct an angle of 30°.

Ex. 3. Hence construct an angle of 15°.

Proposition XXIX. Problem

279. *Through a given point without a given straight line to draw a line parallel to a given line.*

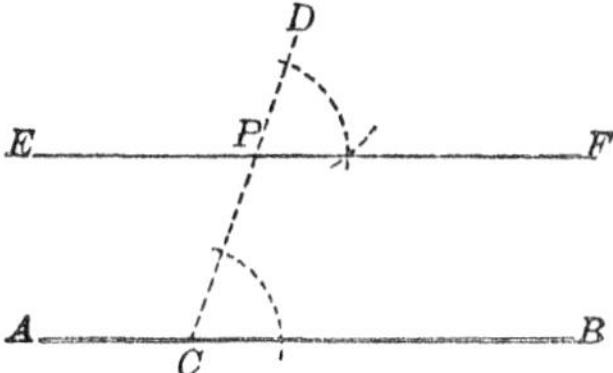

Given any point P without the line AB.

To construct a line through $P \parallel AB$.

Construction. Through P draw any convenient line CD meeting AB in C. Post. 1.

At P in the line CD construct $\angle DPF$ equal to $\angle PCB$. Art. 278.

Then EF is the line required.

Proof. Let the pupil supply the proof.

Ex. 1. Through a given point draw a line parallel to a given line by constructing a parallelogram of which the given point is one vertex.

Ex. 2. On a given line as a diameter, construct a circle.

Ex. 3. Construct an arc of 60° having a radius of 1 in.

Ex. 4. On the figure, p. 148, if the point Q be constructed below the line AB, will the perpendicular required be likely to be more accurately, or less accurately constructed?

Ex. 5. From a given point on a given circumference, how many equal chords can be drawn?

Ex. 6. Through a given point within a given circumference, how many equal chords can be drawn?

Ex. 7. From a given point external to a given circle, how many equal secants can be drawn?

PROPOSITION XXX. PROBLEM

280. *To divide a given straight line into any required number of equal parts.*

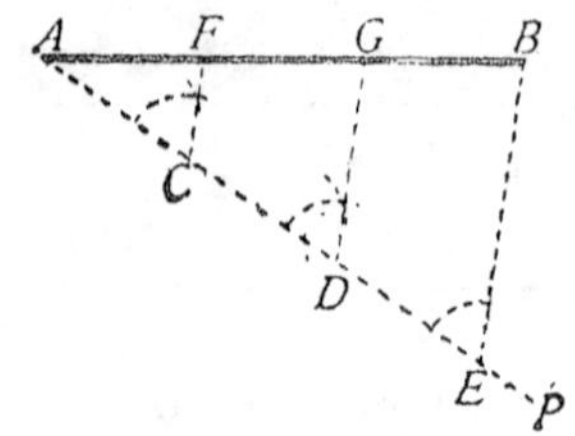

Given the line AB.

To divide AB into a required number (as three) equal parts.

Construction. From A draw the line AP making a convenient angle with AB. Post. 1.

Take AC any line of convenient length and apply it to AP a number of times equal to the number of parts into which AB is to be divided. Post. 2.

From E, the end of the measure when last applied to AP, draw EB. Post. 1.

Through the other points of division on AP, viz., C and D, draw lines $\parallel$ EB and meeting AB at F and G. Art. 279.

Then AB is divided into the required number of parts at F and G.

Proof. $AC = CD = DE$. Constr.

$\therefore AF = FG = GB$, Art. 176.

(*if three or more parallels intercept equal parts on one transversal, they intercept equal parts on every transversal*).

Ex. 1. Construct a right triangle whose legs are 1 in. and $1\frac{1}{2}$ in.

Ex. 2. Construct a rectangle whose base is 2 in. and altitude 1 in.

PROPOSITION XXXI. PROBLEM

281. *To construct a triangle, given two sides and the included angle.*

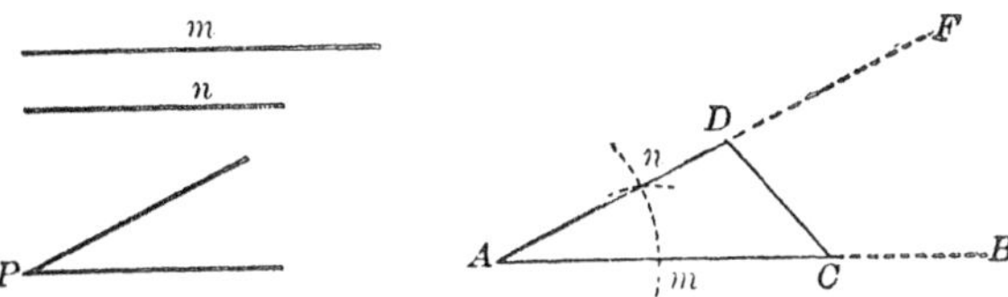

Given m and n two sides of a triangle and P the angle included by them.

To construct the triangle.

Construction. At the point A in the line AB construct $\angle A$ equal to the given $\angle P$. Art. 278.

On the line AB lay off AC equal to m. Post. 2.

On the line AF lay off AD equal to n. Post. 2.

Draw DC. Post. 1.

The $\triangle ADC$ is the triangle required.

Q. E. F.

Ex. 1. Construct a triangle in which two of the sides are 1 in. and $1\frac{1}{4}$ in., and the included angle is 135°.

Ex. 2. Construct an isosceles triangle, in which the base shall be $1\frac{1}{2}$ in. and the altitude 2 in.

Ex. 3. Construct the complement of a given acute angle.

Ex. 4. Construct the supplement of a given angle.

Ex. 5. How is the figure on p. 154 constructed with the fewest adjustments of the compasses?

Ex. 6. Draw a line (segment) and mark off three-fifths of it

PROPOSITION XXXII. PROBLEM

282. *To construct a triangle, given two angles and the included side.*

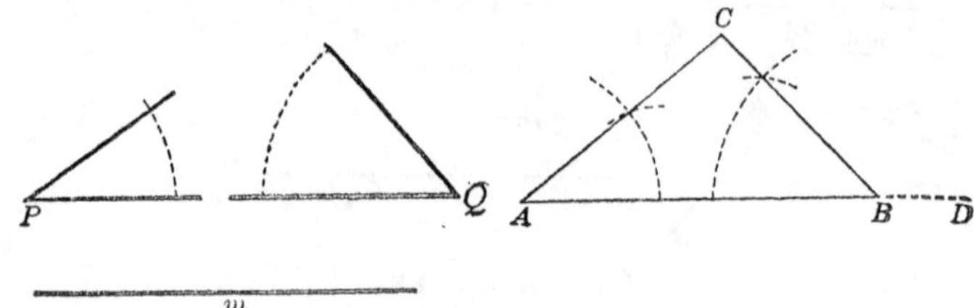

Given the ∠s P and Q and the included side m.

To construct the triangle.

Construction. Take any line AD and on it mark AB equal to m. Post. 2.

At A construct an angle equal to $\angle P$. Art. 278.

At B construct an angle equal to $\angle Q$. Art. 278.

Produce the sides of the ∠s A and B to meet at C. Post. 2.

Then $\triangle ACB$ is the triangle required. Q. E. F.

Discussion. Is it possible to construct the triangle if the sum of the given angles is two right angles? Why? Is it possible if this sum is greater than two right angles? Why?

Ex. 1. Construct a triangle in which two of the angles are 30° and 45°, and the included side is $1\frac{3}{4}$ in.

Ex. 2. Construct the complement of half a given angle.

Ex. 3. Construct an angle of 120°; of 150°; of 135°.

Ex. 4. Trisect a given right angle.

PROPOSITION XXXIII. PROBLEM

283. *To construct a triangle, given the three sides.*

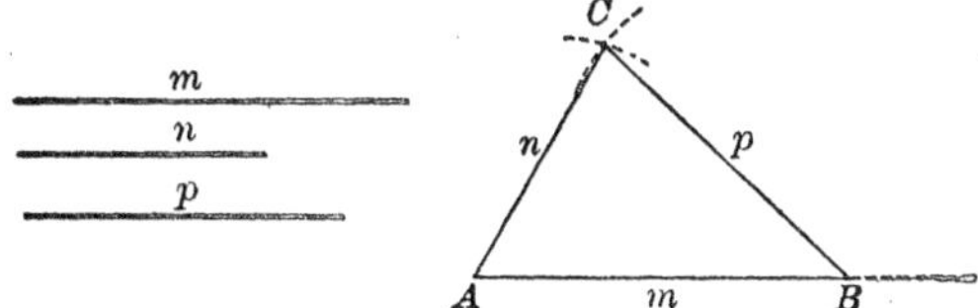

Given m, n and p, the three sides of a triangle.

To construct the triangle.

Construction. Take the line AB equal to m. Post. 2.

With A as a center and with a radius equal to n describe an arc, and with B as a center and with a radius equal to p describe another arc. Post. 3.

Let the two arcs intersect at the point C.

Draw CA and CB. Post. 1.

Then $\triangle ABC$ is the triangle required.

Discussion. Is it possible to construct the triangle if one of the sides is greater than the sum of the other two sides? Why?

What kind of a figure is obtained if one side equals the sum of the other two sides?

Ex. 1. Construct an angle of $22\frac{1}{2}°$.

Ex. 2. Divide a given circumference into four quadrants.

Ex. 3. Construct the figure on page 82, using the concurrence of the three altitudes as a test of the accuracy of the work.

Proposition XXXIV. Problem

284. *To construct a triangle, given two sides and an angle opposite one of them.*

Given m and n two sides of a $\triangle$ and $\angle P$ opposite n.

To construct the triangle.

Construction. Several cases occur, according to the relative size of the given sides and the size of the given angle.

Case I. *When* $n > m$ *(and* $\angle P$ *is acute).*

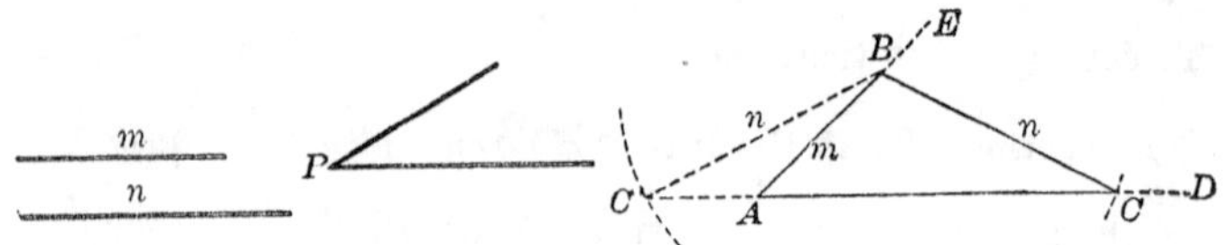

At the point A construct $\angle EAD$ equal to $\angle P$. Art. 278.

On AE take AB equal to m. Post. 2.

With B as a center and with a radius equal to n, describe an arc intersecting AD at C and C'. Post. 3.

Draw BC and BC'. Post. 1.

Two $\triangle$, ABC and ABC', are obtained, containing the sides m and n; but only one of them, $\triangle ABC$, contains the $\angle P$.

$\therefore$ $\triangle ABC$ is the triangle required.

Case II. *When* $n = m$ *(and* $\angle P$ *is acute).*

Make the same construction as in the preceding case.

The arc drawn intersects the line AD in the points A and C.

Hence the isosceles $\triangle ABC$ is the triangle required.

CASE III. *When $n < m$ (and $\angle P$ is acute).*

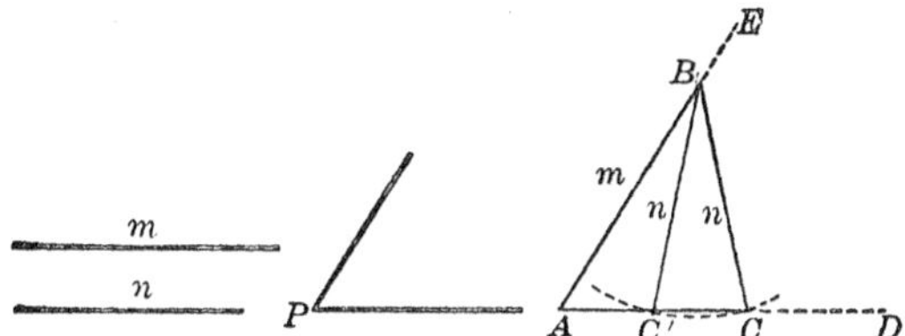

Make the construction in the same way as in Case I.

Two ▵, ABC and ABC', are obtained, each of which contains the sides m and n and an angle equal to $\angle P$ opposite the side n.

$\therefore$ ▵ ABC and ABC' are the triangles required.

Discussion. In Case I, if $\angle P$ is a right $\angle$, let the pupil construct the figure and show that there are two ▵ answering the given conditions. If $\angle P$ is an obtuse angle, let him construct the figure and show that there is but one answer.

In Case II, if $\angle P$ is right, or obtuse, what results are obtained?

In Case III, if $\angle P$ is acute and $n =$ the $\perp$ from B to AD, how many answers are there? also, if $n <$ this $\perp$, how many?

If $\angle P$ is right, or obtuse, what result is obtained?

Ex. 1. Construct a triangle in which two of the sides are 1 in. and $1\frac{1}{2}$ in., and the angle opposite the latter side is 45°.

Ex. 2. Construct a triangle in which two of the sides are $1\frac{1}{2}$ in. and 1 in., and the angle opposite the latter side is 45°.

Ex. 3. Construct a triangle in which two of the sides are $1\frac{1}{2}$ in. and $\frac{3}{4}$ in., and the angle opposite the latter side is 30°.

Ex. 4. Construct the figure of page 80, using the concurrence of the three bisectors as a test of the accuracy of the work.

PROPOSITION XXXV. PROBLEM

285. *To construct a parallelogram, given two sides and the included angle.*

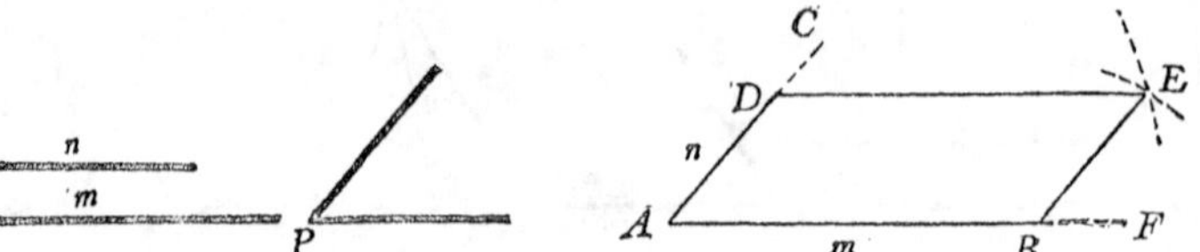

Given m and n two sides, and P the included ∠ of a ▱.

To construct the parallelogram.

Construction. Take line AB equal to m. Post. 2.

At the point A construct $\angle CAB$ equal to $\angle P$. Art. 278.

On the side AC lay off AD equal to n.

From D as a center and with a radius equal to m, and from B as a center with a radius equal to n, describe arcs intersecting in E. Post. 3.

Draw ED and EB. Post. 1.

Then $ABED$ is the parallelogram required.

Proof. Let the pupil supply the proof.

PROPOSITION XXXVI. PROBLEM

286. *To circumscribe a circle about a given triangle.*

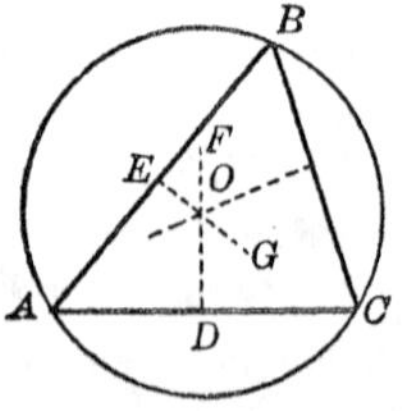

Given the △ ABC.

To construct a circumscribed ⊙ about ABC.

Construction. Erect ⊥s DF and EG at the midpoints of the sides AC and AB, respectively. Art. 274.

From O, the point of intersection of these ⊥s, with a radius equal to OA, describe the ⊙ ABC. Post. 3, Art. 197.

Then ⊙ ABC is the circle required.

Proof. Let the pupil supply the proof.

Q. E. F.

PROPOSITION XXXVII. PROBLEM

287. *To inscribe a circle in a given triangle.*

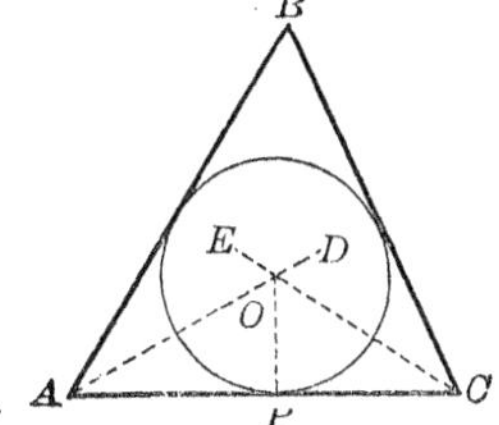

Given the △ ABC.

To construct an inscribed circle in the triangle.

Construction. Draw the line AD bisecting $\angle BAC$, and CE bisecting $\angle BCA$. Art. 277.

From O, the intersection of AD and CE, draw the line $OP \perp AC$. Art. 273.

From O as a center with a radius OP, describe a ⊙. Post. 3.

Then circle O is the circle required.

Proof. Let the pupil supply the proof.

Q. E. F.

Ex. 1. Find the center of a given circumference.

Ex. 2. Construct the figure on page 81, using the concurrence of the three perpendicular bisectors as a test of the accuracy of the work.

PROPOSITION XXXVIII. PROBLEM

288. *Through a given point on the circumference to draw a tangent to a circle.*

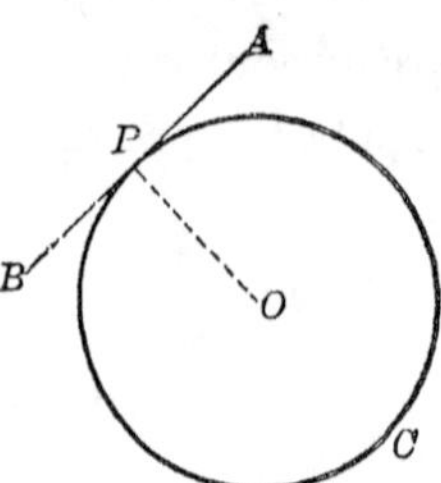

Given any point P on the circumference of the circle O.

To construct a tangent to the circle at the point P.

Construction. Draw the radius OP.

At the point P construct a line $AB \perp OP$. Art. 274.

Then AB is the tangent required.

Proof. Let the pupil supply the proof.

289. An **escribed circle** is a circle tangent to one side of a triangle and to the other two sides produced. Thus the circle O is an escribed circle of the $\triangle ABC$. A center of an escribed circle, as O, is called an **ex-center** of the triangle.

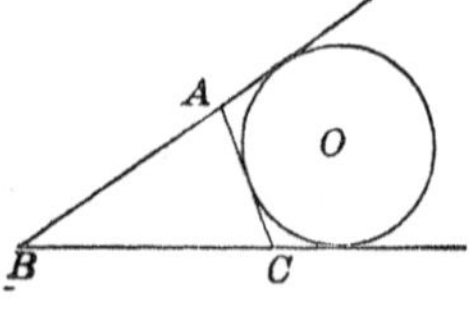

Ex. 1. Draw a triangle and all of its escribed circles.

Ex. 2. Construct the figure on page 83, using the concurrence of the three medians as a test of the accuracy of the work.

Proposition XXXIX. Problem

290. *From a given point without a circle to draw a tangent to the circle.*

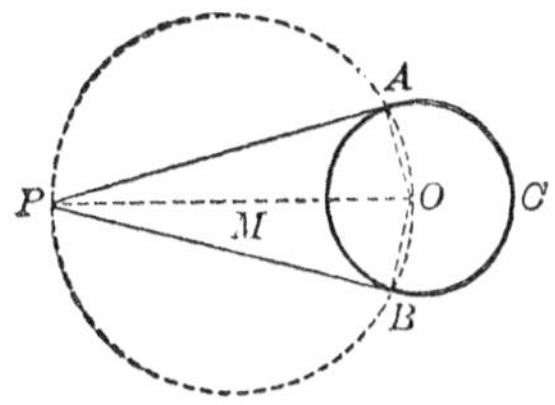

Given P any point without the circle O.

To construct a line through P tangent to the circle O.

Construction. Draw the line PO. Post. 1.

Bisect the line PO at M. Art. 275.

From M as a center with MP as a radius, describe a circumference intersecting the given circumference at A and B. Post. 3.

Draw PA and PB. Post. 1.

Then PA and PB are the tangents required.

Proof. $\angle PAO$ is inscribed in a semicircle. Constr.

$\therefore$ $\angle PAO$ is a right angle. (Why?)

$\therefore$ PA is tangent to the circle O. (Why?)

In like manner PB is tangent to the circle O.

Q. E. F.

Proposition XL. Problem

291. *Upon a given straight line to describe a segment which shall contain a given inscribed angle.*

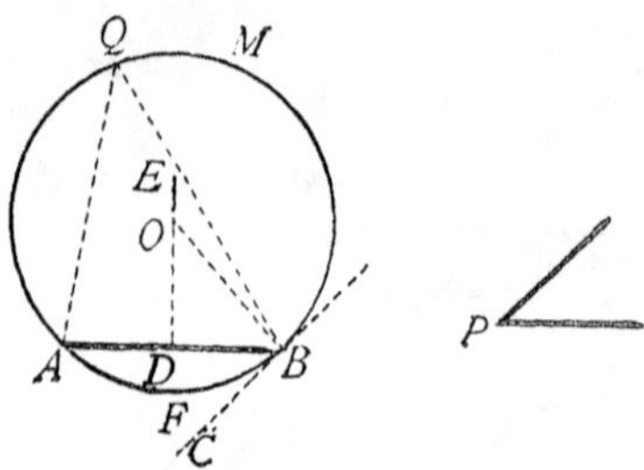

Given the straight line AB and the $\angle P$.

To construct on the line AB a segment of a circle such that any angle inscribed in the segment shall equal $\angle P$.

Construction. At the point B in the line AB construct $\angle ABC$ equal to $\angle P$. Art. 278.

Construct DE the $\perp$ bisector of line AB. Art. 274.

At B construct $BO \perp BC$, and intersecting DE at O. Arts. 274, 122.

From O as a center with OB as a radius, describe the circle AMB. Post. 3.

Then AMB is the required segment.

Proof. Let AQB be any $\angle$ inscribed in the segment AQB.

Then $\angle AQB$ is measured by $\frac{1}{2}$ arc AFB. Art. 258.

But BC is tangent to the circle AMB. (Why?)

$\therefore$ $\angle ABC$ is measured by $\frac{1}{2}$ arc AFB. (Why?)

$\therefore$ $\angle Q = \angle ABC$, or $\angle P$. (Why?)

$\therefore$ any $\angle$ inscribed in segment $AMB = \angle P$.

Q. E. F.

PROPOSITION XLI. THEOREM.

292. *To find the common unit of measure of two commensurable straight lines, and hence the ratio of the lines.*

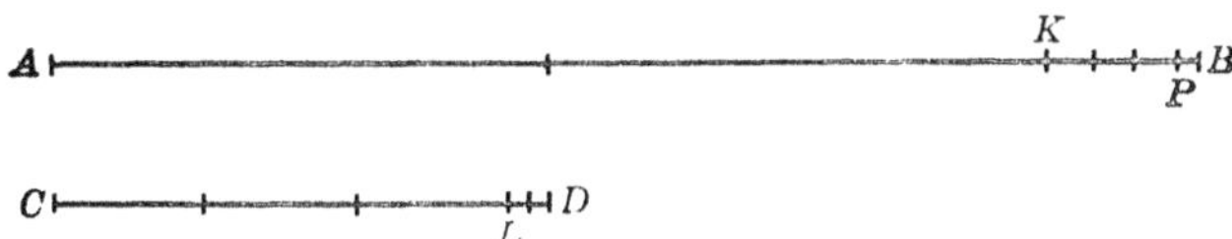

Given two lines AB and CD, of which CD is the shorter.

To construct a common unit of measure of AB and CD, and hence obtain the ratio of AB and CD.

Construction. Apply CD to AB as many times as possible; Post. 2.

Say twice, with a remainder KB.

Then apply KB to CD as many times as possible; Post. 2.

Say three times, with a remainder LD.

Apply LD to KB as many times as possible; Post. 2.

Say three times, with a remainder PB.

Apply PB to LD as many times as possible; Post. 2.

Say it is contained in LD exactly twice.

Then PB is the common unit of measure of AB and CD.

Proof. $$LD = 2PB.$$

$$KB = KP + PB = 3LD + PB = 7PB.$$ Axs. 6, 8.

$$CD = CL + LD = 3KB + LD = 23PB.$$

$$AB = AK + KB = 2CD + KB = 53PB.$$

Hence $$\frac{AB}{CD} = \frac{53\ PB}{23\ PB} = \frac{53}{23}.$$

Q. E. F.

EXERCISES IN CONSTRUCTION PROBLEMS

293. Analysis of problems. The method of analysis (see Art. 196) is of especial value in the solution of construction problems. In general, to investigate the solution of a problem by this method:

Draw a figure in which the required construction is assumed as made ;

Draw auxiliary lines, if necessary ;

Observe the relations between the parts of this figure, in order to discover a known relation on which the required construction depends ;

Having discovered the required relation, construct another figure by the direct use of this relation.

Ex. Through a given point within a circle draw a chord which shall be bisected by the given point.

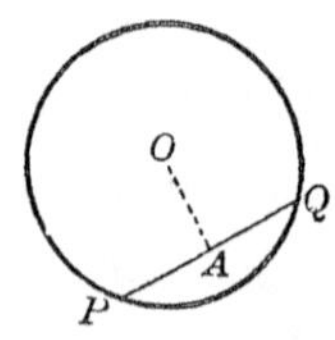

ANALYSIS. Let A be the given point within the given circle O, and let PQ be a chord bisected at the point A. A bisected chord suggests a line OA joining the point of bisection with the center O, and that (Art. 224) $OA \perp PQ$.

SYNTHESIS, or DIRECT SOLUTION. Taking another figure containing the data of the problem, connect the point A with the center of the circle by the line OA.

Through A draw a line $\perp$ OA (Art. 274), and meeting the circumference at the points P and Q. PQ is the chord required.

EXERCISES. GROUP 22

CONSTRUCTION OF STRAIGHT LINES.

Ex. 1. Draw a line parallel to a given line, and tangent to a given circle.

[SUG. Suppose the required line drawn; then the radius to the point of tangency, if produced, is $\perp$ given line, etc.]

Ex. 2. Draw a line perpendicular to a given line, and tangent to a given circle.

Ex. 3. From two points in the circumference of a circle, draw two equal and parallel chords.

Ex. 4. Through a given point draw a line which shall make a given angle with a given line.

Ex. 5. Through a given point draw a line which shall make equal angles with the sides of a given angle.

[SUG. The bisector of the given ∠ will be ⊥ the required line, etc.]

Ex. 6. Through a given point between two given parallel lines, draw a line of given length with its extremities in the two parallel lines.

Ex. 7. Through a given point A within a circle, draw a chord equal to a given line.

Ex. 8. From a given point in the circumference of a circle, draw a chord at a given distance from the center.

Ex. 9. Through a given point on the circumference of a circle, draw a chord which shall be bisected by another given chord.

[SUG. Draw the radius to the given point and on it as a diameter describe a circle, etc. When is the solution impossible?]

294. Construction of points and of loci. In constructing a point to meet certain given conditions, it is often helpful to *construct the locus of a point answering one of the given conditions and observe in what point or points it meets a given line, or meets another locus answering another given condition.*

EXERCISES. GROUP 23

CONSTRUCTION OF POINTS AND LOCI

Ex. 1. Find a point P in a given line AB equidistant from two given points C and D.

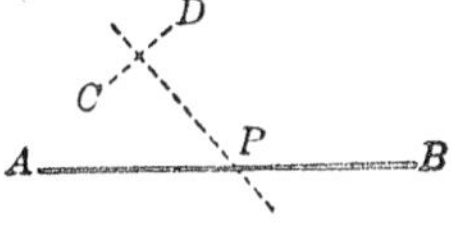

[SUG. Construct the locus of all points equidistant from C and D and observe where it intersects the given line AB.]

Ex. 2. Find a point P in a given circumference, which is equidistant from two given points, C and D.

Ex. 3. Find a point P in a given line, which is equidistant from two given intersecting lines.

Ex. 4. Find a point in a given line, which is at a given distance, d, from a given point.

Ex. 5. Find a point which is at a given distance, a, from a given point A, and at another distance, b, from another given point B.

Discuss the limitations of this problem.

Ex. 6. Find a point equidistant from two given points, and at a given distance from a given straight line.

[SUG. Draw the locus of all points equidistant from the two given points, and also the locus of all points at the given distance from the given straight line, etc.]

Ex. 7. Find a point equidistant from two given points, and at a given distance from another given point.

Ex. 8. Find a point equidistant from two given points, and also from two given intersecting lines.

On the other hand, the **determination of certain loci** *is equivalent to the construction of all points which satisfy one or more given conditions.*

Ex. 9. Find the locus of the center of a circle, which touches a given line at a given point.

[SUG. Construct a number of circles touching the given line at the given point and observe the relation of their centers.]

Ex. 10. Find the locus of the center of a circumference with a given radius, r, which passes through a given fixed point.

Ex. 11. Find the locus of the center of a circle, touching two given intersecting lines.

Ex. 12. Find the locus of the center of a circle, touching two given parallel lines.

Ex. 13. Find the locus of the center of a circle of given radius, r, which touches a given straight line.

Ex. 14. Find the locus of the center of a circle of given radius, r, which touches a given circle.

EXERCISES. GROUP 24

CONSTRUCTION OF RECTILINEAR FIGURES

Construct

Ex. 1. An equilateral triangle, given the perimeter.

Ex. 2. An equilateral triangle, given the altitude.

Ex. 3. An isosceles triangle, given the base and altitude.

Ex. 4. An isosceles triangle, given the base and an angle at the base.

Ex. 5. An isosceles triangle, given the vertex angle and the altitude.

Ex. 6. A right triangle, given a leg and the acute angle adjacent.

Ex. 7. A right triangle, given a leg and the acute angle opposite.

Ex. 8. A right triangle, given the hypotenuse and an acute angle.

Ex. 9. A right triangle, given the hypotenuse and a leg.

Ex. 10. A triangle, given the altitude and the sides including the vertical angle.

[SUG. Through the foot of the altitude draw a line ⊥ altitude and of indefinite length, etc.]

Ex. 11. A triangle, given two sides and the altitude upon one of them.

Ex. 12. A square, given the diagonal.

Ex. 13. A rhombus, given the two diagonals.

Ex. 14. A rhombus, given one angle and one diagonal.

Ex. 15. A parallelogram, given two adjacent sides and an altitude.

Ex. 16. A parallelogram, given a side, the altitude upon that side and an angle.

Ex. 17. A parallelogram, given the diagonals and an angle included by them.

Ex. 18. A quadrilateral, given the sides and one angle.

295. Use of auxiliary lines in constructing rectilinear figures. In constructing polygons, auxiliary lines are frequently of service. Thus it is often of especial value to *construct, first, either the inscribed or the circumscribed circle, and afterward the required triangle or quadrilateral.*

Ex. Construct an isosceles triangle, given the base, b, and the radius, r, of the inscribed circle.

CONSTRUCTION. Draw a circle O with radius equal to r. Draw a tangent at any point A. On this tangent mark off AB and AC each equal $\frac{1}{2}b$. From B and C draw tangents BF and CF to the circle. Then BCF is the required triangle.

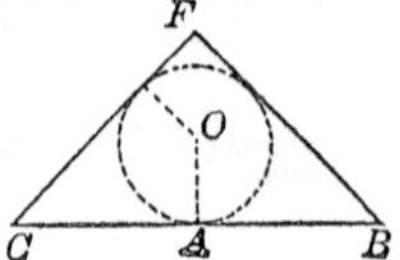

EXERCISES. GROUP 25

CONSTRUCTIONS. AUXILIARY LINES

Construct

Ex. 1. An isosceles triangle, given the base and the radius of the circumscribed circle.

Ex. 2. A right triangle, given the radius of the circumscribed circle and one leg.

Ex. 3. A right triangle, given the radius of the circumscribed circle and an acute angle.

Ex. 4. A right triangle, given the radius of the inscribed circle and an acute angle.

[SUG. Draw the inscribed circle and at its center construct an angle equal to the supplement of the given angle.]

Ex. 5. A triangle, given the base, the altitude and the vertex angle.

[SUG. On the given base construct a segment which shall contain the given vertex angle. See Art. 291.]

Ex. 6. A triangle, given the base, the median to the base, and the vertex angle.

Ex. 7. A triangle, given one side, an adjacent angle, and the radius of the inscribed circle. (See Ex. 4.)

Ex. 8. A triangle, given one side, an adjacent angle, and the radius of the circumscribed circle.

The use of *auxiliary straight lines* may be illustrated as follows:

Ex. 9. Construct a triangle, given the perimeter and two angles.

ANALYSIS. Suppose the required triangle ABC already constructed and let $\measuredangle ABC$ and ACB be the given angles. Produce BC to D and E, making $DB = AB$ and $CE = AC$. Then DE= given perimeter. Also $\angle D = \angle DAB$ $\therefore$ $\angle ABC = 2\angle D$. Similarly $\angle ACB = 2\angle E$. Hence

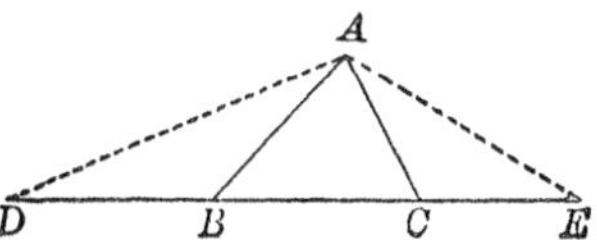

CONSTRUCTION. Take DE the given perimeter; at D construct an angle $= \frac{1}{2}$ of one given angle; at E construct an angle $= \frac{1}{2}$ of the other given angle. Produce the sides of these angles to meet at A. Construct $\angle DAB = \angle D$ and $\angle CAE = \angle E$. Then $\triangle ABC$ is the required triangle, etc.

Construct

Ex. 10. An isosceles triangle, given the perimeter and the altitude.

[SUG. Bisect the perimeter and construct the altitude $\perp$ to it at its midpoint.]

Ex. 11. An isosceles triangle, given the perimeter and the vertex angle.

[SUG. If the vertex $\angle$ is known, the base $\measuredangle$ may be obtained.]

Ex. 12. A right triangle, given an acute angle and the sum of the legs.

[SUG. Given AB the sum of the legs, construct $\angle A = 45°$, etc.]

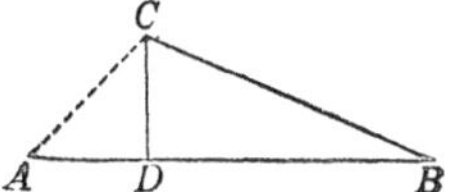

Ex. 13. A right triangle, given an acute angle and the difference of the legs.

Ex. 14. A right triangle, given the hypotenuse and the sum of the legs.

Ex. 15. A right triangle, given an acute angle and the sum of the hypotenuse and one leg.

Ex. 16. A triangle, given an angle, a side and the sum of th other two sides.

Ex. 17. A triangle, given an angle A, the sum of the sides A and BC and the altitude upon AB.

296. Reduction of problems. In many cases a problem may be solved *by reducing the problem to a problem alread solved.* (This is a special kind of analysis.)

Ex. Construct a parallelogram, given the diagonals and one side.

ANALYSIS: Suppose the ▱ $ABCF$ to be the required ▱ alread constructed. Let AF be the given side. If the diagonals ar given, half each diagonal is given (Art. 161). Hence in the △ AOF the three sides are given. Hence the required problem reduces to the problem of constructing a triangle whose three sides are given (Art. 283). Hence

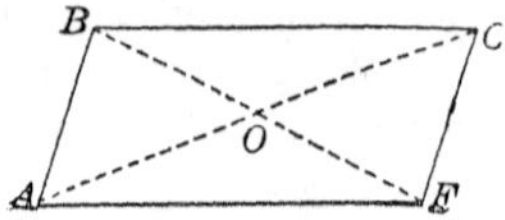

CONSTRUCTION. Let the pupil supply the direct construction.

EXERCISES. GROUP 26

REDUCTION OF CONSTRUCTION PROBLEMS

Construct

Ex. 1. A right triangle, given the altitude upon the hypotenuse and the median upon the same.

Ex. 2. A rectangle, given the perimeter and a diagonal (see Ex. 14, p. 171).

Ex. 3. A rectangle, given the perimeter and an angle made by the diagonals.

Ex. 4. A triangle, given the three angles and the radius of the circumscribed circle.

[SUG. The sides of the △ are the chords of the segments of the ⊙ containing the given ∡.]

Ex. 5. A triangle, given two sides and the median to the third side.

Ex. 6. A triangle, given the three medians.

[SUG. Reduce this to the preceding Ex. See Art. 187.]

Ex. 7. An isosceles trapezoid, given the bases and an angle.

Ex. 8. An isosceles trapezoid, given the bases and a diagonal.

Ex. 9. A trapezoid, given the four sides.

Ex. 10. A trapezoid, given the bases and the two diagonals.

[SUG. Reduce to Art. 283 by producing the lower base a distance equal to the upper base, etc.]

297. Construction of circles. The construction of a required circle is frequently a good illustration of the preceding method of reducing one construction problem to another. For the construction of a circle frequently *reduces to the problem of finding a point (the center of the circle) which answers given conditions.* (See Art. 294.)

Ex. Construct a circle which shall touch two given intersecting lines and have its center in another given line.

This problem is equivalent to the problem of finding a point which shall be in a given line and be equidistant from two other given lines. (See Ex. 3, p. 168.) In some cases, however, the construction of a required circle must be made by an independent method.

EXERCISES. GROUP 27

CONSTRUCTION OF CIRCLES

Construct a circle with given radius, r,

Ex. 1. Which passes through a given point and touches a given line.

Ex. 2. Which has its center in a given line and touches another given line.

Ex. 3. Which passes through two given points.

Construct a circle

Ex. 4. Which touches two given parallel lines and passes through a given point.

Ex. 5. Which passes through two given points and has its center on a given line.

Ex. 6. Which touches three given lines, two of which are parallel.

Ex. 7. Which passes through a given point A and touches a given line BC at a given point B.

[Sug. Draw AB and at B construct a $\perp$ to BC.]

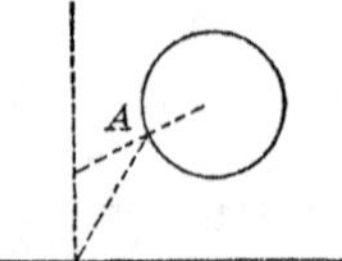

Ex. 8. Which touches a given line and also touches a given circle at a given point A.

Ex. 9. Which touches a given line AB at a given point A and touches a given circle.

EXERCISES. GROUP 28

PROBLEMS SOLVED BY VARIOUS METHODS

Ex. 1. Through a given point draw a line which shall cut two given intersecting lines so as to form an isosceles triangle.

Ex. 2. Construct an isosceles triangle, given the altitude and one leg.

Ex. 3. In a given circumference find a point equidistant from two given intersecting lines.

Ex. 4. Draw a circle which shall touch two given intersecting lines, one of them at a given point.

Ex. 5. Draw a line which shall be terminated by the sides of a given angle, shall equal a given line, and be parallel to another given line.

Ex. 6. Construct a triangle, given one side, an adjacent angle, and the difference of the other two sides.

Ex. 7. Find a point in a given circumference at a given distance from a given point.

Ex. 8. Construct a parallelogram, given a side, an angle, and a diagonal.

Ex. 9. Through a given point within an angle, draw a straight line terminated by the sides of the angle and bisected by the given point.

[SUG. Draw a line from the vertex of the angle to the given point and produce it its own length through the point.]

Ex. 10. Construct a triangle, given the vertex angle and the segments of the base made by the altitude.

[SUG. Use Art. 291.]

Ex. 11. Construct an isosceles triangle, given the angle at the vertex and the base.

Ex. 12. Draw a circle with given radius which shall touch a given circle at a given point.

Ex. 13. Construct a right triangle, given the hypotenuse and the altitude upon the hypotenuse.

Ex. 14. Construct a triangle, given the base and the altitudes upon the other two sides.

[SUG. Construct a semicircle on the given base as a diameter.]

Ex. 15. Find a point in one side of a triangle equidistant from the other two sides.

Ex. 16. Construct a triangle, given the altitude and the angles at the extremities of the base.

Ex. 17. Construct a rhombus, given an angle and a diagonal.

Ex. 18. Draw a circle which shall pass through two given points and have its center equidistant from two given parallel lines.

Ex. 19. Construct a triangle, given one side, an adjacent angle and the radius of the circumscribed circle.

Ex. 20. In a given circle draw a chord equal to a given line and parallel to another given line.

[SUG. Find the distance of the given chord from the center, by constructing a right triangle of which the hypotenuse and one leg are given.]

Ex. 21. Construct a triangle, given an angle, the bisector of that angle, and the altitude from another vertex.

Ex. 22. Find the locus of the points of contact of tangents drawn from a given point to a series of circles having a given center.

[SUG. Use Arts. 229 and 261.]

Ex. 23. Given a line AB and two points, C and D, on the same side of AB. Find a point P in AB such that $\angle APC = \angle BPD$.

·C ·D A P B

[SUG. Draw a ⊥ from C to AB and produce it its own length, etc.]

Ex. 24. Given a line AB and two points C and D on the same side of AB; find a point P in AB such that $CP + PD$ shall be a minimum.

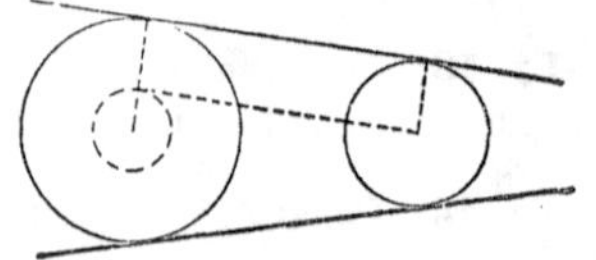

Ex. 25. Draw a common external tangent to two given circles.

Ex. 26. Draw a common internal tangent to two given circles.

Book III

PROPORTION. SIMILAR POLYGONS

THEORY OF PROPORTION

298. **Ratio** has been defined, and its use briefly indicated in Arts. 245, 246.

299. A **proportion** is an expression of the equality of two or more equal ratios. As,

$$\frac{a}{b}=\frac{c}{d}, \text{ or } a : b = c : d.$$

This reads, "the ratio of a to b equals the ratio of c to d," or, "a is to b as c is to d."

300. The **terms of a proportion** are the four quantities used in the proportion. In a proportion

the **antecedents** are the *first* and *third* terms;
the **consequents** are the *second* and *fourth* terms;
the **extremes** are the *first* and *last* terms;
the **means** are the *second* and *third* terms.

A **fourth proportional** is the last term of a proportion (provided the means are not equal).

Thus, in $a : b=c : d$, d is a fourth proportional.

301. A **continued proportion** is a proportion in which each consequent and the next antecedent are the same.

Thus, $a : b=b : c=c : d=d : e$ is a continued proportion.

A **mean proportional** is the middle term in a continued proportion containing but two ratios.

A **third proportional** is the last term in a continued proportion containing but two ratios.

Thus, in $a : b = b : c$, b is a mean proportional, and c is a third proportional.

Proposition I. Theorem

302. *In any proportion, the product of the extremes is equal to the product of the means.*

Given the proportion $a : b = c : d$.

To prove $ad = bc$.

Proof. $\frac{a}{b} = \frac{c}{d}$. Hyp.

Multiply each member by bd.

$ad = bc$. Ax. 4.

Q. E. D.

Proposition II. Theorem

303. *The mean proportional between two quantities is equal to the square root of their product.*

Given the proportion $a : b = b : c$.

To prove $b = \sqrt{ac}$.

Proof. $a : b = b : c$. Hyp.

$\therefore b^2 = ac$, Art. 302.

(*in any proportion, the product of the extremes equals the product of the means*).

$\therefore b = \sqrt{ac}$.

Q. E. D.

Ex. 1. Find the fourth proportional to 2, 3 and 6; also to 3, $\frac{1}{2}$, $\frac{3}{2}$.

Ex. 2. Find the mean proportional between 3 and 6.

Ex. 3. Find the third proportional to 3 and 5.

Proposition III. Theorem

304. *If the product of two quantities is equal to the product of two other quantities, one pair may be made the extremes and the other pair the means of a proportion.*

Given $ad = bc$.

To prove $a : b = c : d$.

Proof. $ad = bc$. Hyp.

Divide each member by bd.

Then $\frac{a}{b} = \frac{c}{d}$. Ax. 5.

Or $a : b = c : d$.

Q. E. D.

305. Cor. 1. *If the antecedents of a proportion are equal, the consequents are equal.*

Thus, if $a : x = a : y$, then $x = y$.

Let the pupil supply the proof.

306. Cor. 2. *If three terms of one proportion are equal to the corresponding three terms of another proportion, the fourth terms of the two proportions are equal.*

Thus, if $a : b = c : x$, and $a : b = c : y$, then $x = y$.

Let the pupil supply the proof.

Ex. 1. Write $ab = pq$ as a proportion in as many different ways as possible.

Ex 2. Write $x(x+1) = 6$ as a proportion; also $x^2 = 15$.

Proposition IV. Theorem

307. *If four quantities are in proportion, they are in proportion by* **alternation**; *that is, the first term is to the third as the second is to the fourth.*

Given the proportion $a : b = c : d$.

To prove $a : c = b : d$.

Proof. $a : b = c : d$. Hyp.

$\therefore ad = bc$, Art. 302.

(*in any proportion, the product of the extremes equals the product of the means*).

Writing a and d as the extremes, and c and b as the means of a proportion,

$a : c = b : d$, Art. 304.

(*if the product of two quantities is equal to the product of two other quantities, one pair may be made the extremes and the other pair the means of a proportion*).

Q. E. D.

Proposition V. Theorem

308. *If four quantities are in proportion, they are in proportion by* **inversion**; *that is, the second term is to the first as the fourth is to the third.*

Given the proportion $a : b = c : d$.

To prove $b : a = d : c$.

Proof. $a : b = c : d$. Hyp.

$\therefore ad = bc$. (Why?)

Writing b and c as the extremes, and a and d as the means,

$b : a = d : c$. (Why?)

Q. E. D.

Ex. Transform $x : a = b : c$ so that x shall be the last term.

Proposition VI. Theorem

309. *If four quantities are in proportion, they are in proportion by* **composition**; *that is, the sum of the first two terms is to the second term as the sum of the last two terms is to the last term.*

Given the proportion $a : b = c : d$.

To prove $a + b : b = c + d : d$.

Proof. $\frac{a}{b} = \frac{c}{d}$. Hyp.

Add 1 to each member of the equality. Ax. 2.

$$\frac{a}{b} + 1 = \frac{c}{d} + 1,$$

Or $\frac{a+b}{b} = \frac{c+d}{d}$.

That is $a + b : b = c + d : d$.

Let the pupil show also that $a + b : a = c + d : c$.

Q. E. D.

Proposition VII. Theorem

310. *If four quantities are in proportion, they are in proportion by* **division**; *that is, the difference of the first two is to the second as the difference of the last two is to the last.*

Given the proportion $a : b = c : d$.

To prove $a - b : b = c - d : d$.

Proof. $\frac{a}{b} = \frac{c}{d}$. Hyp.

Subtract 1 from each member of the equality.

$$\therefore \frac{a}{b} - 1 = \frac{c}{d} - 1. \qquad \text{Ax. 3.}$$

Or $$\frac{a-b}{b} = \frac{c-d}{d}.$$

That is $a - b : b = c - d : d.$ Q. E. D.

Let the pupil show also that $a - b : a = c - d : c$.

Proposition VIII. Theorem

311. *If four quantities are in proportion, they are in proportion by* **composition and division**; *that is, the sum of the first two is to their difference as the sum of the last two is to their difference.*

Given the proportion $a : b = c : d$.

To prove $a + b : a - b = c + d : c - d$.

Proof. $a : b = c : d$. Hyp.

By composition $$\frac{a+b}{b} = \frac{c+d}{d}. \qquad \text{Art. 309.}$$

Also by division $$\frac{a-b}{b} = \frac{c-d}{d}. \qquad \text{Art. 310.}$$

Dividing the corresponding members of the two equalities,

$$\frac{a+b}{a-b} = \frac{c+d}{c-d}. \qquad \text{Ax. 5.}$$

That is $a + b : a - b = c + d : c - d$. Q. E. D.

Ex. 1. What does the proportion $12 : 3 = 8 : 2$ become by composition? also by division?

Ex. 2. What does $2x - 5 : 5 = 3x - 7 : 7$ become by composition?

Proposition IX. Theorem

312. *In a series of equal ratios, the sum of all the antecedents is to the sum of all the consequents as any one antecedent is to its consequent.*

Given $a : b = c : d = e : f = g : h$.

To prove $a + c + e + g : b + d + f + h = a : b$.

Proof. Denote each of the equal ratios by r.

Then $\frac{a}{b} = r, \therefore a = br$.

Similarly $c = dr, e = fr, g = hr$. Ax. 4.

Hence $a + c + e + g = (b + d + f + h)\, r$. Ax. 2.

Dividing by $b + d + f + h$, $\frac{a + c + e + g}{b + d + f + h} = r = \frac{a}{b}$. Ax. 5.

That is $a + c + e + g : b + d + f + h = a : b$.

Q. E. D.

Proposition X. Theorem

313. *The products of the corresponding terms of two or more proportions are in proportion.*

Given $a : b = c : d$, $e : f = g : h$, and $j : k = l : m$.

To prove $aej : bfk = cgl : dhm$.

Proof. $\frac{a}{b} = \frac{c}{d}, \frac{e}{f} = \frac{g}{h}, \frac{j}{k} = \frac{l}{m}$. Hyp.

Multiplying together the corresponding terms of these equalities,

$$\frac{aej}{bfk} = \frac{cgl}{dhm}.$$ Ax. 4.

That is $aej : bfk = cgl : dhm$.

Q. E. D.

Proposition XI. Theorem

314. *Like powers or like roots of the terms of a proportion are in proportion.*

Given the proportion $a : b = c : d$.

To prove $a^n : b^n = c^n : d^n$, and $a^{\frac{1}{n}} : b^{\frac{1}{n}} = c^{\frac{1}{n}} : d^{\frac{1}{n}}$.

Proof. $$\frac{a}{b} = \frac{c}{d}.$$ Hyp.

Raising both members to the nth power,

$$\frac{a^n}{b^n} = \frac{c^n}{d^n}.$$

That is $a^n : b^n = c^n : d^n$.

In like manner $a^{\frac{1}{n}} : b^{\frac{1}{n}} = c^{\frac{1}{n}} : d^{\frac{1}{n}}$.

Q. E. D.

Proposition XII. Theorem

315. *Equimultiples of two quantities have the same ratio as the quantities themselves.*

Given the two quantities a and b.

To prove $ma : mb = a : b$.

Proof. $$\frac{a}{b} = \frac{a}{b}.$$ Ident.

Multiply each term of the first fraction by m.

$$\therefore \frac{ma}{mb} = \frac{a}{b}.$$

That is $ma : mb = a : b$.

Q. E. D.

Ex. 1. Transform $p : q = x : y$ in all possible ways by the use of the properties of a proportion.

Ex. 2. Transform $7 : 3 = 28 : 12$ by composition and division.

Ex. 3. Also $2x + 5 : 2x - 5 = x^2 + 1 : x^2 - 1$.

316. Note. In the theory of proportion as just presented, the quantities used are assumed to be commensurable, but the same theorems may also be proved for proportions whose terms are incommensurable by use of the method of limits. For each incommensurable ratio may be made the limit of a corresponding commensurable ratio; then, by showing that the variable commensurable ratios are equal, it may be proved that the limiting incommensurable ratios are equal.

It is also to be noted, that, in the above theorems, the terms of a ratio must be of the same kind of quantity; that is, both be lines, or both be surfaces, etc. Hence, in order that a proportion be treated by alternation, for instance, all four of the terms must be of the same kind.

Ex. 1. Find the fourth proportional to a, $2a$, $3x$.

Ex. 2. Find the third proportional to $a+b$ and $a-b$.

Ex. 3. Find the value of x in the proportion, $4 : 5 = x : 15$.

Ex. 4. Find a short method of determining whether a given proportion is true or not. Use this method to determine whether the following proportions are true: (1) $4 : 6 = 3 : 9$. (2) $5a : 2a = 15 : 6$.

Ex. 5. Transform the following proportion by composition and division, and afterward find the value of x; $x+1 : x-1 = 7 : 5$.

Ex. 6. Construct exactly the figure of page 77 with the fewest possible adjustments of the compasses.

Ex. 7. Draw an obtuse triangle and construct its three altitudes, using the concurrence of the altitudes as a test of the accuracy of the work.

Ex. 8. Arrange nine points in a plane in such a way that the greatest number of straight lines may pass through them, each line to pass through three points and only three.

PROPOSITION XIII. THEOREM

317. *A line parallel to one side of a triangle and meeting the other two sides, divides these sides proportionally.*

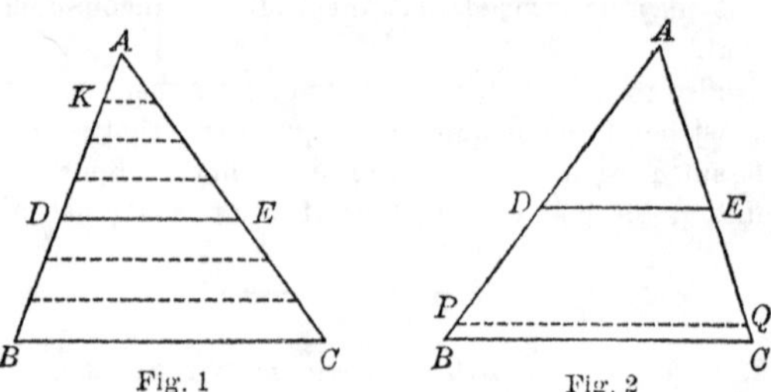

Fig. 1 Fig. 2

Given the triangle ABC and the line $DE \parallel$ base BC and intersecting the sides AB and AC in the points D and E, respectively.

To prove $DB : AD = EC : AE$.

CASE I. *When DB and AD* (Fig. 1) *are commensurable.*

Proof. Take any common unit of measure of DB and AD, as AK, and let it be contained in DB a certain number of times, as n times, and in AD, m times.

Then $$\frac{DB}{AD} = \frac{n}{m}.$$ Art. 245.

Through the points of division of DB and AD draw lines $\parallel BC$.

These lines will divide EC into n, and AE into m parts, all equal. Art. 176.

$$\therefore \frac{EC}{AE} = \frac{n}{m}.$$ (Why?)

$$\therefore \frac{DB}{AD} = \frac{EC}{AE}.$$ (Why?)

CASE II. *When DB and AD* (Fig. 2) *are incommensurable.*

Let the line AD be divided into any number of equal parts and let one of those parts be applied to DB. It will be contained in DB a certain number of times, with a line PB, less than the unit of measure, as a remainder.

Draw $PQ \parallel BC$.

Then DP and AD are commensurable. Constr.

$$\therefore \frac{DP}{AD}=\frac{EQ}{AE}. \qquad \text{Case I.}$$

If now the unit of measure be indefinitely diminished, the line PB, which is less than the unit of measure, will be indefinitely diminished.

Hence $DP \doteq DB$, and $EQ \doteq EC$ as a limit. Art. 251.

Hence $\frac{DP}{AD}$ becomes a variable with $\frac{DB}{AD}$ as its limit. Art. 253, 3.

Also $\frac{EQ}{AE}$ becomes a variable with $\frac{EC}{AE}$ as its limit. Art. 253, 3.

But the variable $\frac{DP}{AD}=$ the variable $\frac{EQ}{AE}$ always. Case I.

$\therefore$ the limit $\frac{DB}{AD}=$ the limit $\frac{EC}{AE}$. Art. 254.

Q. E. D.

318. COR. 1. By composition, Art. 309.

$$DB + AD : AD = EC + AE : AE.$$

Or $AB : AD = AC : AE$.

In like manner $AB : DB = AC : EC$,

or, in general language, *if a line parallel to the base cut the sides of a triangle, a side is to a segment of that side as the other side is to the corresponding segment of the second side.*

319. COR. 2. Using Fig. 2,

$$\frac{PB}{QC}=\frac{AP}{AQ}; \text{ also } \frac{DP}{EQ}=\frac{AP}{AQ} \therefore \frac{PB}{QC}=\frac{DP}{EQ}.$$

Hence, *if two lines are cut by a number of parallels, the corresponding segments are proportional.*

PROPOSITION XIV. THEOREM (CONVERSE OF PROP. XIII)

320. *If a straight line divides two sides of a triangle proportionally, it is parallel to the third side.*

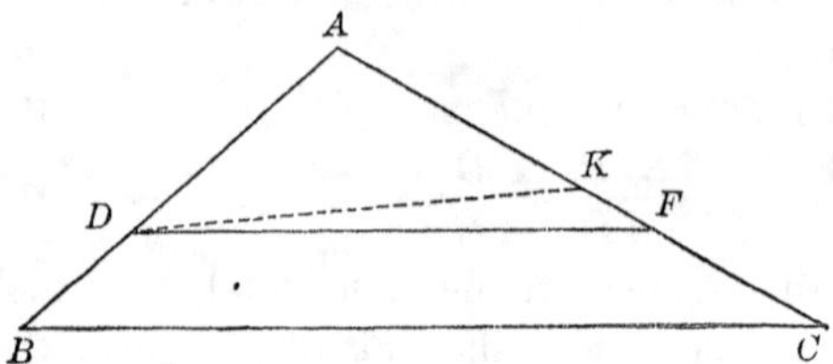

Given the $\triangle ABC$ and the line DF intersecting AB and AC so that $AB : AD = AC : AF$.

To prove $DF \parallel BC$.

Proof. Through D draw the line $DK \parallel BC$ and meeting the side AC in K.

Then $AB : AD = AC : AK$, Art. 318.

(*if a line* $\parallel$ *base cut the sides of a* $\triangle$, *a side is to a segment of that side as the other side is to the corresponding segment of the second side*).

But $AB : AD = AC : AF$. Hyp.

$\therefore AF = AK$, Art. 306.

(*if three terms of one proportion are equal to the corresponding three terms of another proportion, the fourth terms of the two proportions are equal*).

Hence the point K falls on F, and the line DK coincides with the line DF. Art. 66.

But the line $DK \parallel BC$. Constr.

$\therefore DF \parallel BC$,

(*for DF coincides with DK, which is* $\parallel$ *BC*).

Q. E. D.

Ex. 1. In the figure of Prop. XIII, if $AD=6$, $DB=4$, and $AE=9$, find EC.

Ex. 2. Also, if $AD=12$, $DB=8$, and $AC=15$, find AE and EC.

EXERCISES. GROUP 29

REVIEW EXERCISES

Make a list of the properties of

Ex. 1. One straight line (in connection with such points as may be related to the line).

Ex. 2. Two straight lines that meet, or intersect (in connection with such angles as may be formed by the given lines).

Ex. 3. Two or more parallel lines (in connection with transversals and angles formed).

Ex. 4. Right angles.

Ex. 5. Complementary angles.

Ex. 6. Supplementary angles.

Ex. 7. A single triangle (in connection with lines within the triangle, as altitudes, medians, etc.).

Ex. 8. A right triangle.

Ex. 9. An isosceles triangle.

Ex. 10. Two triangles.

Ex. 11. Two right triangles.

Ex. 12. A quadrilateral.

Ex. 13. A trapezoid.

Ex. 14. An isosceles trapezoid.

Ex. 15. A parallelogram.

Ex. 16. A rhombus.

Ex. 17. A polygon.

Ex. 18. A single circle, or circumference (in connection with related points).

Ex. 19. Arcs of a circle (in connection with related lines and angles).

Ex. 20. Chords in a circle (in connection with related lines or arcs).

Ex. 21. Central angles in a circle.

Ex. 22. Tangents to a circle.

Ex. 23. Secants to a circle.

Ex. 24. Two circles.

SIMILAR POLYGONS

321. Def. **Similar polygons** are polygons having their homologous angles equal and their homologous sides proportional.

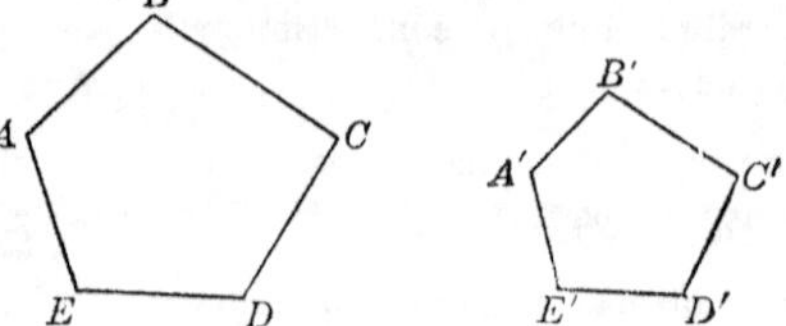

Thus, if the figures $ABCDE$ and $A'B'C'D'E'$ are similar, the angles A, B, C, etc., must equal the angles A', B', C', etc., respectively; also $AB : A'B' = BC : B'C' = CD : C'D'$, etc.

Hence it is constantly to be borne in mind that similarity in shape or form of rectilinear figures involves two distinct properties:

1. *The homologous angles are equal.*
2. *The homologous sides are proportional.*

It should also be clearly realized that one of these properties may be true of two figures, and not the other.

Thus, in the rectangle A and the rhomboid B, the corresponding sides are proportional but the corresponding angles are not equal.

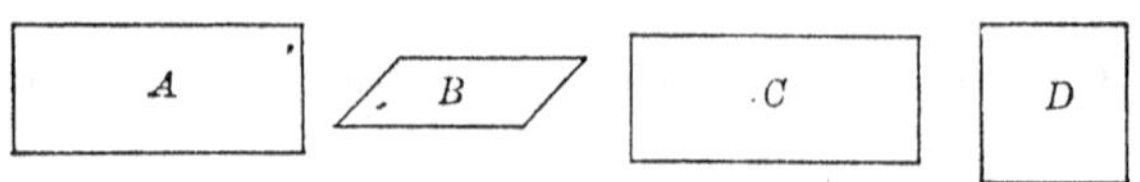

Also, in the rectangle C and the square D, the corresponding angles are equal but the corresponding sides are not proportional.

However, it will be found that, in the case of *triangles*, if one of the two properties is true, the other must be true also.

322. Def. The **ratio of similitude** in two similar figures is the ratio of any two homologous sides in those figures.

PROPOSITION XV. THEOREM

323. *If two triangles are mutually equiangular, they are similar.*

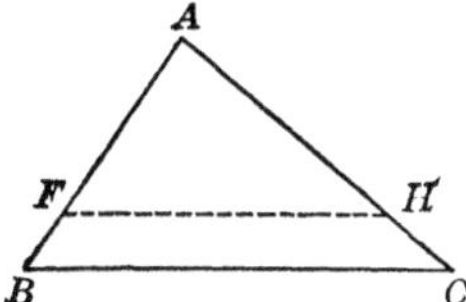

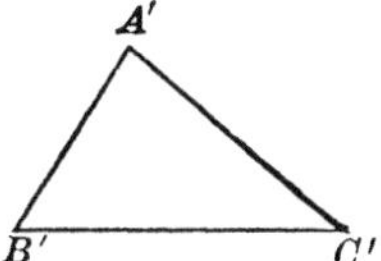

Given the $\triangle$ ABC and $A'B'C'$ with $\angle A = \angle A'$, $\angle B = \angle B'$, and $\angle C = \angle C'$.

To prove the $\triangle$ ABC and $A'B'C'$ similar.

Proof. The given $\triangle$ are mutually equiangular. Hyp.

Hence it only remains to prove that their homologous sides are proportional. Art. 321.

Place the $\triangle A'B'C'$ upon the $\triangle ABC$, so that $\angle A'$ shall coincide with its equal, the $\angle A$, and $B'C'$ take the position FH.

Then $\angle AFH = \angle B$. Hyp.

$\therefore FH \parallel BC$. (Why?)

$\therefore AB : AF = AC : AH$. Art. 317.

Or $AB : A'B' = AC : A'C'$. Ax. 8.

In like manner, by placing the $\triangle A'B'C'$ upon the $\triangle ABC$ so that the $\angle B'$ shall coincide with its equal, the $\angle B$, it may be proved that $AB : A'B' = BC : B'C'$.

Hence the $\triangle$ ABC and $A'B'C'$ are similar. Art. 321.

324. COR. 1. *If two triangles have two angles of one equal to two angles of the other, the triangles are similar;* also,

If two right triangles have an acute angle of one equal to an acute angle of the other, the triangles are similar.

325. COR. 2. *If two triangles are each similar to the same triangle, they are similar to each other.*

PROPOSITION XVI. THEOREM

326. *If two triangles have their homologous sides proportional, they are similar.*

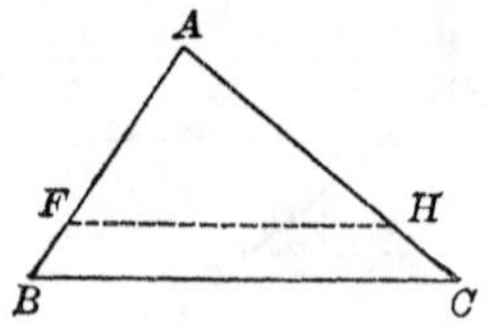

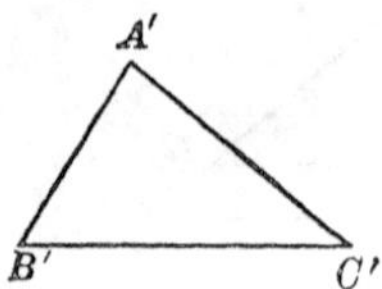

Given the $\triangle$ ABC and $A'B'C'$, in which $AB : A'B' = AC : A'C' = BC : B'C'$.

To prove the $\triangle$ ABC and $A'B'C'$ similar.

Proof. The given $\triangle$ have their homologous sides proportional. Hyp.

Hence it only remains to prove that the $\triangle$ are mutually equiangular. Art. 321.

1. On AB take AF equal to $A'B'$, and on AC take AH equal to $A'C'$. Draw FH.

Then $AB : AF = AC : AH$. Hyp.

$\therefore FH \parallel BC$, Art. 320.

(*if a straight line divides two sides of a* $\triangle$ *proportionally, it is* $\parallel$ *the third side*).

$\therefore \angle AFH = \angle B$, and $\angle AHF = \angle C$. (Why?)

$\therefore$ $\triangle$ AFH and ABC are similar, Art. 323.

(*if two* $\triangle$ *are mutually equiangular, they are similar*).

2. $\therefore AB : AF = BC : FH$. Art. 321.

Or $AB : A'B' = BC : FH$. Ax. 8.

But $AB : A'B' = BC : B'C'$. Hyp.

$\therefore FH = B'C'$. Art. 306.

Hence the $\triangle$ AFH and $A'B'C'$ are equal. Art. 101.

$\therefore \triangle A'B'C'$ is similar to $\triangle ABC$, Ax. 8.
(*for its equal $\triangle AFH$ is similar to $\triangle ABC$*).

Q. E. D.

Proposition XVII. Theorem

327. *If two triangles have an angle of one equal to an angle of the other, and the including sides proportional, the triangles are similar.*

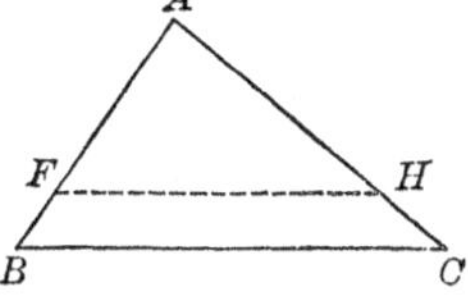

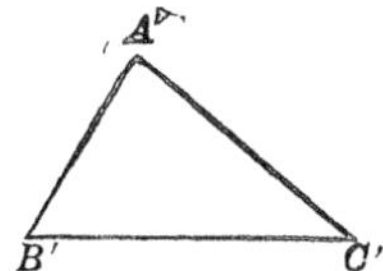

Given the $\triangle$ ABC and $A'B'C'$, in which $\angle A = \angle A'$ and $AB : A'B' = AC : A'C'$.

To prove the $\triangle$ ABC and $A'B'C'$ similar.

Proof. Place the $\triangle A'B'C'$ upon the $\triangle ABC$ so that $\angle A'$ shall coincide with its equal, the $\angle A$, and $B'C'$ take the position FH.

Then $AB : AF = AC : AH$. Hyp.

Hence $FH \parallel BC$. (Why?)

$\therefore \angle AFH = \angle B$, and $\angle AHF = \angle C$. (Why?)

$\therefore$ $\triangle$ ABC and AFH are similar. Art. 323.

Or $\triangle$ ABC and $A'B'C'$ are similar. Ax. 8.

Q. E. D.

Ex. In the $\triangle ABC$, $AB = 6$, $AC = 8$, $BC = 10$; in the similar $\triangle A'B'C'$, $A'B' = 9$. Find $A'C'$ and $B'C'$.

PROPOSITION XVIII. THEOREM

328. *If two triangles have their sides parallel, or perpendicular, each to each, the triangles are similar.*

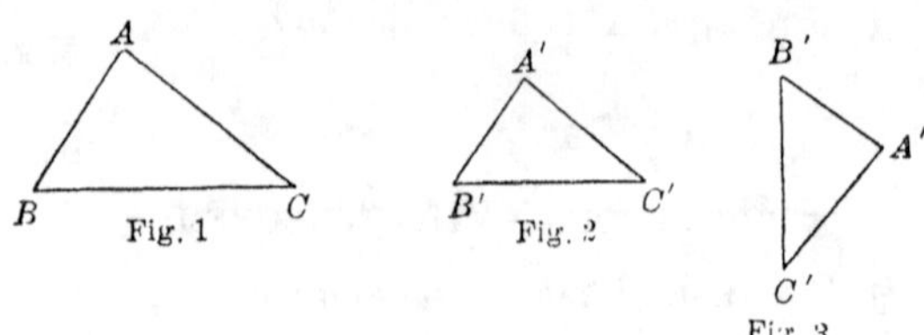

Fig. 1 Fig. 2 Fig. 3

Given the $\triangle A'B'C'$ (Fig. 2) with its sides $\parallel$, and the $\triangle A'B'C'$ (Fig. 3) with its sides $\perp$, to corresponding sides of the $\triangle ABC$.

To prove $\triangle$s ABC and $A'B'C'$ similar.

Proof. The $\angle$s A and A' are either equal or supplementary, (*for the sides forming them are $\parallel$ or $\perp$*). Arts. 130, 132.

Similarly, the $\angle$s B and B', and C and C' are either equal or supplementary.

Hence one of the three following statements must be true concerning the angles of the $\triangle$s: either

1. The $\triangle$s contain three pairs of supplementary $\angle$s and $\angle A + \angle A' = 2$ rt. $\angle$s, $\angle B + \angle B' = 2$ rt. $\angle$s, $\angle C + \angle C' = 2$ rt. $\angle$s; or

2. The $\triangle$s contain two pairs of supplementary $\angle$s, as $\angle A = \angle A'$, $\angle B + \angle B' = 2$ rt. $\angle$s, $\angle C + \angle C' = 2$ rt. $\angle$s; or

3. The $\triangle$s contain three pairs of equal $\angle$s and $\angle A = \angle A'$, $\angle B = \angle B'$, $\angle C = \angle C'$, Art. 139.
(*if two $\angle$s of a $\triangle$ = two $\angle$s of another $\triangle$, the third $\angle$s are equal*).

The first two of these statements are impossible, for the sum of the $\angle$s of two $\triangle$s cannot exceed four rt. $\angle$s. Art. 134.

Hence the third statement is true, and the $\triangle$s ABC and $A'B'C'$ are mutually equiangular, and therefore similar. Art. 323

Q. E. D.

PROPOSITION XIX. THEOREM

329. *If two polygons are similar, they may be separated into the same number of triangles, similar, each to each, and similarly placed.*

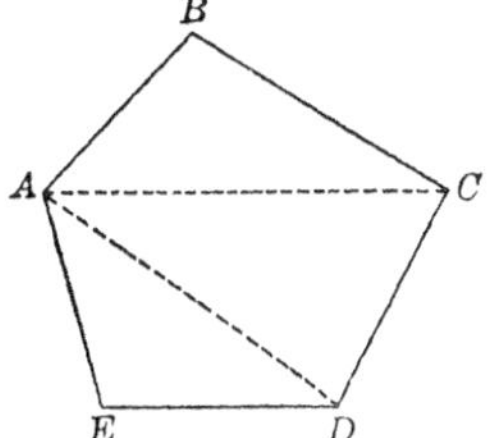

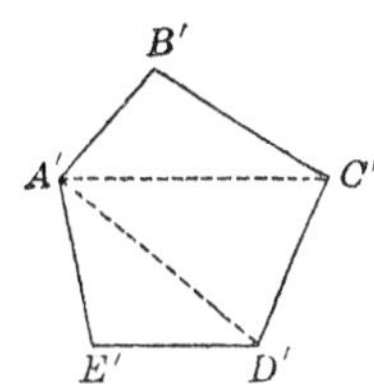

Given the similar polygons $ABCDE$ and $A'B'C'D'E'$, divided into triangles by the diagonals AC, AD, and $A'C'$, $A'D'$ drawn from the corresponding vertices A and A'.

To prove that △ ABC, ACD, ADE are similar to the △ $A'B'C'$, $A'C'D'$, $A'D'E'$, respectively.

Proof. 1. $\angle B = \angle B'$. Art. 321.

Also $AB : A'B' = BC : B'C'$. Art. 321.

$\therefore$ △ ABC and $A'B'C'$ are similar, Art. 327.

(*if two △ have an ∠ of one = an ∠ of the other and the including sides proportional, the △ are similar*).

2. Again $\angle BCD = \angle B'C'D'$, Art. 321.

(*homologous ∠ of similar polygons*).

Also $\angle BCA = \angle B'C'A'$, Art. 321.

(*homologous ∠ of the similar △ ABC and A'B'C'*).

Subtracting $\angle ACD = \angle A'C'D'$. Ax. 3.

But $BC : B'C' = CD : C'D'$, Art. 321.

(*homologous sides of similar polygons are proportional*).

And $BC : B'C' = AC : A'C'$, Art. 321.

(*being homologous sides of the similar △ ABC and A'B'C'*).

Hence $AC : A'C' = CD : C'D'$. Ax. 1.

$\therefore$ the △ ACD and $A'C'D'$ are similar. Art. 327.

3. In like manner it can be shown that the △ ADE and $A'D'E'$ are similar. Q. E. D.

PROPOSITION XX. THEOREM (CONVERSE OF PROP. XIX)

330. *If two polygons are composed of the same number of triangles, similar, each to each, and similarly placed, the polygons are similar.*

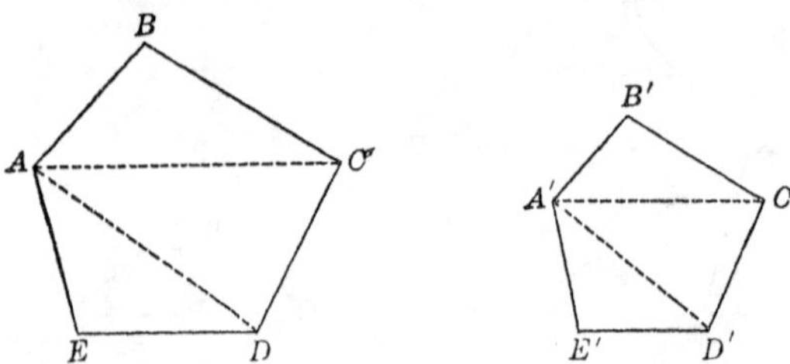

Given the two polygons $ABCDE$ and $A'B'C'D'E'$, in which the $\triangle$ ABC, ACD, ADE are similar, respectively, to the $\triangle$ $A'B'C'$, $A'C'D'$, $A'D'E'$, and are similarly placed.

To prove the polygons $ABCDE$ and $A'B'C'D'E'$ similar.

Proof. $\angle B = \angle B'$, Art. 321.

(*being homologous* $\angle$ *of similar* $\triangle$).

Also $\angle BCA = \angle B'C'A'$. (Why?)

And $\angle ACD = \angle A'C'D'$. (Why?)

Adding equals, $\angle BCD = \angle B'C'D'$. (Why?)

In like manner it may be shown that $\angle CDE = \angle C'D'E'$, $\angle BAE = \angle B'A'E'$, etc.

Hence the polygons are mutually equiangular.

Also $\frac{AB}{A'B'} = \frac{BC}{B'C'}$ (*homologous sides of similar* $\triangle$). Art. 321.

Again $\frac{BC}{B'C'} = \frac{AC}{A'C'}$, and $\frac{CD}{C'D'} = \frac{AC}{A'C'}$ $\therefore \frac{BC}{B'C'} = \frac{CD}{C'D'}$. (Why?)

In like manner $\frac{CD}{C'D'} = \frac{DE}{D'E'} = \frac{AE}{A'E'}$.

Hence the homologous sides of the given polygons are proportional.

$\therefore$ the polygons $ABCDE$ and $A'B'C'D'E'$ are similar. Art. 321

Q. E. D.

331. NOTE. It is often convenient to write a series of equal ratios, like those used in the above proof, as follows:

$$\frac{AB}{A'B'}=\frac{BC}{B'C'}=\left(\frac{AC}{A'C'}\right)=\frac{CD}{C'D'}=\left(\frac{AD}{A'D'}\right)=\frac{DE}{D'E'}=\frac{AE}{A'E'},$$

a ratio which is used merely to show the equality of two other ratios being inclosed in parenthesis.

PROPORTIONAL LINES AND NUMERICAL PROPERTIES OF LINES

PROPOSITION XXI. THEOREM

332. *In any triangle, the bisector of an angle divides the opposite side into segments which are proportional to the other two sides.*

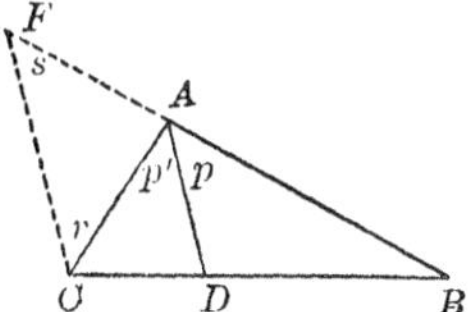

Given the $\triangle ABC$, with the line AD bisecting the $\angle BAC$, and meeting BC at D.

To prove $DC : DB = AC : AB$.

Proof. Draw the line $CF \parallel AD$, and meeting AB produced at F.

Then	$DC : DB = AF : AB$.	Art. 317.
But	$\angle r = \angle p'$.	Art. 124.
And	$\angle s = \angle p$.	Art. 126.
Also	$\angle p' = \angle p$.	Hyp.
	$\therefore \angle r = \angle s$.	(Why?)

$\therefore \triangle ACF$ is isosceles, and $AC = AF$. (Why?)

Substituting AC for its equal, AF, in the above proportion,

$DC : DB = AC : AB$. Ax. 8.

Q. E. D.

Ex. If, in the above figure, $AB=16$, $AC=12$, and $BC=14$, find DC and DB.

[SUG. Let $DC=x$, $DB=14-x$, etc.]

333. A **line divided internally** is a line divided into two parts by a point taken between its extremities.

Thus, the line AB is divided internally at the point P into the segments PA and PB.

To divide a given line internally in a given ratio (as 2 : 7), divide it into a number of equal parts, equal to the *sum* of the terms of the ratio (as 2 + 7, or 9 equal parts).

334. A **line divided externally** is a line whose parts are considered to be the segments included between a point on the line produced and the extremities of the given line.

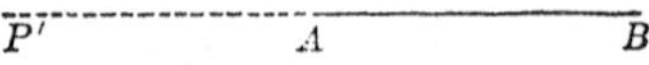

Thus, the line AB is divided externally at the point P' into the two segments $P'A$ and $P'B$.

To divide a given line externally in a given ratio (as 2 : 7), divide it into a number of equal parts, equal to the *difference* of the terms of the ratio (as 7 — 2, or 5 parts) and produce it till the produced part equals the smaller term of the ratio times the unit line found.

335. A **line divided harmonically** is a line divided both internally and externally in the same ratio.

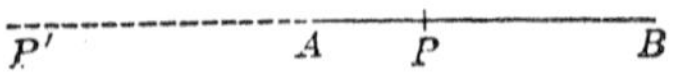

Thus, if the line AB is divided internally, so that $PA : PB = 1 : 2$, and externally, so that $P'A : P'B = 1 : 2$, then $PA : PB = P'A : P'B$, and the line is divided harmonically

---

Ex. 1. Divide a given line harmonically in the ratio 1 : 3.

Ex. 2. Divide a given line harmonically in the ratio 2 : 5.

PROPOSITION XXII. THEOREM

336. *In any triangle, the bisector of an exterior angle divides externally the opposite side produced into segments which are proportional to the other two sides.*

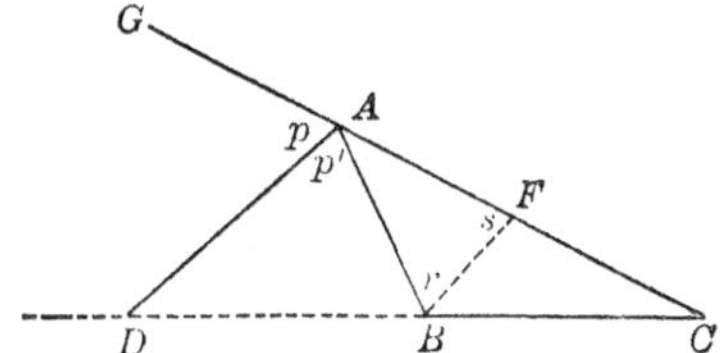

Given the $\triangle ABC$ with its exterior $\angle BAG$ bisected by the line AD meeting the opposite side produced at D.

To prove $DB : DC = AB : AC$.

Proof. Draw the line $BF \parallel AD$ and meeting AC at F.

Then $DB : DC = AF : AC$. (Why?)

But $\angle r = \angle p'$. (Why?)

And $\angle s = \angle p$. (Why?)

Also $\angle p' = \angle p$. (Why?)

$\therefore \angle r = \angle s$. (Why?)

$\therefore \triangle ABF$ is isosceles, and $AB = AF$. (Why?)

Substituting AB for its equal, AF, in the above proportion,

$DB : DC = AB : AC$. Ax. 8.

Q. E. D.

337. COR. *The lines bisecting the interior and exterior angles of a triangle at a given vertex, divide the opposite side harmonically.*

PROPOSITION XXIII. THEOREM

338. *The homologous altitudes of two similar triangles have the same ratio as any two homologous sides.*

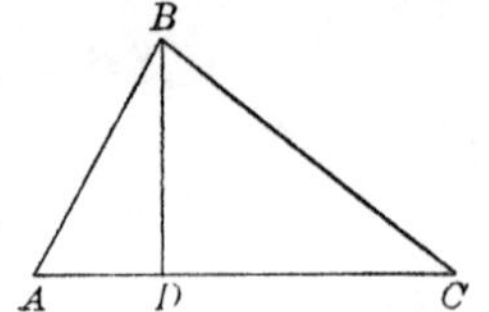

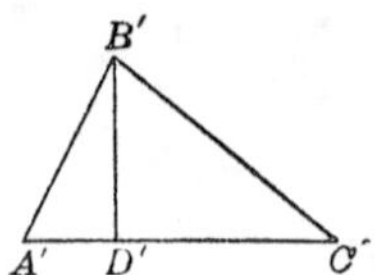

Given the similar △ ABC, $A'B'C'$, and BD, $B'D'$ any two homologous altitudes in these △.

To prove $$\frac{BD}{B'D'}=\frac{BA}{B'A'}=\frac{BC}{B'C'}=\frac{AC}{A'C'}.$$

Proof. In the right △ ABD and $A'B'D'$,

$$\angle A = \angle A', \qquad \text{Art. 321.}$$

(*being homologous* ∠ *of the similar* △ ABC *and* $A'B'C'$).

$\therefore$ △ ABD and $A'B'D'$ are similar, Art. 324.

(*if two rt.* △ *have an acute* ∠ *of one* = *an acute* ∠ *of the other, the* △ *are similar*).

$$\therefore \frac{BD}{B'D'}=\frac{BA}{B'A'}. \qquad \text{Art. 321.}$$

But, in the similar △ ABC and $A'B'C'$,

$$\frac{BA}{B'A'}=\frac{BC}{B'C'}=\frac{AC}{A'C'}. \qquad \text{Art. 321.}$$

Hence $$\frac{BD}{B'D'}=\frac{BA}{B'A'}=\frac{BC}{B'C'}=\frac{AC}{A'C'}. \qquad \text{Ax. 1.}$$

Q. E. D.

Ex. If, in the above figure, $AC=18$, $A'C'=12$, and $BD=15$, find $B'D'$.

PROPOSITION XXIV. THEOREM

339. *If three or more lines pass through the same point and intersect two parallel lines, they intercept proportional segments on the parallel lines.*

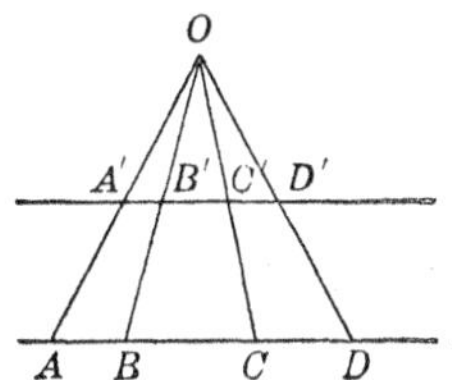

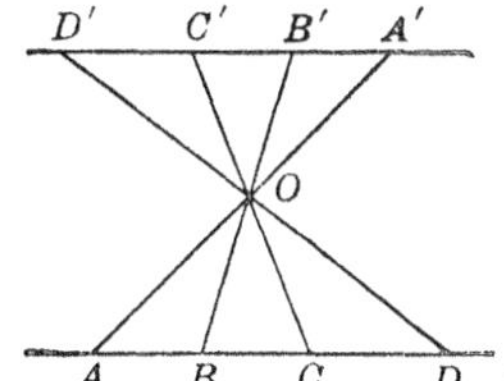

Given the transversals OA, OB, OC, OD intersecting the parallel lines AD and $A'D'$ in the points A, B, C, D and A', B', C', D', respectively.

To prove $$\frac{AB}{A'B'} = \frac{BC}{B'C'} = \frac{CD}{C'D'}.$$

Proof. $A'D' \parallel AD$. Hyp.

Therefore the base ∠s of the △s $A'B'O$, $B'C'O$, etc., are equal to corresponding base ∠s of the △s ABO, BCO, etc.
Art. 126 or Art. 124.

Hence the △s $A'B'O$, $B'C'O$, etc., are similar to the △s ABO, BCO, etc., respectively. Art. 324.

$$\therefore \frac{AB}{A'B'} = \left(\frac{BO}{B'O}\right) = \frac{BC}{B'C'} = \left(\frac{CO}{C'O}\right) = \frac{CD}{C'D'}.$$ Art. 321.

Q. E. D.

Ex. If the sides of a polygon are 2, 3, 4, 5, 6 feet, and, in a similar polygon, the side homologous to 6 is 9, find all the other sides of the second polygon.

PROP. XXV. THEOREM (CONVERSE OF PROP. XXIV)

340. *If three or more non-parallel straight lines intercept proportional segments on two parallels, they pass through the same point (that is, are concurrent).*

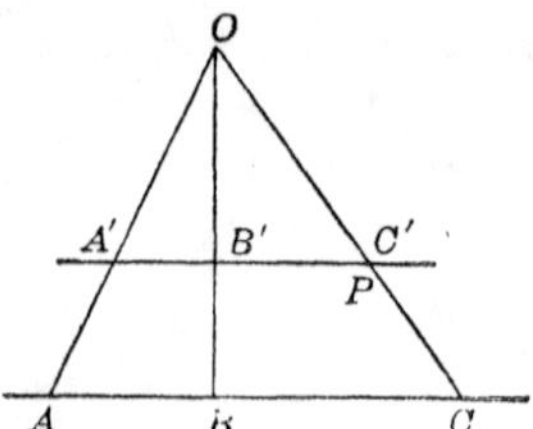

Given the transversals AA', BB', CC' intersecting the parallel lines AC and $A'C'$ so that $\frac{AB}{A'B'}=\frac{BC}{B'C'}$.

To prove that the lines AA', BB', CC' are concurrent.

Proof. Produce the lines AA' and BB' to meet at some point O.

Draw the line OC and let it intersect the line $A'C'$ at some point P.

Then $\frac{AB}{A'B'}=\frac{BC}{B'P}$. Art. 339.

But $\frac{AB}{A'B'}=\frac{BC}{B'C'}$. Hyp.

Hence $B'P=B'C'$. Art. 306.

$\therefore$ point P coincides with point C'.

$\therefore$ line CC' coincides with line CP. Art. 64

$\therefore$ the line CC', if produced, passes through O, (*for it coincides with the line CP, which passes through O*).

$\therefore$ AA', BB', CC' meet in O.

Q. E. D.

PROPOSITION XXVI. THEOREM

341. *The perimeters of two similar polygons have the same ratio as any two homologous sides.*

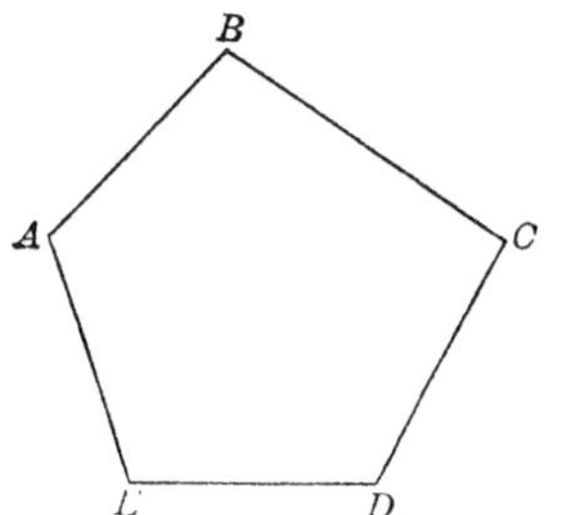

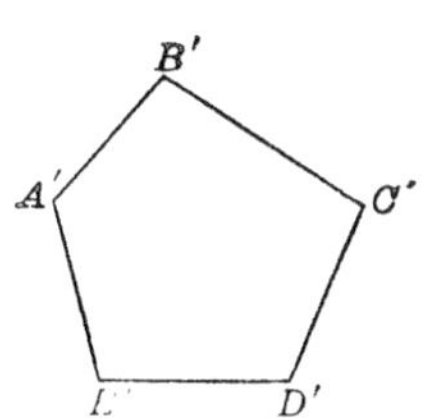

Given the two similar polygons $ABCDE$ and $A'B'C'D'E'$, with their perimeters denoted by P and P' and with AB and $A'B'$ any two homologous sides.

To prove that $P : P' = AB : A'B'$.

Proof. $AB : A'B' = BC : B'C' = CD : C'D' =$ etc. Art. 321.

Hence $AB+BC+CD+$, etc. $: A'B'+B'C'+C'D'+$, etc., $= AB : A'B'$, Art. 312.

(*in a series of equal ratios, the sum of all the antecedents is to the sum of all the consequents as any one antecedent is to its consequent*).

Or $$P : P' = AB : A'B'.$$

Q. E. D.

341 (*a*). COR. *In two similar polygons, any two homologous lines are to each other as any other two homologous lines; and the perimeters are to each other as any two homologous lines.*

Ex. If the perimeter of a given field is 210 rods and a side of this field is to a homologous side of a similar field as 3 : 2, find the perimeter of the second field.

Proposition XXVII. Theorem

342. *In a right triangle,*

I. *The altitude to the hypotenuse is a mean proportional between the segments of the hypotenuse;*

II. *Each leg is a mean proportional between the hypotenuse and the segment of the hypotenuse adjacent to the given leg.*

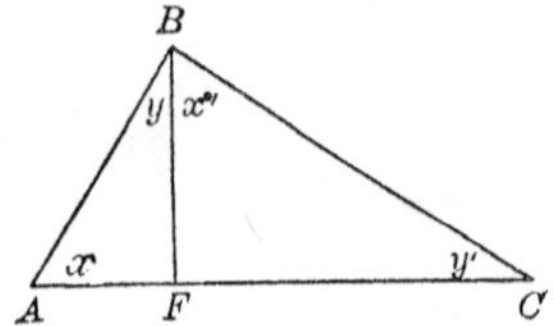

Given the right $\triangle$ ABC and BF the altitude upon the hypotenuse AC.

To prove I. $AF : BF = BF : FC$.

II. $\begin{cases} AC : AB = AB : AF \\ AC : BC = BC : FC. \end{cases}$

Proof. The $\angle A$ is common to the rt. $\triangle$s ABF and ABC.

$\therefore$ $\triangle$s ABF and ABC are similar. Art. 324.

Similarly $\angle C$ is common to the rt. $\triangle$s BFC and ABC, and $\triangle$s BFC and ABC are similar.

$\therefore$ $\triangle$s ABF and BFC are similar, Art. 325.
(*if two $\triangle$s are similar to the same $\triangle$, they are similar to each other*).

Hence, I. In the $\triangle$s ABF and BFC,

$AF : BF = BF : FC$, Art. 321.
(*homologous sides of similar $\triangle$s are proportional*).

II. In the similar $\triangle$ ABC and ABF,

$AC : AB = AB : AF$. Art. 321.

Also, in the similar $\triangle$ ABC and BFC,

$AC : BC = BC : FC$. Art. 321.

Q. E. D.

343. Cor. *The perpendicular to the diameter from any point in the circumference of a circle is a mean proportional between the segments of the diameter; and the chord joining the point to an extremity of the diameter is a mean proportional between the diameter and the segment of the diameter adjacent to the chord.*

344. Def. The **projection of a point** upon a line is the foot of the perpendicular drawn from the point to the line.

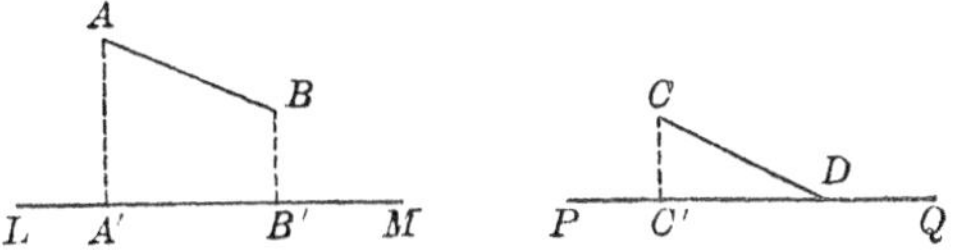

Thus, if AA' be perpendicular to LM, A' is the projection of the point A on the line LM.

345. The **projection of a line** upon another given line is that part of the second line which is included between perpendiculars drawn from the extremities of the first line upon the second line. Thus, the projection of AB on LM is $A'B'$; of CD on PQ, is $C'D$.

Ex. 1. If the segments of the hypotenuse of a right triangle are 3 and 12, find the altitude on the hypotenuse; also find the legs of the triangle.

Ex. 2. If, in the figure of Art. 343, the diameter is 20 and the longer chord 16, find the segment of the diameter adjacent to the chord.

Proposition XXVIII. Theorem

346. *In a right triangle, the square of the hypotenuse is equal to the sum of the squares of the legs.*

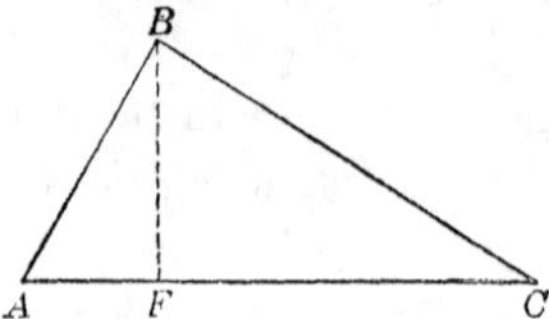

Given the right triangle ABC with AC the hypotenuse.

To prove $\overline{AC}^2 = \overline{AB}^2 + \overline{BC}^2$.

Proof. Draw the line $BF \perp AC$. Then

$AC : AB = AB : AF \therefore AC \times AF = \overline{AB}^2$. Arts. 342, 302.

Also $AC : BC = BC : FC \therefore AC \times FC = \overline{BC}^2$. (Why?)

Adding equals, $AC\ (AF + FC) = \overline{AB}^2 + \overline{BC}^2$. Ax. 2.

Or $\overline{AC}^2 = \overline{AB}^2 + \overline{BC}^2$. Axs. 6, 8.

Q. E. D.

347. Cor. 1. *In a right triangle, the square of either leg is equal to the square of the hypotenuse minus the square of the other leg.*

348. Cor. 2. In the square $ABCD$, the diagonal divides the square into two right triangles.

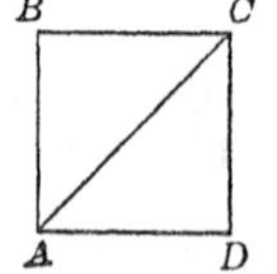

Hence $\overline{AC}^2 = \overline{AB}^2 + \overline{BC}^2 = 2\ \overline{AB}^2$.

$\therefore AC = AB\sqrt{2}$, or $\frac{AC}{AB} = \frac{\sqrt{2}}{1}$.

Hence *the diagonal and the side of a square are incommensurable.*

Ex. 1. If the legs of a rt. △ are $1\frac{1}{2}$ in. and 2 in., find the hypotenuse.

Ex. 2. In the figure on p. 204, show that $\overline{AB}^2 : \overline{BC}^2 = AF : FC$.

Proposition XXIX. Theorem

349. *In any oblique triangle, the square of a side opposite an acute angle is equal to the sum of the squares of the other two sides, diminished by twice the product of one of those sides by the projection of the other side upon it.*

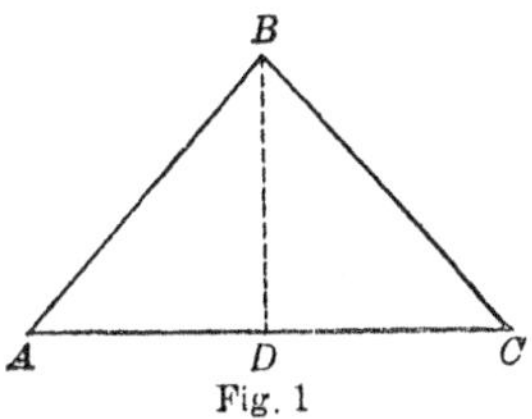

Fig. 1

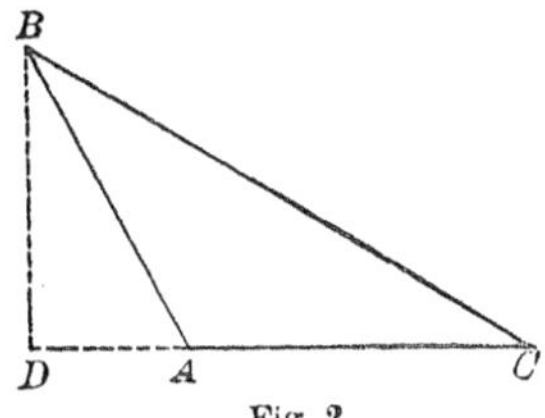

Fig. 2

Given acute $\angle C$ in $\triangle ABC$, and DC the projection of the side BC on the side AC.

To prove $\overline{AB}^2 = \overline{BC}^2 + \overline{AC}^2 - 2\, AC \times DC$.

Proof. If D falls on AC (Fig. 1), $AD = AC - DC$.

If D falls on AC produced (Fig. 2), $AD = DC - AC$.

In either case, $\overline{AD}^2 = \overline{AC}^2 + \overline{DC}^2 - 2\, AC \times DC$. Ax. 4.

Adding $\overline{BD}^2$ to each of these equals,

$$\overline{AD}^2 + \overline{BD}^2 = \overline{AC}^2 + \overline{DC}^2 + \overline{BD}^2 - 2\, AC \times DC. \quad \text{Ax. 2.}$$

But, in the rt. $\triangle ABD$, $\overline{AD}^2 + \overline{BD}^2 = \overline{AB}^2$, Art. 346.

and, in the rt. $\triangle DBC$, $\overline{DC}^2 + \overline{BD}^2 = \overline{BC}^2$. (Why?)

Substituting these values in the above equality,

$$\overline{AB}^2 = \overline{BC}^2 + \overline{AC}^2 - 2\, AC \times DC. \quad \text{Ax. 8.}$$

Q. E. D.

Ex. 1. A line 10 in. long makes an angle of 45° with a second line; find the projection of the first line on the second.

Ex. 2. Find the same, if the angle is 60°.

Ex. 3. If the side of an equilateral triangle is a, find its projection on the base.

Ex. 4. If, in Fig. 1, $BC = 10$, $AC = 12$, and $\angle C = 60°$, find AB.

PROPOSITION XXX. THEOREM

350. *In any obtuse triangle, the square of the side opposite an obtuse angle is equal to the sum of the squares of the other two sides, increased by twice the product of one of the sides by the projection of the other side upon that side.*

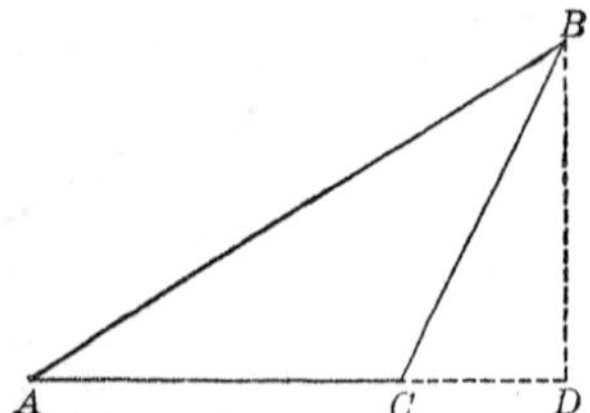

Given the obtuse $\angle ACB$ in the $\triangle ABC$, and CD the projection of BC on AC produced.

To prove $\overline{AB}^2 = \overline{AC}^2 + \overline{BC}^2 + 2\,AC \times CD.$

Proof. $AD = AC + CD.$ Ax. 6

$\therefore \overline{AD}^2 = \overline{AC}^2 + \overline{CD}^2 + 2\,AC \times CD.$ Ax. 4.

Adding $\overline{BD}^2$ to each of these equals,

$\overline{AD}^2 + \overline{BD}^2 = \overline{AC}^2 + \overline{CD}^2 + \overline{BD}^2 + 2\,AC \times CD.$ Ax. 2.

But, in the rt. $\triangle ABD$, $\overline{AD}^2 + \overline{BD}^2 = \overline{AB}^2.$ (Why ?)

And, in the rt. $\triangle CBD$, $\overline{CD}^2 + \overline{BD}^2 = \overline{BC}^2.$ (Why ?)

Substituting these values in the above equality,

$\overline{AB}^2 = \overline{AC}^2 + \overline{BC}^2 + 2\,AC \times CD.$ Ax. 8.

Q. E. D.

351. COR. *If the square on one side of a triangle equals the sum of the squares on the other two sides, the angle opposite the first side is a right angle;* for it cannot be acute (Art. 349), or obtuse (Art. 350).

Ex. 1. If, in the above figure, $BC = 10$, $AC = 2$, and $\angle BCA = 120°$, find AB.

Ex. 2. If $AB = 20$, $BC = 14$, and $AC = 12$, find CD.

Proposition XXXI. Theorem

352. *If, in any triangle, a median be drawn to one side,*

I. *The sum of the squares of the other two sides is equal to twice the square of half the given side, increased by twice the square of the median upon that side;* and

II. *The difference of the squares of the other two sides is equal to twice the product of the given side by the projection of the median upon that side.*

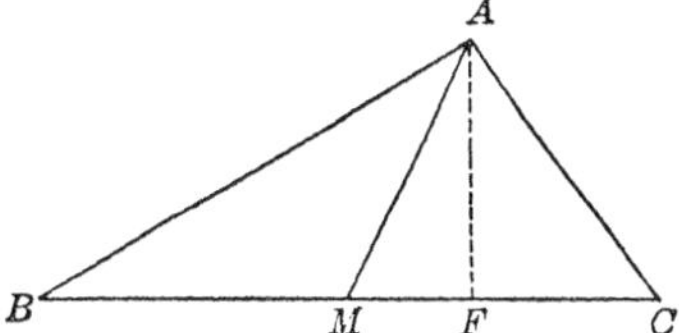

Given the $\triangle ABC$, $AB > AC$, AM the median upon BC, and MF the projection of AM on BC.

To prove I. $\overline{AB}^2 + \overline{AC}^2 = 2\,\overline{BM}^2 + 2\,\overline{AM}^2$.

II. $\overline{AB}^2 - \overline{AC}^2 = 2\,BC \times MF$.

Proof. In the $\triangle$s BMA and AMC, $BM = MC$, $AM = AM$, and $AB > AC$. Hyp.

$\therefore \angle AMB$ is greater than $\angle AMC$. Art. 108.

$\therefore \angle AMB$ is obtuse (*for it is greater than half a straight* $\angle$).

In obtuse $\triangle ABM$, $\overline{AB}^2 = \overline{BM}^2 + \overline{AM}^2 + 2\,BM \times MF$. Art. 350.

In acute $\triangle ACM$, $\overline{AC}^2 = \overline{MC}^2 + \overline{AM}^2 - 2\,MC \times MF$. Art. 349.

Adding, $\overline{AB}^2 + \overline{AC}^2 = 2\,\overline{BM}^2 + 2\,\overline{AM}^2$, Ax. 2.
(*for* $MC = BM$).

Subtracting, $\overline{AB}^2 - \overline{AC}^2 = 2\,BC \times MF$, Axs. 2, 6.
(*for* $BM + MC = BC$).

Q. E. D.

353. Formula for median of a triangle in terms of its sides. In the Fig., p. 209, denoting AB by c, AC by b, BC by a, and AM by m, by Art. 352, $b^2 + c^2 = 2m^2 + 2\left(\frac{a}{2}\right)^2$, whence $m = \frac{1}{2}\sqrt{2(b^2 + c^2) - a^2}$.

PROPOSITION XXXII. THEOREM

354. *If two chords in a circle intersect, the product of the segments of one chord is equal to the product of the segments of the other chord.*

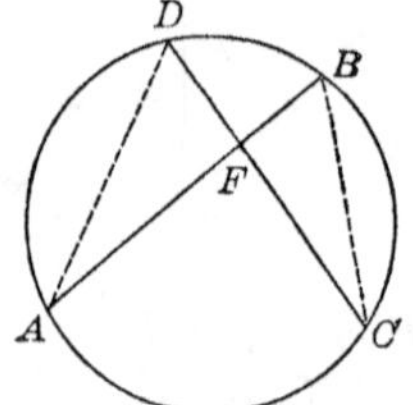

Given the ⊙ $ADBC$ with the chords AB and CD intersecting at the point F.

To prove $AF \times FB = CF \times FD$.

Proof. Draw AD and CB.

Then, in the △ AFD and CFB, $\angle A = \angle C$, Art. 258.
(each being measured by ½ arc DB).

Also $\angle D = \angle B$. (Why?)

Hence the △ AFD and CFB are similar. (Why?)

$\therefore AF : CF = FD : FB$. (Why?)

And $AF \times FB = CF \times FD$. (Why?)

Q. E. D.

355. COR. 1. *If through a fixed point within a circle a chord be drawn, the product of the segments of the chord is constant, in whatever direction the chord be drawn.*

356. DEF. **Four directly proportional quantities** are four quantities in proportion in such a way that both the antecedents belong to one figure and both the consequents to another figure.

Four reciprocally proportional quantities are four quantities in proportion in such a way that the means belong to one figure and the extremes to another figure.

357. COR. 2. *The segments of two chords intersecting in a circle are reciprocally proportional* (the segments being considered as parts of the chords, not of the $\triangle$s).

PROPOSITION XXXIII. THEOREM

358. *If, from a given point, a secant and a tangent be drawn to a circle, the tangent is the mean proportional between the whole secant and its external segment.*

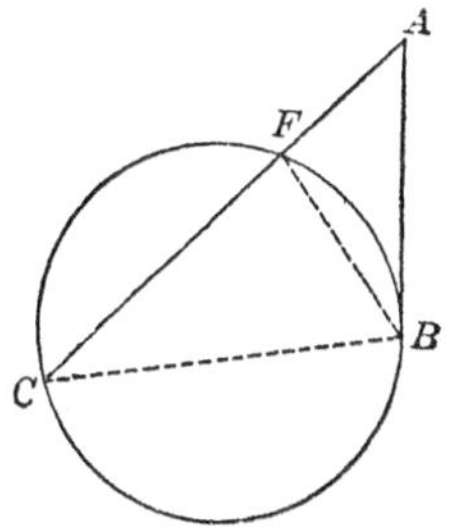

Given AB, a tangent, and AC, a secant, to the circle BCF, and AF the external segment of the secant.

To prove $AC : AB = AB : AF$.

Proof. Draw the chords BC and BF.

Then, in the $\triangle$s ABC and ABF, $\angle A = \angle A$. Ident.

$\angle C = \angle ABF$. Arts. 258, 264.

$\therefore$ the $\triangle$s ABC and ABF are similar. (Why?)

$\therefore AC : AB = AB : AF$. (Why?)

Q. E. D.

359. Cor. 1. *If, from a given point, a tangent and a secant be drawn to a circle, the product of the whole secant and its external segment is equal to the square of the tangent.*

360. Cor. 2. *If, from a given point without a circle, a secant be drawn, the product of the secant and its external segment is constant, in whatever direction the secant be drawn.*

For the product of each secant and its external segment equals the square of the tangent, which is constant.

361. Cor. 3. *If two secants be drawn from an external point to a circle, the whole secants and their external segments are reciprocally proportional.*

Proposition XXXIV. Theorem

362. *The square of the bisector of an angle of a triangle is equal to the product of the sides forming the angle, diminished by the product of the segments of the third side formed by the bisector.*

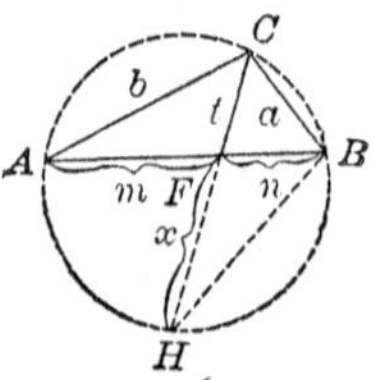

Given the $\triangle ABC$, CF (or t) the bisector of $\angle ACB$, and m and n the segments of AB formed by CF.

To prove $t^2 = ab - mn$.

Circumscribe a $\odot$ about the $\triangle ABC$.

Proof. Produce CF to meet the circumference at H, and draw the chord BH.

Then, in the △ ACF and CHB, $\angle A = \angle H$. (Why?)

And $\angle ACF = \angle HCB$. (Why?)

Hence the △ ACF and CHB are similar. (Why?)

$\therefore b : t = x + t : a$. (Why?)

$\therefore t(x+t) = ab$. (Why?)

Or $tx + t^2 = ab$.

Subtracting tx, $t^2 = ab - tx$. Ax. 3.

But $tx = mn$. Art. 354.

Substituting mn for tx,

$t^2 = ab - mn$. Ax. 8.

Q. E. D.

363. Formula for bisector of an angle of a triangle.

$m : n = b : a$ (Art. 332) $\therefore m + n : m = b + a : b$ (Art. 309), or $c : m = a + b : b$ (Ax. 8) $\therefore m = \frac{bc}{a+b}$ (Art. 302, Ax. 5).

In like manner $n = \frac{ac}{a+b}$. Substituting for m and n,

$$t^2 = ab - \frac{abc^2}{(a+b)^2} = \frac{ab(a+b+c)(a+b-c)}{(a+b)^2}.$$

Let $a + b + c = 2s$; then $a + b - c = 2s - 2c = 2(s - c)$.

Then $$t = \frac{2}{a+b}\sqrt{abs(s-c)}.$$

Ex. 1. On the figure, p. 210, let $AF=10$, $FB=4$, and $FC=8$. Find FD.

Ex. 2. On the figure, p. 211, let $AC=16$ and $AB=12$. Find AF.

Ex. 3. On the figure, p. 212, let $AC=16$, $CB=12$, and $AB=14$. Find CF.

Proposition XXXV. Theorem

364. *In any triangle, the product of any two sides is equal to the product of the diameter of the circumscribed circle by the altitude upon the third side.*

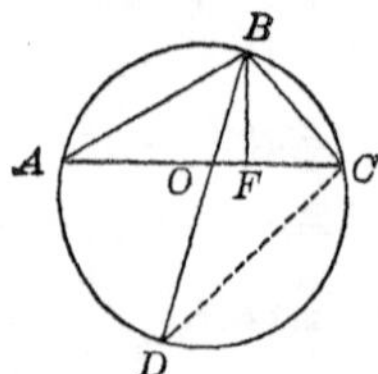

Given the $\triangle$ ABC, $ABCD$ a circumscribed $\odot$, BD the diameter of this $\odot$, and BF the altitude upon AC.

To prove $AB \times BC = BD \times BF$.

Proof. Draw the chord DC.

Then, in the $\triangle$s ABF and DBC, $\angle A = \angle D$. (Why ?)

$\angle DCB$ is a rt. $\angle$. (Why ?)

$\therefore$ $\triangle$s ABF and DBC are similar. (Why ?)

$\therefore AB : BD = BF : BC$. (Why ?)

$\therefore AB \times BC = BD \times BF$. (Why ?)

Q. E. D.

365. Cor. *The diameter of a circle circumscribed about a triangle equals the product of two sides of the triangle divided by the altitude upon the third side.*

Ex. In the above figure, if $AB = 8$, $BC = 6$, and $BF = 4\frac{4}{5}$, find the radius of the circumscribed circle.

CONSTRUCTION PROBLEMS

Proposition XXXVI. Problem

366. *To construct a fourth proportional to three given lines.*

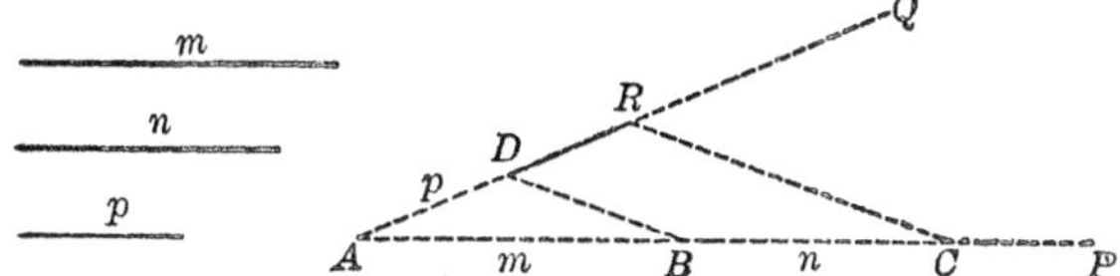

Given the lines m, n, p.

To construct a fourth proportional to m, n, and p.

Construction. Take any two lines, AP and AQ, making any convenient $\angle A$.

On AP take $AB=m$, and $BC=n$.

On AQ take $AD=p$.

Draw BD.*

Through the point C draw $CR \parallel DB$, and meeting AQ at R. Art. 279.

Then DR is the fourth proportional required.

Proof. $AB : BC = AD : DR$. Art. 317.

Or $m : n = p : DR$. Ax. 8.

Q. E. F.

Proposition XXXVII. Problem

367. *To construct a third proportional to two given lines.*

Let the pupil supply the construction and proof.

[Sug. Use the method of Art. 366, making $p=n$.]

Ex. 1. Construct a fourth proportional to three lines, $\frac{3}{4}$, 1 and $1\frac{1}{2}$ in. long.

Ex. 2. Construct a third proportional to two lines, 2 and $1\frac{1}{2}$ in. long.

* From now on, references to the postulates will be omitted where these are not necessary to complete some other reference.

PROPOSITION XXXVIII. PROBLEM

368. *To construct a mean proportional between two given lines.*

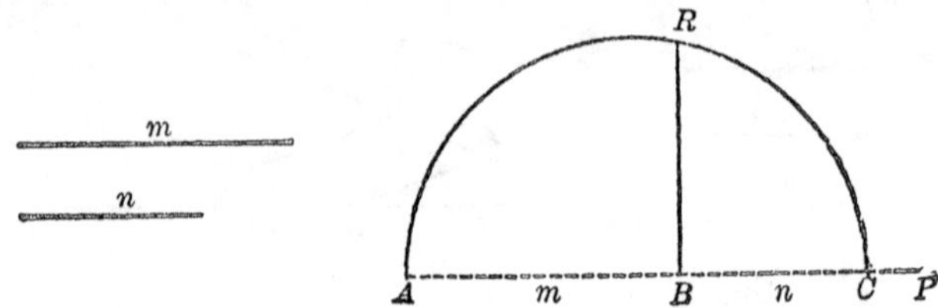

Given the lines m and n.

To construct a mean proportional between m and n.

Construction. On the line AP take $AB = m$, and $BC = n$.

On AC as a diameter construct a semi-circumference. Art. 275, Post. 3.

At B erect a $\perp$ to AC meeting the semi-circumference at R. Art. 274.

Then BR is the mean proportional required.

Proof. $AB : BR = BR : BC$, Art. 343.

(*the $\perp$ to the diameter from any point in the circumference of a circle is a mean proportional between the segments of the diameter*).

Substituting for AB and BC their values m and n,

$$m : BR = BR : n.$$ Ax. 8.

Q. E. F.

Ex. 1. Construct the third proportional to two lines, 1 and $1\frac{1}{2}$ in. long.

Ex. 2. Construct a mean proportional between two lines, 1 and 2 in. long.

Ex. 3. Taking any line as 1, construct $\sqrt{2}$.

PROPOSITION XXXIX. PROBLEM

369. *To divide a given straight line into parts proportional to a number of given lines.*

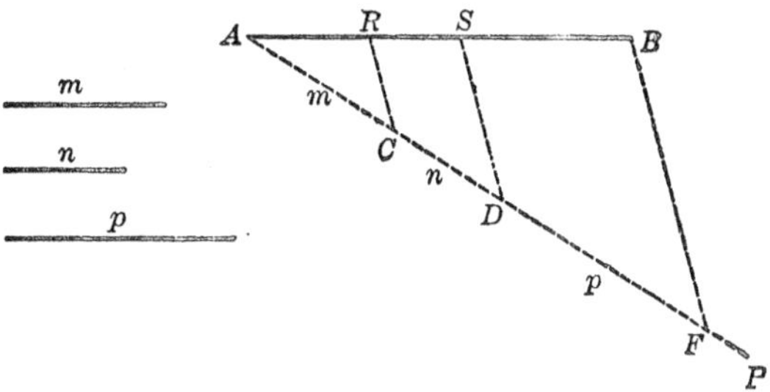

Given the straight lines AB, m, n and p.

To divide AB into parts proportional to m, n and p.

Construction. Draw the line AP, making any convenient angle with AB.

On AP mark off $AC=m$, $CD=n$, and $DF=p$. Draw BF.

Through the points C and D draw lines $\parallel BF$, and meeting AB at R and S. Art. 279.

Then AR, RS and SB are the segments required.

Proof. $$\frac{AR}{AC}=\frac{RS}{CD}=\frac{SB}{DF}.$$ Art. 319.

(*if two lines are cut by a number of parallels, the corresponding segments are proportional*).

For AC, CD and DF substitute their equals m, n and p.

Then $$\frac{AR}{m}=\frac{RS}{n}=\frac{SB}{p}.$$ Ax. 8.

Q. E. F.

370. DEF. **A straight line divided in extreme and mean ratio** is a straight line divided into two segments such that one of the segments is a mean proportional between the whole line and the other segment.

Ex. Divide a given line into parts proportional to 2, 3 and 4.

PROPOSITION XL. PROBLEM

371. *To divide a given straight line in extreme and mean ratio.*

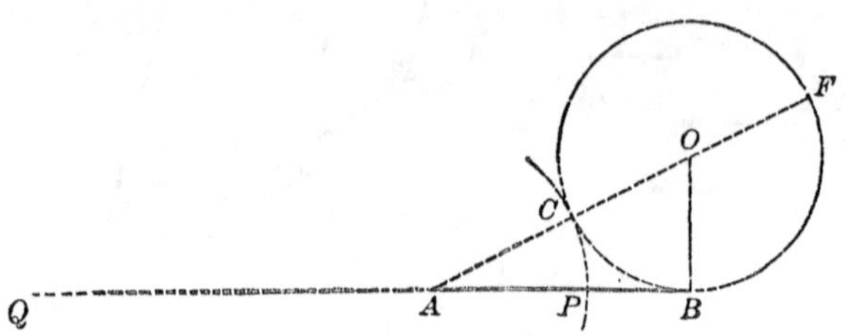

Given the line AB.

To divide AB in extreme and mean ratio.

Construction. At one end of the given line, as B, construct a $\perp$ OB equal to $\frac{1}{2}AB$. Arts. 274, 275.

From O as a center and with OB as a radius, describe a circumference.

Draw AO meeting the circumference at C, and produce it to meet the circumference again at F.

On AB mark off AP equal to AC; on BA produced take $AQ=AF$.

Then AB is divided in extreme and mean ratio internally at P, and externally at Q.

Proof. 1. $AF : AB = AB : AC$. Art. 358.

$\therefore AF - AB : AB = AB - AC : AC$. (Why?)

But $CF = 2OB = AB$, and $AC = AP$.

Hence $AF - AB = AP$, and $AB - AC = PB$.

$\therefore AP : AB = PB : AP$, or $AB : AP = AP : PB$.

Ax. 8. Art. 308.

2. $AF : AB = AB : AC$. (Why?)

$\therefore AF + AB : AF = AB + AC : AB$. (Why?)

But $AF + AB = QB$, and $AB + AC = AF = QA$.

$\therefore QB : QA = QA : AB$. Ax. 8.

Q. E. F.

Proposition XLI. Problem

372. *Upon a given straight line, to construct a polygon similar to a given polygon, and similarly placed.*

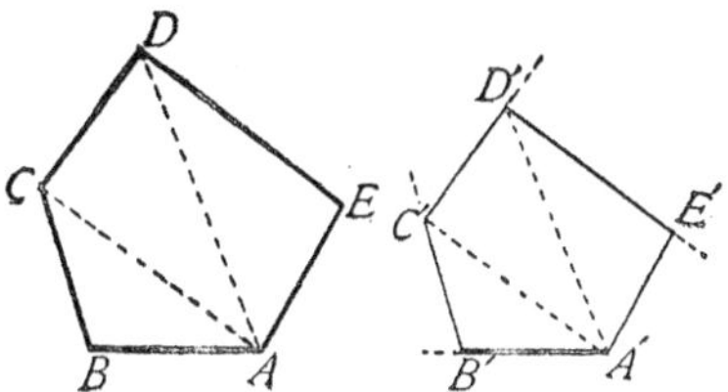

Given the polygon $ABCDE$ and the line $A'B'$.

To construct on $A'B'$ a polygon similar to $ABCDE$ and similarly placed.

Construction. In the given polygon, draw the diagonals AC and AD, dividing the polygon into triangles.

At B', on the line $A'B'$, construct $\angle A'B'C'$ equal to $\angle B$; and at A' construct $\angle B'A'C'$ equal to $\angle BAC$. Art. 278.

Produce the lines $B'C'$ and $A'C'$ to meet at C'.

In like manner, on $A'C'$ construct $\triangle A'C'D'$, equiangular with $\triangle ACD$ and similarly placed; and on $A'D'$ construct the $\triangle A'D'E'$, equiangular with $\triangle ADE$ and similarly placed.

Then $A'B'C'D'E'$ is the polygon required.

Proof. The $\triangle$s $A'B'C'$, $A'C'D'$, etc., are similar to the $\triangle$s ABC, ACD, etc., respectively. Art. 324.

Hence the polygons $A'B'C'D'E'$ and $ABCDE$ are similar. Art. 330.

Q. E. F.

EXERCISES. GROUP 30

SIMILAR TRIANGLES

Let the pupil make a list of all the conditions that make two triangles similar (see Arts. 323, 324, 325, etc.).

Ex. 1. Given $AD \perp BC$, and $BF \perp AC$; prove $\triangle$ ADC and BFC similar.

Ex. 2. In the same figure, prove the $\triangle$ AOF and BFC similar. What other triangle on this figure is similar to $\triangle$ BFC?

Ex. 3. Two isosceles triangles are similar if their vertex angles are equal.

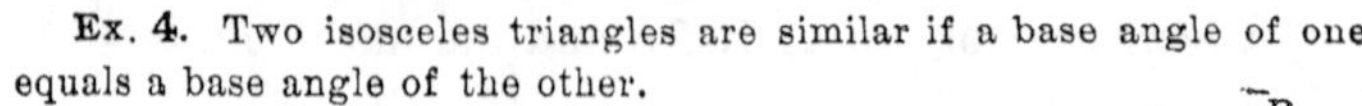

Ex. 4. Two isosceles triangles are similar if a base angle of one equals a base angle of the other.

Ex. 5. Given arc AC=arc BC; prove $\triangle$ APC and AFC similar.

Ex. 6. Prove that the diagonals and bases of a trapezoid together form a pair of similar triangles.

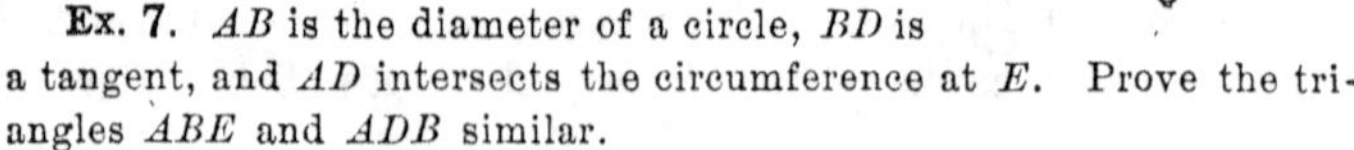

Ex. 7. AB is the diameter of a circle, BD is a tangent, and AD intersects the circumference at E. Prove the triangles ABE and ADB similar.

Ex. 8. BC is a chord in a circle, AQ is the diameter perpendicular to BC and meeting it at N; AP is any chord intersecting BC in M. Prove the $\triangle$ AMN and APQ similar.

Ex. 9. The triangle ABC is inscribed in a circle; the bisector of the angle A meets BC in D and the circumference in P. Prove the triangles BAD and APC similar.

Ex. 10. A pair of homologous medians divide two similar triangles into triangles which are similar each to each.

Ex. 11. Two rectangles are similar if two adjacent sides of one are proportional to the homologous sides of the other.

Ex. 12. Two circles intersect in the points A and B. AC and AD are each a tangent in one circle and a chord in the other. Prove the $\triangle$ ABC and ABD similar.

[SUG. Prove $\angle BAD = \angle ACB$, etc.]

373. Proof that lines are proportional. In order to prove that certain lines are proportional, or have proportional relations, it is usually best to *show that the given lines are homologous sides of similar triangles.*

Sometimes, however, other methods of proof are used (as the theorems of Arts. 354 and 358); but these, if investigated, are usually found to be the method of similar triangles in disguise.

EXERCISES. GROUP 31

PROPORTIONAL LINES

Ex. 1. On the figure of Ex. 1, p. 220, prove $AD \times BC = BF \times AC$, and $BC \times OD = BO \times FC$.

Ex. 2. On the figure of Ex. 5, p. 220, prove $CP : CA = CA : CF$. (Hence as P moves the product of what two lines is constant?)

Ex. 3. The diagonals of a trapezoid divide each other into proportional segments.

Ex. 4. In the isosceles triangle ABC, $AB = AC$, on the side AB, the point P is taken so that PC equals the base. Prove $AB \times PB = \overline{BC}^2$.

Ex. 5. In a triangle the median to the base bisects all lines parallel to the base and terminated by the sides.

Ex. 6. If PQ is any line through F, the midpoint of the line AB, and AP and BQ are perpendicular to PQ, show that the ratio $PF : FQ$ is constant.

Ex. 7. The triangle ABC is inscribed in a circle. F is the midpoint of the arc AC, and BF intersects the line AC in E. Prove $AB : BC = AE : EC$.

[SUG. Use Art. 332.]

Ex. 8. If two circles intersect, the common chord, if produced, bisects the common tangent.

[SUG. Use Art. 358.]

Ex. 9. If two circles intersect, tangents drawn to the two circles from any point in the common chord produced are equal.

Ex. 10. Given AD, PT and BC ∥ ; prove $PQ = RT$.

[SUG. Show that $\frac{PQ}{BC} = \frac{RT}{BC}$, by showing them equal to a common ratio.]

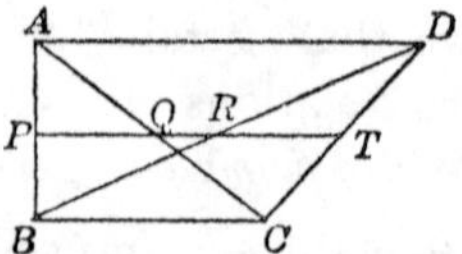

Ex. 11. Lines are drawn from a point O within the triangle, to the vertices of the triangle ABC. From B' any point in OB, $B'A'$ is drawn parallel to BA and meeting OA in A', and $B'C'$ is drawn parallel to BC and meeting OC in C'. Prove $A'B' : AB = B'C' : BC$, and the triangles ABC and $A'B'C'$ similar.

Ex. 12. Given $ABCD$ a ▱, and P any point in BC produced; prove $\overline{AR}^2 = RQ \times RP$.

[SUG. Compare the similar △ ABR and RQD; also the similar △ ARD and RBP.]

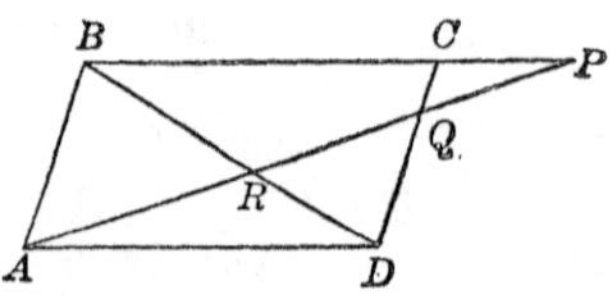

EXERCISES. GROUP 32

NUMERICAL PROPERTIES OF LINES

Ex. 1. If AD is the altitude of the triangle ABC, $\overline{AB}^2 - \overline{AC}^2 = \overline{BD}^2 - \overline{DC}^2$.

Ex. 2. If the diagonals of a quadrilateral are perpendicular to each other, the sum of the squares of one pair of opposite sides equals the sum of the squares of the other pair of sides.

Ex. 3. The square of the altitude of an equilateral triangle is three-fourths the square of one side.

Ex. 4. If AB is the hypotenuse of a right triangle, and the leg BC is bisected at K, $\overline{AB}^2 - \overline{AK}^2 = 3\overline{CK}^2$.

Ex. 5. PQ is a line parallel to the hypotenuse AB of a right triangle ABC, and meeting AC in P and BC in Q. Prove $\overline{AQ}^2 + \overline{BP}^2 = \overline{AB}^2 + \overline{PQ}^2$.

Ex. 6. In the right triangle ABC, BE and CF bisect the legs AC and AB in the points E and F. Prove $4\overline{BE}^2 + 4\overline{CF}^2 = 5\overline{BC}^2$.

EXERCISES. GROUP 33

AUXILIARY LINES

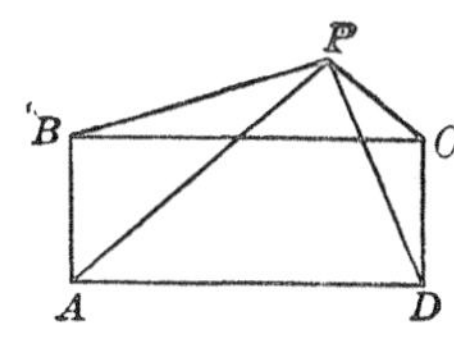

Ex. 1. Given $ABCD$ a rectangle; prove $\overline{PA}^2+\overline{PC}^2=\overline{PB}^2+\overline{PD}^2$.

Ex. 2. The common tangent of two circles divides the line of centers into segments which have the same ratio as the diameters of the circles.

[SUG. Draw radii to the points of contact.]

Ex. 3. AB and AC are the legs of an isosceles triangle and BF is an altitude. Prove $2AC\times FC=\overline{BC}^2$.

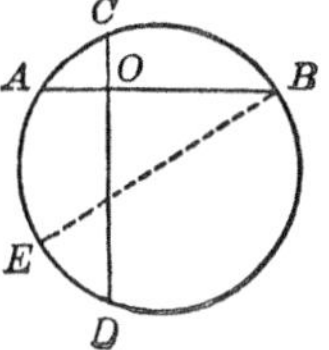

Ex. 4. Given the chords AB and CD perpendicular to each other and intersecting at O; prove $\overline{OA}^2+\overline{OB}^2+\overline{OC}^2+\overline{OD}^2=(\text{diameter})^2$.

[SUG. Draw the diameter BE and the chords AC, BD, DE. Prove $AC=ED$, etc.]

Ex. 5. ABC is an inscribed isosceles triangle of which AB and AC are the legs. AD is a chord meeting BC in E. Prove $\overline{AB}^2=AD\times AE$.

Ex. 6. If C is the vertex of an isosceles triangle ABC, and D is a point in the base produced, then $\overline{CD}^2=\overline{CB}^2+AD\times BD$.

Ex. 7. Two circles touch at the point T. PTP' and QTQ' are lines drawn meeting the circumferences in P, Q and P', Q' respectively. Prove the triangles PTQ and $P'TQ'$ similar.

[SUG. Draw the common tangent at T.]

Ex. 8. Two circles touch at the point T, through T three lines are drawn meeting the circumferences in P, Q, R and P', Q', R', respectively. Prove the triangles PQR and $P'Q'R'$ similar.

Ex. 9. If A is the midpoint of CD, an arc of a circle, and AP is any chord intersecting the chord CD in Q, prove that $AP\times AQ$ is a constant.

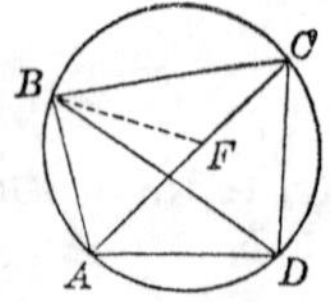

Ex. 10. In an inscribed quadrilateral, the product of the diagonals is equal to the sum of the products of the opposite sides.

[Sug. Draw BF so that $\angle CBF = \angle ABD$ and use similar triangles.]

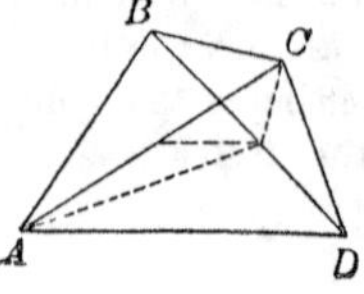

Ex. 11. The sum of the squares of the sides of any quadrilateral is equal to the sum of the squares of the diagonals, plus four times the square of the line joining the midpoints of the diagonals.

EXERCISES. GROUP 34

INDIRECT DEMONSTRATIONS

Ex. 1. If the sum of the squares on two sides of a triangle is greater than the square on the third side, the angle included by the two given sides is an acute angle.

Ex. 2. If D is a point in the side AC of the triangle ABC, and $AD : DC = AB : BC$, then DB bisects angle ABC.

Ex. 3. A given straight line can be divided in a given ratio at but one point.

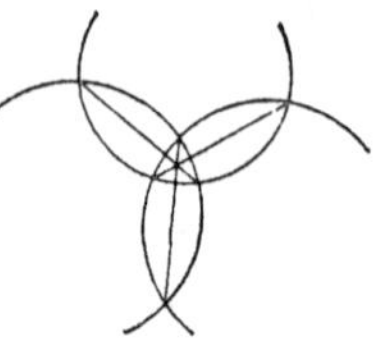

Ex. 4. If the sides of two triangles are parallel, each to each, and a straight line be passed through each pair of homologous vertices, these lines, if produced, will meet in a common point.

Ex. 5. If each of three circles intersects the other two, the three common chords intersect in one point.

EXERCISES. GROUP 35

THEOREMS PROVED BY VARIOUS METHODS

Ex. 1. In the figure on p. 204, show that $AB \times BF = BC \times AF$.

Ex. 2. In the same figure, if $FC = 3AF$, show that $\overline{AB}^2 : \overline{BC}^2 = 1 : 3$.

Ex. 3. AB is the diameter of a circle and PB is a tangent. If AP meets the circumference in the point Q, prove that $AP \times AQ = \overline{AB}^2$.

Ex. 4. If the line bisecting the parallel sides of a trapezoid be produced, it meets the legs produced in a common point.

[Sug. See Art. 340.]

Ex. 5. In similar triangles, homologous medians have the same ratio as homologous sides.

Ex. 6. A diameter AB is produced to the point C; CP is perpendicular to AC; PB produced meets the circumference at Q. Prove the triangles AQB and PCB similar.

Ex. 7. If PA and PB are chords in a circle, and CD is a line parallel to the tangent at P and meeting PA and PB at C and D, the triangles PAB and PCD are similar.

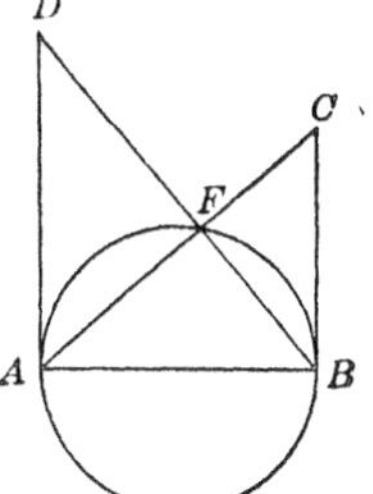

Ex. 8. Given AB a diameter and AD and BC tangents, AC and DB intersecting at any point F on the circumference; prove AB a mean proportional between the sides AD and BC.

Ex. 9. In any isosceles triangle, the square of one of the legs equals the square on a line drawn from the vertex to any point of the base plus the product of the segments of the base.

Ex. 10. A line drawn through the intersection of the diagonals of a trapezoid parallel to the bases and terminated by the legs is bisected by the diagonals.

[Sug. See Ex. 10, p. 222.]

Ex. 11. If a chord is bisected by another chord, each segment of the first chord is a mean proportional between the segments of the second chord.

Ex. 12. In a parallelogram the sum of the squares of the sides equals the sum of the squares of the diagonals.

Ex. 13. If two circles are tangent externally, and a line is drawn through the point of contact terminated by the circumferences, the chords intercepted in the two circles are to each other as the radii.

Ex. 14. Three times the sum of the squares of the sides of a triangle equals four times the sum of the squares of the medians.

Ex. 15. Find the locus of the midpoints of lines in a triangle parallel to the base and terminated by the sides.

Ex. 16. Given AB the diameter, AP, PQR, BR, tangents; prove $PQ \times QR$ a constant (= radius squared).

Ex. 17. O is the center of a circle and A is any point within the circle; OA is produced to B, so that $OA \times OB$ equals the radius squared. If P is any point in the circumference, the angles OPA and OBP are equal.

[SUG. Use Art. 327.]

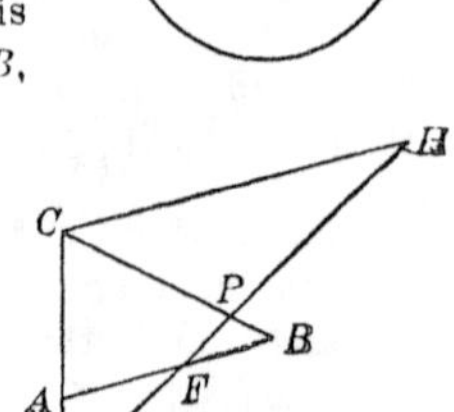

Ex. 18. Given $AF = FB$, and $CH \parallel AB$; prove $HP : FP = HK : FK$.

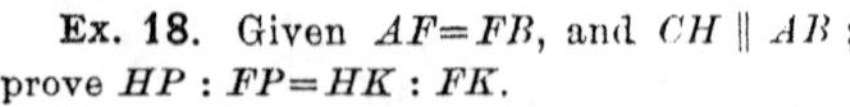

[SUG. $HP : FP = CH : FB$, etc.]

Ex. 19. If from any point P within the triangle ABC the perpendiculars PQ, PR, PT are drawn to the sides AB, AC, BC, respectively, then $\overline{AQ}^2 + \overline{BT}^2 + \overline{RC}^2 = \overline{AR}^2 + \overline{BQ}^2 + \overline{CT}^2$.

EXERCISES. GROUP 36

PROBLEMS

Given three lines a, b, c,

Ex. 1. Construct $x = \frac{ab}{c}$; also $x = \frac{ab}{2c}$.

Ex. 2. Construct $x = \sqrt{a^2 - b^2}$, i. e., $\sqrt{(a+b)(a-b)}$.

Ex. 3. Construct $x = \sqrt{3ab}$, i. e., $\sqrt{(3a)b}$.

Ex. 4. Construct $x = \sqrt{a^2 - bc}$, i. e., $\sqrt{a^2 - (\sqrt{bc})^2}$.

Ex. 5. Given a line denoted by 1, construct $\sqrt{3}$; also $\frac{1}{2}\sqrt{5}$.

Ex. 6. Divide a line into three parts proportional to 2, $\frac{1}{3}$, $\frac{3}{5}$.

Ex. 7. Divide a line harmonically in the ratio 3 : 5.

Ex. 8. Divide one side of a triangle into segments proportional to the other two sides.

Ex. 9. Divide a line into segments in the ratio $1 : \sqrt{2}$.

Ex. 10. Given a point P in the side AB of a triangle ABC, draw a line from P to AC produced so that the line drawn may be bisected by BC.

[SUG. Suppose the required line, PQR, drawn meeting BC in Q and AC in R. From P draw $PL \parallel AC$. Compare the ▵ PLQ and QRC.]

Ex. 11. Through a given point P in the arc subtended by the chord AB draw a chord which shall be bisected by AB.

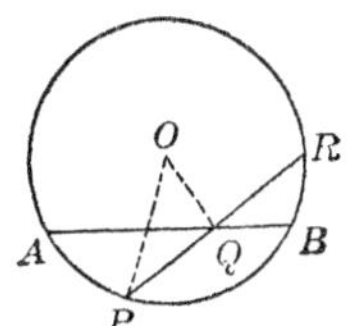

[SUG. Suppose the required chord drawn, viz., PQR. Join the center O with P and Q. What kind of an angle is OQP, etc.?]

Ex. 12. In an obtuse triangle draw a line from the vertex of the obtuse angle to the opposite side which shall be a mean proportional between the segments of the opposite side.

[SUG. Circumscribe a circle about the triangle and reduce the problem to the preceding Ex.]

Ex. 13. Find a point P in the arc subtended by the chord AB such that chord PA : chord $PB = 2 : 3$.

[SUG. Suppose the required construction made, and also the chord AB divided in the ratio $2 : 3$ at the point Q. How do the angles APQ and QPB compare?]

Ex. 14. Given the perimeter, construct a triangle similar to a given triangle.

Ex. 15. Given the altitude of a triangle, construct a triangle similar to a given triangle.

Ex. 16. In a given circle inscribe a triangle similar to a given triangle.

Ex. 17. About a given circle circumscribe a triangle similar to a given triangle.

Ex. 18. By drawing a line parallel to one of the sides of a given rectangle, divide the rectangle into two similar rectangles.

Ex. 19. Inscribe a square in a given triangle.

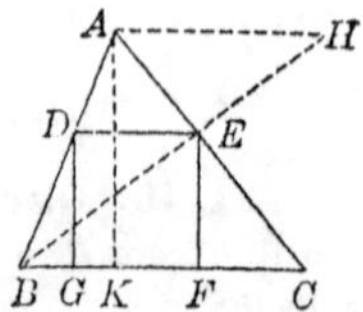

[SUG. If ABC is the given triangle, suppose $DGFE$ the required inscribed square. Join BE and produce it to meet $AH \parallel BC$. Prove $AH = AK$, etc.]

374. The method of similars in solving geometrical problems is best shown by the aid of an example.

Ex. In the side BC of a triangle ABC find a point D such that the perpendiculars from it to the other sides shall be in the ratio 3 : 1.

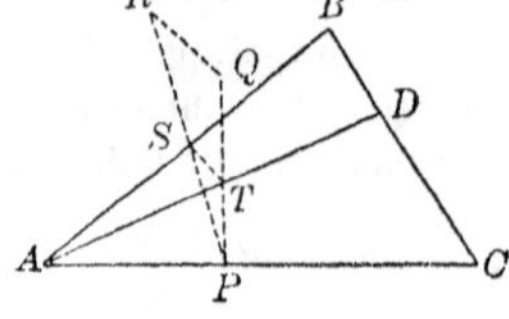

CONSTRUCTION. At any point P in AC erect a $\perp$ PQ of any convenient length. In a direction $\perp$ AB draw $RQ = \frac{1}{3}$ QP. Join RP. From S draw $ST \parallel RQ$. Produce AT to D. Then D is the required point.

Let the pupil supply the proof.

EXERCISES. GROUP 37

PROBLEMS SOLVED BY METHOD OF SIMILARS

Ex. 1. In one side of a triangle find a point such that the perpendiculars from it to the other two sides shall be in the ratio $m : n$.

Ex. 2. Find a point the perpendiculars from which to the three sides of a given triangle shall be in a given ratio.

[SUG. Use Ex. 1 twice.]

Ex. 3. Construct a circle which shall touch two given lines and pass through a given point.

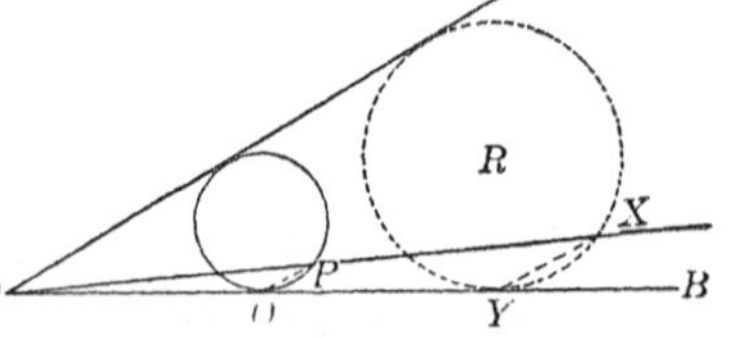

[SUG. Let OA and OB be the given lines and P the given point. Draw any $\odot R$ touching the two lines (OB at Y) and intersecting OP produced at X. Draw the chord XY, etc.]

Ex. 4 Inscribe a square in a given semi-circle.

[SUG. Circumscribe a semi-circle about any given square, by taking the midpoint of the base of the square as a center, and the line from this midpoint to a non-adjacent vertex as a radius, etc.]

Ex. 5. Solve Ex. 19, p. 228, by the method of similars.

375. Algebraic analysis of problems. *The conditions of a problem may often be stated as an algebraic equation; by solving the equation, the length of a desired line in terms of known lines may then be obtained, and the problem solved by constructing the algebraic expression thus obtained.*

Ex. Find a point P in the line AB such that $\overline{AP}^2 = 3\overline{BP}^2$.

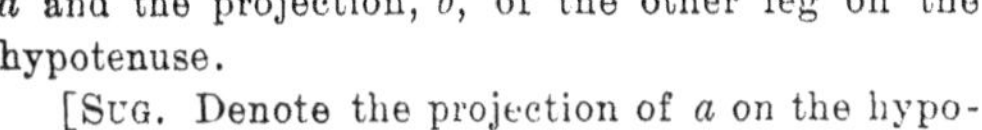

ANALYSIS AND CONSTRUCTION. Denote AB by a, AP by x, and PB by $a - x$. Then $x^2 = 3(a - x)^2$.

$\therefore 2x^2 - 6ax = -3a^2$, and $x = \dfrac{3a \pm a\sqrt{3}}{2}$.

Construct $a\sqrt{3}$, whence construct $\dfrac{3a - a\sqrt{3}}{2}$; lay off the line obtained, as AP, on AB; this gives the point P of internal division. Similarly, the construction of $\dfrac{3a + a\sqrt{3}}{2}$ gives P', the point of external division.

EXERCISES. GROUP 38

PROBLEMS SOLVED BY ALGEBRAIC ANALYSIS

Ex. 1. Find a point P in a given line AB such that $\overline{AP}^2 = 2\overline{BP}^2$.

Ex. 2. Construct a right triangle, given one leg a and the projection, b, of the other leg on the hypotenuse.

[SUG. Denote the projection of a on the hypotenuse by x. Then $a^2 = x(x + b)$, etc.]

Ex. 3. Inscribe a square in a given semicircle.

Ex. 4. From a given line cut off a part which shall be a mean proportional between the remainder of the line and another given line.

Ex. 5. Given AC and CB arcs of $90°$, and a a given line. Draw the chord CQ intersecting AB in P so that $PQ = a$.

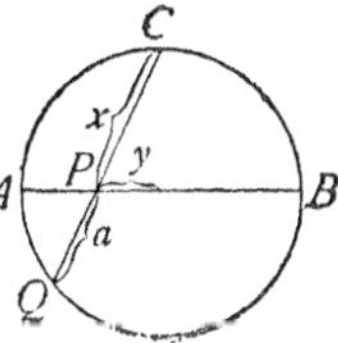

[SUG. $x^2 - y^2 = r^2$. $ax = (r + y)(r - y)$, etc.]

Ex. 6. Given the greater segment of a line divid[ed i]n extreme and mean ratio, construct the line.

EXERCISES. GROUP 39

PROBLEMS SOLVED BY VARIOUS METHODS

Ex. 1. Construct two lines, given their sum (a line AB) and their ratio ($m : n$).

Ex. 2. Construct two lines, given their difference and their ratio.

Ex. 3. Divide a trapezoid into two similar trapezoids by drawing a line parallel to the bases.

[SUG. Conceive the figure drawn, and compare the ratio of the bases in the two trapezoids formed.]

Ex. 4. Construct a mean proportional between two given lines by use of Art. 358.

Ex. 5. Construct a circle which shall pass through two given points and touch a given line.

Ex. 6. From a given point draw a secant to a circle so that the external segment shall equal half the secant.

[SUG. Draw a tangent to the ⊙ and use the algebraic method.]

Ex. 7. From a given external point P, draw a secant meeting a circle in A and B so that $PA : AB = m : n$.

[SUG. Draw a tangent to the circle from the point P and denote its length by t. Denote PA by mx and AB by nx. Then $m(m+n)x^2 = t^2$, or $t : mx = mx : \frac{mt}{m+n}$, etc.]

Ex. 8. Through a given point P draw a straight line so that the parts of it, included between that point and perpendiculars drawn to the line from two other given points, shall be in a given ratio.

[SUG. Join the last two points, and divide the line between them in the given ratio.]

Ex. 9. Construct a straight line so that the perpendiculars on it from three given points shall be in a given ratio.

[SUG. Let P, Q, R, be the given points and $m : n : p$ the given ratio. Divide PQ in the ratio $m : n$ and QR in the ratio $n : p$, etc.]

Ex. 10. Upon a given line as hypotenuse construct a right triangle one leg of which shall be a mean proportional betwe- the other leg and the hypotenuse.

Book IV

AREAS OF POLYGONS

376. **A unit of surface** is a square whose side is a unit of length, as a square inch, a square yard, or a square centimeter.

377. The **area of a surface** is the number of units of surface which the given surface contains.

It is important for the student to grasp firmly the fact that *area* means not mere vague largeness of surface, but that it is a *number*. Being a number, it can be resolved into factors, it may be determined as a product of simpler numbers, and handled with ease and precision in various ways.

378. **Equivalent plane figures** are plane figures having equal areas.

Thus two triangles may have equal areas (be equivalent) and yet not be of the same shape, that is, not be equal (congruent).

379. **Abbreviations.** Instead of "area of a rectangle," for example, it is often convenient to say simply "rectangle." So instead of "the number of linear units in the base," we use simply "the base." In like manner, for "product of the number of linear units in the base by the number of linear units in the altitude," a common abbreviation is "product of the base by the altitude."

COMPARISON OF RECTANGLES

PROPOSITION I. THEOREM

380. *If two rectangles have the same altitude, they are to each other as their bases.*

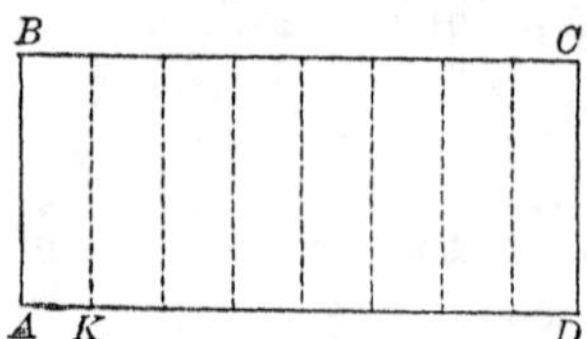

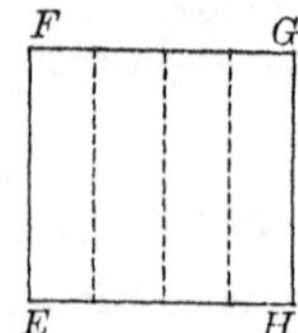

Given the rectangles $EFGH$ and $ABCD$, having their altitudes EF and AB equal.

To prove $EFGH : ABCD = EH : AD.$

CASE I. *When the bases are commensurable.*

Proof. Take some common measure of EH and AD, as AK, and let it be contained in EH n times and in AD m times.

Hence $EH : AD = n : m.$ (Why?)

Through the points of division of the bases of the two rectangles draw lines perpendicular to the bases.

These lines will divide EG into n, and AC into m small rectangles, all equal. Art. 163.

Hence $EFGH : ABCD = n : m.$ (Why?)

$\therefore EFGH : ABCD = EH : AD.$ (Why?)

CASE II. *When the bases are incommensurable.*

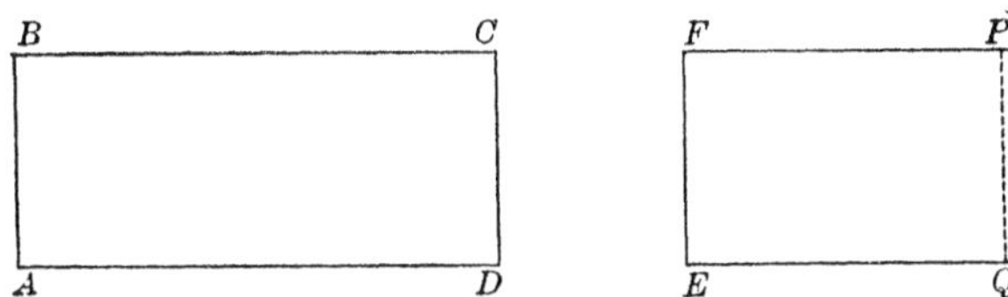

Proof. Divide the base AD into any number of equal parts, and apply one of these parts to EH.

It will be contained in EH a certain number of times with a remainder QH, less than the unit of measure.

Draw $QP \perp EH$, meeting FG at P.

Then EQ and AD are commensurable. Constr.

$$\therefore \frac{EFPQ}{ABCD} = \frac{EQ}{AD}. \qquad \text{Case I.}$$

If now the unit of measure be indefinitely diminished, the line QH, which is less than the unit of measure, will be indefinitely diminished.

$\therefore EQ \doteq EH$ as a limit; $EFPQ \doteq EFGH$ as a limit. Art. 251.

Hence $\frac{EFPQ}{ABCD}$ becomes a variable with $\frac{EFGH}{ABCD}$ as its limit; also $\frac{EQ}{AD}$ becomes a variable with $\frac{EH}{AD}$ as its limit. Art. 253, 3.

But the variable $\frac{EFPQ}{ABCD}$ = the variable $\frac{EQ}{AD}$ always. Case I.

$$\therefore \text{the limit } \frac{EFGH}{ABCD} = \text{the limit } \frac{EH}{AD}. \qquad \text{(Why?)}$$

Q. E. D.

381. COR. *If two rectangles have equal bases, they are to each other as their altitudes.*

PROPOSITION II. THEOREM

382. *The areas of any two rectangles are to each other as the products of their bases by their altitudes.*

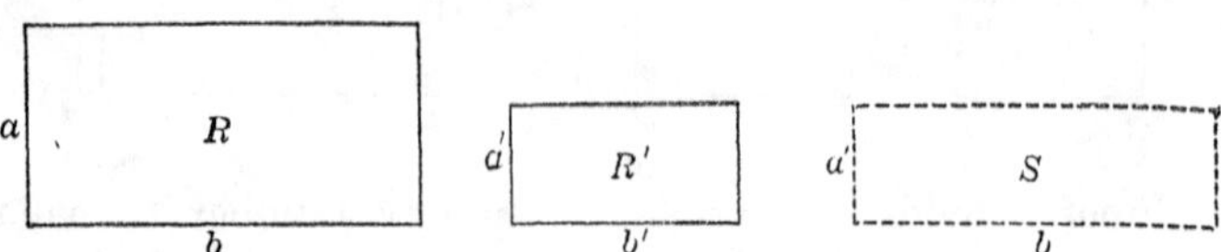

Given the rectangles R and R', having the bases b and b', and the altitudes a and a', respectively.

To prove $$\frac{R}{R'}=\frac{b \times a}{b' \times a'}.$$

Proof. Construct a rectangle, S, having its base equal to that of R, and its altitude equal to that of R'.

Then $$\frac{R}{S}=\frac{a}{a'}.$$ Art. 381.

Also $$\frac{S}{R'}=\frac{b}{b'}.$$ Art. 380.

Taking the product of the corresponding members of the two equalities,

$$\frac{R}{R'}=\frac{b \times a}{b' \times a'}.$$ Ax. 4.

Q. E. D.

Ex. 1. Find the ratio of the area of a rectangle whose dimensions are 12×8 in. to that of one whose dimensions are 9×2 in.

Ex. 2. How many bricks, each 8×5 in., will it take to cover a pavement 60×9 ft.?

AREAS OF POLYGONS

PROPOSITION III. THEOREM

383. *The area of a rectangle is equal to the product of its base by its altitude.*

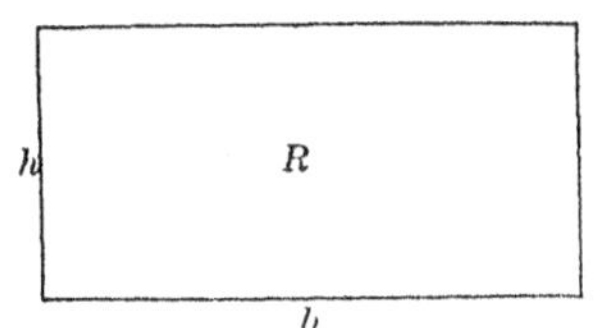

Given the rectangle R, with a base containing b, and an altitude containing h units of linear measure.

To prove area of $R = b \times h$.

Proof. Let U be a square each side of which contains 1 unit of linear measure.

Then U is the unit of surface. Art. 376.

$$\therefore \frac{R}{U} = \frac{b \times h}{1 \times 1} = b \times h.$$ Art. 382.

But $\frac{R}{U}$ is the area of R, Art. 377.

(*by definition of area*).

$$\therefore \text{ area of } R = b \times h.$$

Q. E. D.

384. NOTE. By use of this theorem, the problem of finding the area of a rectangle is reduced to the simpler problem of measuring the two linear dimensions of the rectangle and taking their product. (See Art. 1.)

Ex. 1. Find, in square feet, the area of a rectangle 8 yds. long and 5 ft. wide.

Ex. 2. The area of a rectangle is 60 sq. ft. and its altitude is 5 ft. Find the base.

PROPOSITION IV. THEOREM

385. *The area of a parallelogram is equal to the product of its base by its altitude.*

Given the ▱ $ABCD$ with the base AD (denoted by b) and the altitude DF (denoted by h).

To prove area of $ABCD = b \times h$.

Proof. From A draw $AK \parallel DF$, and meeting CB produced, at K.

Then $AK \perp CK$. Art. 123.

$\therefore AKFD$ is a rectangle with base b and altitude h. (Why?)

In the rt. ⊿s AKB and DFC,

$AB = DC$. (Why?)

$AK = DF$. (Why?)

$\therefore \triangle AKB = \triangle DFC$. (Why?)

To each of these equals add the figure $ABFD$;

Then rectangle $AKFD$ ≎ ▱ $ABCD$. Ax. 2.

But area of the rectangle $AKFD = b \times h$. Art. 383.

$\therefore$ area ▱ $ABCD = b \times h$. (Why?)

Q. E. D.

386. COR. 1. *Parallelograms which have equal bases and equal altitudes are equivalent.*

387. COR. 2. *Parallelograms which have equal bases are to each other as their altitudes;*

Parallelograms which have equal altitudes are to each other as their bases.

388. COR. 3. *Any two parallelograms are to each other as the products of their bases and altitudes.*

PROPOSITION V. THEOREM

389. *The area of a triangle is equal to one-half the product of its base by its altitude.*

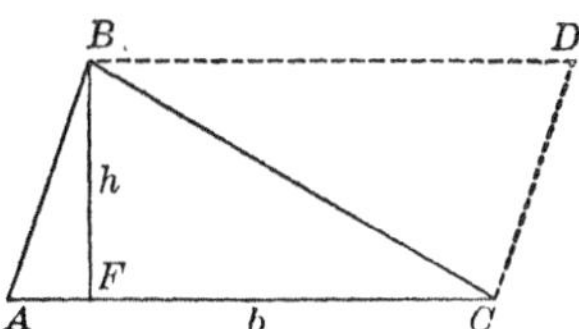

Given the $\triangle ABC$ with the base AC (denoted by b), and the altitude FB (denoted by h).

To prove area of $\triangle ABC = \frac{1}{2} b \times h$.

Proof. Draw $BD \parallel AC$, and $CD \parallel AB$, forming the ▱ $ABDC$.

Then BC is a diagonal of ▱ $ABDC$.

$\therefore \triangle ABC = \frac{1}{2}$ ▱ $ABDC$. Art. 156.

But area ▱ $ABDC = b \times h$. (Why?)

$\therefore$ area $\triangle ABC = \frac{1}{2} b \times h$. Ax. 5.

Q. E. D.

390. COR. 1. *Triangles which have equal bases and equal altitudes (or which have equal bases and their vertices in a line parallel to the base) are equivalent.*

391. COR. 2. *Triangles which have equal bases are to each other as their altitudes;*

Triangles which have equal altitudes are to each other as their bases.

392. COR. 3. *Any two triangles are to each other as the products of their bases and altitudes.*

Ex. 1. Find the area of a parallelogram whose base is 9 ft. 8 in. and whose altitude is 2 ft. 3 in.

Ex. 2. Find the altitude of a triangle whose area is 180 sq. in. and whose base is 1 ft. 3 in.

PROPOSITION VI. THEOREM

393. *If a, b, c denote the sides of a triangle opposite the angles A, B, C, respectively, and $s=\frac{1}{2}(a+b+c)$, the area of the triangle $=\sqrt{s(s-a)(s-b)(s-c)}$.*

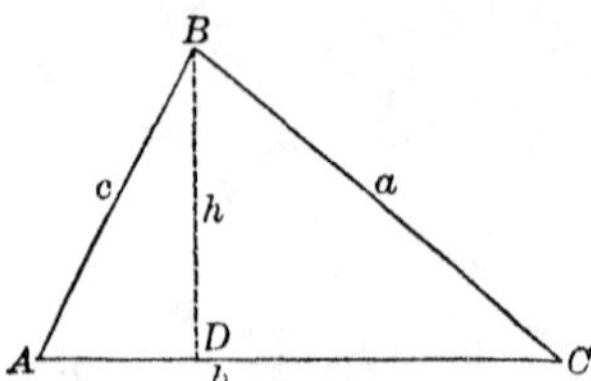

Given the $\triangle ABC$ with the sides opposite $\angle A$, B and C, denoted by a, b and c, respectively, $\frac{1}{2}(a+b+c)$ denoted by s, and A an acute angle.

To prove area $\triangle ABC=\sqrt{s(s-a)(s-b)(s-c)}$.

Proof. Draw the altitude BD and denote BD by h.

Then $a^2=b^2+c^2-2b\times AD$. Art. 349.

$\therefore 2b\times AD=b^2+c^2-a^2$. Axs. 2, 3.

$\therefore AD=\frac{b^2+c^2-a^2}{2b}$. Ax. 4.

But $h^2=c^2-AD^2=(c+AD)(c-AD)$ Art. 347.

$$=\left(c+\frac{b^2+c^2-a^2}{2b}\right)\left(c-\frac{b^2+c^2-a^2}{2b}\right) \quad \text{Ax. 8.}$$

$$=\left(\frac{2bc+b^2+c^2-a^2}{2b}\right)\left(\frac{2bc-b^2-c^2+a^2}{2b}\right)$$

$$=\frac{[(b+c)^2-a^2][a^2-(b-c)^2]}{4b^2}$$

$$=\frac{(b+c+a)(b+c-a)(a+b-c)(a-b+c)}{4b^2}.$$

Now $a+b+c=2s$ $\therefore a+b-c=2s-2c$, etc.

Hyp., Axs. 4, 3.

$$\therefore h^2 = \frac{2s(2s-2a)(2s-2c)(2s-2b)}{4b^2} = \frac{16s(s-a)(s-b)(s-c)}{4b^2}$$

$$\therefore h = \frac{2\sqrt{s(s-a)(s-b)(s-c)}}{b}.$$

But area $\triangle ABC = \frac{1}{2} b \times h$. Art. 390.

$\therefore$ area $\triangle ABC = \sqrt{s(s-a)(s-b)(s-c)}$. Ax. 8.

Q. E. D.

Proposition VII. Theorem

394. *The area of a trapezoid is equal to one-half the sum of its bases multiplied by its altitude.*

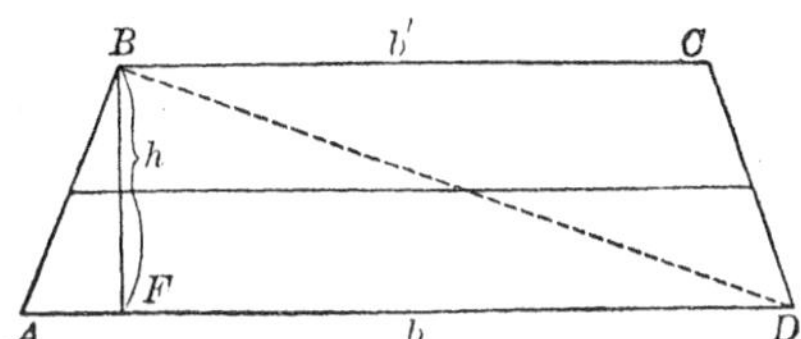

Given the trapezoid $ABCD$ with the bases AD and BC (denoted by b and b'), and the altitude FB (denoted by h).

To prove area of $ABCD = \frac{1}{2}(b + b') \times h$.

Proof. Draw the diagonal BD.

Then area of $\triangle ABD = \frac{1}{2} b \times h$. (Why?)

And area of $\triangle BCD = \frac{1}{2} b' \times h$. (Why?)

Adding, area of $ABCD = \frac{1}{2}(b' + b) h$. (Why?)

Q. E. D.

395. Cor. *The area of a trapezoid equals the product of the median of the trapezoid by the altitude.* For the median of a trapezoid equals one-half the sum of the bases (Art. 179).

396. SCHOLIUM. The *area of a polygon of four or more sides can usually be found in one of several ways;* as, *by dividing the polygon into triangles and taking the sum of the areas of the triangles;* or, by *drawing the longest diagonal of the polygon and drawing perpendiculars to this diagonal from the vertices which it does not meet, and obtaining the sum of the areas of the triangles and trapezoids thus formed.*

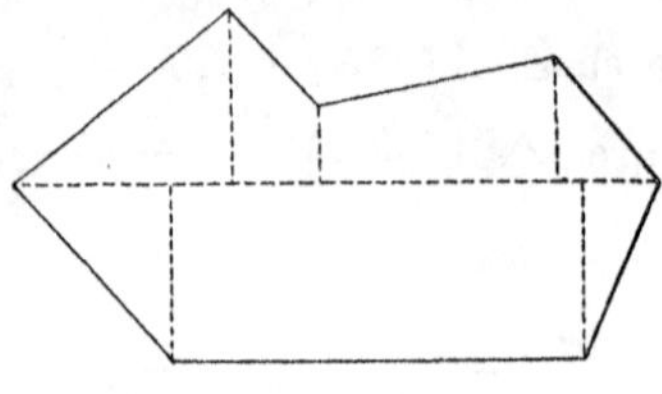

Ex. 1. Find the area of a parallelogram whose base is 1 yd., and whose altitude is 1 ft.

Ex. 2. Find the area of a triangle whose sides are 5, 6, and 7 in.

Ex. 3. Find the area of a trapezoid whose bases are 18 and 10 in., and whose altitude is 6 in.

Ex. 4. If the area of a trapezoid is 135, and its bases are 12 and 18, find its altitude.

Ex. 5. The measurement of the area of a parallelogram reduces to the measurement of what two straight lines?

Ex. 6. The measurement of the area of a triangle reduces to the measurement of what lines?

Ex. 7. The measurement of the area of a trapezoid reduces to the measurement of what lines?

Ex. 8. Prove the theorem of Art. 385 by drawing perpendiculars from B and C, instead of from A and D.

COMPARISON OF POLYGONS

Proposition VIII. Theorem

397. *If two triangles have an angle of one equal to an angle of the other, their areas are to each other as the products of the sides including the equal angles.*

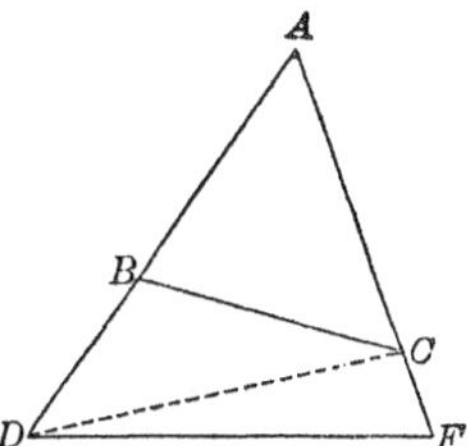

Given the $\triangle$ ABC and ADF having $\angle A$ in common.

To prove $$\frac{\triangle ABC}{\triangle ADF}=\frac{AB\times AC}{AD\times AF}.$$

Proof. Draw the line DC.

Then the $\triangle$ ABC and ADC may be regarded as having their bases in the line AD, and as having the common vertex C.

$$\therefore \frac{\triangle ABC}{\triangle ADC}=\frac{AB}{AD}. \qquad \text{Art. 391.}$$

In like manner $$\frac{\triangle ADC}{\triangle ADF}=\frac{AC}{AF}. \qquad \text{(Why?)}$$

Multiplying the corresponding members of these equalities,

$$\frac{\triangle ABC}{\triangle ADF}=\frac{AB\times AC}{AD\times AF}. \qquad \text{Ax. 4.}$$

Q. E. D.

Ex. In the above figure, if $AB=12$, $AC=18$, $AD=30$, and $AF=32$, find the ratio of the areas of the $\triangle$ ABC and ADF.

PROPOSITION IX. THEOREM

398. *The areas of two similar triangles are to each other as the squares of any two homologous sides.*

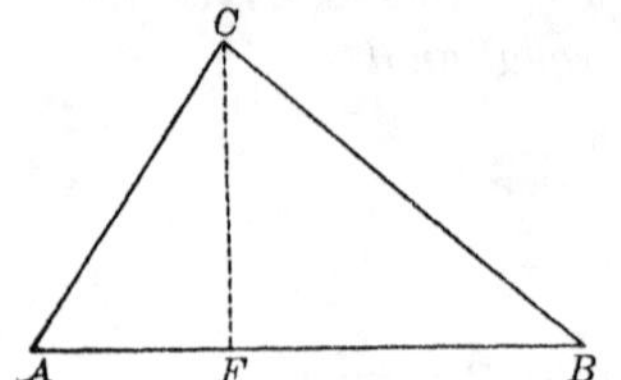

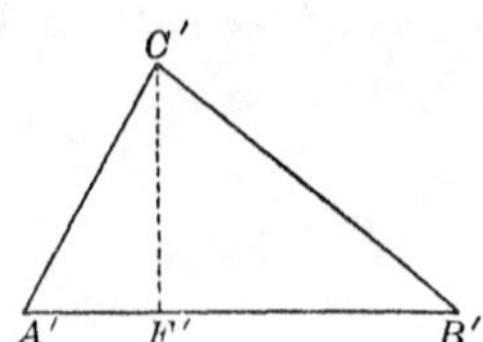

Given the similar $\triangle$ ABC and $A'B'C'$ with AB and $A'B'$ homologous sides.

To prove $$\frac{\triangle ABC}{\triangle A'B'C'}=\frac{\overline{AB}^2}{\overline{A'B'}^2}.$$

Proof. Draw the homologous altitudes CF and $C'F'$.

Then $$\frac{\triangle ABC}{\triangle A'B'C'}=\frac{AB\times CF}{A'B'\times C'F'}=\frac{AB}{A'B'}\times\frac{CF}{C'F'},$$ Art. 392.

(*any two* $\triangle$ *are to each other as the products of their bases and altitudes*).

But $$\frac{CF}{C'F'}=\frac{AB}{A'B'}.$$ Art. 338.

Substituting $\frac{AB}{A'B'}$ for its equal $\frac{CF}{C'F'}$, Ax. 8.

$$\frac{\triangle ABC}{\triangle A'B'C'}=\frac{AB}{A'B'}\times\frac{AB}{A'B'}=\frac{\overline{AB}^2}{\overline{A'B'}^2}.$$

Q. E. D.

Ex. 1. If a pair of homologous sides of two similar triangles are 4 ft. and 5 ft., find the ratio of the areas of the triangles.

Ex. 2. Prove Prop. IX by use of Prop. VIII.

Proposition X. Theorem

399. *The areas of two similar polygons are to each other as the squares of any two homologous sides.*

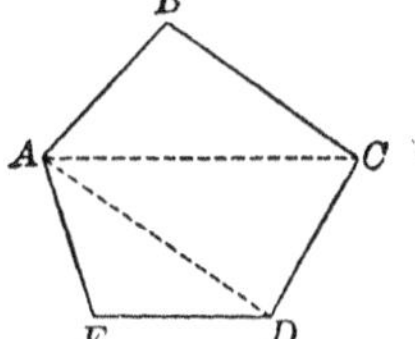

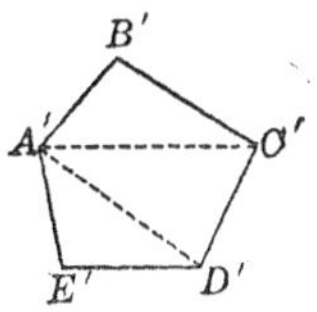

Given the similar polygons $ABCDE$ and $A'B'C'D'E'$, with their areas denoted by S and S', respectively, and with AB and $A'B'$ any pair of homologous sides.

To prove $S : S' = \overline{AB}^2 : \overline{A'B'}^2$.

Proof. Draw the diagonals AC, AD and $A'C'$, $A'D'$ from the homologous vertices A and A'.

These diagonals will divide the polygons into similar $\triangle$s. Art. 329.

$$\therefore \frac{\triangle ABC}{\triangle A'B'C'} = \frac{\overline{AB}^2}{\overline{A'B'}^2}. \qquad \text{Art. 398.}$$

$$\text{But } \frac{\triangle ABC}{\triangle A'B'C'} = \left(\frac{\overline{AC}^2}{\overline{A'C'}^2}\right) = \frac{\triangle ACD}{\triangle A'C'D'} = \left(\frac{\overline{AD}^2}{\overline{A'D'}^2}\right) = \frac{\triangle ADE}{\triangle A'D'E'}. \qquad \text{(Why?)}$$

$$\therefore \frac{\triangle ABC}{\triangle A'B'C'} = \frac{\triangle ACD}{\triangle A'C'D'} = \frac{\triangle ADE}{\triangle A'D'E'}. \qquad \text{(Why?)}$$

$$\therefore \frac{\triangle ABC + \triangle ACD + \triangle ADE}{\triangle A'B'C' + \triangle A'C'D' + \triangle A'D'E'} = \frac{\triangle ABC}{\triangle A'B'C'}. \qquad \text{Art. 312.}$$

$$\therefore \frac{S}{S'} = \frac{\triangle ABC}{\triangle A'B'C'}. \qquad \text{Ax. 6.}$$

$$\therefore \frac{S}{S'} = \frac{\overline{AB}^2}{\overline{A'B'}^2}. \qquad \text{Ax. 1.}$$

Q. E. D.

Ex. If a pair of homologous sides of two similar polygons are 1 and 2 ft., find the ratio of the areas of the polygons.

PROPOSITION XI. THEOREM

400. *In a right triangle, the square on the hypotenuse is equivalent to the sum of the squares on the two legs.*

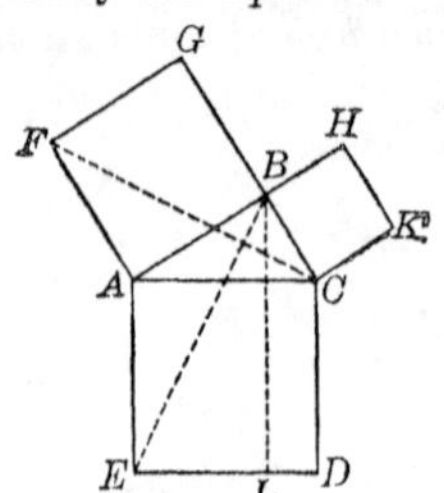

Given AD the square on AC the hypotenuse of the rt. $\triangle ABC$, and BF and BK the squares on the legs BA and BC respectively.

To prove $AD \approx BF + BK$.

Proof. Through B draw $BL \parallel AE$, and meeting ED in L. Draw BE and FC.

Then ∡ ABC and ABG are rt. ∡. (Why?)

$\therefore GBC$ is a straight line. (Why?)

In the ▵ BAE and FAC, $AB = AF$, and $AE = AC$. (Why?)

Also $\angle BAE = \angle FAC$, Ax. 2.

(*for each* $= \angle BAC + a$ *rt.* $\angle$).

$\therefore \triangle BAE = \triangle FAC$. (Why?)

But rectangle $AL \approx 2 \triangle BAE$.

(*for AL has the same base, AE, and the same altitude, EL, as $\triangle BAE$*).

Also square $BF \approx 2 \triangle FAC$. (Why?)

$\therefore$ rectangle $AL \approx$ square BF. Ax. 1.

In like manner rectangle $CL \approx$ square BK.

Adding, $AL + CL$, or $AD \approx BF + BK$. Ax. 2.

Q. E. D.

401. COR. *The square on either leg of a right triangle is equivalent to the square on the hypotenuse diminished by the square on the other leg.*

CONSTRUCTION PROBLEMS

PROPOSITION XII. PROBLEM

402. *To construct a square equivalent to the sum of two given squares.*

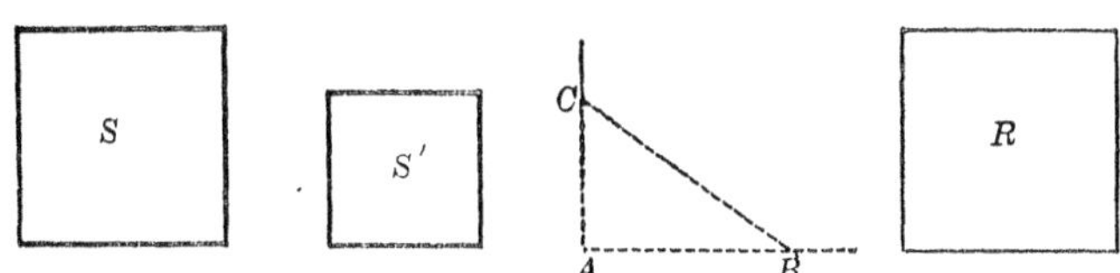

Given two squares S and S'.

To construct a square equivalent to $S + S'$.

Construction. Construct a right angle BAC. Art. 274.

On one side of this angle take AB equal to a side of S, and on the other side take AC equal to a side of S'.

Draw BC.

On a line equal to BC construct the square R.

Then R is the square required.

Proof. $R = \overline{BC}^2 \approx \overline{AB}^2 + \overline{AC}^2$. Art. 400.

$\therefore R \approx S + S'$ Ax. 8.

Q. E. F.

403. COR. *To construct a square equivalent to the sum of three or more given squares.* At C in the above figure erect a line $CD \perp BC$ (Art. 274), and equal to a side of the third given square. Draw DB. DB will be a side of a square equivalent to the sum of three given squares, etc.

Ex. 1. Construct a square equivalent to the sum of two squares whose sides are $\frac{1}{2}$ in. and 1 in., respectively.

Ex. 2. By use of Art. 403, taking a given line as unity, construct $\sqrt{2}$; also $\sqrt{3}$. For example, construct a line $\sqrt{2}$ inches long; also one $\sqrt{3}$ inches long.

Proposition XIII. Problem

404. *To construct a square equivalent to the difference of two given squares.*

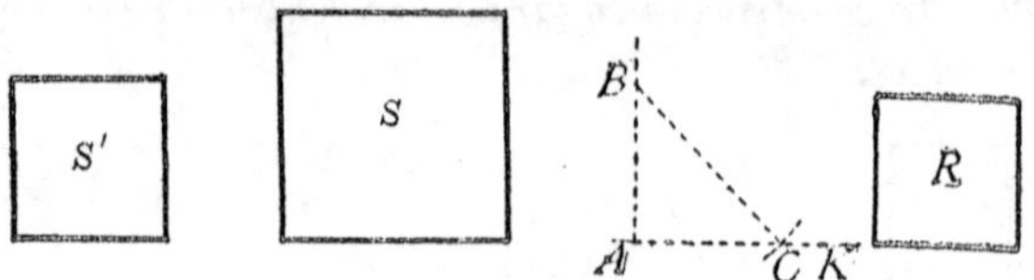

Given the squares S and S'.

To construct a square equivalent to the difference of S and S'.

Construction. Construct a right angle BAK. Art. 274.

On one side AB take AB equal to a side of the smaller given square S'.

From B as a center with a radius BC, equal to a side of the larger square, describe an arc intersecting AK in C.

On a line equal to AC construct the square R.

Then R is the square required.

Proof. $R = \overline{AC}^2 \simeq \overline{BC}^2 - \overline{AB}^2$, Art. 401.

(*the square on either leg of a right triangle is equivalent to the square on the hypotenuse diminished by the square on the other leg*).

Hence $R \simeq S - S'$. Ax. 8.

Q. E. F.

Ex. Construct a square equivalent to the difference of two squares whose sides are 1 in. and ½ in., respectively.

PROPOSITION XIV. PROBLEM

405. *To construct a square equivalent to a given parallelogram.*

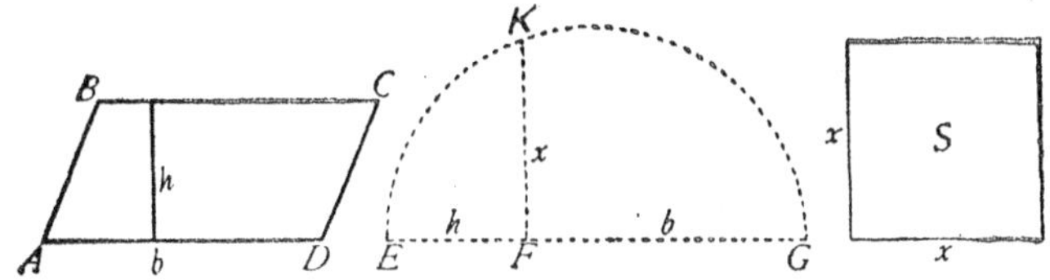

Given the ▱ $ABCD$ with base b and altitude h.

To construct a square equivalent to $ABCD$.

Construction. On the line EG take EF equal to h and FG equal to b.

On EG as a diameter construct a semicircle. Art. 275, Post. 3.

At F erect a ⊥ meeting the semicircumference at K. Art. 274.

On a line equal to FK construct the square S.

Then S is the required square.

Proof. $S = \overline{KF}^2$.

But $\overline{KF}^2 = b \times h$, Art. 343.

(*a ⊥ from any point in a circumference to a diameter is a mean proportional between the segments of the diameter*).

But area ▱ $ABCD = b \times h$. (Why?)

$\therefore S \approx$ area ▱ $ABCD$.

Q. E. F.

406. COR. *To construct a square equivalent to a given triangle*, construct the mean proportional between the base and half the altitude of the triangle and construct a square on this mean proportional.

PROPOSITION XV. PROBLEM

407. *To construct a triangle equivalent to a given polygon.*

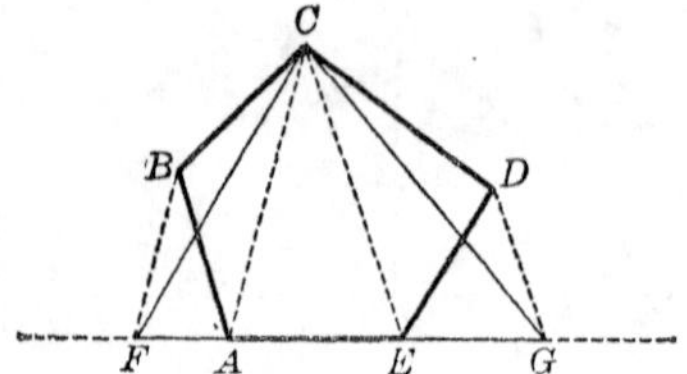

Given the polygon $ABCDE$.

To construct a triangle equivalent to $ABCDE$.

Construction. Let A, B, C be any three consecutive vertices in the given polygon.

Draw the diagonal AC.

Draw $BF \parallel AC$ (Art. 279), and meeting AE produced at F.

Draw FC.

In the polygon $FCDE$ take the three consecutive vertices C, D, E, and draw the diagonal CE.

Draw $DG \parallel CE$, and meeting AE produced at G. Draw CG.

Then $\triangle FCG$ is the triangle required.

Proof. $\triangle ABC \approx \triangle AFC$, Art. 390.

(*having the same base AC, and their vertices in a line $BF \parallel$ the base*).

Also $\triangle ACE = \triangle ACE$. Ident.

And $\triangle ECD \approx \triangle ECG$, Art. 390.

(*having the same base CE and their vertices in line $DG \parallel$ base*).

Adding, $\triangle ABC + \triangle ACE + \triangle ECD \approx \triangle AFC + \triangle ACE + \triangle ECG$. Ax. 2.

Or polygon $ABCDE \approx \triangle FCG$.

Q. E. F.

408. COR. *To construct a square equivalent to a given polygon*, use Arts. 407 and 406.

Ex. Construct a triangle equivalent to a given hexagon.

Proposition XVI. Problem

409. *To construct a rectangle equivalent to a given square, and having the sum of its base and altitude equal to a given line.*

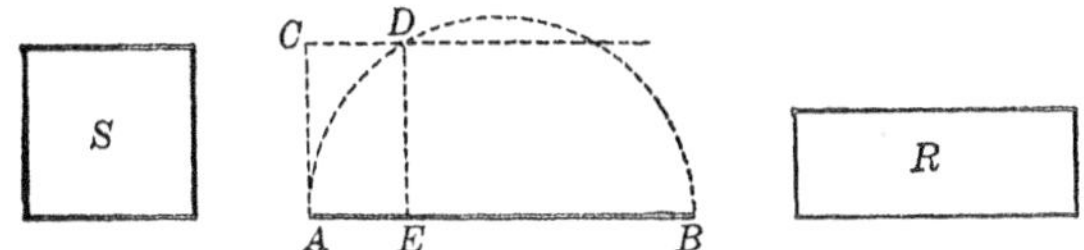

Given the square S and the line AB.

To construct a rectangle equivalent to S, and having the sum of its base and altitude equal to AB.

Construction. On AB as a diameter describe the semicircumference ADB. Art. 275, Post. 3.

At the point A erect a $\perp$, AC, equal to a side of S. Art. 274.

Through C draw a line $\parallel$ AB (Art. 279), and meeting the circumference at D.

Draw $DE \perp AB$. Art. 273.

Construct the rectangle R with a base equal to EB and an altitude equal to AE.

Then R is the rectangle required.

Proof. $\overline{DE}^2 = AE \times EB$. (Why ?)

But $DE = CA$. (Why ?)

$\therefore \overline{CA}^2 = AE \times EB$. (Why ?)

Or $S \approx R$.

Q. E. F.

410. Cor. The above problem is equivalent to the problem: *Given the sum and product of two lines, to construct the lines.*

PROPOSITION XVII. PROBLEM

411. *To construct a rectangle equivalent to a given square, and having the difference of its base and altitude equal to a given line.*

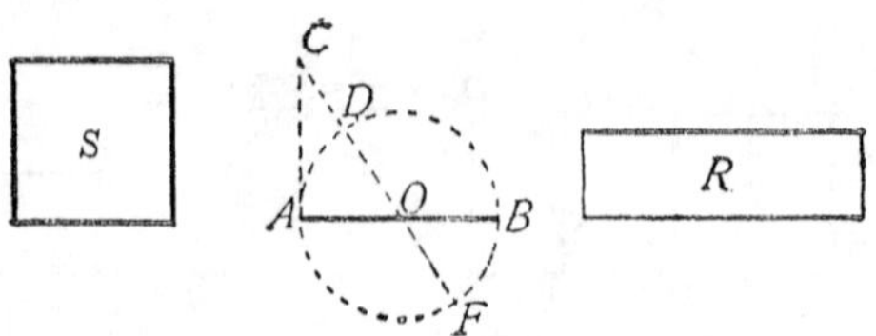

Given the square S and the line AB.

To construct a rectangle equivalent to S, and having the difference of its base and altitude equal to AB.

Construction. On AB as a diameter describe the circumference $ADBF$. Art. 275, Post. 3.

At A erect the $\perp$ AC equal to a side of S. Art. 274.

Draw CF through the center O, and meeting the circumference at the points D and F.

Construct a rectangle R with base equal to CF and altitude equal to CD.

Then R is the rectangle required.

Proof. $CF : CA = CA : CD$. Art. 358.

$\therefore \overline{CA}^2 = CF \times CD$. (Why ?)

$\therefore S \approx R$. (Why ?)

Also the difference of the base and altitude of $R = CF - CD = DF = AB$.

Q. E. F.

412. COR. The above problem is equivalent to the problem: *Given the difference and the product of two lines, to construct the lines.*

PROPOSITION XVIII. PROBLEM

413. *To construct a polygon similar to two given similar polygons, and equivalent to their sum.*

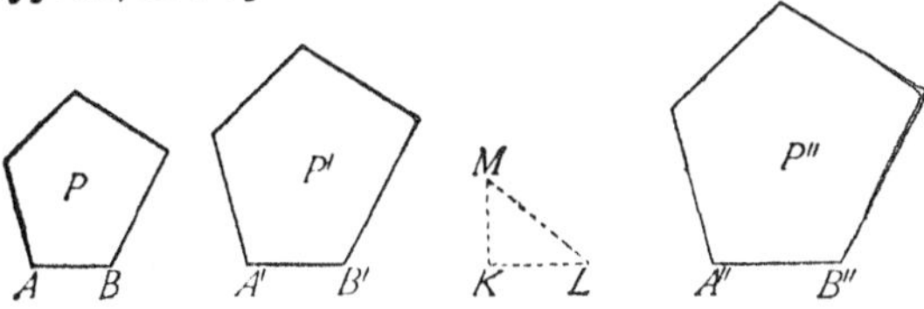

Given the similar polygons P and P'.

To construct a polygon similar to P and P', and equivalent to their sum.

Construction. Take any two homologous sides, AB and $A'B'$, of P and P'.

Draw $MK \perp KL$ (Art. 274), making $MK = AB$, and $KL = A'B'$.

Draw ML.

On $A''B''$, equal to ML, as a side homologous to AB construct the polygon P'' similar to P. Art. 372.

Then P'' is the polygon required.

Proof. $\frac{P}{P''} = \frac{\overline{AB}^2}{\overline{A''B''}^2}$; also $\frac{P'}{P''} = \frac{\overline{A'B'}^2}{\overline{A''B''}^2}$, Art. 399.

(*the areas of two similar polygons are to each other as the squares of their homologous sides*).

Adding, $\therefore \frac{P + P'}{P''} = \frac{\overline{AB}^2 + \overline{A'B'}^2}{\overline{A''B''}^2}$. Ax. 2.

But $\overline{MK}^2 + \overline{KL}^2 = \overline{ML}^2$, Art. 346.

Or $\overline{AB}^2 + \overline{A'B'}^2 = \overline{A''B''}^2$. Ax. 8.

$\therefore \frac{\overline{AB}^2 + \overline{A'B'}^2}{\overline{A''B''}^2} = \frac{\overline{A''B''}^2}{\overline{A''B''}^2} = 1$. Ax. 8.

$\therefore \frac{P + P'}{P''} = 1$ (Ax. 1.) $\therefore P + P' \approx P''$. Ax. 4.

Q. E. F.

PROPOSITION XIX. PROBLEM

414. *To construct a square which shall have a given ratio to a given square.*

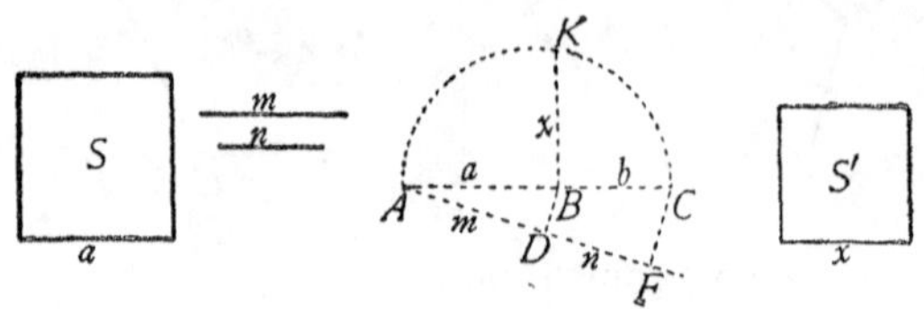

Given the square S and the lines m and n.

To construct a square which shall be to S in the ratio $n : m$.

Construction. Take AB equal to a side of S and draw AF, making a convenient angle with AB.

On AF take AD equal to m, and DF equal to n.

Draw DB. Draw $FC \parallel DB$, meeting AB produced in C. Art. 279.

On AC, as a diameter, construct a semicircumference AKC. Art. 275, Post. 3.

At B erect a $\perp$ BK meeting the semicircumference at K. Art. 274.

Construct a square S' having a side equal to BK, or x.

Then S' is the square required.

Proof. $x^2 = a \times b.$ (Why?)

Also $a : b = m : n.$ (Why?)

Hence $\frac{S}{S'} = \frac{a^2}{x^2} = \frac{a^2}{ab} = \frac{a}{b} = \frac{m}{n}.$ Axs. 8, 5.

$\therefore \frac{S}{S'} = \frac{m}{n}.$ Ax. 1.

Q. E. F.

Proposition XX. Problem

415. *To construct a polygon similar to a given polygon, and having a given ratio to it.*

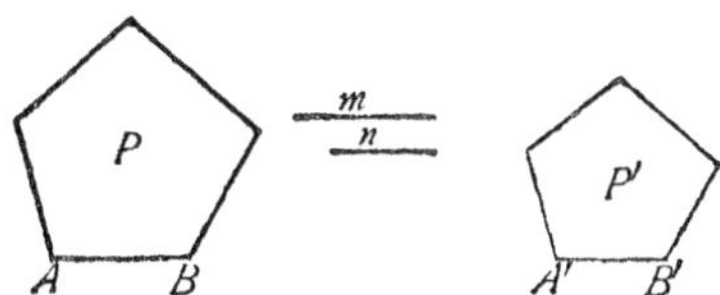

Given the polygon P and the lines m and n.

To construct a polygon P' which shall be similar to P, and be to P in the ratio $n : m$.

Construction. Construct a square which shall be to the square on AB as $n : m$. Art. 414.

Let $A'B'$ be a side of this square.

Upon $A'B'$ as a side homologous to AB construct a polygon P' similar to P. Art. 372.

Then P' is the polygon required.

Proof. $$\frac{P}{P'} = \frac{\overline{AB}^2}{\overline{A'B'}^2}.$$ (Why ?)

But $$\frac{\overline{AB}^2}{\overline{A'B'}^2} = \frac{m}{n}.$$ Constr.

Hence $$\frac{P}{P'} = \frac{m}{n}.$$ (Why ?)

Q. E. F.

PROPOSITION XXI. PROBLEM

416. *To construct a polygon similar to one given polygon, and equivalent to another given polygon.*

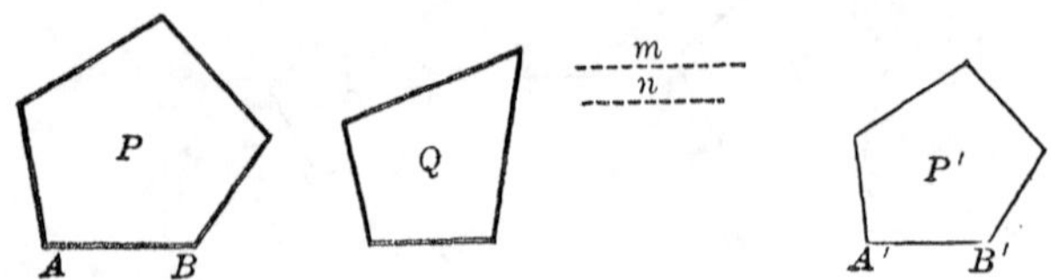

Given the polygons P and Q.

To construct a polygon similar to P, and equivalent to Q.

Construction. Construct a square equivalent to P, and let m be one of its sides. Art. 408.

Construct a square equivalent to Q, and let n be one of its sides. Art. 408.

Construct $A'B'$, the fourth proportional to m, n, and AB. Art. 366.

On $A'B'$, as a side homologous to AB, construct a polygon P' similar to P. Art. 372.

Then P' is the polygon required.

Proof. $\frac{P}{Q}=\frac{m^2}{n^2}=\frac{\overline{AB}^2}{\overline{A'B'}^2}=\frac{P}{P'}$. Constr., Arts. 314, 399.

$\therefore \frac{P}{Q}=\frac{P}{P'}$. Ax. 1.

$\therefore P' \approx Q$. Art. 305.

Q. E. F.

EXERCISES. GROUP 40

THEOREMS CONCERNING AREAS

Ex. 1. The diagonals of a parallelogram divide the parallelogram into four equivalent triangles.

Ex. 2. Any straight line drawn through the point of intersection of the diagonals of a parallelogram divides the parallelogram into two equivalent parts.

Ex. 3. If, in the triangle ABC, D and F are the midpoints of the sides AB and AC, respectively, the area of ADF equals one-fourth the area of ABC.

[SUG. Use Art. 397.]

Ex. 4. If the midpoints of two adjacent sides of a parallelogram be joined, the area of the triangle so formed equals one-eighth the area of the parallelogram.

Ex. 5. If, in the triangle ABC, D and F are the midpoints of the sides AB and AC, respectively, the triangles ADC and AFB are equivalent.

Ex. 6. In a right triangle show, by obtaining expressions for the area of the figure, that the product of the legs equals the product of the hypotenuse by the altitude upon the hypotenuse.

Ex. 7. If two triangles are equivalent, and the altitude of one is three times the altitude of the other, find the ratio of their bases.

Ex 8. If two isosceles triangles have their legs equal, and half of the base of one equivalent to the altitude of the other, the triangles are equivalent.

Ex. 9 If two triangles have an angle of one the supplement of an angle of the other, their areas are to each other as the products of the sides including these angles.

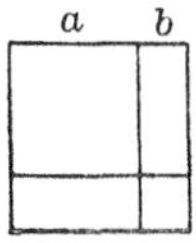

Ex. 10. Prove geometrically that $(a+b)^2=a^2+b^2+2ab$.

Ex. 11. Similarly prove $(a-b)^2=a^2+b^2-2ab$.

Ex 12. Similarly prove $(a+b)(a-b)=a^2-b^2$.

Ex. 13. The line joining the midpoints of the parallel sides of a trapezoid divides the trapezoid into two equivalent parts.

Ex. 14. The lines joining the midpoint of one diagonal of a quadrilateral to the vertices not joined by the diagonal divide the quadrilateral into two equivalent parts.

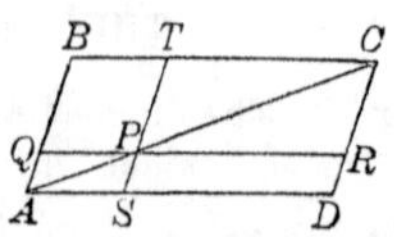

Ex. 15. Given QR and TS passing through P, any point on the diagonal AC of a ▱, $QR \parallel AD$, and $TS \parallel AB$; prove $QBTP \approx PRDS$.

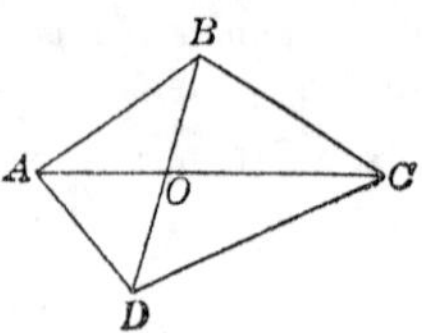

Ex. 16. Given $OB = OD$; prove $\triangle ABC \approx \triangle ADC$. Let the pupil also state this as a theorem in general language.

EXERCISES. GROUP 41

USE OF AUXILIARY LINES

Ex. 1. Given $ABCD$ a ▱ and P any point inside $ABDC$; prove $\triangle PAD + \triangle PBC \approx \triangle PAB + \triangle PCD$.

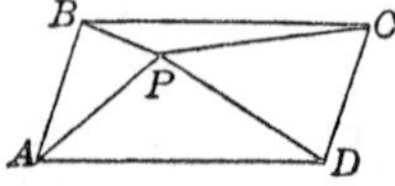

Ex. 2. The area of a triangle is equal to one-half the product of its perimeter by the radius of the inscribed circle.

Ex. 3 If the extremities of one leg of a trapezoid be joined to the midpoint of the other leg, the middle one of the three triangles thus formed is equivalent to half the trapezoid.

Ex. 4. The area of a trapezoid is equal to the product of one leg by the perpendicular on that leg from the midpoint of the other leg.

Ex 5. If the midpoints of the sides of a quadrilateral be joined in order, the parallelogram thus formed is equivalent to one-half the quadrilateral.

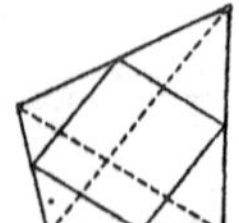

Ex. 6. A quadrilateral is equivalent to a triangle two of whose sides are the diagonals of the quadrilateral, the angle included by these sides being equal to one of the angles formed by the intersection of the diagonals.

EXERCISES. GROUP 42

THEOREMS PROVED BY VARIOUS METHODS

Ex. 1. If through the midpoint of one leg of a trapezoid a line be drawn parallel to the other leg to meet one base and the other base produced, the parallelogram so formed is equivalent to the trapezoid.

Ex. 2. If the midpoints of two sides of a triangle be joined to any point in the base, the quadrilateral so formed is equivalent to half the triangle.

Ex. 3. If P is any point on AC the diagonal of a parallelogram $ABCD$, the triangles APB and APD are equivalent.

Ex. 4. If the side of an equilateral triangle be denoted by a, the area of the triangle equals $\frac{a^2\sqrt{3}}{4}$.

Ex. 5. Find the ratio of the areas of two equilateral triangles, if the altitude of one equals the side of the other.

Ex. 6. If perpendiculars be drawn from any point within an equilateral triangle to the three sides, their sum is equal to the altitude of the triangle.

Ex. 7. If E is the intersection of the diagonals AC and BD of a quadrilateral, and the triangle ADE is equivalent to the $\triangle BEC$, then the lines AB and CD are parallel.

Ex. 8. If, in the quadrilateral $ABCD$, the triangles ABC and ADC are equivalent, the diagonal AC bisects the diagonal BD.

Ex. 9. If two triangles have two sides of one equal to two sides of the other, and the included angles supplementary, the triangles are equivalent.

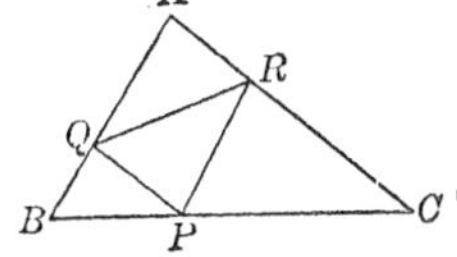

Ex. 10. If, in the parallelogram $ABCD$, F is the midpoint of the side BC, and AF intersects BD in K, the triangle $BKF = \frac{1}{12}$ the parallelogram $ABCD$.

Ex. 11. Given $PQ \parallel AC$, and $PR \parallel AB$; prove $\triangle QAR$ a mean proportional between $\triangle BQP$ and $\triangle PRC$.

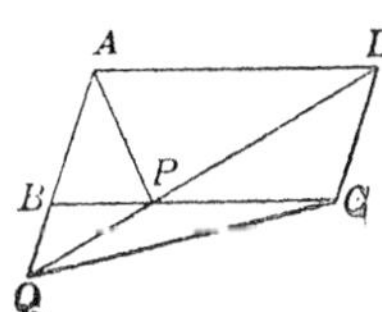

Ex. 12. P is any point in the side BC of the parallelogram $ABCD$ and DP produced meets AB produced in Q. Show that the triangles BPA and CPQ are equivalent.

EXERCISES. GROUP 43

PROBLEMS IN CONSTRUCTING AREAS

Ex. 1. Construct a square having twice the area of a given square.

Ex. 2. Construct a square having three times the area of a given square.

Ex. 3. Construct a square equivalent to the sum of three given squares.

Ex. 4. Transform a given triangle into an equivalent isosceles triangle having the same base.

Ex. 5. Transform a given triangle into an equivalent triangle having the same base, but having a given angle adjacent to the base.

Ex. 6. Transform a triangle into an equivalent triangle with the same base, but naving another given side.

Ex. 7. Transform a parallelogram into an equivalent parallelogram having the same base, but containing a given angle.

Ex. 8. Construct a triangle similar to a given triangle and containing twice the area.

To construct a similar triangle containing five times the area; how is the construction changed?

Ex 9. Bisect a given triangle by a line parallel to the base.

Ex. 10. Construct a polygon similar to two given similar polygons, and equivalent to their difference.

Ex. 11. Draw a line parallel to one side of a given rectangle, and cutting off five-sevenths of the area.

Ex. 12. Bisect a parallelogram by a line perpendicular to the base.

Ex. 13. Through any given point draw a line bisecting a given parallelogram.

Ex. 14. Construct a triangle equivalent to a given triangle, ABC, having a given base AD, but the $\angle BAC$ adjacent to the base unchanged.

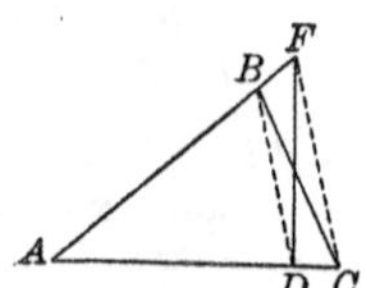

[Sug. Draw $CF \parallel DB$, etc.]

Ex. 15. Transform a parallelogram into an equivalent parallelogram having a given base, but the angle adjacent to the base unchanged.

Ex. 16. Transform a given triangle into an equivalent right triangle having a given leg.

[SUG. Use Ex. 14, then Ex. 5.]

Ex. 17. Transform a given triangle into an equivalent right triangle having a given hypotenuse.

[SUG. Find the altitude upon the hypotenuse of the new triangle by finding the fourth proportional to what three lines ?]

Ex. 18. Transform a re-entrant pentagon into an equivalent triangle.

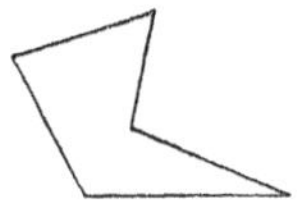

Ex. 19. Transform a given triangle into an equivalent equilateral triangle.

[SUG. See Art. 416.]

Ex. 20. Bisect a triangle by a line perpendicular to one of its sides.

[SUG. See Art. 416.]

Ex. 21. Construct a square equivalent to two-thirds of a given square.

EXERCISES. GROUP 44

PROBLEMS SOLVED BY ALGEBRAIC ANALYSIS

Ex. 1. Transform a given rectangle into an equivalent rectangle with a given base.

Ex. 2. Transform a given square into a right triangle having a given leg.

Ex. 3. Transform a given triangle into an equivalent isosceles right triangle.

Ex. 4. Draw a line cutting off from a given triangle an isosceles triangle equivalent to one-half the given triangle.

[SUG. Use Art. 397.]

Ex. 5. Through a given point in one side of a given triangle draw a line bisecting the area of the triangle.

Ex. 6. Transform a given square into a rectangle which shall have three times the perimeter of the given square.

EXERCISES. GROUP 45

PROBLEMS SOLVED BY VARIOUS METHODS

Ex. 1. Bisect a given parallelogram by a line parallel to the base.

Ex. 2. Transform a parallelogram into an equivalent parallelogram having the same base and a given side adjacent to the base.

Ex. 3. Construct a square which shall contain four-sevenths of the area of a given square.

Ex. 4. In two different ways construct a square having three times the area of a given square.

Ex. 5. Trisect a given triangle by lines parallel to the base.

Ex. 6. Find a point within a triangle such that lines drawn from it to the vertices trisect the area.

Ex. 7. Find a point within a triangle such that lines drawn from it to the three vertices divide the area into parts which shall have the ratio 2 : 3 : 4.

[SUG. If one of the small $\triangle$s contains $\frac{2}{9}$ the area of original $\triangle$, a line through its vertex cuts off $\frac{2}{9}$ the altitude, etc.]

Ex. 8. Divide a triangle into three equivalent parts by lines through a given vertex.

Ex. 9. Divide a triangle into three equivalent parts by lines drawn through a given point P in one of the sides.

[SUG. Use Art. 397.]

Ex. 10. Divide a given quadrilateral into three equivalent parts by lines drawn through a given vertex.

Ex. 11. Through a given point in the base of a trapezoid draw a line bisecting the area of the trapezoid.

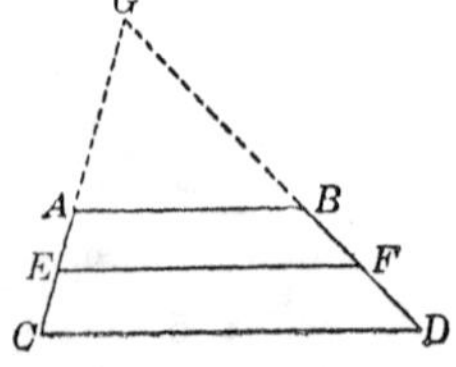

Ex. 12. Bisect the area of a trapezoid by a line drawn parallel to the bases.

[SUG. Construct $\triangle$ GEF similar to $\triangle$ ABG and equivalent to $\frac{1}{2}$ sum of what two $\triangle$s?]

BOOK V

REGULAR POLYGONS. MEASUREMENT OF THE CIRCLE

417. DEF. A **regular polygon** is a polygon that is both equilateral and equiangular.

PROPOSITION I. THEOREM

418. *An equilateral polygon that is inscribed in a circle is also equiangular and regular.*

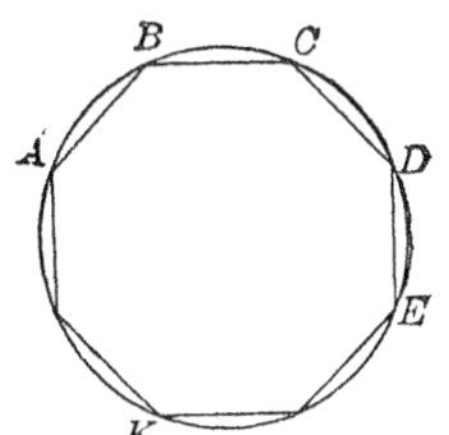

Given $ABC \ldots K$ an inscribed polygon, with its sides AB, BC, CD, etc., equal.

To prove the polygon $ABC \ldots K$ equiangular and regular.

Proof. Arc AB = arc BC = arc CD, etc. Art. 218.

$\therefore$ arc ABC = arc BCD = arc CDE, etc. Ax. 2.

$\therefore \angle ABC = \angle BCD = \angle CDE$, etc., Art. 260.

(*all* $\measuredangle$ *inscribed in the same segment, or in equal segments, are equal*).

$\therefore$ the polygon $ABC \ldots K$ is equiangular.

$\therefore$ the polygon $ABC \ldots K$ is regular. Art. 417.

Q. E. D.

419. Cor. 1. *If the arcs subtended by the sides of a regular inscribed polygon be bisected, and each point of bisection be joined to the nearest vertices of the polygon, a regular inscribed polygon of double the number of sides is formed.*

420. Cor. 2. *The perimeter of an inscribed polygon is less than the perimeter of an inscribed polygon of double the number of sides.*

Proposition II. Theorem

421. *A circle may be circumscribed about, and a circle may be inscribed in, any regular polygon.*

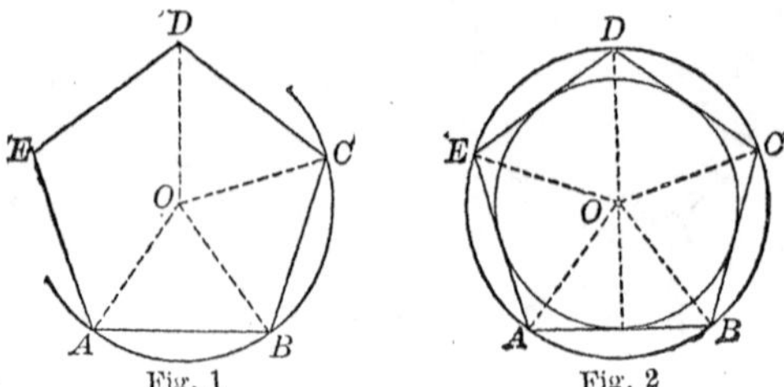

Fig. 1 Fig. 2

Given the regular polygon $ABCDE$.

To prove that a ⊙ may be circumscribed about, or inscribed in, $ABCDE$.

Proof. I. Through A, B and C (Fig. 1), any three successive vertices of the polygon $ABCDE$, pass a circumference. Art. 235.

Let O be the center of this circumference.

Draw the radii OA, OB, OC. Also draw the line OD.

Then, in $\triangle\ OBC$, $OB = OC$. (Why?)

$\therefore\ \angle OBC = \angle OCB$. (Why?)

But $\angle ABC = \angle BCD$, Art. 417.
(*being* ∡ *of a regular polygon*).

Subtracting, $\angle OBA = \angle OCD$. (Why?)

Hence, in the △ OAB and OCD,

$OB = OC$. (Why?)

$AB = CD$. (Why?)

$\angle OBA = \angle OCD$,
(*just proved*).

$\therefore \triangle ABO = \triangle OCD$. (Why?)

$\therefore OD = OA$. (Why?)

Hence the circumference which passes through the vertices A, B and C, will also pass through the vertex D.

In like manner, it may be proved that this circumference will pass through the vertex E.

Hence a circle described with O as a center, and OA as a radius, will be circumscribed about the given polygon.

II. The sides of the polygon $ABCDE$ (Fig. 2) are equal chords in the circle O.

Hence they are equidistant from the center. Art. 226.

∴ a circle described with O as a center, and the distance from O to one of the sides of the polygon as a radius, will be inscribed in the given polygon.

Q. E. D.

422. Def. The **center of a regular polygon** is the common center of the inscribed and circumscribed circles, as the point O in the above figure.

423. Def. The **radius of a regular polygon** is the radius of the circumscribed circle, as OA in the above figure.

424. Def. The **apothem** of a regular polygon is the radius of the inscribed circle.

425. DEF. **The angle at the center of a regular polygon** is the angle between two radii drawn to the extremities of any side, as the angle AOB.

426. COR. *The angle at the center of a regular polygon is equal to four right angles divided by the number of sides.*

Hence, if n denote the number of sides in the polygon, the *angle at the center of a regular polygon equals* $\frac{4 \text{ rt. } \angle s}{n}$; *also the angle between an apothem and the nearest radius, in a regular polygon of n sides, equals* $\frac{2 \text{ rt. } \angle s}{n}$.

PROPOSITION III. THEOREM

427. *If the circumference of a circle be divided into any number of equal arcs,*

I. *The chords of these arcs form a regular inscribed polygon;*

II. *Tangents to the circumference at the points of division form a regular circumscribed polygon.*

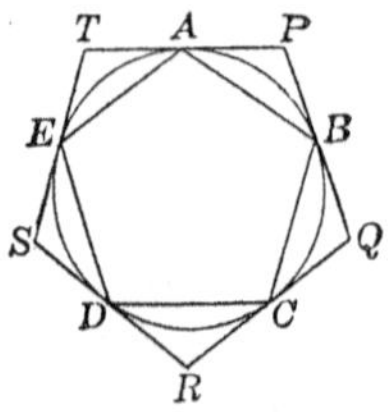

Given the circumference ABC, divided into the equal arcs AB, BC, CD, etc., the chords AB, BC, etc., and PQ, QR, etc., lines tangent to the circle at B, C, etc.

To prove $ABCDE$ a regular inscribed polygon, and $PQRST$ **a regular circumscribed polygon.**

Proof. I. The chords AB, BC, CD, etc., are equal. (Why ?)

$\therefore$ polygon $ABCDE$ is equilateral and regular. Art. 418.

II. In the $\triangle$ APB, BQC, CRD, etc.,

$AB = BC = CD$, etc. (Why ?)

Also $\angle PAB = \angle PBA = \angle QBC = \angle QCB = \angle RCD$, etc. Art. 264.

(*each being measured by half of one of the equal arcs* AB, BC, CD, *etc.*).

$\therefore$ $\triangle$ APB, BQC, CRD, etc., are equal, isosceles triangles. (Why ?)

$\therefore$ $\angle P = \angle Q = \angle R$, etc. (Why ?)

And $AP = PB = BQ = QC$, etc. (Why ?)

$\therefore$ $PQ = QR = RS$, etc. Ax. 4.

$\therefore$ $PQRST$ is a regular polygon. Art. 417.

Q. E. D.

428. COR. 1. *If the arcs AB, BC, CD, etc., be bisected, and a tangent be drawn at each point of bisection, a circumscribed regular polygon of double the number of sides of $PQRST$ will be formed.*

429. COR. 2. *The perimeter of a circumscribed regular polygon is greater than that of a circumscribed regular polygon of double the number of sides.*

Ex. 1. Find the number of degrees in the central angle of a regular pentagon. Of a regular hexagon. Of a square.

Ex. 2. What is the short name for an inscribed equilateral quadrilateral ?

Proposition IV. Theorem

430. *Tangents to a circle at the midpoints of the arcs subtended by the sides of a regular inscribed polygon form a regular circumscribed polygon whose sides are parallel to the corresponding sides of the inscribed polygon.*

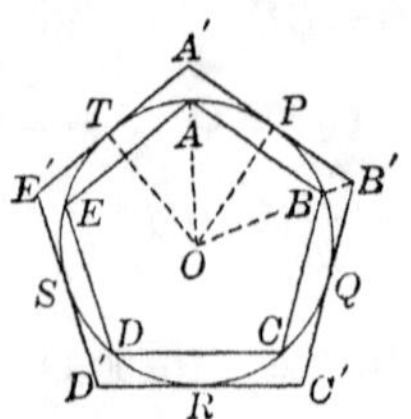

Given the regular polygon $ABCDE$ inscribed in the ⊙ ACD; P, Q, R, etc., the midpoints of the arcs AB, BC, CD, etc.; and $A'B'$, $B'C'$, $C'D'$, etc., tangents to the circle at P, Q, R, etc.

To prove $A'B'C'D'E'$ a regular polygon with its sides $\parallel$ corresponding sides of the polygon $ABCDE$.

Proof. The arcs AB, BC, CD, etc., are equal. Art. 218.

$\therefore$ the arcs AP, PB, BQ, QC, etc., are equal. Ax. 5.

$\therefore$ the arcs PQ, QR, RS, etc., are equal. Ax. 4.

$\therefore$ $A'B'C'D'E'$ is a regular polygon, Art. 427.

(*if the circumference of a ⊙ be divided, etc.*).

Side $AB \perp OP$. Art. 113.

$A'B' \perp OP$. (Why?)

$\therefore AB \parallel A'B'$. (Why?)

In like manner, each pair of homologous sides in the two polygons is parallel.

Q. E. D.

431. Cor. *Homologous radii of an inscribed and a circumscribed regular polygon, whose sides are parallel, coincide in direction.* Thus, in the above figure ∠s POA and POA' each $= \frac{2 \text{ rt. } \angle s}{n}$ $\therefore OA$ and OA' coincide in direction. Art. 426.

PROPOSITION V. THEOREM

432. *Two regular polygons of the same number of sides are similar.*

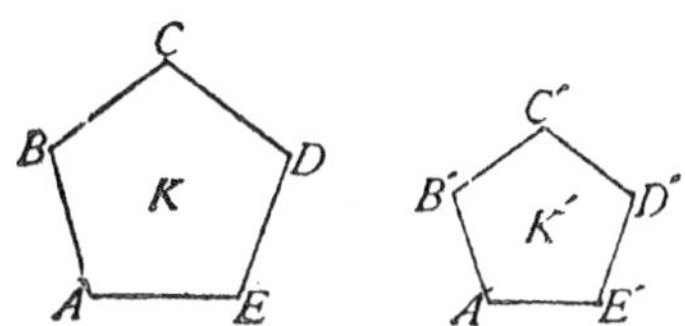

Given K and K' two regular polygons, each of n sides.

To prove K and K' similar.

Proof. Each $\angle$ of $K=\frac{(n-2)\ 2\text{ rt. }\angle s}{n}$, Art. 174.

(*in an equiangular polygon of n sides, each* $\angle=\frac{(n-2)\ 2\text{ rt. }\angle s}{n}$).

Similarly each $\angle$ of $K'=\frac{(n-2)\ 2\text{ rt. }\angle s}{n}$.

Hence K and K' are mutually equiangular. Ax. 1, Art. 169.

Also $AB=BC\quad\therefore\quad\frac{AB}{BC}=1.$ Art. 417, Ax. 5.

And $A'B'=B'C'\quad\therefore\quad\frac{A'B'}{B'C'}=1.$ (Why?)

$\therefore\ \frac{AB}{BC}=\frac{A'B'}{B'C'}$, or $\frac{AB}{A'B'}=\frac{BC}{B'C'}$. (Why?)

In like manner $\frac{BC}{B'C'}=\frac{CD}{C'D'}=\frac{DE}{D'E'}$, etc.

Hence K and K' have their homologous sides proportional.

Hence K and K' are similar. Art. 321.

Q. E. D.

433. COR. *The areas of two regular polygons of the same number of sides are to each other as the squares of any two homologous sides.*

Proposition VI. Theorem

434. I. *The perimeters of two regular polygons of the same number of sides are to each other as the radii of their circumscribed circles, or as the radii of their inscribed circles;*

II. *Their areas are to each other as the squares of these radii.*

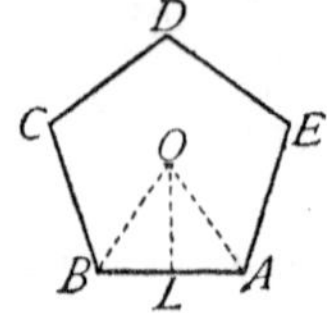

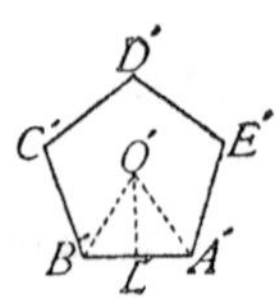

Given AC and $A'C'$ two regular polygons, each of n sides, with centers O and O', and with perimeters denoted by P and P', radii by R and R', and apothems by r and r', respectively.

To prove. I. $P : P' = R : R' = r : r'$.

II. Area AC : area $A'C' = R^2 : R'^2 = r^2 : r'^2$.

Proof. I. The polygons AC and $A'C'$ are similar. Art. 432.

Hence $P : P' = AB : A'B'$. Art. 341.

But, in the △ OAB and $O'A'B'$,

$\angle AOB = \angle A'O'B'$, (*for each* $\angle = \frac{4 \text{ rt. } \angle s}{n}$). Art. 426.

Also $OA : OB = O'A' : O'B'$, (*for each* △ *is isosceles*).

∴ △ OAB and $O'A'B'$ are similar. Art. 327.

∴ $AB : A'B' = OA : O'A'$. Art. 321.

And $AB : A'B' = OL : O'L'$. Art. 338.

∴ $P : P' = OA : O'A' = OL : O'L'$. Ax. 1.

Or $P : P' = R : R' = r : r'$.

II. Area AC : area $A'C' = \overline{AB}^2 : \overline{A'B'}^2$. Art. 399.

But $\overline{AB}^2 : \overline{A'B'}^2 = R^2 : R'^2 = r^2 : r'^2$. Art. 314.

∴ area AC ; area $A'C' = R^2 : R'^2 ; r^2 : r'^2$. Ax. 1.

Q. E. D.

PROPOSITION VII. THEOREM

435. *If the number of sides of a regular inscribed polygon be indefinitely increased, the apothem of the polygon approaches the radius as a limit.*

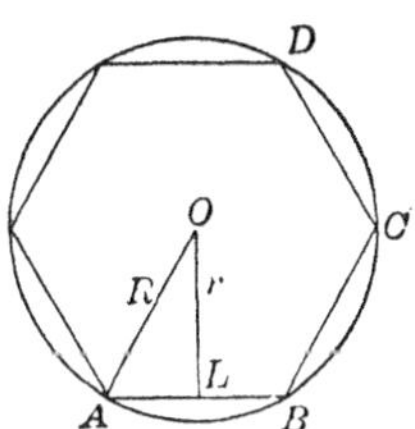

Given the regular inscribed polygon $AB \ldots D$ of n sides, with radius OA and apothem OL.

To prove that, as n is indefinitely increased, OL approaches OA as a limit.

Proof. In the $\triangle OAL$, $AL > OA - OL$, Art. 93.

(*any side of a $\triangle$ is greater than the difference between the other two sides.*)

But, as $n \doteq \infty$, $AB \doteq 0 \quad \therefore \quad AL \doteq 0$. Art. 253, 3.

Hence $OA - OL \doteq 0$.

$\therefore \quad OL \doteq OA$, or $r \doteq R$.

Or the limit of the apothem OL is the radius OA.

Q. E. D.

436. COR. As $n \doteq \infty$, $R^2 - r^2 \doteq 0$.

For, $R^2 - r^2 = (R + r)\ (R - r)$. But, as $n \doteq \infty$,

$R + r \doteq R + R$ or $2\ R$, and $R - r \doteq 0$. Ax. 8.

$\therefore \quad R^2 - r^2 \doteq 2R \times 0$ or 0. Ax. 8.

Ex. A pair of homologous sides of two regular pentagons are 2 and 3 ft. Find the ratio of the areas of the polygons.

Proposition VIII. Theorem

437. *The length of any line inclosing the circumference of a circle, and not passing within the circumference, is greater than the length of the circumference.*

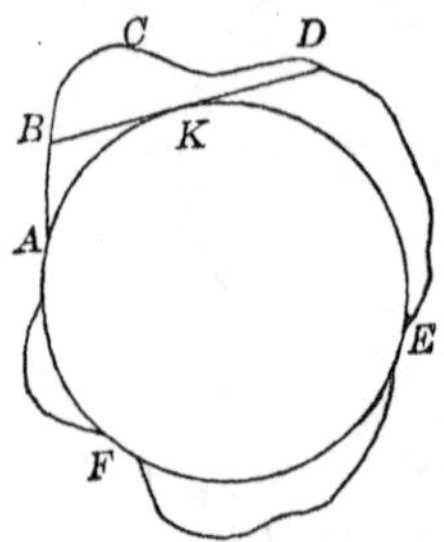

Given the circumference $AKEF$, and $ABCDEF$ any line which does not pass within $AKEF$.

To prove circumference $AKEF <$ perimeter $ABCDEF$.

Proof. Let K be any point on the circumference $AKEF$ not touched by the line $ACDEF$.

At K draw a tangent to the given circle, meeting $ACEF$ at B and D.

Then the straight line $BKD <$ line BCD. Art. 15.

To each of these unequals add the line $DEFAB$.

Then line $ABKDEF <$ line $ACDEF$. Ax. 9.

Hence every line enveloping the circular area $AKEF$, except the circumference $AKEF$, may be shortened.

Hence the circumference $AKEF$ is shorter than any line enveloping it.

Q. E. D.

438. Cor. 1. *The circumference of a circle is less than the perimeter of any polygon circumscribed about the circle.*

439. Cor. 2. *The circumference of a circle is greater than the perimeter of any polygon inscribed in the circle,* for each side of an inscribed polygon is less than the arc subtended by it.

440. Cor. 3. *The difference between the perimeters of an inscribed and a circumscribed polygon is greater than the difference between either perimeter and the circumference of the circle.*

Proposition IX. Theorem

441. *If the number of sides of a regular inscribed, or of a regular circumscribed polygon be indefinitely increased,*

I. *The perimeter of each polygon approaches the circumference as a limit;*

II. *The area of each polygon approaches the area of the circle as a limit.*

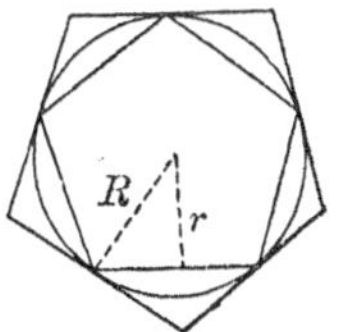

Given a circle of circumference C and area A, with regular inscribed and circumscribed polygons, each of n sides, with their perimeters denoted by P and P', and their areas by K and K', respectively.

To prove that as n is indefinitely increased, P and P'

each approaches C as a limit, and K and K' each approaches A as a limit.

Proof I. Denote the apothems of the two polygons by R and r.

Then $$\frac{P'}{P}=\frac{R}{r}.$$ Art. 434.

Hence $$\frac{P'-P}{P}=\frac{R-r}{r}.$$ Art. 310.

Let $n \doteq \infty$; then $R-r \doteq 0$. Art. 435.

$$\therefore \frac{R-r}{r} \doteq 0,$$ Art. 253, 4.

(*for the denominator, r, increases as n increases*).

Hence $\frac{P'-P}{P} \doteq 0$ (Ax. 8). $\therefore P'-P \doteq 0$, (Art. 253, 4).

But $P'-C < P'-P$ (Art. 440) $\therefore P'-C \doteq 0$, or $P' \doteq C$.

Also $C-P < P'-P$ (Art. 440) $\therefore C-P \doteq 0$, or $P \doteq C$.

II. $\frac{K'}{K}=\frac{R^2}{r^2}$ (Art. 434) $\therefore \frac{K'-K}{K}=\frac{R^2-r^2}{r^2}$. Art. 310.

Let $n \doteq \infty$; then $R^2-r^2 \doteq 0$. Art. 436.

$$\therefore \frac{R^2-r^2}{r^2} \doteq 0,$$ Art. 253, 4.

(*for the denominator, r^2, increases*).

Hence $\frac{K'-K}{K} \doteq 0$ (Ax. 8). $\therefore K'-K \doteq 0$, (Art. 253, 4).

But $K'-A < K'-K$ (Ax. 7) $\therefore K'-A \doteq 0$, or $K' \doteq A$.

Also $A-K < K'-K$ (Ax. 7) $\therefore A-K \doteq 0$, or $K \doteq A$.

Q. E. D.

Ex. Find the perimeter and area of a square field, one of whose sides is 10 rods. Find the same in a square field, one of whose sides is 20 rods. Is it more economical, therefore, to fence land in large or small fields?

PROPOSITION X. THEOREM

442. *Two circumferences have the same ratio as their radii, or as their diameters.*

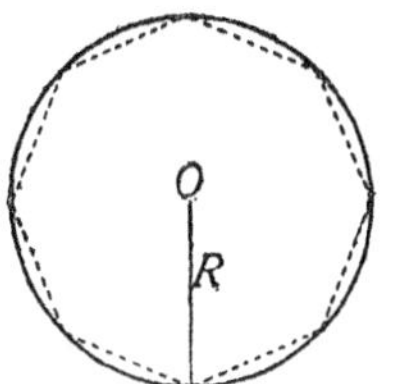

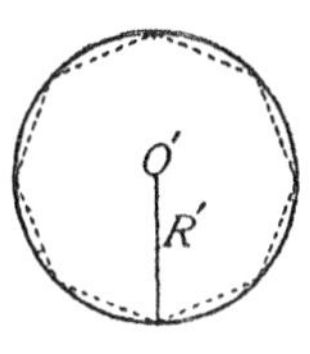

Given the circles O and O', with circumferences denoted by C and C', radii by R and R', diameters by D and D', respectively.

To prove $C : C' = R : R' = D : D'$.

Proof. In the given ⊙ let regular polygons of the same number of sides be inscribed. Denote the perimeters of the inscribed polygons by P and P', respectively.

Then $\frac{P}{P'} = \frac{R}{R'}$. Art. 434.

Hence, by alternation, $\frac{P}{R} = \frac{P'}{R'}$. Art. 307.

If, now, the number of sides of the similar inscribed polygons be increased indefinitely, P becomes a variable approaching C as a limit. Art. 441.

Hence $\frac{P}{R}$ becomes a variable approaching $\frac{C}{R}$ as a limit. Art. 253, 3.

Similarly, the variable $\frac{P'}{R'}$ approaches $\frac{C'}{R'}$ as a limit.

But the variable $\frac{P}{R}$ = the variable $\frac{P'}{R'}$ always.

$\therefore \frac{C}{R} = \frac{C'}{R'}$. Art. 254.

$\therefore C : C' = R : R'$. Art. 307.

Also $\frac{R}{R'} = \frac{2R}{2R'}$ (Art. 315.) $\therefore \frac{C}{C'} = \frac{2R}{2R'} = \frac{D}{D'}$. Ax. 8.

Q. E. D.

PROPOSITION XI. THEOREM

443. *In every circle, the ratio of the circumference to the diameter (that is, the number of times the diameter is contained in the circumference) is the same, and may be denoted by an appropriate symbol* (π).

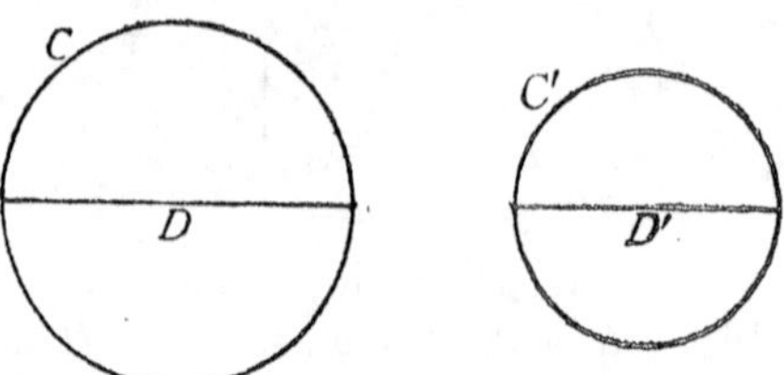

Given two ⊙ with circumferences denoted by C and C', and diameters by D and D', respectively.

To prove $\frac{C}{D}=\frac{C'}{D'}=\pi$.

Proof. $\frac{C}{C'}=\frac{D}{D'}$. Art. 442.

By alternation $\frac{C}{D}=\frac{C'}{D'}$. Art. 307.

Hence, in any given circle, $\frac{C}{D}$ has a value equal to that of $\frac{C}{D}$ in some standard circle.

$\therefore$ the value of $\frac{C}{D}$ is the same in all circles.

Denoting this constant by π, in every circle $\frac{C}{D}=\pi$.

Q. E. D.

444. Formula for the circumference in terms of the radius.

We have $\frac{C}{D}=\pi \quad \therefore \; C=\pi D=\pi(2R)$.

$$\therefore \; C=2\pi R.$$

Ex. Find the circumference of a circle whose radius is 10 inches.

445. Formula for length of arc of a circle.

$$\frac{\text{arc}}{\text{circumference}}=\frac{\text{central}\ \angle}{360°}\text{(Art. 255.)}\ \therefore\ \text{arc}=\frac{\text{central}\ \angle}{360°}\times 2\pi R.$$

$$\therefore\ \text{arc}=\frac{\text{central}\ \angle}{180°}\times \pi R.$$

PROPOSITION XII. THEOREM

446. *The area of a regular polygon is equal to one-half the product of its perimeter by its apothem.*

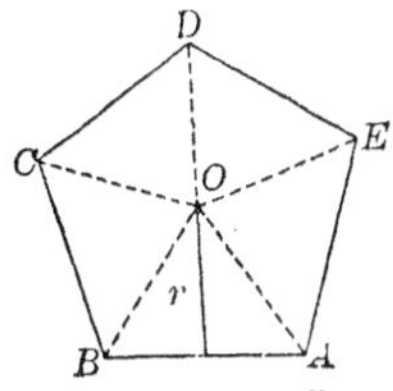

Given the regular polygon $ABCDE$ with area denoted by K, perimeter by P, and apothem by r.

To prove $K=\frac{1}{2}P\times r$.

Proof. Draw the radii OA, OB, OC, etc., dividing the polygon into as many ⚠ as the polygon has sides.

All the ⚠ thus formed have the same altitude, r. Art. 208.

∴ the area of each △ = ½ product of its base by r. Art. 389.

Hence the sum of the areas of the ⚠ = ½ product of the sum of the bases of the ⚠ by r, or $=\frac{1}{2}P\times r$.

But the sum of the areas of the ⚠ equals the area of the polygon. Ax. 6.

Hence $K=\frac{1}{2}P\times r$. Ax. 8.

Q. E. D.

447. DEF. **Similar sectors** are sectors in different circles which have equal angles at the center.

448. DEF. **Similar segments** are segments in different circles whose arcs subtend equal angles at the center.

PROPOSITION XIII. THEOREM

449. *The area of a circle is equal to one-half the product of its circumference by its radius.*

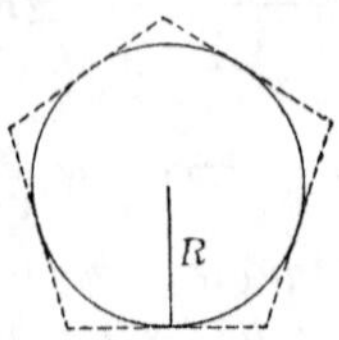

Given a ⊙ with circumference denoted by C, radius by R, and area by K.

To prove $K = \frac{1}{2} C \times R$.

Proof. Circumscribe a regular polygon about the given circle, and denote its perimeter by P and its area by K'.

In this case the apothem of the regular polygon is R.

Hence $K' = \frac{1}{2} P \times R$. Art. 446.

Let the number of sides of the circumscribed polygon be increased indefinitely; then

K' becomes a variable approaching K as its limit; Art. 441.

P becomes a variable approaching C as its limit; Art. 441.

And variable $\frac{1}{2} P \times R$ approaches $\frac{1}{2} C \times R$ as a limit. Art. 253, 2.

But $K' = \frac{1}{2} P \times R$ always. Art. 446.

Hence $K = \frac{1}{2} C \times R$. Art. 254.

Q. E. D.

450. **Formula for the area of a circle in terms of the radius R.**

$K = \frac{1}{2} C \times R$; but $C = 2\pi R$. Art. 444.

$\therefore = \frac{1}{2}(2\pi R) R$. Or $K = \pi R^2$. Ax. 8.

Again $R = \frac{1}{2} D \therefore K = \frac{\pi D^2}{4}$. Ax. 8.

451. COR. 1. *The area of a circle is equal to the square of the radius, multiplied by π; or to one-fourth the square of the diameter, multiplied by π.*

452. Cor. 2. *The areas of two circles are to each other as the squares of their radii, or as the squares of their diameters.*

For $\frac{K}{K'} = \frac{\pi R^2}{\pi R'^2} = \frac{R^2}{R'^2}$; also $\frac{K}{K'} = \frac{\frac{1}{4}\pi D^2}{\frac{1}{4}\pi D'^2} = \frac{D^2}{D'^2}$.

453. **The area of a sector** *is equal to one-half the product of its radius by its arc.*

Hence area of sector $= \frac{1}{2}\left(\frac{\text{central } \angle}{180} \times \pi R\right) R$. Art. 445.

Or area of sector $= \frac{\text{central } \angle}{360} \times \pi R^2$.

454. Cor. 3. *Similar sectors are to each other as the squares of their radii.*

Ex. 1. The measurement of the area of a circle can be reduced to the measurement of the length of what single straight line? Can it be reduced to the measurement of any other single straight line? to the measurement of a single curved line?

Find the area of a circle,

Ex. 2. Whose radius is 10 ft.

Ex. 3. Whose diameter is 10 ft.

Ex. 4. Whose radius is b ft.

Ex. 5. Whose radius is $\frac{1}{2}b$ ft.

Ex. 6. Whose radius is $2b$ ft.

Ex. 7. Whose radius is R; $\frac{1}{2}R$; $2R$.

Ex. 8. Whose radius is $R\sqrt{3}$.

Ex. 9. Whose diameter is $\frac{1}{2}R\sqrt{2}$.

Ex. 10. If the radius of one circle is 10 times as great as the radius of another circle, how do their areas compare? Also how do their circumferences compare?

Ex. 11. A wheel with 6 cogs is geared to a wheel with 48 cogs. How many revolutions will the smaller wheel make while the larger wheel revolves once?

Ex. 12. A 2 in. pipe will discharge how much more water in a given time than a 1 in. pipe?

PROPOSITION XIV. THEOREM

455. *The areas of two similar segments are to each other as the squares of their radii.*

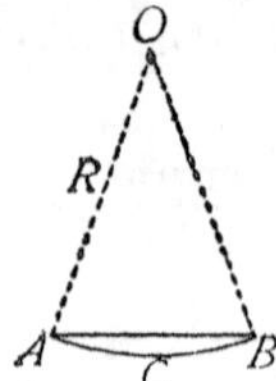

Given the similar segments ABC and $A'B'C'$ in circles whose radii are R and R'.

To prove segment ABC : segment $A'B'C' = R^2 : R'^2$.

Proof. Draw the radii OA, OB, $O'A'$, $O'B'$.

Then the sectors OAB and $O'A'B'$ are similar. Arts. 448, 447.

And $\triangle$s OAB and $O'A'B'$ are similar. Art. 327.

$$\therefore \frac{\text{sector } OAB}{\text{sector } O'A'B'} = \frac{R^2}{R'^2}; \text{ and } \frac{\triangle\ OAB}{\triangle\ O'A'B'} = \frac{R^2}{R'^2}.$$ Art. 454, 398.

$$\therefore \frac{\text{sector } OAB}{\text{sector } O'A'B'} = \frac{\triangle\ OAB}{\triangle\ O'A'B'}.$$ (Why?)

$$\therefore \frac{\text{sector } OAB}{\triangle\ OAB} = \frac{\text{sector } O'A'B'}{\triangle\ O'A'B'}.$$ (Why?)

$$\therefore \frac{\text{sector } OAB - \triangle\ OAB}{\triangle\ OAB} = \frac{\text{sector } O'A'B' - \triangle\ O'A'B'}{\triangle\ O'A'B'}.$$ (Why?)

Or $$\frac{\text{segment } ABC}{\triangle\ OAB} = \frac{\text{segment } A'B'C'}{\triangle\ O'A'B'}.$$

$$\therefore \frac{\text{segment } ABC}{\text{segment } A'B'C'} = \left(\frac{\triangle\ OAB}{\triangle\ O'A'B'}\right) = \frac{R^2}{R'^2}.$$ Arts. 307, 398.

Q. E. D.

CONSTRUCTION PROBLEMS

Proposition XV. Problem

456. *To inscribe a square in a given circle.*

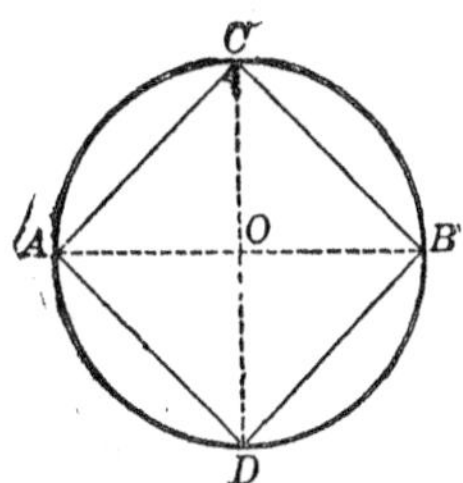

Let the pupil supply a solution.

457. Cor. *By bisecting the arcs AC, CB, BD, etc., and drawing chords, a regular octagon may be inscribed in the circle; by repeating the process, regular polygons of* 16, 32, 64, . . . *and* 2^n *sides may be inscribed, where n is a positive integer greater than* 1.

How can a regular polygon of 2^n sides be circumscribed about a given circle?

Proposition XVI. Problem

458. *To inscribe a regular hexagon in a given circle.*

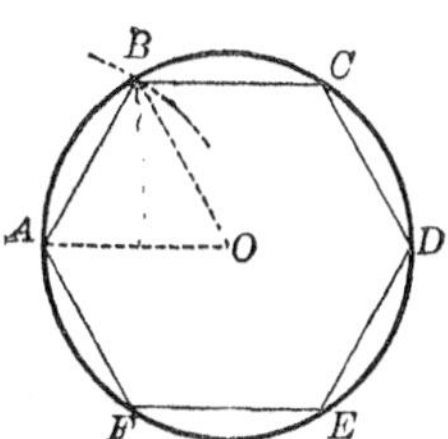

Let the pupil supply a solution.

459. Cor. 1. *The side of a regular inscribed hexagon equals the radius of the circle.*

460. Cor. 2. *By joining the alternate vertices of a regular inscribed hexagon, an equilateral triangle can be inscribed in a given circle.*

461. Cor. 3. *By bisecting the arcs AB, BC, CD, etc., and drawing chords, a regular polygon of 12 sides can be inscribed in a given circle; by repeating the process, regular polygons of* 24, 48, . . . 3×2^n *sides can be inscribed.*

Similarly, how can a regular polygon of 3×2^n sides be circumscribed about a given circle?

Proposition XVII. Problem

462. *To inscribe a regular decagon in a given circle.*

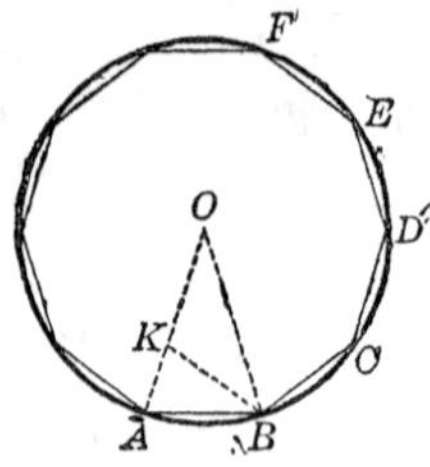

Given the circle O.

To inscribe a regular decagon in the given ⊙.

Construction. Draw any radius OA, and divide OA in extreme and mean ratio at K, OK being the greater segment. Art. 371.

With A as a center and OK as a radius, describe an arc cutting the given circumference at B.

Then the chord AB is a side of the decagon required.

Proof. Draw OB and KB.

Then $OA : OK = OK : KA$. Constr.

But $AB = OK$, $\therefore OA : AB = AB : KA$. Ax. 8.

Also $\angle OAB = \angle KAB$. Ident.

$\therefore$ △s OAB and KAB are similar. Art. 327.

$\therefore \angle O = \angle ABK$ (*homolog.* ∠s *of similar* △s).

And △ AKB is isosceles, Art. 321.

(*being similar to* △ AOB, *which is isosceles*).

$\therefore AB = KB = KO$. Ax. 1.

Hence $\angle O = \angle KBO$. (Why?)

$\therefore \angle ABO = 2 \angle O$. Ax. 2.

$\therefore \angle OAB = 2 \angle O$. Art. 99.

$\angle O = \angle O$.

Adding, $\angle ABO + \angle OAB + \angle O = 5 \angle O$.

But $\angle ABO + \angle OAB + \angle O = 2$ rt. ∠s. Art. 134.

$\therefore 5 \angle O = 2$ rt. ∠s. Ax. 1.

$\therefore \angle O = \frac{1}{5}$ of 2 rt. ∠s, or $\frac{1}{10}$ of 4 rt. ∠s.

$\therefore$ arc AB is $\frac{1}{10}$ of the circumference.

Hence, if the chord AB be applied ten times in succession to the circumference, a regular decagon will be inscribed in the given circle. Art. 418.

Q. E. F.

463. Cor. 1. *By joining the alternate vertices of a regular inscribed decagon, a regular pentagon can be inscribed in a given circle.*

464. Cor. 2. *By bisecting the arcs AB, BC . . . and drawing chords, a regular polygon of 20 sides can be inscribed in a given circle; by repeating the process a regular polygon of 40, 80, . . . 5×2^n sides can be inscribed.*

How can a regular polygon of 5×2^n sides be circumscribed about a given circle?

Proposition XVIII. Problem

465. *To inscribe a regular polygon of fifteen sides (pentedecagon) in a given circle.*

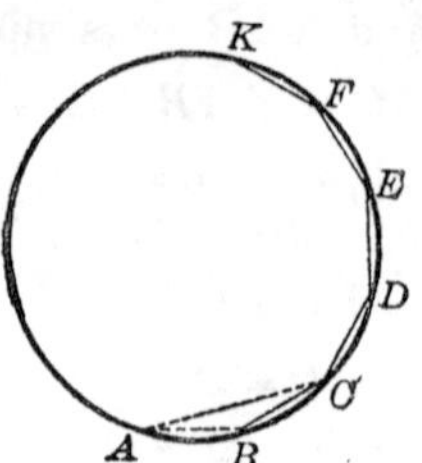

Given the ⊙ AK.

To inscribe a regular pentedecagon in the given ⊙.

Construction. Draw the chord AC equal to the radius of the given ⊙, and the chord AB equal to a side of a regular decagon inscribed in the circle. Art. 462.

Draw the chord BC.

Then BC is a side of the required pentedecagon.

Proof. AC is a side of a regular inscribed hexagon. Art. 459.

$\therefore$ arc $AC = \frac{1}{6}$ of the circumference.

In like manner arc $AB = \frac{1}{10}$ of the circumference. Art. 462.

$\therefore$ arc $BC = \frac{1}{6} - \frac{1}{10}$, or $\frac{1}{15}$ of the circumference. Ax. 3.

Hence, if chord BC be applied fifteen times in succession to the circumference, a regular pentedecagon will be inscribed in the given circle. Art. 418.

Q. E. F.

466. Cor. *By bisecting the arcs BC, CD, . . . etc., drawing chords, and repeating the process, regular polygons of* 30, 60, . . . 15×2^n *sides can be inscribed in a given circle.*

How can a regular polygon of 15×2^n sides be circumscribed about a given circle?

COMPUTATION PROBLEMS

PROPOSITION XIX. PROBLEM

467. ***Given the side and radius of a regular inscribed polygon, to find the side of a regular inscribed polygon of double the number of sides, in terms of the given quantities.***

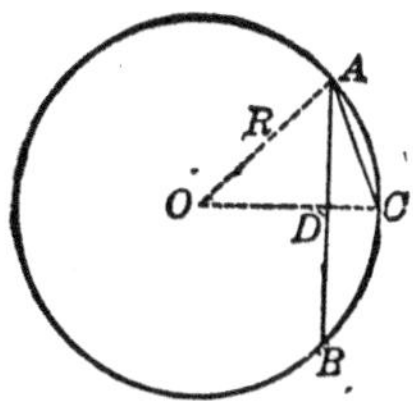

Given the circle O with radius R, AB a side of a regular inscribed polygon, and AC a side of the regular inscribed polygon of double the number of sides.

To determine AC in terms of AB and R.

Solution. Draw the radii OA and OC.

Then OC is the $\perp$ bisector of AB. (Why?)

But arc $ACB <$ semicircumference. (Why?)

$\therefore AC <$ a quadrant. Ax. 10.

$\therefore \angle AOC$ is an acute $\angle$. Art. 257.

Hence, in $\triangle OAC$, $\overline{AC}^2 = \overline{OA}^2 + \overline{OC}^2 - 2\,OC \times OD$. Art. 349.

Or $AC^2 = 2\,R^2 - 2\,R \times OD$

But, in the rt. $\triangle OAD$, $\overline{OD}^2 = \overline{OA}^2 - \overline{AD}^2$, Art. 347.

Or $\overline{OD}^2 = R^2 - (\frac{1}{2}\,AB)^2$. Ax. 8.

$$\therefore OD = \sqrt{R^2 - \tfrac{1}{4}\,\overline{AB}^2} = \tfrac{1}{2}\sqrt{4\,R^2 - \overline{AB}^2}.$$

Substituting for OD its value thus obtained,

$$\overline{AC}^2 = 2\,R^2 - R\sqrt{4\,R^2 - \overline{AB}^2}. \quad \text{Ax. 8.}$$

Or $$AC = \sqrt{R\,(2\,R - \sqrt{4\,R^2 - \overline{AB}^2})},$$

468. Special Formulas. If the radius R, be taken as 1, $AC = \sqrt{2 - \sqrt{4 - \overline{AB}^2}}$.

If a side of the regular inscribed polygon of n sides be denoted by S_n, and $R = 1$, then $S_{2n} = \sqrt{2 - \sqrt{4 - S_n^2}}$.

PROPOSITION XX. PROBLEM

469. *To compute approximately the numerical value of π.*

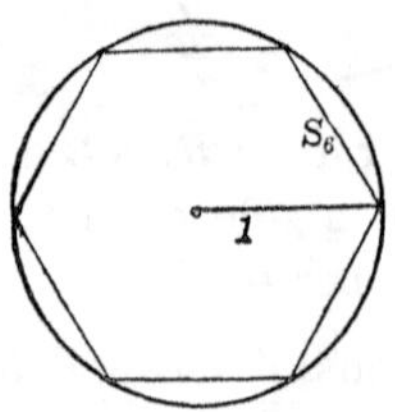

Given a ⊙ whose radius is 1, and whose circumference is denoted by C.

To compute C, i. e., 2π, and hence find the numerical value of π approximately.

Computation. 1. Inscribe a regular hexagon in the given circle, and denote its side by S_6.

Then $S_6 = 1$. Art. 459.

$\therefore$ perimeter of the inscribed hexagon $= 6$.

2. Inscribe a regular polygon of double the number (12) sides.

Then, by the second formula of Art. 468,

$$S_{12} = \sqrt{2 - \sqrt{4 - 1}} = 0.51763809 +.$$

Denoting the perimeter of a regular inscribed polygon of n sides by P_n,

$$P_{12} = 12\,(0.51763809 +) = 6.21165708.$$

3. In the second formula of Art. 468, let $n=12$.

$\therefore S_{24}=\sqrt{2-\sqrt{4-(0.51763809+)^2}}=0.26108238+$, etc.
Hence $P_{24}=6.26525772$.

Computing S_{48}, P_{48}, etc., in like manner, the following results are obtained:

$S_{12}=\sqrt{2-\sqrt{4-1}}=.51763809. \quad \therefore P_{12}=6.21165708.$

$S_{24}=\sqrt{2-\sqrt{4-(.5176809)^2}}=.26105238. \quad \therefore P_{24}=6.26525722.$

$S_{48}=\sqrt{2-\sqrt{4-(.26105238)^2}}=.13080626. \quad \therefore P_{48}=6.27870041.$

$S_{96}=\sqrt{2-\sqrt{4-(.13080626)^2}}=.06543817. \quad \therefore P_{96}=6.28206396.$

$S_{192}=\sqrt{2-\sqrt{4-(.06543817)^2}}=.03272346. \quad \therefore P_{192}=6.28290510.$

$S_{384}=\sqrt{2-\sqrt{4-(.03272346)^2}}=.01636228. \quad \therefore P_{384}=6.28311544.$

$S_{768}=\sqrt{2-\sqrt{4-(.01636228)^2}}=.00818126. \quad \therefore P_{768}=6.28316941.$

By continuing the computation it is found that the first six decimal figures in the value of the perimeter of the inscribed polygon remain unchanged.

$$\therefore C, \text{ or } 2\pi=6.283169 \text{ approximately.}$$
$$\therefore \pi=3.14159 \text{ approximately.}$$

MAXIMA AND MINIMA

470. Def. **A maximum** (see Art. 268) is the greatest of a group of magnitudes, all of which satisfy certain given conditions.

Thus, the diameter is the maximum chord which can be drawn in a circle.

471. Def. **A minimum** is the smallest of a group of magnitudes, all of which satisfy certain given conditions.

Thus, of all lines which can be drawn from a given point to a given line the perpendicular is the minimum.

Certain maxima and minima have already been studied, and we now proceed to investigate more particularly those relating to regular polygons and the circle.

PROPOSITION XXI. THEOREM

472. *Of all triangles which have two sides equal, that triangle in which these sides include a right angle is the maximum.*

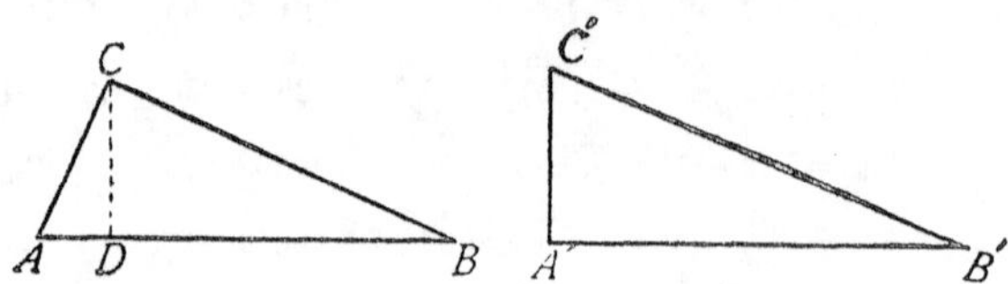

Given the ⟁ ABC and $A'B'C'$ in which $AB = A'B'$, $CA = C'A'$, and $\angle A'$ is a rt. $\angle$.

To prove $\triangle A'B'C' > \triangle ABC$.

Proof. In the $\triangle ACB$, draw $CD \perp AB$.

Then $CD < CA$. (Why?)

$\therefore CD < C'A'$. Ax. 8.

But the ⟁ ABC and $A'B'C'$ have equal bases. Hyp.

∴ these ⟁ are to each other as their altitudes, CD and $C'A'$. Art. 391.

$\therefore \triangle A'B'C' > \triangle ABC$.
(*for* $C'A' > CD$).

Q. E. D.

473. Isoperimetric figures are figures having equal perimeters.

Ex. 1. Find the minimum line that can be drawn between two given parallel lines.

Ex. 2. What is the largest stick that can be placed on a rectangular table 12 x 5 ft., and not have an end projecting over a side of the table?

Proposition XXII. Theorem

474. *Of all isoperimetric triangles which have the same base, the isosceles triangle is the maximum.*

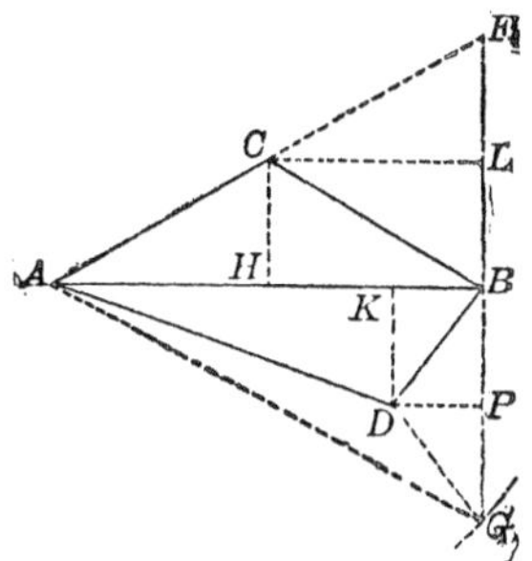

Given the $\triangle$ ABC and ABD having the same base AB and equal perimeters, $AC=CB$, and AD and DB unequal.

To prove $\triangle ACB > \triangle ADB$.

Proof. Produce AC to F, making $CF=AC$.

Draw FB, and from D as a center, with a radius equal to DB, describe an arc cutting FB produced in G.

Draw DG and AG.

Draw CH and $DK \perp AB$; also CL and $DP \perp FG$.

Then $\angle ABF$ is a right $\angle$, Art. 261.

(*for it may be inscribed in a semicircle whose center is C and whose diameter is ACF*).

Also ADG is not a straight line,

(*for, if it were, the* $\angle$ *DAB and DBA would be complements of the* $=\angle$ *DGB and DBG, respectively, and hence would be equal, and* $\therefore \triangle DAB$ *would be isosceles, which is contrary to the hypothesis*).

$\therefore AF=AC+CB=AD+DB=AD+DG$. Constr. Hyp.

But $AD+DG > AG$ (Art. 92). $\therefore AF > AG$. Ax. 8.

$\therefore BF > BG$. Art. 111.

$\therefore \frac{1}{2} BF > \frac{1}{2} BG$, or $CH > KD$. Ax. 10.

$\therefore \triangle ACB > \triangle ADB$. Art. 391.

Q. E. D.

PROPOSITION XXIII. THEOREM

475. *Of isoperimetric polygons having the same number of sides, the maximum is equilateral.*

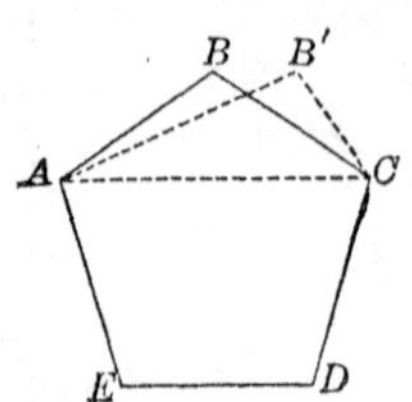

Given $ABCDE$ the maximum of all polygons having a given perimeter, and a given number of sides.

To prove $ABCDE$ equilateral.

Proof. If $ABCDE$ is not equilateral, at least two of its sides, as AB and BC, must be unequal,

If this is possible, on the diagonal AC as a base, construct a triangle having the same perimeter as ABC, and having side $AB'=B'C$.

Then $\triangle AB'C > \triangle ABC$. Art. 474.

To each of these unequals add the polygon $ACDE$.

$\therefore AB'CDE > ABCDE$. Ax. 9.

But this is contrary to the hypothesis that $ABCDE$ is the maximum of the class of polygons considered.

Hence $AB=BC$, and $ABCDE$ is equilateral.

Q. E. D.

Ex. Of all circles which are described on a given line as chord, which is the minimum?

PROPOSITION XXIV. THEOREM

476. *Of all polygons having all sides given but one, the maximum can be inscribed in a semicircle having the undetermined side as a diameter.*

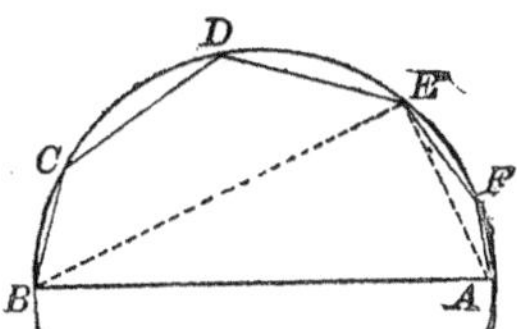

Given the polygon $ABCDEF$, the maximum of all polygons having the sides BC, CD, DE, EF, FA, in common, and AB undetermined.

To prove that AB is the diameter of a semicircle in which $ABCDEF$ can be inscribed.

Proof. Draw lines from any vertex, E, to A and B.

The $\triangle$ BEA must be the maximum of all $\triangle$s having the sides BE and EA,

(*for, if it is not, by increasing or decreasing the angle BEA, the $\triangle$ BEA can be changed till it is a maximum, the rest of figure, $BCDE$ and EFA, meantime remaining unchanged; thus the area of the polygon $ABCDEF$ would be increased, which is contrary to the hypothesis that $ABCDEF$ is a maximum*).

$\therefore$ $\angle BEA$ is a right $\angle$. Art 472.

$\therefore$ E is on the semicircumference of which AB is the diameter.

In like manner the other vertices, C, D and F, must be on the semicircumference which has AB for a diameter.

Q. E. D.

S

Proposition XXV. Theorem

477. *Of all polygons formed with the same given sides, that which can be inscribed in a circle is the maximum.*

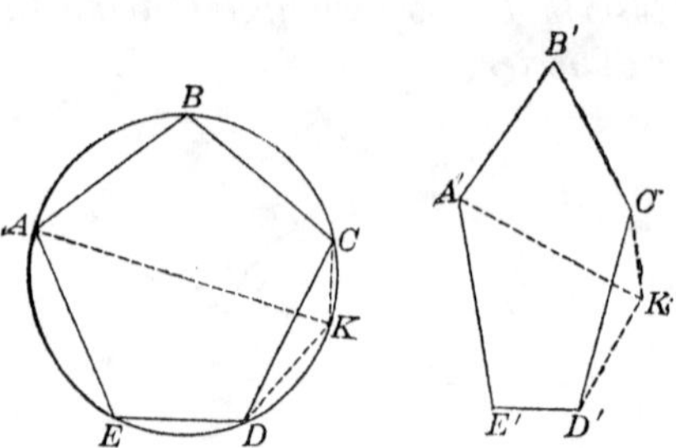

Given $ABCDE$ a polygon which can be inscribed in a $\odot$, and $A'B'C'D'E'$ a polygon which has the same sides as $ABCDE$, but which cannot be inscribed in a $\odot$.

To prove $ABCDE > A'B'C'D'E'$.

Proof. From any vertex, A, of $ABCDE$ draw the diameter AK, and join K to the adjacent vertices C and D.

Upon $C'D'$ construct the triangle $C'K'D'$ equal to $\triangle\ CDK$, and draw $A'K'$.

Then	area $ABCK >$ area $A'B'C'K'$.	Art. 476.
Also	area $AEDK >$ area $A'E'D'K'$.	(Why ?)
Adding,	$ABCKDE > A'B'C'K'D'E'$.	(Why ?)
But	$\triangle\ CKD = \triangle\ C'K'D'$.	Constr.
Subtracting,	$ABCDE > A'B'C'D'E'$.	(Why ?)

Q. E. D.

478. Note. It might happen that one of the parts of the second figure formed by the diameter $A'K'$, as $A'B'C'K'$, could be inscribed in a semicircle, and $\therefore = ABCK$. How, then, would the above proof be modified ?

479. Cor. *Of all isoperimetric polygons of a given number of sides the maximum polygon is regular.*

For it is equilateral (Art. 475), and can be inscribed in a circle (Art. 477), and is, therefore, regular (Art. 417).

PROPOSITION XXVI. THEOREM

480. *Of two isoperimetric regular polygons, that which has the greater number of sides has the greater area.*

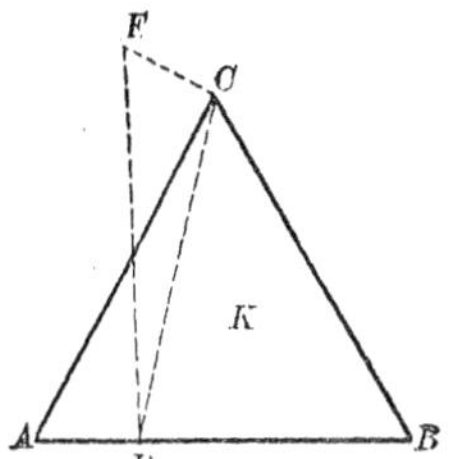

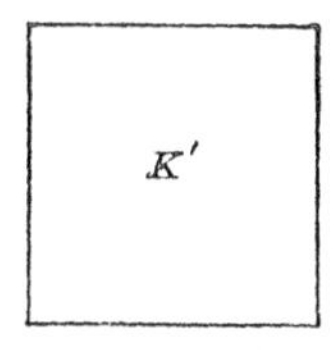

Given K a regular polygon of any number of sides, as three, and K' a regular polygon of one more, or four sides, and let K and K' have equal perimeters.

To prove $K' > K$.

Proof. From any vertex, C, of K, draw a line CD to any point D of the side AB, which meets one of the sides of $\angle C$.

Construct the $\triangle DCF$, having $CF = DA$, and $DF = CA$.

$\therefore \triangle DCF = \triangle CDA$. Art. 101.

Adding $\triangle CBD$, $DFCB \approx K$. Ax. 2.

Hence the polygon $DFCB$ has the same perimeter as K', and the same area as K.

But $DFCB$ is an irregular, while K' is a regular polygon of four sides.

$\therefore K' > DFCB$. Art. 479.

$\therefore K' > K$. Ax. 8.

In like manner it may be shown that a regular polygon of one more, or five sides, $> K$, and so on.

Q. E. D.

481. COR. *Of isoperimetric plane figures, the circle is the maximum.* That is, the area of a circle is greater than the area of any polygon with equal perimeter.

Proposition XXVII. Theorem

482. *Of two equivalent regular polygons that which has the less number of sides has the greater perimeter.*

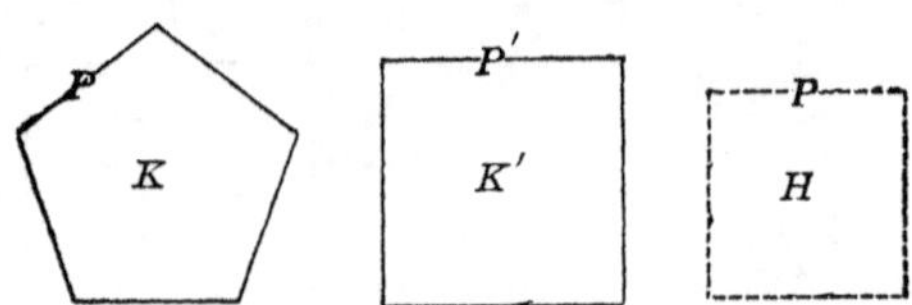

Given K and K' two regular polygons having the same area, and having their perimeters denoted by P and P', but K' having the less number of sides.

To prove $P' > P$.

Proof. Let H be a regular polygon having the same number of sides as K', and the same perimeter as K.

Then $K > H$. Art. 480.

$\therefore K' > H$. Ax. 8.

$\therefore P' > P$. Art. 399.

Q. E. D.

Ex. 1. Find the area of a triangle in which two of the sides are 6 and 12 in., and the included angle is 90°. Find the area of another triangle having two sides of 6 and 12 in., and the included angle 60°.

Ex. 2. How long is the fence about a garden 60 x 40 ft.? How many square feet in the area of the garden? Find also the length of fence and area of a garden 50 ft. square.

Ex. 3. Find the area of an equilateral triangle, a square, a hexagon, and a circle, in each of which the perimeter is 1 ft.

Ex. 4. Find the perimeter of an equilateral triangle, a square, and a circle, in each of which the area is 24 sq. in.

Ex. 5. What principle of maxima and minima is illustrated in each of the four preceding Exs.?

SYMMETRY

483. Symmetry of polygons. Many of the properties of regular figures can be obtained in a simple and expeditious way by the use of the ideas of symmetry.

484. An **axis of symmetry** is a line such that, if part of a figure be folded over upon it as an axis, the part folded over will coincide with the remaining part of the figure.

EXERCISES. GROUP 46

Ex. 1. How many axes of symmetry has an isosceles triangle? an equilateral triangle?

Ex. 2. How many has a square? a regular pentagon?

Ex. 3. How many has a regular hexagon? a regular heptagon?

Ex. 4. How many has a regular octagon? a regular polygon of n sides?

Ex. 5. How many has a circle?

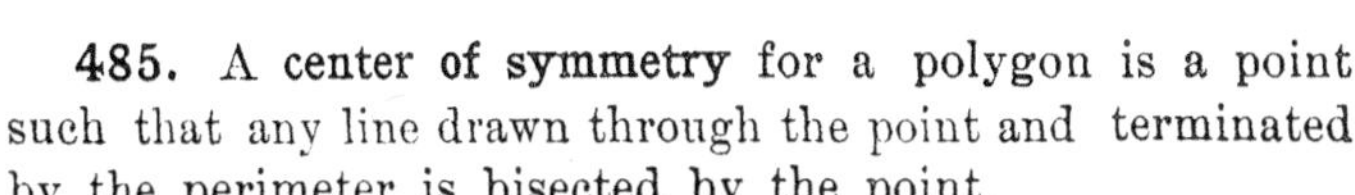

485. A **center of symmetry** for a polygon is a point such that any line drawn through the point and terminated by the perimeter is bisected by the point.

EXERCISES. GROUP 47

Ex. 1. Has an equilateral triangle a center of symmetry? Has a square?

Ex. 2. Has a regular pentagon a center of symmetry? Has a regular hexagon?

Ex. 3. In general, which regular polygons have a center of symmetry, and which do not?

Ex. 4. Has a circle a center of symmetry?

Ex. 5. Which is the most symmetrical plane figure studied thus far?

SYMMETRY WITH RESPECT TO A LINE OR AXIS

486. Two points symmetrical with respect to a line or axis are points such that the straight line joining them is bisected by the given line at right angles.

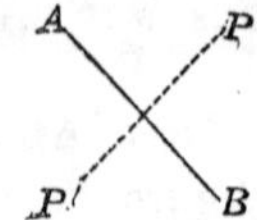

Thus, if PP' is bisected by AB, and PP' is $\perp AB$, the points P and P' are symmetrical with respect to the axis AB.

487. A figure symmetrical with respect to an axis is a figure such that each point in the one part of the figure has a point in the other part symmetrical to the given point, with respect to an axis.

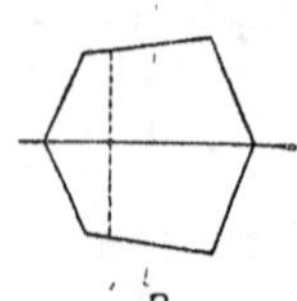

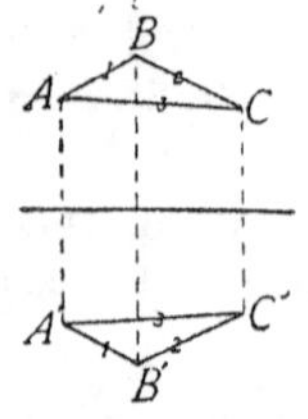

488. Two figures symmetrical with respect to an axis are two figures such that each point in the one figure has a point in the other figure symmetrical to the given point, with respect to an axis.

SYMMETRY WITH RESPECT TO A POINT OR CENTER

489. Two points symmetrical with respect to a point or center are points such that the straight line joining them is bisected by the point or center.

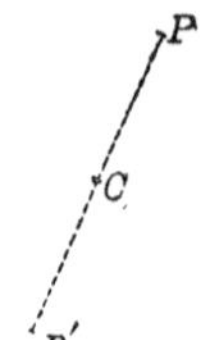

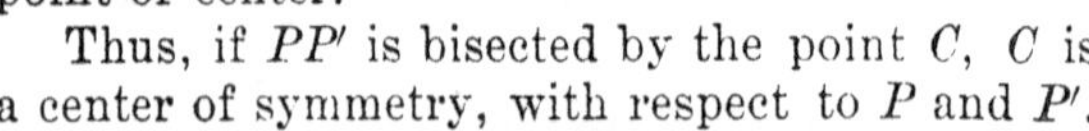

Thus, if PP' is bisected by the point C, C is a center of symmetry, with respect to P and P'.

490. A figure symmetrical with respect to a point or center is a figure such that each point in the figure has another point in the figure symmetrical to the given point with respect to the center.

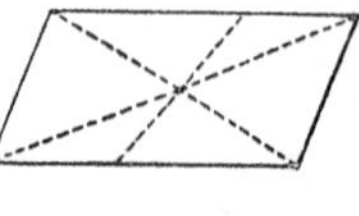

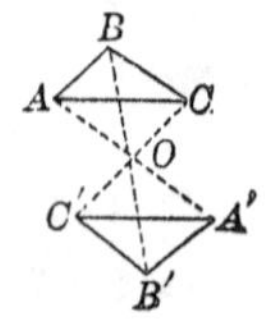

491. Two figures symmetrical with respect to a center are figures such that each point in one figure has a point in the other figure symmetrical to it with respect to the center.

PROPOSITION XXVIII. THEOREM

492. *If a figure is symmetrical with respect to two axes which are perpendicular to each other, it is symmetrical with respect to their point of intersection as a center.*

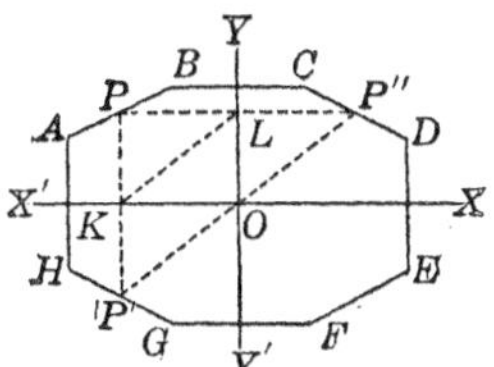

Given the figure $ABC \ldots H$ symmetrical with respect to the two axes XX' and YY'; and $XX' \perp YY'$ and intersecting it at O.

To prove $ABC \ldots H$ symmetrical with respect to O.

Proof. Take any point P in the perimeter of the figure, and determine the points P' and P'', symmetrical with respect to P, by drawing $PKP' \perp XX'$, and $PLP'' \perp YY'$.

Draw KL, $P'O$, OP''.

Then $PP' \parallel YY'$, and $PP'' \parallel XX'$. Art. 121.

But $PK = KP'$. Art. 486.

Also $PK =$ and $\parallel LO$. Art. 157.

$\therefore KP' =$ and $\parallel LO$. Ax. 1.

$\therefore KLOP'$ is a $\square$, and $KL =$ and $\parallel P'O$. Art. 160.

In like manner it may be shown that $KL =$ and $\parallel OP''$.

$\therefore P'O =$ and $\parallel OP''$. Ax. 1, Art. 122.

$\therefore P'OP''$ is a straight line bisected by point O. Geom. Ax. 3.

Hence any straight line drawn through O, and terminated by the perimeter, is bisected at O,

(*for P is any point on the perimeter*).

$\therefore O$ is a center of symmetry for the given figure. Art. 490.

Q. E. D.

EXERCISES. GROUP 48

THEOREMS CONCERNING REGULAR POLYGONS AND THE CIRCLE

Ex. 1. The diagonals of a regular pentagon are equal.

Ex. 2. If $ABCDE$ is a regular pentagon, the triangles ABK and ABC are similar.

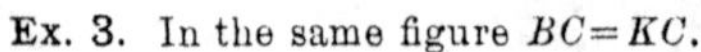

Ex. 3. In the same figure $BC=KC$.

Ex. 4. Also $KCDE$ is a parallelogram.

Ex. 5. Also $AC=AB+BK$.

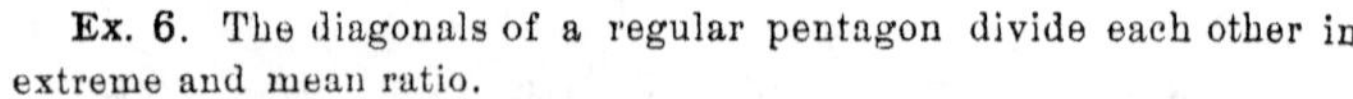

Ex. 6. The diagonals of a regular pentagon divide each other in extreme and mean ratio.

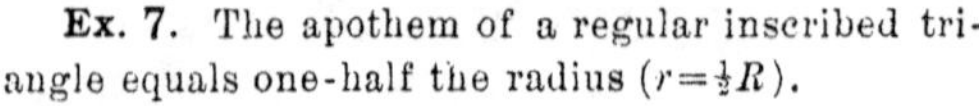

Ex. 7. The apothem of a regular inscribed triangle equals one-half the radius ($r=\frac{1}{2}R$).

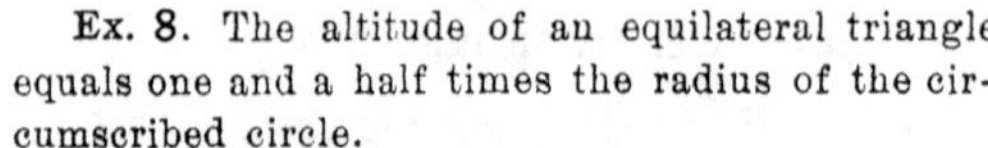

Ex. 8. The altitude of an equilateral triangle equals one and a half times the radius of the circumscribed circle.

Ex. 9. The altitude of an equilateral triangle is to the diameter of the circumscribed circle as 3 : 4.

Ex. 10. The side of an equilateral triangle equals $R\sqrt{3}$, the radius of the circumscribed circle being denoted by R.

Ex. 11. The apothem of a regular inscribed hexagon equals one-half the side of a regular inscribed triangle ($=\frac{R}{2}\sqrt{3}$).

Ex. 12. The side of a regular circumscribed triangle is double the side of the regular inscribed triangle.

Ex. 13. Find the side of an inscribed square in terms of R; also the side of a circumscribed square.

Ex. 14. Find the apothem of an inscribed square, and also of a circumscribed square, in terms of R.

Ex. 15. Find the areas of the inscribed and circumscribed squares in terms of R, and show that one of these is double the other.

Ex. 16. Show that the area of a regular inscribed triangle is $\frac{3\sqrt{3}R^2}{4}$.

Ex. 17. Show that the area of a regular inscribed hexagon is $\frac{3\sqrt{3}R^2}{2}$. What, then, is the ratio of the area of a regular inscribed triangle to that of a regular inscribed hexagon?

Ex. 18. The area of a regular inscribed hexagon is three-fourths the area of the regular circumscribed hexagon.

Ex. 19. The area of a regular inscribed hexagon is a mean proportional between the areas of a regular inscribed and a regular circumscribed triangle.

[SUG. On a figure similar to that of Prop. IV, p. 266, let OP intersect AB in K, and compare the ⚠ OKA, OAP, and OPA'.]

Ex. 20. The area of a regular inscribed polygon of $2n$ sides is a mean proportional between the areas of regular inscribed and circumscribed polygons of n sides.

Ex. 21. The area of a regular inscribed octagon equals the area of the rectangle whose base and altitude are the sides of the circumscribed and inscribed squares respectively.

Ex. 22. The area of an inscribed regular dodecagon equals three times the square of the radius.

Ex. 23. An angle of a regular polygon is the supplement of the angle at the center.

Ex. 24. Diagonals drawn from a vertex of a regular polygon of n sides divide the angle at that vertex into $n-2$ equal parts.

Ex. 25. The diagonals formed by joining the alternate vertices of a regular hexagon form another regular hexagon. Find also the ratio of the areas of the two hexagons.

Ex. 26. If squares be erected on the sides of a regular hexagon, the lines joining their exterior vertices form a regular dodecagon. Find also the area of this dodecagon in terms of b, a side of the hexagon.

Ex. 27. The square of a side of an inscribed equilateral triangle equals the square of a side of an inscribed square added to the square of a side of an inscribed regular hexagon.

Ex. 28. The area of a circle equals four times the area of a circle described on its radius as a diameter.

Ex. 29. The area of a circular ring equals the area of a circle whose diameter is the chord of the outer circle tangent to the inner circle.

Ex. 30. Given *AaBbC*, *AcB*, *BdC*, semicircles; prove that the sum of the two crescents *AcBa* and *BdCb* equals the area of the right triangle *ABC*.

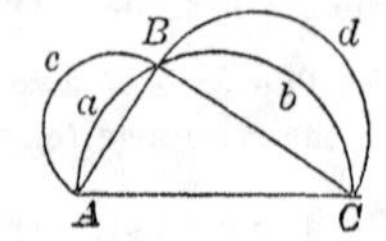

Ex. 31. An equiangular polygon inscribed in a circle is regular if the number of its sides be odd.

[SUG. In the figure to Prop. III, p. 264, arc *AEDC*=arc *EDCB* ∴ arc *AE*=arc *BC*, ∴ side *AE*=side *BC*, etc.]

EXERCISES. GROUP 49

MAXIMA AND MINIMA

Ex. 1. Of isoperimetric parallelograms with the same base the rectangle is the maximum.

Ex. 2. Of equivalent parallelograms with the same base, which has the minimum perimeter?

Ex. 3. Of isoperimetric rectangles which is the maximum?

Ex. 4. Divide a given line into two parts such that their product is a maximum.

Ex. 5. Find a point in the hypotenuse of a right triangle such that the sum of the squares of the perpendiculars drawn from the point to the legs shall be a minimum.

Ex. 6. How shall a mile of wire fence be stretched so as to contain the maximum area?

Ex. 7. Find the area in acres included by a mile of wire fence if it be stretched as a square, a regular hexagon, and a circle respectively.

Ex. 8. Of all triangles with the same base and equal altitudes, the isosceles triangle has the least perimeter.

Ex. 9. Of all polygons of a given number of sides inscribed in a given circle, the maximum is regular.

[SUG. Prove the maximum polygon (1) equilateral, (2) equiangular.]

Ex. 10. Find the maximum rectangle inscribed in a circle.

Ex. 11. Find the maximum rectangle that can be inscribed in a semicircle.

[SUG. Inscribe a square in the circle.]

Ex. 12. Of trapezoids inscribed in a semicircle (having the diameter as one base), find the maximum.

Ex. 13. Divide a given straight line into two parts such that the sum of the squares of these parts shall be a minimum.

EXERCISES. GROUP 50

SYMMETRY

Ex. 1. A rhombus has how many axes of symmetry? Has it a center of symmetry?

Ex. 2. What axis of symmetry has a quadrilateral which has two pairs of equal adjacent sides? Has such a figure a center of symmetry?

Ex. 3. A parallelogram is symmetrical with respect to the point of intersection of its diagonals.

Ex. 4. A segment of a circle is symmetrical with respect to what axis?

Ex. 5. Has a trapezium a center of symmetry? An axis of symmetry?

Ex. 6. How many axes of symmetry have two equal circles taken as one figure? Have they a center of symmetry?

Ex. 7. What axis of symmetry have any two circles?

Ex. 8. How must two equilateral triangles be placed so as to have a center of symmetry? So as to have an axis of symmetry?

Ex. 9. If two polygons are symmetrical with reference to a center, any two homologous sides are equal and parallel and drawn in opposite directions.

EXERCISES. GROUP 81

VARIOUS THEOREMS

Ex. 1. The square on the diameter of a circle equals twice the square inscribed in the circle.

Ex. 2. The altitude of an inscribed equilateral triangle is to the radius of the circle as 3 : 2.

Ex. 3. The diagonal joining any two opposite vertices of a regular hexagon passes through the center.

Ex. 4. The radius of an inscribed regular polygon is a mean proportional between its apothem and the radius of the circumscribed regular polygon of the same number of sides.

Ex. 5. If the sides of a regular hexagon be produced, their points of intersection are the vertices of another regular hexagon. Also find the ratio of the areas of the two hexagons.

Ex. 6. Of all lines drawn through a given point within an angle, and terminated by the sides of the angle, the line which is bisected at the given point cuts off the minimum area.

Ex. 7. Each angle of a regular polygon of $n+2$ sides contains $\frac{2n}{n+2}$ right angles.

Ex. 8. The diagonals from a vertex of a regular polygon of $n+2$ sides divide the angle at that vertex into n equal parts.

Ex. 9. The sum of the perpendiculars drawn to the sides of a regular polygon of n sides from any point within the polygon equals n times the apothem.

Ex. 10. An equiangular polygon circumscribed about a circle is regular.

[SUG. Draw radii from the points of contact, and lines from the vertices of the polygon to the center.]

Ex. 11. An equilateral polygon circumscribed about a polygon is regular if the number of its sides is odd.

[SUG. See Figure of Prop. III, p. 264. Prove $AT=BQ$. Draw radii and prove $\angle T=\angle Q$, etc.]

Ex. 12. How must two equal isosceles triangles be placed so as to have a center of symmetry? How must they be placed so as to have an axis of symmetry?

If, in a regular inscribed polygon of n sides, S_n denotes a side and r_n denotes the apothem, show that

Ex. 13. For the triangle, $S_3 = R\sqrt{3}$, $r_3 = \frac{1}{2}R$.

Ex. 14. For the square, $S_4 = R\sqrt{2}$, $r_4 = \frac{1}{2}R\sqrt{2}$.

Ex. 15. For the hexagon, $S_6 = R$, $r_6 = \frac{1}{2}R\sqrt{3}$.

Ex. 16. For the decagon, $S_{10} = \dfrac{R(\sqrt{5}-1)}{2}$, $r_{10} = \dfrac{R\sqrt{10+2\sqrt{5}}}{4}$.

Ex. 17. For the pentagon, $S_5 = \dfrac{R\sqrt{10-2\sqrt{5}}}{2}$, $r_5 = \dfrac{R(\sqrt{5}+1)}{4}$.

Ex. 18. For the octagon, $S_8 = R\sqrt{2-\sqrt{2}}$.

Ex. 19. For the dodecagon, $S_{12} = R\sqrt{2-\sqrt{3}}$.

Ex. 20. Prove that $S_5^2 = R^2 + S_{10}^2$.

Ex. 21. Prove that $S_5^2 = S_6^2 + S_{10}^2$.

Ex. 22. If ADB, AaC, CbB are semicircles and $DC \perp AB$, prove that the area bounded by the three semicircumferences equals the area described on DC as a diameter.

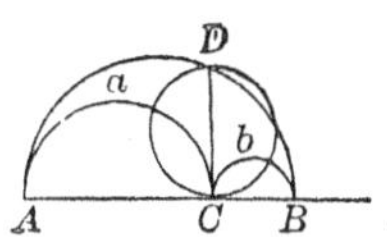

Ex. 23. If p_n denotes the perimeter of an inscribed polygon of n sides and P_n the perimeter of a circumscribed polygon of n sides, prove that $P_{2n} = \dfrac{2p_n P_n}{p_n + P_n}$.

EXERCISES. GROUP 52

PROBLEMS

Circumscribe about a given circle

Ex. 1. An equilateral triangle.

Ex. 2. A square.

Ex. 3. A regular pentagon.

Ex. 4. A regular hexagon.

Ex. 5. A regular octagon.

Ex. 6. A regular decagon.

Ex. 7. Construct a regular pentagram, or 5-pointed star.

Ex. 8. Construct a hexagram, or 6-pointed star.

Ex. 9. Construct an 8-pointed star.

Ex. 10. Construct an angle of 36°.

Ex. 11. Construct angles of 18°, 9°, 72°.

Ex. 12. Construct angles of 24°, 12°, 6°, 48°.

Ex. 13. Divide a given circumference into two parts which shall be in the ratio of 3 : 7.

Ex. 14. Construct a regular pentagon which shall have twice the perimeter of a given regular pentagon.

Ex. 15. Construct a regular pentagon whose perimeter shall equal the sum of the perimeters of two given regular pentagons.

Ex. 16. Construct a regular pentagon whose area shall be twice the area of a given regular pentagon.

Ex. 17. Construct a regular pentagon whose area shall be equal to the sum of the areas of two given regular pentagons.

Ex. 18. Construct a circumference which shall be twice the circumference of a given circle.

Ex. 19. Construct a circle whose area shall be three times the area of a given circle.

Ex. 20. Construct a circumference equivalent to the sum, and another equivalent to the difference, of two given circumferences.

Ex. 21. Construct a circle equivalent to the sum, and another equivalent to the difference, of two given circles.

Ex. 22. Construct a circle whose area shall be two-thirds the area of a given circle.

Ex. 23. Bisect the area of a given circle by a concentric circumference.

Ex. 24. Divide the area of a given circle into five equal parts by drawing concentric circumferences.

On a given line construct

Ex. 25. A regular pentagon.

Ex. 26. A regular hexagon.

Ex. 27. A regular dodecagon.

Ex. 28. A circle equivalent to a given semicircle.

Ex. 29. Inscribe a regular octagon in a given square.

Ex. 30. Inscribe a circle in a given sector.

Ex. 31. Inscribe a square in a given segment.

Ex. 32. In a given equilateral triangle inscribe three equal circles, each of which touches the other two circles and a side of the triangle.

Ex. 33. In a given circle inscribe three equal circles which shall touch each other and the given circumference.

NUMERICAL APPLICATIONS OF PLANE GEOMETRY

METHODS OF NUMERICAL COMPUTATIONS

493. Cancellation. In numerical work in geometry, as elsewhere, the labor of computations may frequently be economized. Those methods of abbreviating work, which are particularly applicable in the ordinary numerical applications of geometry, may be briefly indicated, as follows:

To simplify numerical work by cancellation, *group together as a whole all the numerical processes of a given problem, and make all possible cancellations before proceeding to a final numerical reduction.*

Ex. Find the ratio of the area of a rectangle, whose base and altitude are 42 and 24 inches, to the area of a trapezoid, whose bases are 21 and 35 and altitude 12.

By Arts. 383, 394

$$\frac{\text{area of rectangle}}{\text{area of trapezoid}} = \frac{42 \times 24}{6(21+35)} = \frac{\overset{\overset{3}{\cancel{6}}}{\cancel{42}} \times \overset{4}{\cancel{24}}}{\cancel{6} \times \underset{\underset{4}{\cancel{8}}}{\cancel{56}}} = 3, \textit{ Ratio.}$$

494. Use of radicals and of π. Where radicals enter in the course of the solution of a numerical problem, it frequently saves labor *not to extract the root of the radical till the final answer is to be obtained.*

Ex. 1. Find the area of a circle circumscribed about a square whose side is 8.

The diagonal of the square must be $8\sqrt{2}$ (Art. 346).

$\therefore$ the radius of $\odot = 4\sqrt{2}$.

$\therefore$ by Art. 449, area of $\odot = \pi(4\sqrt{2})^2 = 32\pi = 100.6$, *Area.*

Similarly in the use of π, it frequently saves labor *not to substitute its numerical value for π till late in the process of solution.*

Ex. 2. Find the radius of a circle whose area is equal to the sum of the areas of two circles whose radii are 6 and 8 inches, respectively.

Denote the radius of the required circle by x.

Then, by Art. 449, $\pi x^2 = 36\pi + 64\pi$.

$$\therefore \pi x^2 = 100\pi.$$

$$\therefore x^2 = 100, \text{ and } x = 10, \textit{ Radius.}$$

495. Use of x, y, etc., as symbols for unknown quantities. In some cases *a numerical computation is greatly facilitated by the use of a specific symbol for an unknown quantity.*

Ex. In a triangle whose sides are 12, 18, and 25, find the segments of the side 25 made by the bisector of the angle opposite.

Denote the required segments of side 25 by x and $25 - x$.

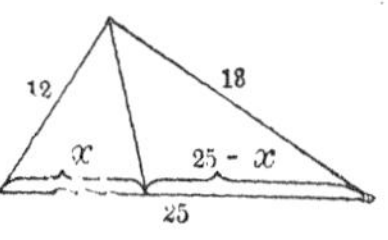

Then $12 : 18 = x : 25 - x$ (Art. 332)

$\therefore 18x = 12(25 - x)$ (Art. 302)

$\therefore x = 10.$ }

And $25 - x = 15.$ } *Segments.*

496. Limitations of numerical computations. Owing to the limitations of human eyesight and of the instruments used in making measurements, no measurement can be accurate beyond the fifth or sixth figure; and in ordinary work, such as is done by a carpenter, measurements are not accurate beyond the third figure. As all numerical applications of geometry are based on practical measurements, *it is not necessary to carry arithmetical work beyond the fifth or sixth significant digit.*

Other methods of facilitating numerical computations, as by the **use of logarithms,** are beyond the scope of this book.

T

NUMERICAL PROPERTIES OF LINES

EXERCISES. GROUP 53

THE RIGHT TRIANGLE

Ex. 1. Find the hypotenuse of a right triangle whose legs are 12 and 35.

Ex. 2. The hypotenuse of a right triangle is 29, and one leg is 20. Find the other leg.

Ex. 3. If a window is 15 ft. from the ground and the foot of a ladder is to be 8 feet from the house, how long a ladder is necessary to reach the window?

Ex. 4. Find the diagonals of a rectangle whose sides are 5 and 12.

Ex. 5. If the base of an isosceles triangle is 8 and a leg is 5, find the altitude.

Ex. 6. Find the diagonal of a square whose side is 1 ft. 6 in.

Ex. 7. The diagonals of a rhombus are 24 and 10. Find a side.

Ex. 8. One side of a rhombus is 17 and one diagonal is 30. Find the other diagonal.

Ex. 9. In a circle whose radius is 5, find the length of the longest and shortest chords through a point at a distance 3 from the center.

Ex. 10. In a circle whose radius is 25 in., find the distance from the center to a chord 48 in. long.

Ex. 11. If a chord 12 in. long is 5 in. from the center of a circle, how far from the center of the same circle is a chord 10 in. long?

Ex. 12. A ladder 40 ft. long reaches a window 20 ft. high on one side of a street and, if turned on its foot, reaches a window 30 ft. high on the other side. How wide is the street?

Ex. 13. If one leg of a right triangle is 10 and the hypotenuse is twice the other leg, find the hypotenuse.

Ex. 14. Find the altitude of an equilateral triangle whose side is 6.

Ex. 15. Find the side of an equilateral triangle whose altitude is 8.

Ex. 16. Find the side of a square whose diagonal is 15.

Ex. 17. One leg of a right triangle is 3, and the sum of the hypotenuse and the other leg is 9. Find the sides.

Ex. 18. A tree 90 ft. high is broken off 40 ft. from the ground. How far from the foot of the tree will the top strike?

Ex. 19. The radii of two circles are 1 and 6 in., and their centers are 13 in. apart. Find the length of the common external tangent.

Ex. 20. The sides of a triangle are 10, 11, 12. Find the length of the projection of the side whose length is 10, on the side 12.

EXERCISES. GROUP 54

TRIANGLES IN GENERAL

Ex. 1. The sides of a triangle are 12, 18 and 20. Find the segments of the side 20, made by the bisector of the angle opposite.

Ex. 2. In the same triangle, find the segments of the side 20, made by the bisector of the exterior angle opposite.

Ex. 3. If the legs of a right triangle are 6 and 8, find the hypotenuse, the altitude on the hypotenuse, and the projections of the legs on the hypotenuse.

Ex. 4. Is a triangle acute, obtuse, or right, if the three sides are 5, 12, 14; 5, 11, 12; 5, 12, 13; 4, 5, 6?

Ex. 5. If the sides of a triangle are 6, 7 and 8, compute the length of the altitude on 8.

Ex. 6. Also the length of the median on the same side.

Ex. 7. Also the length of the bisector of the angle opposite the side 8.

Ex 8. If two sides and a diagonal of a parallelogram are 8, 12 and 10, find the other diagonal.

[SUG. Use Art. 352.]

Ex. 9. Two sides of a triangle are 10 and 18, and the median to the third side is 12. Find the third side.

Ex. 10. Two sides of a triangle are 17 and 16, and the altitude on the third side is 15. Find the third side.

Ex. 11. The hypotenuse of a right triangle is 10, and the altitude on the hypotenuse is 4. Find the segments of the hypotenuse and the legs.

Ex. 12. Find the three medians, the three bisectors, and the three altitudes of a triangle whose sides are 13, 14, 15.

EXERCISES. GROUP 56

CIRCUMFERENCES AND ARCS

Using $\pi = \frac{22}{7}$,

Ex. 1. Find the circumference of a circle whose radius is 1 ft. 9 in.

Ex. 2. Find the radius of a circle whose circumference is 121 ft.

Ex. 3. A bicycle wheel 28 in. in diameter makes, in an afternoon, 3,000 revolutions. How far does the bicycle travel?

Ex. 4. What is the diameter of a wheel which makes 1,400 revolutions in going 8,800 yds.?

Ex. 5. If the diameter of a circle is 20, find the length of an arc of 60°; also of 83°.

Ex. 6. If the length of an arc is 14 and the radius is 6, find the number of degrees in the arc.

Ex. 7. If the arc of a quadrant is 1 ft. in length, find the diameter.

Ex. 8. Two concentric circumferences are 88 and 132 in. in length, respectively. Find the width of the circular ring between them.

Ex. 9. If the year be taken as 365¼ da., and the earth's orbit a circle whose radius is 93,250,000 miles, find the velocity of the earth in its orbit per second.

Find the radius and circumference of a circle circumscribed about

Ex. 10. A square whose side is 5.

Ex. 11. An equilateral triangle whose side is 4.

Ex. 12. A rectangle whose sides are 12 and 5.

Ex. 13. Find the central angle of a sector whose perimeter equals one-half the circumference.

Ex. 14. Find the diameter of a circle circumscribed about a triangle whose sides are 7, 15, and 20.

Ex. 15. Find the radius of a circle whose circumference equals the perimeter of a square whose diagonal is 10.

EXERCISES. GROUP 56

CHORDS, TANGENTS, AND SECANTS

Ex. 1. Two intersecting chords of a circle are 11 and 14 in., and the segments of the first chord are 8 and 3 in. Find the segments of the second chord.

[SUG. Denote the required segments by x and $14 - x$.]

Ex. 2. In a circle whose radius is 12 in., a chord 16 in. long is passed through a point 9 in. from the center. Find the segments of the chord.

Ex. 3. Two secants drawn from a point to a circle are 24 and 27 in. long. If the external segment of the first is 6 in., find the external segment of the second.

Ex. 4. From a given point a secant whose external and internal segments are 9 and 16 is drawn to a circle. Find the length of the tangent drawn from the same point to the circle.

Ex. 5. From a given point a tangent 24 in. long is drawn to a circle whose radius is 18 in. Find the distance of the point from the center.

Ex. 6. If a diameter 60 in. long is divided into 5 equal parts by chords perpendicular to it, find the length of the chords.

Ex. 7. If a mountain 3 miles high is visible 150 miles at sea, what is the diameter of the earth ?

Ex. 8. If the earth is a sphere of radius 4,000 miles, how far will the light of a lighthouse 100 ft. high be visible at sea ?

EXERCISES. GROUP 57

LINES IN SIMILAR FIGURES

Ex. 1. If the sides of a triangle are 6, 7 and 8, and the shortest side of a similar triangle is 18, find the other sides of the second triangle.

Ex. 2. If a post 5 ft. high casts a shadow 3 ft. long, find the height of a steeple which casts a shadow 90 ft. long.

Ex. 3. In a triangle whose base is 14 and altitude 12, a line is drawn parallel to the base and at a distance 2 from the base. Find the length of the line thus drawn.

Ex. 4. The upper and lower bases of a trapezoid are 12 and 20 and the altitude is 8. If the legs are produced till they meet, find the altitude of each of the two triangles thus formed.

Ex. 5. If the upper and lower bases of a trapezoid are b_1 and b_2 and the altitude is h, find the altitude of each of the triangles formed by producing the legs.

Ex. 6. If the perimeters of two similar polygons are 300 and 400 and a side of the first is 27, find the homologous side of the second.

Ex. 7. If the perimeter of a regular polygon is three times the perimeter of a regular polygon of the same number of sides, what is the ratio of their apothems ?

Ex. 8. If the circumferences of two circles are 600 and 400 ft., what is the ratio of their diameters ?

Ex. 9. In the preceding example, if a chord of the first circle is 30, what is the length of a chord in the second circle, subtending the same number of degrees of arc ?

COMPUTATION OF AREAS

EXERCISES. GROUP 58

AREAS OF TRIANGLES

Ex. 1. Find the area in acres of a triangular field whose base is 300 ft. and altitude 200 ft.

Ex. 2. Find the area of a triangle whose sides are 10, 17, and 21.

Ex. 3. Find the area in acres of a field whose sides are 60, 70, and 80 chains.

Ex. 4. Find the area in acres of a triangular field each of whose sides is 10 chains.

Find the area of

Ex. 5. An isosceles triangle whose base is 16, and each of whose legs is 34.

Ex. 6. An equilateral triangle whose altitude is 8.

Ex. 7. A right triangle in which the segments of the hypotenuse made by the altitude upon it are 12 and 3; also, in one in which the segments are a and b.

Ex. 8. An isosceles right triangle whose hypotenuse is 12.

Ex. 9. A right triangle in which the hypotenuse is 41 and one leg is 9.

Ex. 10. Find in two ways the area of a triangle whose sides are 6, 5, 5.

Ex. 11. A side of a given equilateral triangle is 4 ft. longer than the altitude. Find the area of the triangle.

Ex. 12. The area of an isosceles triangle is 144 and a leg is 24. Find the base.

Ex. 13. The area of an equilateral triangle is $4\sqrt{3}$. Find a side.

Ex. 14. The area of a triangle is 1125, and $a : b : c = 2 : 3 : 4$. Find a, b, c.

Ex. 15. The area of a triangle is 6 sq. in., and two of its sides are 3 and 5 in. Find the remaining side.

EXERCISES. GROUP 59

AREAS OF OTHER RECTILINEAR FIGURES

Find the area of

Ex. 1. A parallelogram whose base is 24 ft. 6 in. and whose altitude is 12 ft. 9 in.

Ex. 2. A trapezoid whose bases are 12 and 20 in. and whose altitude is $1\frac{1}{2}$ ft.

Ex. 3. A rhombus whose diagonals are 9 ft. and 2 yds.

Ex. 4. A quadrilateral in which the sides AB, BC, CD, DA are 12, 13, 14, 15 and the diagonal AC is 17.

Ex. 5. A quadrilateral in which the sides are 27, 36, 30, 25 and the angle included between the first two sides is a right angle.

Ex. 6. A square whose diagonal is 12 in.

Ex. 7. Find the number of boards, each 4 yds. long and 6 in. wide, which are necessary to cover a floor 48×24 ft.

Ex. 8. How many persons can stand in a room 15×9 ft., if each person requires 27×18 in.?

Ex. 9. The baseball diamond is a square each side of which is 90 ft. What fraction of an acre is its area?

Ex. 10. A rectangular garden contains 4,524 sq. yds. and is 20 yds. longer than wide. Find its dimensions.

Ex. 11. Each side of a rhombus is 24 ft. and each of the larger angles is double a smaller one. Find the area.

Ex. 12. Find the area of a rhombus one of whose sides is 17, and one of whose diagonals is 30.

Ex. 13. The area of a trapezoid is 4 acres, one base is 120 yds., and the altitude is 100 yds. Find the other base.

Ex. 14. The bases of an isosceles trapezoid are 20 and 36 and the legs are 17. Find the area.

Ex. 15. The base of a triangle is 20 and the altitude 18. Find the length of a line parallel to the base which cuts off a trapezoid whose area is 80 sq. ft.

[SUG. Denote the altitude of trapezoid by $18 - x$ and find its upper base by similar triangles.]

Ex. 16. The perimeter of a polygon, circumscribed about a circle whose radius is 20, is 340. Find the area of the polygon.

Ex. 17. The area of a rectangle is 144 and the base is three times the altitude. Find the dimensions.

Ex. 18. Find the area of a regular hexagon one of whose sides is 10.

Ex. 19. Find the area of a regular decagon inscribed in a circle whose radius is 20.

Ex. 20. Find a side of a regular hexagon whose area is 200 sq. in.

EXERCISES. GROUP 60

AREAS OF CIRCULAR FIGURES

Ex. 1. Find the area in acres of a circle whose radius is 100 yds.

Ex. 2. Find the radius in inches of a circle whose area is 1 sq. yd.

Ex. 3. Find the area of a circle whose circumference is p.

Ex. 4. Find the radius of a circle whose area equals the sum of the areas of two circles whose radii are 9 and 40 in.

Ex. 5. Find the radius of a circle whose area equals the sum of the areas of three circles whose radii are 20, 28, 29.

Ex. 6. In a circle of radius 50 find the area of a sector of 80°.

Ex. 7. Also of a segment of 60°; of a segment of 300°; of a segment of 240°.

Ex. 8. In a circle whose radius is 7, the area of a sector is 45 sq. ft. Find the number of degrees in its angle.

Ex. 9. In a circle whose radius is 10, find the sum of the segments formed by an inscribed square.

Ex. 10. A circular mill-pond, $\frac{1}{4}$ mile in diameter, contains a circular island, 100 yds. in diameter. Find the water surface of the pond in acres.

Ex. 11. The same pond is surrounded by a driveway 30 ft. wide. Find the area of the driveway.

Ex. 12. Two tangents to a circle, whose radius is 15, include an angle of 60°. Find the area included between the tangents and the radii to the points of contact.

Ex. 13. Find the length of the tether by which a cow must be tied, in order to graze over exactly one acre.

Ex. 14. Three equal circles touch each other externally. Show that the area included between them is $R^2(\sqrt{3} - \frac{\pi}{2})$.

Ex. 15. If the area included between three equal circles which touch each other externally is a square foot, find the radius of each circle in inches.

EXERCISES. GROUP 61

AREAS OF SIMILAR FIGURES

Ex. 1. The homologous sides of two similar triangles are 3 and 5. Find the ratio of their areas.

Ex. 2. The homologous sides of two similar polygons are 4 and 7, and the area of the first polygon is 112. Find the area of the second polygon.

Ex. 3. The radius of a circle is 6. Find the radius of a circle having three times the area of the given circle.

Ex. 4. The areas of two circles are as 16 to 9, and the radius of the first is 8. Find the radius of the second.

Ex. 5. The sides of a triangle are 5, 6, 7. Find the sides of a similar triangle containing 9 times the area of the given triangle.

Ex. 6. If, in finding the area of a circle, a student uses $D=50$ as $R=50$, how will the area as computed differ from the correct area?

Ex. 7. In a triangle whose base is 24 in. and altitude is 18 in., the altitude is bisected by a line parallel to the base. Find the area of the triangle cut off.

Ex. 8. In the triangle of Ex. 7, what part of the altitude must be cut off in order that the area of the triangle be bisected?

Ex. 9. In a circle whose diameter is 30 in., what are the diameters of concentric circumferences which divide the area into three equivalent parts?

Ex. 10. If a circle be constructed on the radius of a given circle, and segments, one in each circle, be formed by a line drawn from the point of contact, find the ratio of the segments.

EXERCISES. GROUP 62

GENERAL NUMERICAL EXERCISES IN PLANE GEOMETRY

Ex. 1. The leg of an isosceles triangle is 10 and the base is 16. Find the altitude and the area.

Ex. 2. Find the area of a triangle whose sides are 25, 39, 40. Also find the radius of a circle equivalent to this triangle.

Ex. 3. Find the area of a regular hexagon inscribed in a circle whose radius is 2.

Ex. 4. The sides of a triangle are 7, 8 and 9 inches. Find the sides of a triangle of four times the area. Also, of twice the area.

Ex. 5. The temple of Herod is said to have accommodated 210,000 people at one time. If each person required 27×18 in., and one-third the space inside the temple be allowed for walls, sanctuaries, etc., what were the dimensions of the temple, if it was a square?

Ex. 6. If the sides of a triangle are 12, 16 and 21, what are the segments of the side 21 made by the bisector of the angle opposite?

Ex. 7. The sides of a quadrilateral in order are 5, 5, 4, 3, and the first two of these sides contain an angle of 60°. Find the area.

Ex. 8. Find the diameter of a wheel which, in a mile, makes 480 revolutions.

Ex. 9. The area of a trapezoid is 112 and the two bases are 12 and 16. Find the altitude.

Find the radius of a circle equivalent to

Ex. 10. A square whose side is 10.

Ex. 11. An equilateral triangle whose side is 12.

Ex. 12. A trapezoid whose bases are 16 and 18 and altitude 9.

Ex. 13. A semicircle whose radius is 15.

Ex. 14. Find the diameter of a circle whose area shall be equivalent to the sum of two circles whose diameters are 144 and 17.

Ex. 15. A circle, a square, and an equilateral triangle each have a perimeter of 12 yds. Find the area of each figure.

Ex. 16. In a circle whose area is 400, the area of a sector is 125. Find the angle of the sector.

Ex. 17. How many acres are included within a half-mile running track, if the track is in the shape of a rectangle twice as long as it is wide?

Ex. 18. In a square whose side is 6 in., find the area of the inscribed and of the circumscribed circles.

Ex. 19. Find the area of the circle circumscribed about a rectangle whose sides are 40 and 9.

Ex. 20. One leg of a right triangle is 12, and the difference between the hypotenuse and the other leg is 8. Find the area.

Ex. 21. Find the area of an isosceles right triangle whose hypotenuse is 20 ft.

Ex. 22. In a triangle whose sides are 16, 18, 20, find the length of the altitude, median, and bisector of the angle opposite the longest side.

Ex. 23. A line 16 inches long is divided internally in the ratio of 3 : 5; find the segments. Also find the segments when the line is divided externally in the same ratio.

Ex. 24. A line 16 inches long is divided in extreme and mean ratio. Find the segments.

Ex. 25. If a line is divided in extreme and mean ratio and the smaller segment is 4, find the whole line.

Ex. 26. A triangle whose altitude is 20 is bisected by a line parallel to the base. Into what segments is the altitude divided?

Ex. 27. Find the area of a square inscribed in a circle whose radius is 5.

Ex. 28. In a circle whose diameter is 20, a chord is passed through a point at a distance 6 from the center, perpendicular to the diameter through that point. Find the length of this chord, and of the chords drawn from its extremities to the ends of the diameter.

Ex. 29. Each leg of an isosceles trapezoid is 10, and one base exceeds the other by 16. Find the altitude.

Ex. 30. If three arcs, each of 60° and having 10 for a radius, are each concave to the other two, find the area included by them.

Ex. 31. Find the area of a trapezoid whose legs are 4 and 5, and whose bases are 8 and 11.

Ex. 32. A square piece of land and a circular piece each contain 1 acre. How many more feet of fence does one require than the other?

Ex. 33. If the base of a triangle is doubled and the altitude remains unchanged, how is the area affected? If the altitude is doubled and the base remains unchanged? If both the base and the altitude are doubled?

Ex. 34. The side of an equilateral triangle is 12; find the area of the inscribed, and of the circumscribed circle.

Ex. 35. In a given circle a chord of 60° is 16. Find the chord of 120°. Also of 30°.

Ex. 36. In a given triangle, equivalent to a rectangle whose sides are 40 and 20, the base is 32. Find the altitude.

Ex. 37. Find the side of an equilateral triangle equivalent to a circle whose diameter is 10.

Ex. 38. The area of a rhombus is 156 sq. in., and one side is 1 ft. 1 in. Find the diagonals.

Ex. 39. The sides of a triangle are 8, 10, 12. Find the areas of the triangles made by the bisector of the angle opposite the side 12.

Ex. 40. In a circle of area 275 sq. ft., a rectangle of area 150 sq. ft. is inscribed. Show how to find the sides of the rectangle.

EXERCISES. GROUP 63

EXERCISES INVOLVING THE METRIC SYSTEM

Ex. 1. Find the area of a triangle of which the base is 16 dm. and the altitude 80 cm.

Ex. 2. Find the area of a triangle whose sides are 6 m., 70 dm., 800 cm.

Ex. 3. Find the area in square meters of a circle whose radius is 14 dm.

Ex. 4. If the hypotenuse of a right triangle is 17 dm. and one leg is 150 cm., find the other leg and the area.

Ex. 5. If the circumference of a circle is 1 m., find the area of the circle in square decimeters.

Ex. 6. Find the area in hectares, and also in acres, of a circle whose radius is 100 m.

Ex. 7. If the diagonal of a rectangle is 35 dm. and one side is 800 mm., find the area in square meters, and also in square inches.

Ex. 8. Find the area of a trapezoid whose bases are 600 cm. and 2 m., and whose altitude is 80 dm.

Ex. 9. If a rectangular field is 700 dm. long and 200 m. wide, find its area in hectares and in acres.

Ex. 10. In a given circle two chords, whose lengths are 15 dm. and 13 dm., intersect. If the segments of the first chord are 12 dm. and 3 dm., find the segments of the second chord.

Ex. 11. Find in decimeters the radius of a circle equivalent to a square whose side is 1 ft. 6 in.

Ex. 12. Find in feet the diameter of a wheel which, in going 10 kilometers, makes 5,000 revolutions.

SOLID GEOMETRY

BOOK VI

LINES, PLANES AND ANGLES IN SPACE

DEFINITIONS AND FIRST PRINCIPLES

497. Solid Geometry treats of the properties of space of three dimensions.

Many of the properties of space of three dimensions are determined by use of the plane and of the properties of plane figures already obtained in Plane Geometry.

498. A **plane** is a surface such that, if any two points in it be joined by a straight line, the line lies wholly in the surface.

499. A **plane is determined** by given points or lines, if no other plane can pass through the given points or lines without coinciding with the given plane.

500. Fundamental property of a plane in space. *A plane is determined by any three points not in a straight line.*

For, if through a line connecting two given points, A and B, a plane be passed, the plane, if rotated, can pass through a third given point, C, in but one position.

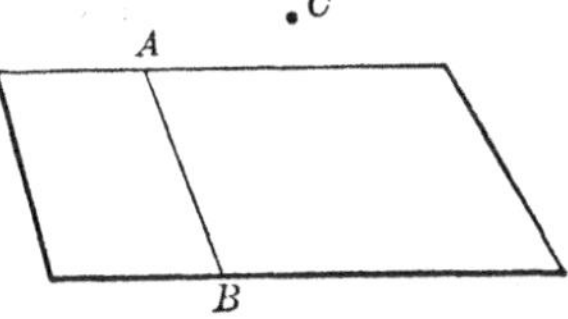

The importance of the above principle is seen from the fact that it reduces an unlimited surface to three points, thus making a vast economy to the attention. It also enables us to connect different planes, and treat of their properties systematically.

501. Other modes of determining a plane. *A plane may also be determined* by any equivalent of three points not in a straight line, as by

a straight line and a point outside the line; or by
two intersecting straight lines; or by
two parallel straight lines.

It is often more convenient to use one of these latter methods of determining a plane than to reduce the data to three points and use Art. 500.

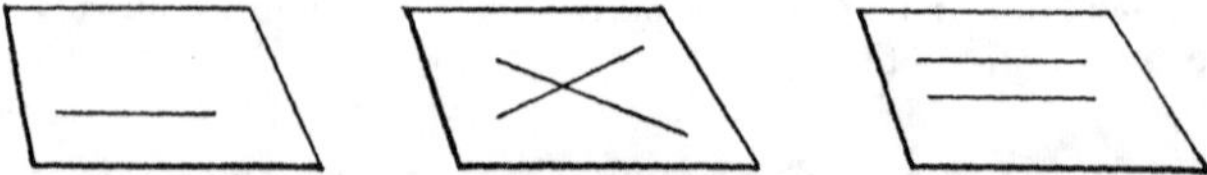

502. Representation of a plane in geometric figures. In reasoning concerning the plane, it is often an advantage to have the plane represented in all directions. Hence, in drawing a geometric figure, a plane is usually represented to the eye by a small parallelogram.

This is virtually a double use of two intersecting lines, or of two parallel lines, to determine a plane (Art. 501).

503. Postulate of Solid Geometry. The principle of Art. 499 may also be stated as a postulate, thus:

Through any three points not in a straight line (or their equivalent) a plane may be passed.

504. The **foot** of a line is the point in which the line intersects a given plane.

505. A straight line perpendicular to a plane is a line perpendicular to every line in the plane drawn through its foot.

A straight line perpendicular to a plane is sometimes called a **normal** to the plane.

506. A **parallel straight line and plane** are a line and plane which cannot meet, however far they be produced.

507. Parallel planes are planes which cannot meet, however far they be produced.

508. Properties of planes inferred immediately.

1. *A straight line, not in a given plane, can intersect the given plane in but one point.*

For, if the line intersect the given plane in two or more points, by definition of a plane, the line must lie in the plane. Art. 498.

2. *The intersection of two planes is a straight line.*

For, if two points common to the two planes be joined by a straight line, this line lies in each plane (Art. 498); and no other point can be common to the two planes, for, through a straight line and a point outside of it only one plane can be passed. Art. 501.

Ex. 1. Give an example of a plane surface; of a curved surface; of a surface, part plane and part curved; of a surface composed of different plane surfaces.

Ex. 2. Four points, not all in the same plane determine how many different planes? how many different straight lines?

Ex. 3. Three parallel straight lines, not in the same plane, determine how many different planes?

Ex. 4. Four parallel straight lines can determine how many different planes?

Ex. 5. Two intersecting straight lines and a point, not in their plane, determine how many different planes?

Proposition I. Theorem

509. *If a straight line is perpendicular to each of two other straight lines at their point of intersection, it is perpendicular to the plane of those lines.*

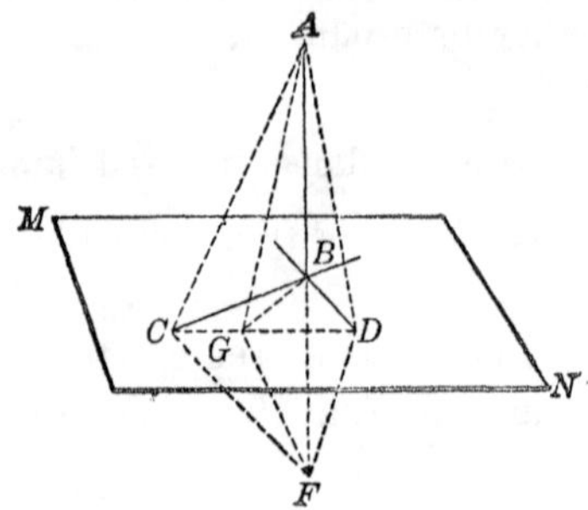

Given $AB \perp$ lines BC and BD, and the plane MN passing through BC and BD.

To prove $AB \perp$ plane MN.

Proof. Through B draw BG, any other line in the plane MN.

Draw any convenient line CD intersecting BC, BG and BD in the points C, G and D, respectively.

Produce the line AB to F, making $BF = AB$.

Connect the points C, G, D with A, and also with F.

Then, in the ⚠ ACD and FCD, $CD = CD$. **Ident.**

$AC = CF$, and $AD = DF$. **Art. 112.**

$\therefore \triangle ACD = \triangle FCD$. **(Why?)**

$\therefore \angle ACD = \angle FCD$. **(Why?)**

Then, in the ⚠ ACG and FCG, $CG = CG$, **(Why?)**

$AC = CF$, and $\angle ACG = \angle FCG$. **(Why?)**

$\therefore \triangle ACG = \triangle FCG$. **(Why?)**

$\therefore AG = GF$. (Why?)

$\therefore$ B and G are each equidistant from the points A and F.

$\therefore$ BG is $\perp AF$; that is, $AB \perp BG$. Art. 113.

$\therefore AB \perp$ plane MN, Art. 505.

(*for it is* $\perp$ *any line*, BG, *in the plane* MN, *through its foot*).

Q. E. D.

Proposition II. Theorem

510. *All the perpendiculars that can be drawn to a given line at a given point in the line lie in a plane perpendicular to the line at the given point.*

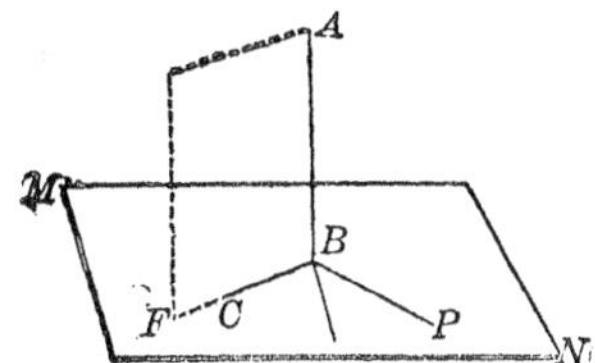

Given the plane MN and the line BC both $\perp$ line AB at the point B.

To prove that BC lies in the plane MN.

Proof. Pass a plane AF through the intersecting lines AB and BC. Art. 503.

This plane will intersect the plane MN in a straight line BF. Art. 508, 2.

But $AB \perp$ plane MN (Hyp.) $\therefore AB \perp BF$. Art. 505.

Also $AB \perp BC$. Hyp.

$\therefore$ in the plane AF, BC and $BF \perp AB$ at B.

$\therefore$ BC and BF coincide. Art. 71.

But BF is in the plane MN.

$\therefore$ BC must be in the plane MN,

(*for* BC *coincides with* BF, *which lies in the plane* MN).

Q. E. D.

511. COR. 1. *At a given point B in the straight line AB, to construct a plane perpendicular to the line AB.* Pass a plane AF through AB in any convenient direction, and in the plane AF at the point B construct $BF \perp AB$ (Art. 274). Pass another plane through AB, and in it construct $BP \perp AB$. Through the lines BF and BP pass the plane MN (Art. 503). MN is the required plane (Art. 509).

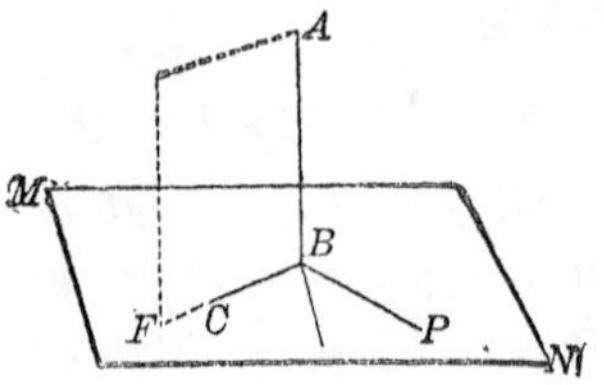

512. COR. 2. *Through a given external point, P, to pass a plane perpendicular to a given line, AB.* Pass a plane through AB and P (Art. 503), and in this plane draw $PB \perp AB$ (Art. 273). Pass another plane through AB, as AF, and in AF draw $BF \perp AB$ at B (Art. 274). Pass a plane through BP and BF (Art. 503). This will be the plane required (Art. 509).

513. COR. 3. *Through a given point but one plane can be passed perpendicular to a given line.*

Ex. 1. Five points, no four of which are in the same plane, determine how many different planes? how many different straight lines?

Ex. 2. A straight line and two points, not all of which are in the same plane, determine how many different planes?

Ex. 3. In the figure, p. 322, prove that the triangles GAD and GDF are equal.

Ex. 4. In the same figure, if $AB = 8$ and $BC = 6$, find FC.

Proposition III. Problem

514. *At a given point in a plane, to erect a perpendicular to the plane.*

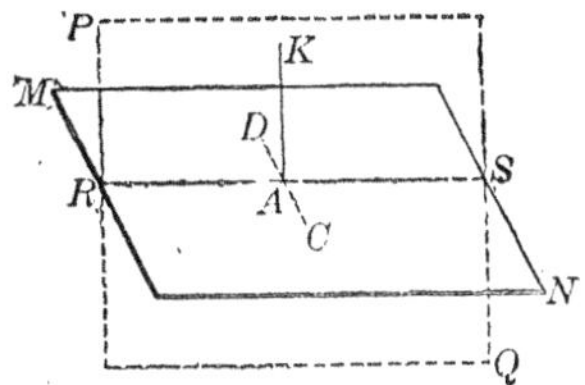

Given the point A in plane MN.

To construct a line perpendicular to MN at the point A.

Construction. Through the point A draw any line CD in the plane MN.

Also through the point A pass the plane $PQ \perp CD$ (Art. 511), intersecting the plane MN in the line RS. Art. 508, 2.

In the plane PQ draw $AK \perp$ line RS at A. Art. 274.

Then AK is the $\perp$ required.

Proof. $CD \perp$ plane PQ. Constr.

$\therefore CD \perp AK$. Art. 505.

Hence $AK \perp CD$.

But $AK \perp RS$. Constr.

$\therefore AK \perp$ plane MN. Art. 509.

Q. E. F.

515. Cor. *At a given point in a plane but one perpendicular to the plane can be drawn.* For, if two $\perp$s could be drawn at the given point, a plane could be passed through them intersecting the given plane. Then the two $\perp$s would be in the new plane and $\perp$ to the same line (the line of intersection of the two planes, Art. 505); which is impossible (Art. 71).

PROPOSITION IV. PROBLEM

516. *From a given point without a plane, to draw a line perpendicular to the plane.*

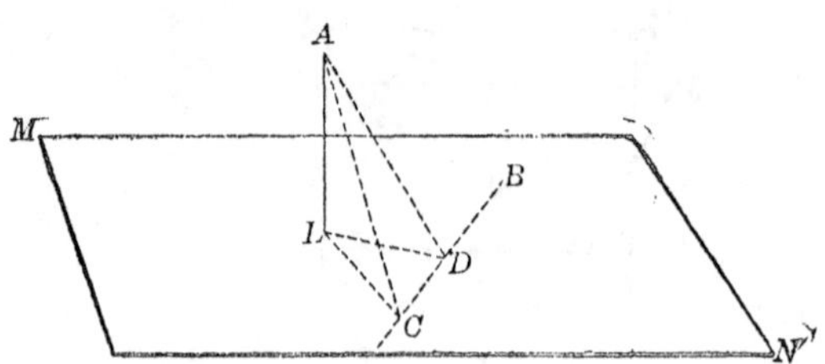

Given the plane MN and the point A external to it.

To construct from A a line $\perp$ plane MN.

Construction. In the plane MN draw any convenient line BC. Pass a plane through BC and A (Art. 503), and in this plane draw $AD \perp BC$. Art. 273.

In the plane MN draw $LD \perp BC$. Art. 274.

Pass a plane through AD and LD (Art. 503), and in that plane draw $AL \perp LD$. Art. 273.

Then AL is the $\perp$ required.

Proof. Take any point C in BC except D, and draw LC and AC.

Then $\triangle$s ADC, ADL and LDC are right $\triangle$s. Constr.

$\therefore \overline{AC}^2 = \overline{AD}^2 + \overline{DC}^2$. Art. 400.

$\therefore \overline{AC}^2 = \overline{AL}^2 + \overline{LD}^2 + \overline{DC}^2$. Art. 400, Ax. 8.

$\therefore \overline{AC}^2 = \overline{AL}^2 + \overline{LC}^2$. Art. 400, Ax. 8.

$\therefore \angle ALC$ is a right $\angle$. Art. 351.

But $AL \perp LD$. Constr.

$\therefore AL \perp MN$. Art. 509.

Q. E. F.

517. COR. *But one perpendicular can be drawn from a given external point to a given plane.*

PROPOSITION V. THEOREM

518. I. *Oblique lines drawn from a point to a plane, meeting the plane at equal distances from the foot of the perpendicular, are equal;*

II. *Of two oblique lines drawn from a point to a plane, but meeting the plane at unequal distances from the foot of the perpendicular, the more remote is the greater.*

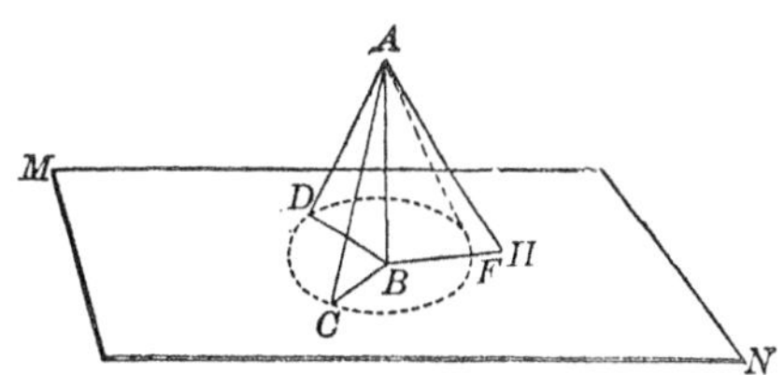

Given $AB \perp$ plane MN, $BD = BC$, and $BH > BC$.

To prove $AD = AC$, and $AH > AC$.

Proof. I. In the right ⊿s ABD and ABC,

$AB = AB$, and $BD = BC$. (Why ?)

$\therefore \triangle ABD = \triangle ABC$. (Why ?)

$\therefore AD = AC$. (Why ?)

II. On BH take $BF = BC$, and draw AF.

Then $AF = AC$ (*by part of theorem just proved*).

But $AH > AF$. (Why ?)

$\therefore AH > AC$. Ax. 8.

Q. E. D.

519. COR. 1. CONVERSELY: *Equal oblique lines drawn from a point to a plane meet the plane at equal distances from the foot of the perpendicular drawn from the same point to the plane; and, of two unequal lines so drawn, the greater line meets the plane at the greater distance from the foot of the perpendicular.*

520. COR. 2. *The locus of a point in space equidistant from all the points in the circumference of a circle is a straight line passing through the center of the circle and perpendicular to its plane.*

521. COR. 3. *The perpendicular is the shortest line that can be drawn from a given point to a given plane.*

522. DEF. The **distance from a point to a plane** is the perpendicular drawn from the point to the plane.

PROPOSITION VI. THEOREM

523. *If from the foot of a perpendicular to a plane a line be drawn at right angles to any line in the plane, the line drawn from the point of intersection so formed to any point in the perpendicular, is perpendicular to the line of the plane.*

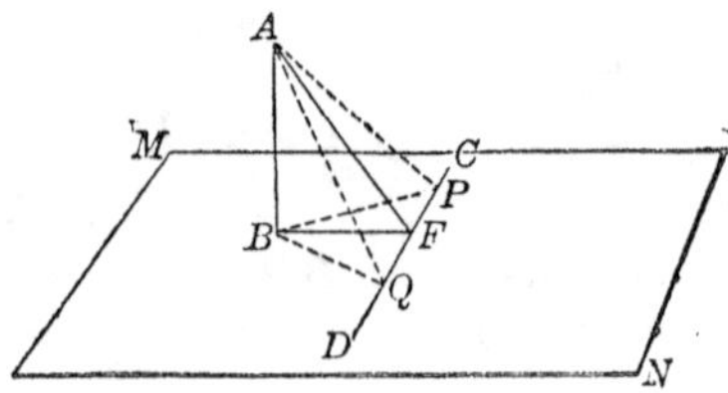

Given $AB \perp$ plane MN, and $BF \perp CD$, any line in MN.

To prove $AF \perp CD$.

Proof. On CD take FP and FQ equal segments.

Draw AP, BP, AQ, BQ.

Then $BP = BQ$. Art. 112.

Hence $AP = AQ$. Art. 518.

$\therefore$ in the line AF, the point A is equidistant from P and Q, and F is equidistant from P and Q.

$\therefore AF \perp CD$. (Why?)

Q. E. D.

Ex. In the above figure, if $AB=6$, $AF=8$, and $AQ=10$, find QF, BF and BQ.

PROPOSITION VII. THEOREM

524. *Two straight lines perpendicular to the same plane are parallel.*

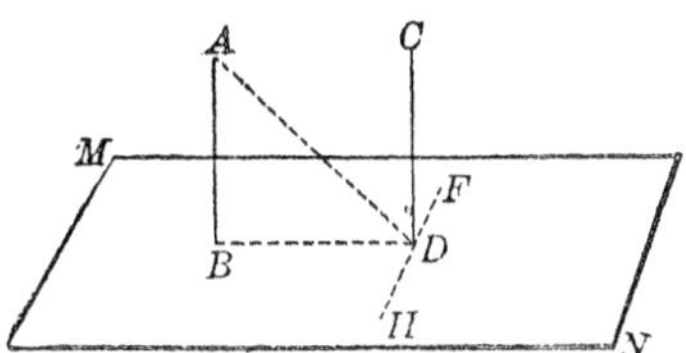

Given the lines AB and $CD \perp$ plane MN.

To prove $AB \parallel CD$.

Proof. Draw BD, and through D, in the plane MN, draw $FH \perp BD$.

Draw AD.

Then $BD \perp FH$. Constr.

$AD \perp FH$. Art. 523.

$CD \perp FH$. Art. 505.

$\therefore$ BD, AD and CD are all $\perp$ FH at the point D.

$\therefore$ BD, AD and CD all lie in the same plane. Art. 510.

$\therefore$ AB and CD are in the same plane. (Why ?)

But AB and CD are $\perp$ BD. Art. 505.

$\therefore$ AB and CD are $\parallel$. Art. 121.

Q. E. D.

525. COR. 1. *If one of two parallel lines is perpendicular to a plane, the other is perpendicular to the plane also.*

For, if AB and CD be $\parallel$, and $AB \perp$ plane PQ, a line drawn from $C \perp PQ$ must be $\parallel AB$. Art. 524.

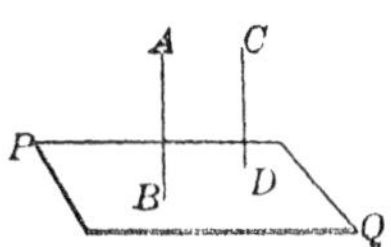

But CD must coincide with the line so drawn (Art. 47, 3); $\therefore$ $CD \perp PQ$.

526. Cor. 2. *If two straight lines are each parallel to a third straight line, they are parallel to each other.* For, if a plane be drawn ⊥ to the third line, each of the two other lines must be ⊥ to it (Art. 525), and therefore be ∥ to each other (Art. 524).

Proposition VIII. Theorem

527. *If a straight line external to a given plane is parallel to a line in the plane, then the first line is parallel to the given plane.*

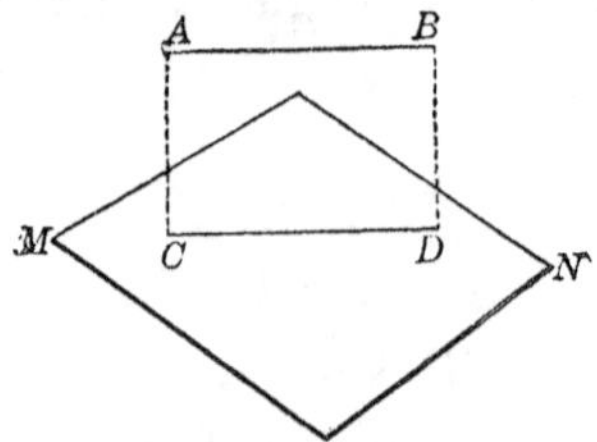

Given the straight line AB ∥ line CD in the plane MN.

To prove AB ∥ plane MN.

Proof. Pass a plane through the ∥ lines AB and CD.

If AB meets MN it must meet it in the line CD.

But AB and CD cannot meet, for they are ∥. **Art. 120.**

∴ AB and MN cannot meet and are parallel. **Art. 506.**

Q. E. D.

528. Cor. 1. *If a straight line is parallel to a plane, the intersection of the plane with any plane passing through the given line is parallel to the given line.*

529. COR. 2. *Through a given line* (*CD*) *to pass a plane parallel to another given line* (*AB*).

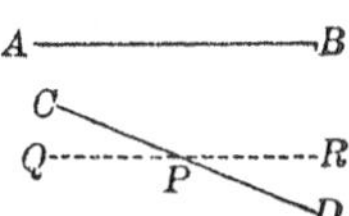

Through P, any point in CD, draw $QR \parallel AB$ (Art. 279). Through CD and QR pass a plane (Art. 503). This will be the plane required (Art. 527).

If AB and CD are not parallel, but one plane can be drawn through $CD \parallel AB$.

PROPOSITION IX. THEOREM

530. *Two planes perpendicular to the same straight line are parallel.*

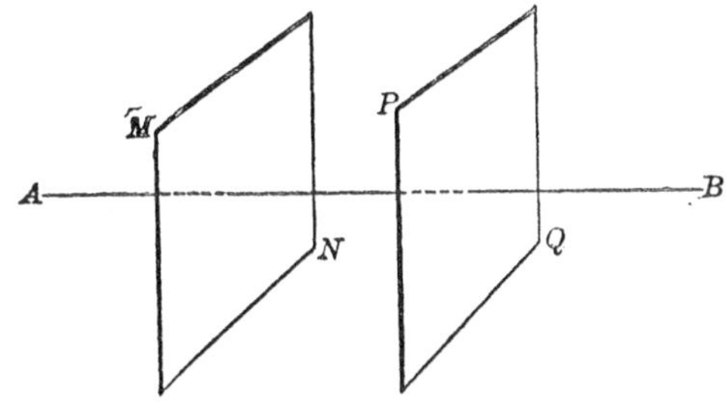

Given the planes MN and $PQ \perp$ line AB.

To prove $MN \parallel PQ$.

Proof. If MN and PQ are not parallel, on being produced they will meet.

We shall then have two planes drawn from a point perpendicular to a given line, which is impossible. Art. 513.

$\therefore$ MN and PQ are parallel. Art. 507.

Q. E. D.

PROPOSITION X. THEOREM

531. *If two parallel planes are cut by a third plane, the intersections are parallel lines.*

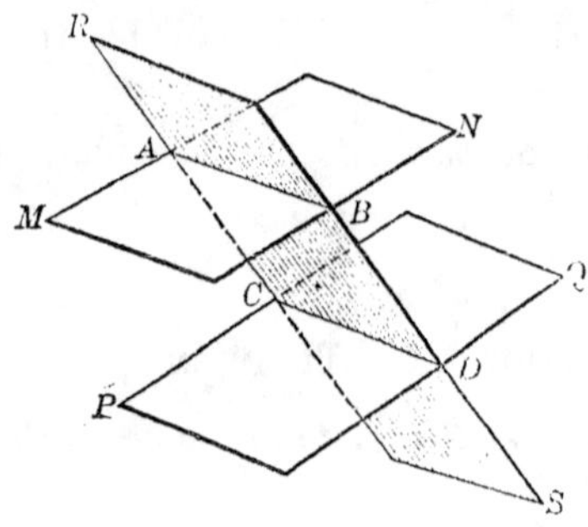

Given *MN* and *PQ* two || planes intersected by the plane *RS* in the lines *AB* and *CD*.

To prove *AB* || *CD*.

Proof. *AB* and *CD* lie in the same plane *RS*.

Also *AB* and *CD* cannot meet; for if they did meet the planes *MN* and *PQ* would meet, which is impossible. Art. 507.

∴ *AB* and *CD* are parallel. Art. 41.

Q. E. D.

532. COR. 1. *Parallel lines included between parallel planes are equal.* For, if *AC* and *BD* are two parallel lines, a plane may be passed through them (Art. 503), intersecting *MN* and *PQ* in the || lines *AB* and *CD*. Art. 531.

∴ *ABDC* is a parallelogram. Art. 147.

∴ $AC = BD$. Art. 155.

533. Cor. 2. *Two parallel planes are everywhere equidistant.*

For lines ⊥ to one of them are || (Art. 524). Hence the segments of these lines included between the || planes are equal (Art. 532).

Proposition XI. Theorem

534. *If two intersecting lines are each parallel to a given plane, the plane of these lines is parallel to the given plane.*

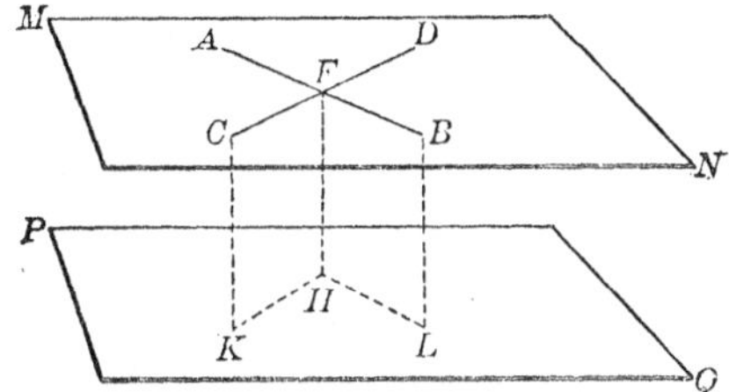

Given the lines AB and CD, each || plane PQ, and intersecting in the point F; and MN a plane through AB and CD.

To prove $MN \parallel PQ$.

Proof. From the point F draw $FH \perp PQ$.

Pass a plane through FC and FH, intersecting PQ in HK; also pass a plane through FB and FH, intersecting PQ in HL.

Then	$HK \parallel FC$, and $HL \parallel FB$.	Art. 528.
But	$FH \perp HK$ and HL.	Art. 505.
	$\therefore FH \perp FC$ and FB.	Art. 123.
	$\therefore FH \perp MN$.	Art. 509.
	$\therefore MN \parallel PQ$.	Art. 530.

Q. E. D.

PROPOSITION XII. THEOREM

535. *A straight line perpendicular to one of two parallel planes is perpendicular to the other also.*

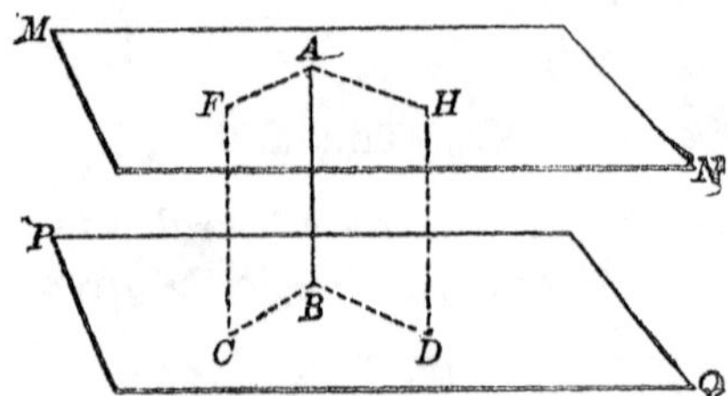

Given the plane $MN \parallel$ plane PQ, and $AB \perp PQ$.

To prove $AB \perp MN$.

Proof. Through AB pass a plane intersecting PQ and MN in the lines BC and AF, respectively; also through AB pass another plane intersecting PQ and MN in BD and AH, respectively.

Then	$BC \parallel AF$, and $BD \parallel AH$.	Art. 531.
But	$AB \perp BC$ and BD.	Art. 505.
	$\therefore AB \perp AF$ and AH.	Art. 123.
	$\therefore AB \perp$ plane MN.	Art. 509.

Q. E. D.

536. COR. 1. *Through a given point to pass a plane parallel to a given plane.*

Let the pupil supply the construction.

537. COR. 2. *Through a given point but one plane can be passed parallel to a given plane.*

Proposition XIII. Theorem

538. *If two angles not in the same plane have their corresponding sides parallel and extending in the same direction, the angles are equal and their planes are parallel.*

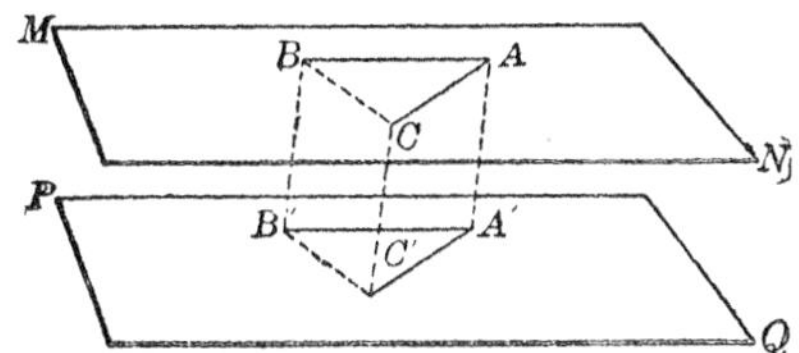

Given the $\angle BAC$ in the plane MN, and the $\angle B'A'C'$ in the plane PQ; AB and $A'B'$ $\parallel$ and extending in the same direction; and AC and $A'C'$ $\parallel$ and extending in the same direction.

To prove $\angle BAC = \angle B'A'C'$, and plane $MN \parallel$ plane PQ.

Proof. Take $AB = A'B'$, and $AC = A'C'$.

Draw AA', BB', CC', BC, $B'C'$.

Then $ABB'A'$ is a ▱, — Art. 160.

(*for AB and A'B' are = and* $\parallel$).

$\therefore$ BB' and AA' are = and $\parallel$. — Art. 155.

In like manner CC' and AA' are = and $\parallel$.

$\therefore$ BB' and CC' are = and $\parallel$. — (Why ?)

$\therefore$ $BCC'B'$ is a ▱, and $BC = B'C'$. — (Why ?)

$\therefore$ $\triangle ABC = \triangle A'B'C'$. — (Why ?)

$\therefore$ $\angle A = \angle A'$. — (Why ?)

Also $AB \parallel A'B'$, $\therefore$ $AB \parallel$ plane PQ. — Art. 527.

Similarly $AC \parallel$ plane PQ.

$\therefore$ plane $MN \parallel$ plane PQ. — Art. 534.

Q. E. D.

PROPOSITION XIV. THEOREM

539. *If two straight lines are intersected by three parallel planes, the corresponding segments of these lines are proportional.*

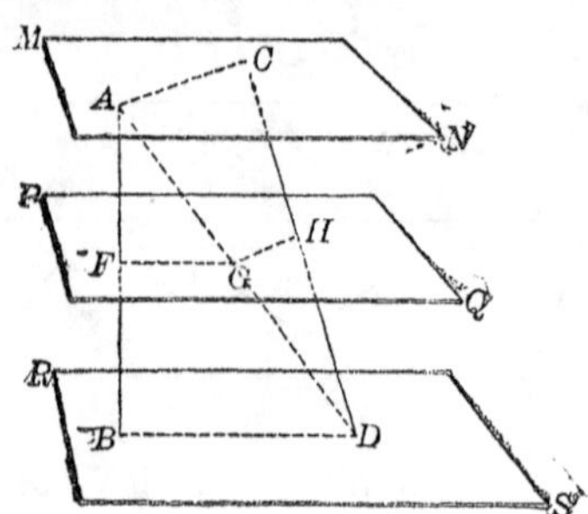

Given the straight lines AB and CD intersected by the $\parallel$ planes MN, PQ and RS in the points A, F, B, and C, H, D, respectively.

To prove $$\frac{AF}{FB}=\frac{CH}{HD}.$$

Proof. Draw the line AD intersecting the plane PQ in G.

Draw FG, BD, GH, AC.

Then $FG \parallel BD$, and $GH \parallel AC$. Art. 531.

$$\therefore \frac{AF}{FB}=\frac{AG}{GD}.$$ Art. 317.

And $$\frac{CH}{HD}=\frac{AG}{GD}.$$ (Why?)

$$\therefore \frac{AF}{FB}=\frac{CH}{HD}.$$ (Why?)

Q. E. D.

Ex. 1. In above figure, if $AF=2$, $FB=5$, and $CH=3$, find CD.

Ex. 2. If $CH=3$, $HD=4$, and $AB=10$, find AF and BF.

DIHEDRAL ANGLES

540. A **dihedral angle** is the opening between two intersecting planes.

From certain points of view, a dihedral angle may be regarded as a wedge or slice of space cut out by the planes forming the dihedral angle.

541. The **faces** of a dihedral angle are the planes forming the dihedral angle.

The **edge** of a dihedral angle is the straight line in which the faces intersect.

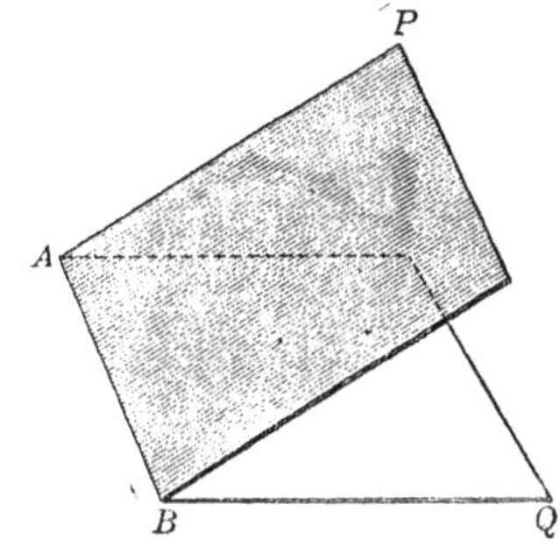

542. Naming dihedral angles. A dihedral angle may be named, or denoted, by naming its edge, as the dihedral angle AB; or by naming four points, two on the edge and one on each face, those on the edge coming between the points on the faces, as P–AB–Q. The latter method is necessary in naming two or more dihedral angles which have a common edge.

543. Equal dihedral angles are dihedral angles which can be made to coincide.

544. Adjacent dihedral angles are dihedral angles having the same edge and a face between them in common.

545. Vertical dihedral angles are two dihedral angles having the same edge, and the faces of one the prolongations of the faces in the other.

546. A **right dihedral angle** is one of two equal adjacent dihedral angles formed by two planes.

547. A **plane perpendicular** to a given plane is a plane forming a right dihedral angle with the given plane.

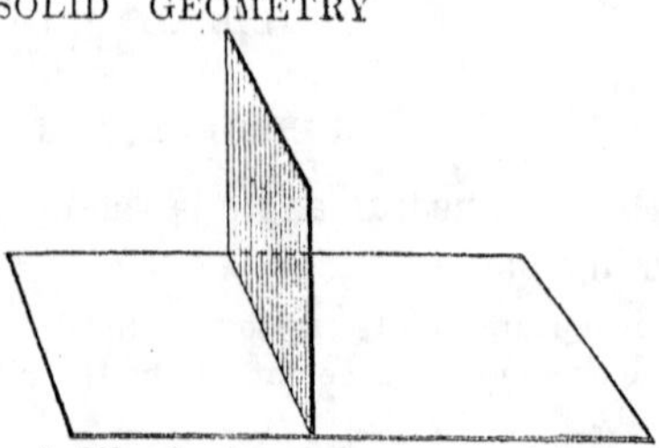

Many of the properties of dihedral angles are obtained most conveniently by using a plane angle to represent the dihedral angle.

548. The **plane angle of a dihedral angle** is the angle formed by two lines drawn one in each face, perpendicular to the edge at the same point.

Thus, in the dihedral angle *C–AB–F*, if *PQ* is a line in the face *AD* perpendicular to the edge *AB* at *P*, and *PR* is a line in face *AF* perpendicular to the edge *AB* at *P*, the angle *QPR* is the plane angle of the dihedral angle *C–AB–F*.

549. Property of plane angles of a dihedral angle.

The magnitude of the plane angle of a dihedral angle is the same at every point of the edge. For let *EAC* be the plane ∠ of the dihedral ∠ *E–AB–D* at the point *A*.
Then *PR* ∥ *AE*, and *PQ* ∥ *AC* (Art. 121.)
∴ ∠*RPQ* = ∠*EAC* (Art. 538).

550. The **projection of a point upon a plane** is the foot of a perpendicular drawn from the point to the plane.

551. The **projection of a line upon a plane** is the locus of the projections of all the points of the line on the plane. Thus *A'B'* is the projection of *AB* on the plane *MN*.

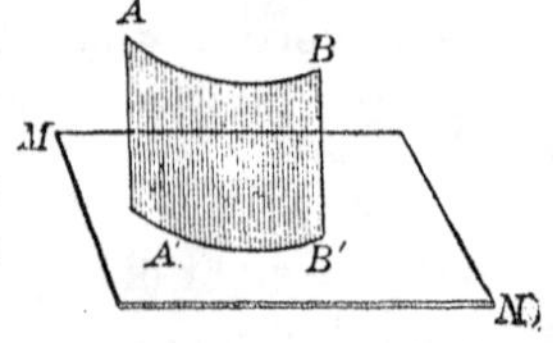

PROPOSITION XV. THEOREM

552. *Two dihedral angles are equal if their plane angles are equal.*

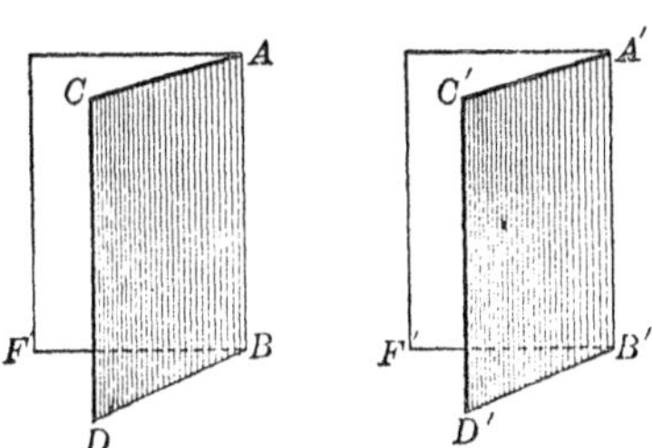

Given $\angle DBF$ the plane $\angle$ of the dihedral $\angle C\text{-}AB\text{-}F$, $\angle D'B'F'$ the plane $\angle$ of the dihedral $\angle C'\text{-}A'B'\text{-}F'$, and $\angle DBF = \angle D'B'F'$.

To prove $\angle C\text{-}AB\text{-}F = \angle C'\text{-}A'B'\text{-}F'$.

Proof. Apply the dihedral $\angle C'\text{-}A'B'\text{-}F'$ to $\angle C\text{-}AB\text{-}F$ so that $\angle D'B'F'$ coincides with its equal, $\angle DBF$. Geom. Ax. 2.

Then line $A'B'$ must coincide with AB, Art. 515.

(*for $A'B'$ and AB are both $\perp$ plane DBF at the point B*).

Hence the plane $A'B'D'$ will coincide with plane ABD, Art. 501.

(*through two intersecting lines only one plane can be passed*).

Also the plane $A'B'F'$ will coincide with the plane ABF,

(*same reason*).

$\therefore$ $\angle C'\text{-}A'B'\text{-}F'$ coincides with $\angle C\text{-}AB\text{-}F$ and is equal to it. Art. 47.

Q. E. D.

553. COR. *The vertical dihedral angles formed by two intersecting planes are equal.*

In like manner, many other properties of plane angles are true of dihedral angles.

Proposition XVI. Theorem

554. *Two dihedral angles have the same ratio as their plane angles.*

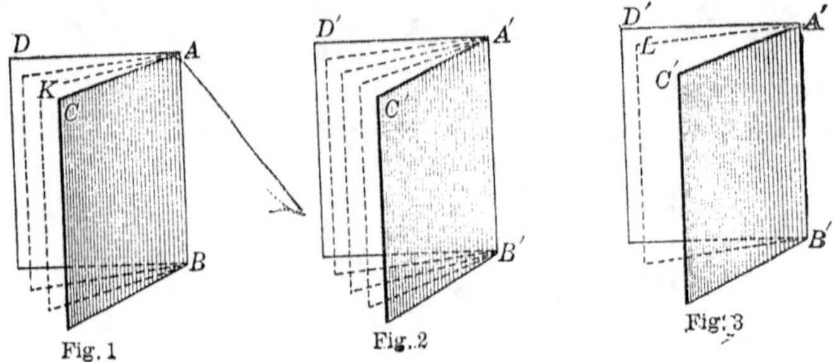

Given the dihedral $\angle s$ $C\text{-}AB\text{-}D$ and $C'\text{-}A'B'\text{-}D'$ having the plane $\angle s$ CAD and $C'A'D'$, respectively.

To prove $\angle C'\text{-}A'B'\text{-}D' : \angle C\text{-}AB\text{-}D = \angle C'A'D' : \angle CAD$.

Case I. *When the plane* $\angle s$ $C'A'D'$ *and* CAD (Figs. 2 and 1), *are commensurable.*

Proof. Find a common measure of the $\angle s$ $C'A'D'$ and CAD, as $\angle CAK$, and let it be contained in $\angle C'A'D'$ n times, and in $\angle CAD$ m times.

Then $$\angle C'A'D' : \angle CAD = n : m.$$

Through $A'B'$ and the lines of division of $\angle C'A'D'$ pass planes, and through AB and the lines of division of $\angle CAD$ pass planes. These planes will divide the dihedral $\angle C'\text{-}A'B'\text{-}D'$ into n, and $\angle C\text{-}AB\text{-}D$ into m parts, all equal. Art. 552.

$$\therefore \angle C'\text{-}A'B'\text{-}D' : \angle C\text{-}AB\text{-}D = n : m.$$

Hence $\angle C'\text{-}A'B'\text{-}D' : \angle C\text{-}AB\text{-}D = \angle C'A'D' : \angle CAD$. (Why?)

CASE II. *When the plane angles $C'A'D'$ and CAD* (Figs. 3 and 1) *are incommensurable.*

Proof. Divide the $\angle CAD$ into any number of equal parts, and apply one of these parts to the $\angle C'A'D'$. It will be contained a certain number of times with a remainder, as $\angle LA'D'$, less than the unit of measure.

Hence the $\angle s\ C'A'L$ and CAD are commensurable.

$\therefore \angle C'-A'B'-L : \angle C-AB-D = \angle C'A'L : \angle CAD$. Case I.

If now we let the unit of measure be indefinitely diminished, the $\angle LA'D'$, which is less than the unit of measure, will be indefinitely diminished.

$\therefore \angle C'A'L \doteq \angle C'A'D'$ as a limit, and

$\angle C'-A'B'-L \doteq \angle C'-A'B'-D'$ as a limit. Art. 251.

Hence $\dfrac{\angle C'-A'B'-L}{\angle C-AB-D}$ becomes a variable, with $\dfrac{\angle C'-A'B'-D'}{\angle C-AB-D}$ as its limit; Art. 253, 3.

Also $\dfrac{\angle C'A'L}{\angle CAD}$ becomes a variable with $\dfrac{\angle C'A'D'}{\angle CAD}$ as its limit. Art. 253, 3.

But the variable $\dfrac{\angle C'-A'B'-L}{\angle C-AB-D} =$ the variable $\dfrac{\angle C'A'L}{\angle CAD}$ always. Case I.

$\therefore$ the limit $\dfrac{\angle C'-A'B'-D'}{\angle C-AB-D} =$ the limit $\dfrac{\angle C'A'D'}{\angle CAD}$. Art. 254.

Q. E. D.

Ex. 1. How many straight lines are necessary to indicate a dihedral angle (as $\angle E-AB-D$, p. 338)? How many straight lines are necessary to indicate the plane angle of a dihedral angle? Hence, what is the advantage of using a plane angle of a dihedral angle instead of the dihedral angle itself?

Ex. 2. Give three additional properties of dihedral angles analogous to properties of plane angles given in Book I.

PROPOSITION XVII. THEOREM

555. *If a straight line is perpendicular to a plane, every plane drawn through that line is perpendicular to the plane.*

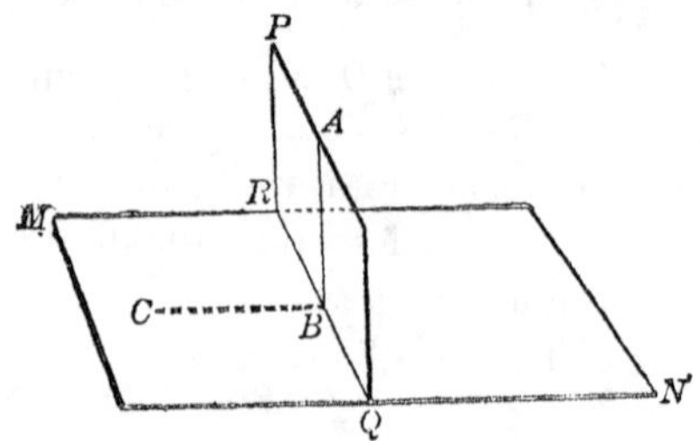

Given the line $AB \perp$ plane MN, and the plane PQ passing through AB and intersecting MN in RQ.

To prove $PQ \perp MN$.

Proof. In the plane MN draw $BC \perp RQ$ at B.

But $AB \perp RQ$. Art. 505.

$\therefore \angle ABC$ is the plane $\angle$ of the dihedral $\angle P-RQ-M$. Art. 548.

But $\angle ABC$ is a right $\angle$, Art. 505.
(*for $AB \perp MN$ by hyp.*).

$\therefore PQ \perp MN$. Art. 547.

Q. E. D.

556. COR. *A plane perpendicular to the edge of a dihedral angle is perpendicular to each of the two faces forming the dihedral angle.*

PROPOSITION XVIII. THEOREM

557. *If two planes are perpendicular to each other, a straight line drawn in one of them perpendicular to their line of intersection is perpendicular to the other plane.*

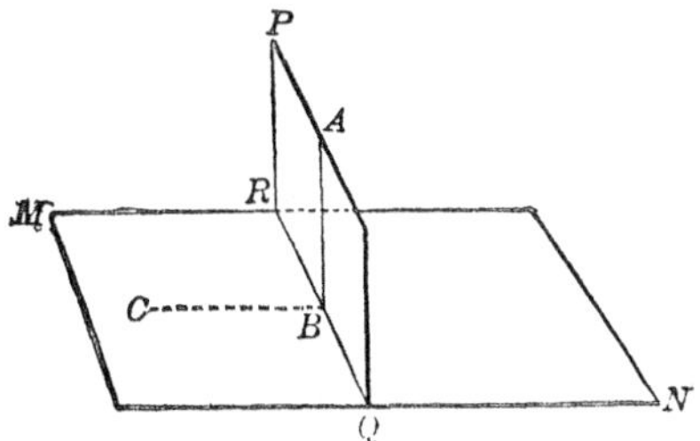

Given the plane $PQ \perp$ plane MN and intersecting it in the line RQ; and AB a line in $PQ \perp RQ$.

To prove $AB \perp$ plane MN.

Proof. In the plane MN draw $BC \perp RQ$.

$\therefore \angle ABC$ is the plane $\angle$ of the dihedral $\angle P-RQ-M$. Art. 548.

$\therefore \angle ABC$ is a rt. $\angle$, Art. 554.

(*for* $P-RQ-M$ *is a right dihedral* $\angle$).

$\therefore AB \perp BC$ and RQ at their intersection.

$\therefore AB \perp$ plane MN. (Why ?)

Q. E. D.

558. COR. 1. *If two planes are perpendicular to each other, a perpendicular to one of them at any point of their intersection will lie in the other plane.*

For, in the above figure, a $\perp$ erected at the point B in the plane MN must coincide with AB lying in the plane PQ and $\perp MN$, for at a given point in a plane only one $\perp$ can be drawn to that plane (Art. 515).

559. COR. 2. *If two planes are perpendicular to each other, a perpendicular to one plane, from a point in the other plane, will lie in the other plane.*

Proposition XIX. Theorem

560. *If two intersecting planes are each perpendicular to a third plane, their line of intersection is perpendicular to the third plane.*

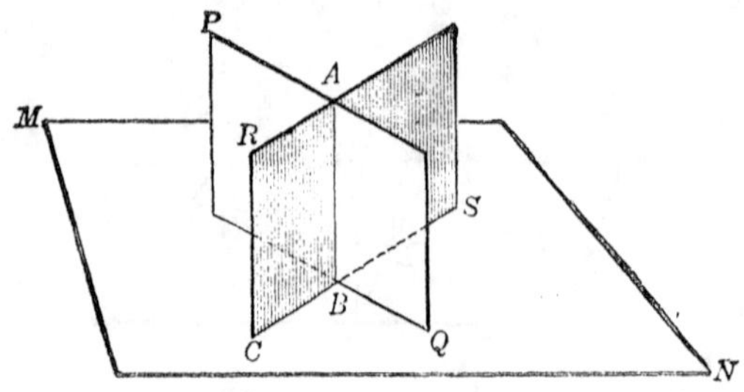

Given the planes PQ and $RS \perp$ plane MN, and intersecting in the line AB.

To prove $AB \perp$ plane MN.

Proof. At the point B in which the three planes meet erect a $\perp$ to the plane MN. This $\perp$ must lie in the plane PQ, and also in the plane RS. Art. 558.

Hence this $\perp$ must coincide with AB, the intersection of PQ and RS. Art. 508, 2.

$\therefore AB \perp$ plane MN.

Q. E. D.

561. Cor. *If two planes, including a right dihedral angle, are each perpendicular to a third plane, the intersection of any two of the planes is perpendicular to the third plane, and each of the three lines of intersection is perpendicular to the other two.*

Ex. 1. Name all the dihedral angles on the above figure.

Ex. 2. If $\angle CBQ = 30°$, find the ratio of each pair of dihedral $\angle$s.

Proposition XX. Theorem

562. *Every point in the plane which bisects a given dihedral angle is equidistant from the faces of the dihedral angle.*

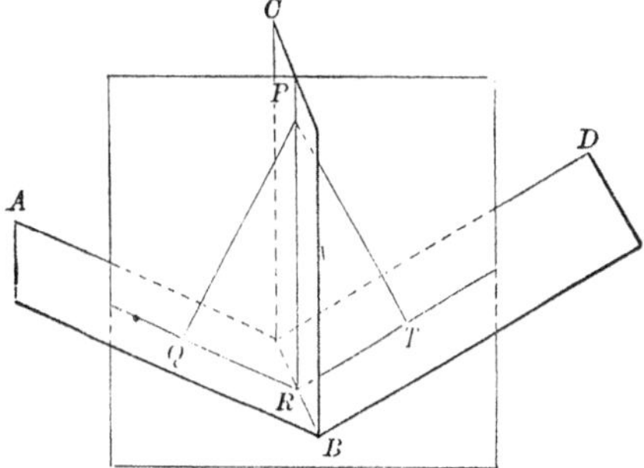

Given plane CB bisecting the dihedral $\angle A\text{–}BR\text{–}D$, P any point in plane BC, PQ and $PT \perp$ faces BA and BD, respectively.

To prove $PQ = PT$.

Proof. Through PQ and PT pass a plane intersecting AB in QR, BD in RT, and BC in PR.

Then plane $PQT \perp$ planes AB and BD. **Art. 555.**

$\therefore$ plane $PQT \perp$ line RB, the intersection of the planes AB and BD. **Art. 560.**

$\therefore RB \perp RQ$, RP and RT. **Art. 505.**

$\therefore \angle s$ QRP and PRT are the plane $\angle s$ of the dihedral $\angle s$ $A\text{–}BR\text{–}P$ and $P\text{–}BR\text{–}D$. **Art. 548.**

But these dihedral $\angle s$ are equal. **Hyp.**

$\therefore \angle QRP = \angle PRT$. **Art. 554.**

$\therefore$ rt. $\triangle PQR =$ rt. $\triangle PRT$. **(Why?)**

$\therefore PQ = PT$. **(Why?)**

Q. E. D.

563. *The locus of all points equidistant from the faces of a dihedral angle is the plane bisecting the dihedral angle.*

PROPOSITION XXI. PROBLEM

564. *Through any straight line not perpendicular to a given plane, to pass a plane perpendicular to the given plane*

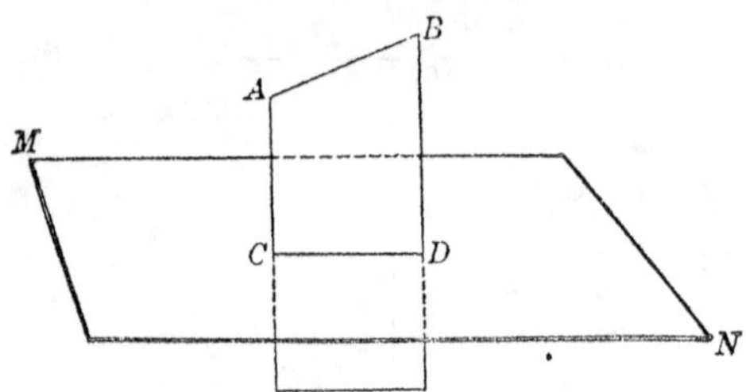

Given the line AB not ⊥ plane MN.

To construct a plane passing through AB and ⊥ MN.

Construction. From a point A in the line AB draw a ⊥ AC to the plane MN. Art. 516.

Through the intersecting lines AB and AC pass the plane AD. Art. 503.

Then AD is the plane required.

Proof. The plane AD passes through AB. Constr.

Also plane AD ⊥ plane MN, Art. 555.

(*for it contains AC, which is ⊥ MN*).

Q. E. F.

565. COR. 1. *Through a straight line not perpendicular to a given plane only one plane can be passed perpendicular to that plane.*

For, if two planes could be passed through AB ⊥ plane MN, this intersection AB would be ⊥ MN (Art. 560), which is contrary to the hypothesis.

566. COR. 2. *The projection upon a plane of a straight line not perpendicular to that plane is a straight line.*

For, if a plane be passed through the given line ⊥ to the given plane, the foot of a ⊥ from any point in the line to the given plane will be in the intersection of the two planes (Art. 559).

PROPOSITION XXII. THEOREM

567. *The acute angle which a line makes with its projection on a plane is the least angle which it makes with any line of the plane through its foot.*

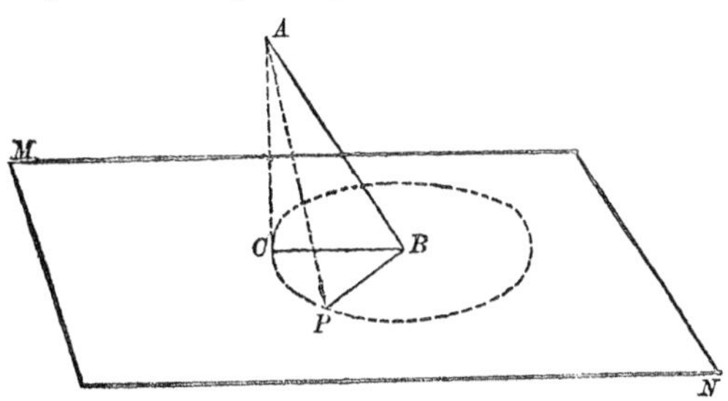

Given line AB meeting the plane MN in the point B, BC the projection of AB on MN, and PB any other line in the plane MN through B.

To prove that $\angle ABC$ is less than $\angle ABP$.

Proof. Lay off PB equal to CB, and draw AC and AP.

Then, in the $\triangle$ ABC and ABP,

$$AB = AB. \qquad \text{(Why?)}$$

$$BC = BP. \qquad \text{(Why?)}$$

But $$AC < AP. \qquad \text{Art. 521.}$$

$$\therefore \angle ABC \text{ is less than } \angle ABP \qquad \text{Art. 108.}$$

Q. E. D.

568. DEF. The **inclination of a line to a plane** is the acute angle which the given line makes with its projection upon the given plane.

Ex. 1. A plane has an inclination of 47° to each of the faces of a dihedral angle and is parallel to the edge of the dihedral angle; how many degrees are in the plane angle of the dihedral angle?

Ex. 2. In the figure on page 345, if $PT = QT$, how large is the dihedral $\angle A\text{-}BR\text{-}D$? if $PT = RT$, how large is it?

Proposition XXIII. Problem

569. *To draw a common perpendicular to any two lines not in the same plane.*

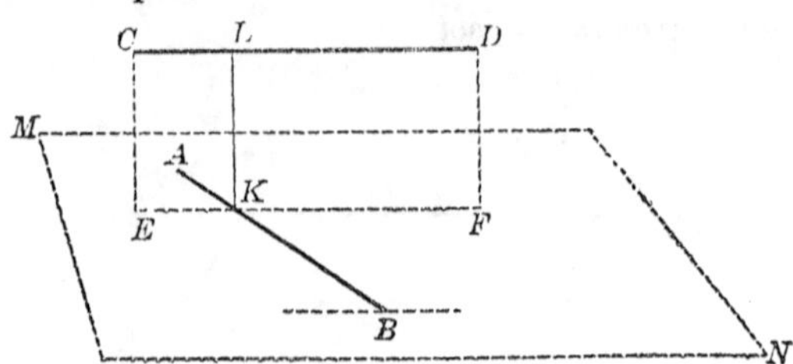

Given the lines AB and CD not in the same plane.

To construct a line perpendicular to both AB and CD.

Construction. Through AB pass a plane $MN \parallel$ line CD. Art. 529

Through CD pass a plane $CF \perp$ plane MN (Art. 564), and intersecting plane MN in the line EF.

Then $EF \parallel CD$ (Art. 528), $\therefore EF$ must intersect AB (which is not $\parallel CD$ by hyp.) in some point K.

At K in the plane CF draw $LK \perp EF$. Art. 274

Then LK is the perpendicular required.

Proof. $LK \perp EF$. Constr.

$\therefore LK \perp CD$. Art. 123.

Also $LK \perp$ plane MN. Art. 557

$\therefore LK \perp$ line AB. (Why?)

$\therefore LK \perp$ both CD and AB.

Q. E. F.

570. *Only one perpendicular can be drawn between two lines not in the same plane.*

For, if possible, in the above figure let another line BD be drawn $\perp AB$ and CD. Then, if a line be drawn through $B \parallel CD$, $BD \perp$ this line (Art. 123), and $\therefore \perp$ plane MN (Art. 509). Draw $DF \perp$ line EF; then $DF \perp$ plane MN (Art. 557). Hence from the point D two $\perp$s, DB and DF, are drawn to the plane MN, which is impossible (Art. 517).

POLYHEDRAL ANGLES

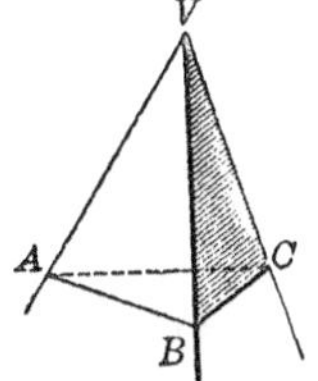

571. A **polyhedral angle** is the amount of opening between three or more planes meeting at a point.

Such an angle may be regarded as a portion of space cut out by the planes forming the angle.

572. The **vertex** of a polyhedral angle is the point in which the planes forming the angle meet; the **edges** are the lines in which the planes intersect; the **faces** are the portions of the planes forming the polyhedral angle which are included between the edges; the **face angles** are the angles formed by the edges.

Each two adjacent faces of a polyhedral angle form a *dihedral angle*.

The **parts of a polyhedral angle** are its face angles and dihedral angles taken together.

573. Naming a polyhedral angle. A polyhedral angle is named either by naming the vertex, as V; or by naming the vertex and a point on each edge, as V-ABC.

In case two or more polyhedral angles have the same vertex, the latter method is necessary.

In the above polyhedral angle, the vertex is V; the edges are VA, VB, VC; the face angles are AVB, BVC, AVC.

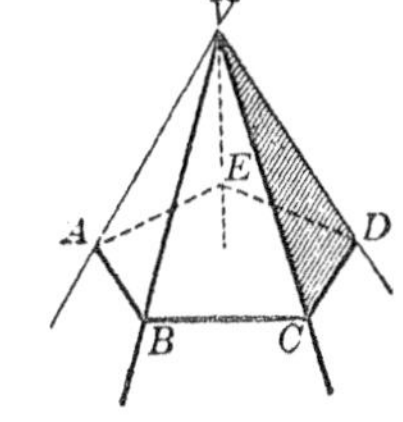

574. A **convex polyhedral angle** is a polyhedral angle in which a section made by a plane cutting all the edges is a convex polygon, as V-$ABCDE$.

575. A **trihedral angle** is a polyhedral angle having three faces; a **tetrahedral angle** is one having four faces, etc.

576. A **trihedral angle is rectangular, birectangular, or trirectangular,** according as it contains *one*, *two*, or *three* right dihedral angles.

577. An **isosceles trihedral angle** is a trihedral angle two of whose face angles are equal.

578. Vertical polyhedral angles are polyhedral angles having the same vertex and the faces of one the faces of the other produced.

579. Two equal polyhedral angles are polyhedral angles having their corresponding parts equal and arranged in the same order, as $V\text{-}ABC$ and $V'\text{-}A'B'C'$.

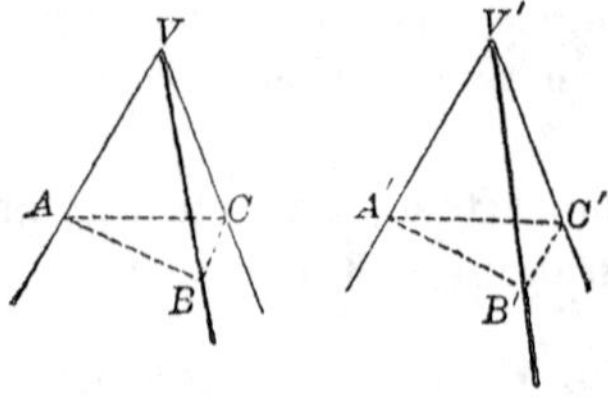

Two equal polyhedral angles may be made to coincide.

580. Two symmetrical polyhedral angles are polyhedral angles having their corresponding parts equal but arranged in reverse order.

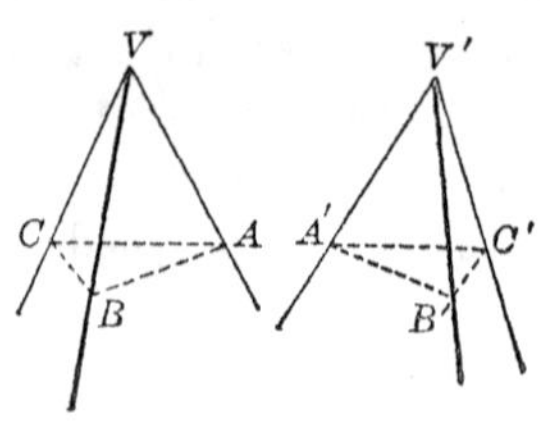

If the faces of a trihedral angle, $V\text{-}ABC$, be produced, they will form a vertical trihedral angle, $V\text{-}A'B'C'$, which is symmetrical to $V\text{-}ABC$. For, if $V\text{-}A'B'C'$ be rotated forward about a horizontal axis through V, the two trihedral angles are seen to have their corresponding parts equal but arranged in reverse order.

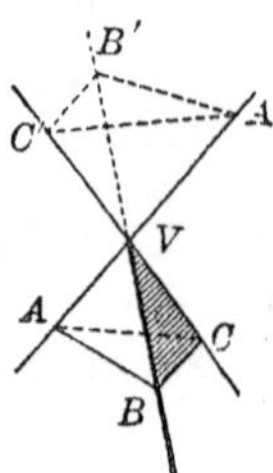

Similarly, any two vertical polyhedral angles are symmetrical.

581. Equivalence of symmetrical polyhedral angles. It has been shown in Plane Geometry (Art. 488) that two triangles (or polygons) symmetrical with respect to an axis have their corresponding parts equal and arranged in reverse order. By sliding two such figures about in a plane they cannot be made to coincide, but by lifting one of them up from the plane in which it lies and turning it over it may be made to coincide with the other figure.

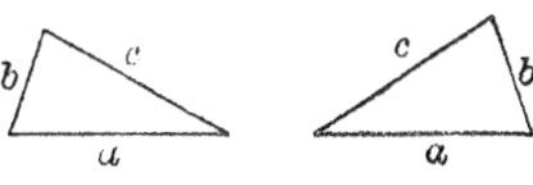

Symmetrical polyhedral angles, however, cannot be made to coincide in any way; hence some indirect method of showing their equivalence is necessary. See Ex. 29, p. 358, and Arts. 789–792.

Ex. 1. Name the trihedral angles on the figure to Prop. XX. If $\angle PRQ=90°$, what kind of trihedral angles are those on the figure? If $\angle PRQ=30°$, what kind are they?

Ex. 2. Are two trirectangular trihedral angles necessarily equal? Prove this.

Ex. 3. Are two lines which are perpendicular to the same plane necessarily parallel? Are two planes which are perpendicular to the same plane necessarily parallel? Are two planes which are perpendicular to the same line necessarily parallel?

Ex. 4. Let the pupil cut out three pieces of pasteboard of the form ndicated in the accompanying figures; cut them half through where :he lines are dotted; fold them and fasten the edges so as to form hree trihedral angles, two of which (Figs. 1 and 2) shall be equal and two (Figs. 1 and 3) symmetrical. By experiment, let the pupil ind which pair may be made to coincide, and which not.

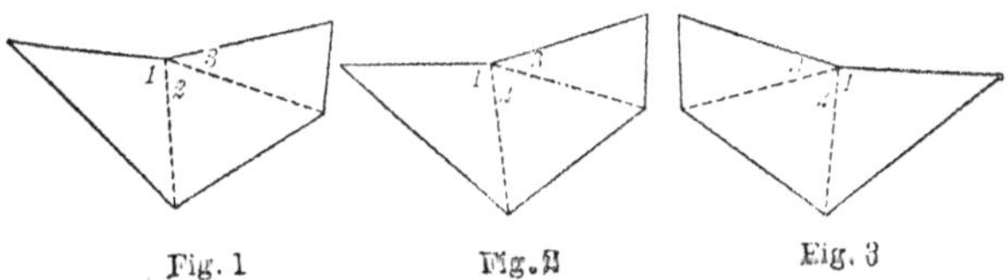

Fig. 1 Fig. 2 Fig. 3

PROPOSITION XXIV. THEOREM

582. *The sum of any two face angles of a trihedral angle is greater than the third face angle.*

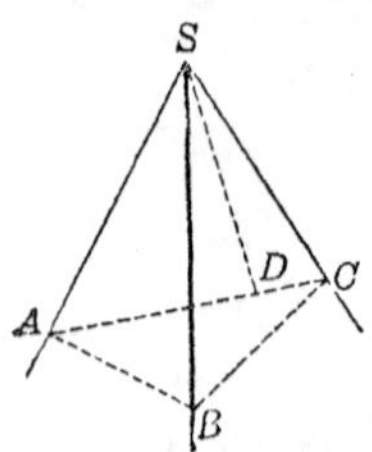

Given the trihedral angle S–ABC, with angle ASC its greatest face angle.

To prove $\angle ASB + \angle BSC$ greater than $\angle ASC$.

Proof. In the face ASC draw SD, making $\angle ASD = \angle ASB$.

Take $SD = SB$.

In the face ASC draw the line ADC in any convenient direction, and draw AB and BC.

Then, in the $\triangle$ ASB and ASD, $SA = SA$. (Why ?)

$SB = SD$, and $\angle ASB = \angle ASD$. (Why ?)

$\therefore \triangle ASB = \triangle ASD$. (Why ?)

$\therefore AB = AD$. (Why ?)

Also $AB + BC > AC$. (Why ?)

Hence, subtracting the equals AB and AD.

$BC > DC$. (Why ?)

Hence, in the $\triangle$ BSC and DSC, $SC = SC$, $SB = SD$, and $BC > DC$. (Why ?)

$\therefore \angle BSC$ is greater than $\angle DSC$. Art. 108.

To each of these unequals add the equals $\angle ASB$ and $\angle ASD$.

$\therefore \angle ASB + \angle BSC$ is greater than $\angle ASC$. (Why ?)

Q. E. D.

Ex. In the above figure, if $\angle ASC$ equals one of the other face angles at S, as $\angle ASB$, how is the theorem proved ?

PROPOSITION XXV. THEOREM

583. *The sum of the face angles of any convex polyhedral angle is less than four right angles.*

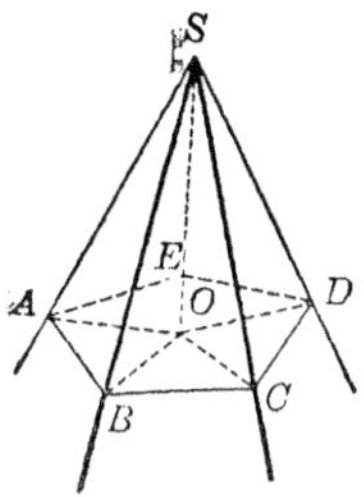

Given the polyhedral angle $S\text{-}ABCDE$.

To prove the sum of the face $\angle s$ at S less than 4 rt. $\angle s$.

Proof. Pass a plane cutting the edges of the given polyhedral angle in the points A, B, C, D, E.

From any point O in the polygon $ABCDE$ draw OA, OB, OC, OD, OE.

Denote the $\triangle s$ having the common vertex S as the S $\triangle s$, and those having the common vertex O as the O $\triangle s$.

Then the sum of the $\angle s$ of the S $\triangle s$ = the sum of $\angle s$ of the O $\triangle s$. Art. 134.

But $\angle SBA + \angle SBC$ is greater than $\angle ABC$,
$\angle SCB + \angle SCD$ is greater than $\angle BCD$, etc. } Art 582.

$\therefore$ the sum of the base $\angle s$ of the S $\triangle s >$ the sum of the base $\angle s$ of the O $\triangle s$. Ax. 9.

$\therefore$ the sum of the vertex $\angle s$ of the S $\triangle s <$ the sum of the vertex $\angle s$ of the O $\triangle s$, Ax. 11.

(*if unequals be subtracted from equals, the remainders are unequal in reverse order*).

But the sum of the $\angle s$ at O = 4 rt. $\angle s$. (Why?)

$\therefore$ the sum of face $\angle s$ at $S <$ 4 rt. $\angle s$. Ax. 8.

Q. E. D.

W

Proposition XXVI. Theorem

584. *If two trihedral angles have the three face angles of one equal to the three face angles of the other, the trihedral angles have their corresponding dihedral angles equal, and are either equal or symmetrical, according as their corresponding face angles are arranged in the same or in reverse order.*

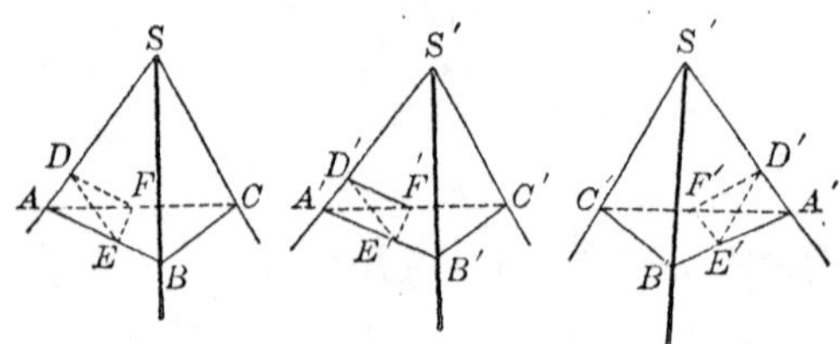

Given the trihedral ∡ S–ABC and S'–$A'B'C'$, having the face ∡ ASB, ASC and BSC equal to the face ∡ $A'S'B'$, $A'S'C'$ and $B'S'C'$, respectively.

To prove that the corresponding dihedral ∡ of S–ABC and S'–$A'B'C'$ are equal, and that ∡ S–ABC and S'–$A'B'C'$ are either equal or symmetrical.

Proof. On the edges of the trihedral ∡ take SA, SB, SC, $S'A'$, $S'B'$, $S'C'$ all equal.

Draw AB, AC, BC, $A'B'$, $A'C'$, $B'C'$.

Then, 1. In the ⊿ ASB and $A'S'B'$, $SA = S'A'$, $SB = S'B'$, and $\angle ASB = \angle A'S'B'$. (Why?)

$\therefore \triangle ASB = \triangle A'S'B'$. (Why?)

$\therefore AB = A'B'$. (Why?)

2. In like manner $AC = A'C'$, and $BC = B'C'$.

$\therefore \triangle ABC = \triangle A'B'C'$. (Why?)

3. Take D a convenient point in SA, and draw DE in the face ASB, and DF in the face ASC, each $\perp SA$.

DE and DF meet AB and AC in points E and F, respectively, (*for* $\angle$s *SAB and SAC are acute*).

Similarly, take $S'D'=SD$ and construct $\triangle D'E'F'$.

Then, in the rt. $\triangle$s ADE and $A'D'E'$, $AD=A'D'$, and $\angle DAE=D'A'E'$. (Why?)

$\therefore \triangle ADE=\triangle A'D'E'$. (Why?)

$\therefore AE=A'E'$, and $DE=D'E'$. (Why?)

4. In like manner it may be shown that $AF=A'F'$, and $DF=D'F'$.

$\therefore \triangle AEF=\triangle A'E'F'$. (Why?)

And $EF=E'F'$ (Why?)

5. Hence, in the $\triangle$s DEF and $D'E'F'$, $DE=D'E'$, $DF=D'F'$ and $EF=E'F'$. (Why?)

$\therefore \triangle DEF=\triangle D'E'F'$. (Why?)

$\therefore \angle EDF=\angle E'D'F'$. (Why?)

But these $\angle$s are the plane $\angle$s of the dihedral $\angle$s whose edges are SA and $S'A'$.

$\therefore$ dihedral $\angle B\text{-}AS\text{-}C=$ dihedral $\angle B'\text{-}A'S'\text{-}C'$ Art. 552.

In like manner it may be shown that the dihedral $\angle$s at SB and $S'B'$ are equal; and that those at SC and $S'C'$ are equal.

$\therefore$ the trihedral $\angle$s S and S' are either equal or symmetrical. Arts. 579, 580.

Q. E. D.

EXERCISES. GROUP 64

THEOREMS CONCERNING THE LINE AND PLANE IN SPACE

Ex. 1. A segment of a line not parallel to a plane is longer than its projection in the plane.

Ex. 2. Equal straight lines drawn from a point to a plane are equally inclined to the plane.

Ex. 3. A line and plane perpendicular to the same plane are parallel.

Ex. 4. If three planes intersecting in three straight lines are perpendicular to a plane, their lines of intersection are parallel.

Ex. 5. If a plane bisects any line at right angles, any point in the plane is equidistant from the ends of the line.

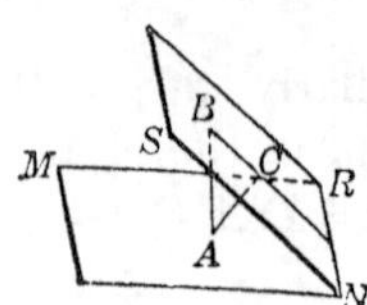

Ex. 6. Given $AB \perp$ plane MN,
and $AC \perp$ plane RS;
prove $BC \perp NR$.

Ex. 7. Given $PQ \perp$ plane MN,
$PR \perp$ plane BL,
and $RS \perp$ plane MN;
prove $QS \perp AB$.

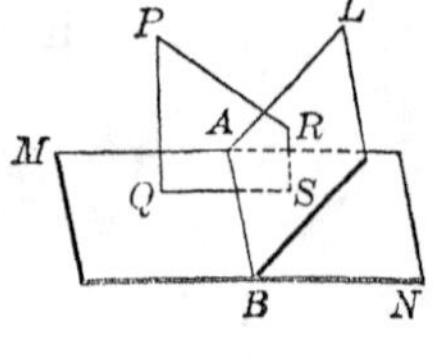

Ex. 8. If a line is perpendicular to one of two intersecting planes, its projection on the other plane is perpendicular to the line of intersection of the two planes.

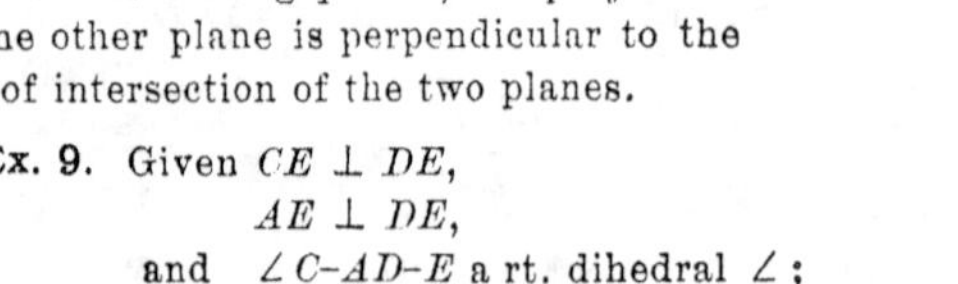

Ex. 9. Given $CE \perp DE$,
$AE \perp DE$,
and $\angle C\text{-}AD\text{-}E$ a rt. dihedral $\angle$;
prove $CA \perp$ plane DAE.

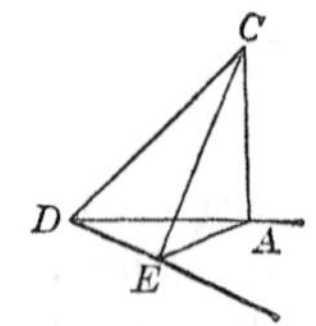

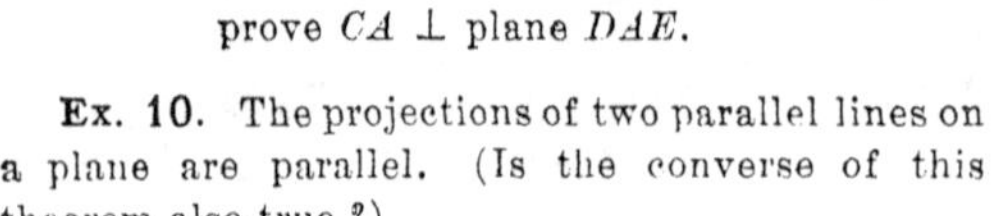

Ex. 10. The projections of two parallel lines on a plane are parallel. (Is the converse of this theorem also true?)

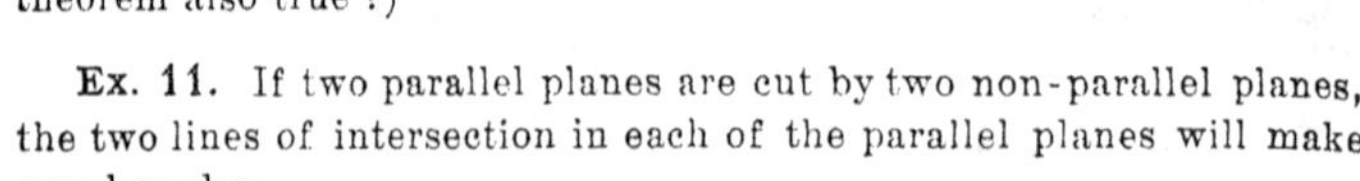

Ex. 11. If two parallel planes are cut by two non-parallel planes, the two lines of intersection in each of the parallel planes will make equal angles.

Ex. 12. If a line is perpendicular to a plane, any plane parallel to the line is perpendicular to the plane. (Is the converse true?)

Ex. 13. In the figure to Prop. VI, given $AB \perp MN$ and $AF \perp DC$; prove $BF \perp DC$.

Ex. 14. Two planes parallel to a third plane are parallel to each other.

[SUG. Draw a line $\perp$ third plane.]

Ex. 15. The projections upon a plane of two equal and parallel straight lines are equal and parallel.

Ex. 16. A line parallel to two planes is parallel to their intersection.

Ex. 17. In the figure to Prop. XXII, if angle CBP is obtuse, prove the angle ABP obtuse.

Ex. 18. In a quadrilateral in space (i. e., a quadrilateral whose vertices are not all in the same plane), show that the lines joining the midpoints of the sides form a parallelogram.

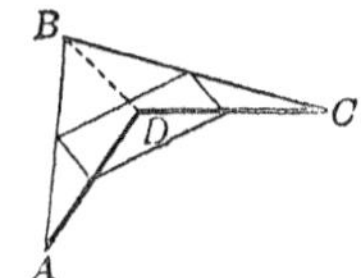

Ex. 19. The lines joining the midpoints of the opposite sides of a quadrilateral in space bisect each other.

Ex. 20. The planes bisecting the dihedral angles of a trihedral angle meet in a line every point of which is equidistant from the three faces.

[SUG. See Art. 562.]

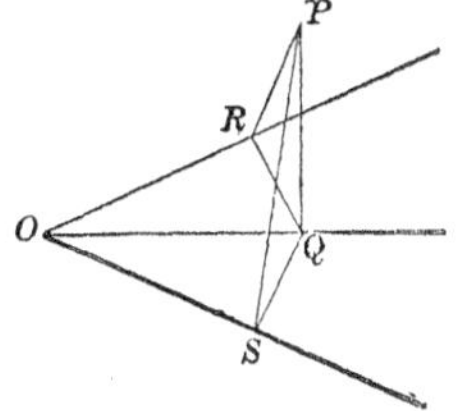

Ex. 21. Given OQ bisecting $\angle ROS$,
$PQ \perp$ plane ROS,
$QR \perp OR$,
and $QS \perp OS$;
prove $PR = PS$, $PR \perp OR$,
and $PS \perp OS$.

Ex. 22. In a plane bisecting a given plane angle, and perpendicular to its plane, every point is equidistant from the sides of the angle.

[SUG. See Ex. 21; or through P any point in the bisecting plane pass planes $\perp$ to the sides of the $\angle$, etc.]

Ex. 23. In a trihedral angle, the three planes bisecting the three face angles at right angles to their respective planes, intersect in a line every point of which is equidistant from the three edges of the trihedral angle.

Ex. 24. If two face angles of a trihedral angle are equal, the dihedral angles opposite them are equal.

Ex. 25. In the figure to Prop. XXIV, prove that $\angle ASC + \angle BSC$ is greater than $\angle ASD + \angle BSD$.

Ex. 26. The common perpendicular to two lines in space is the shortest line between them.

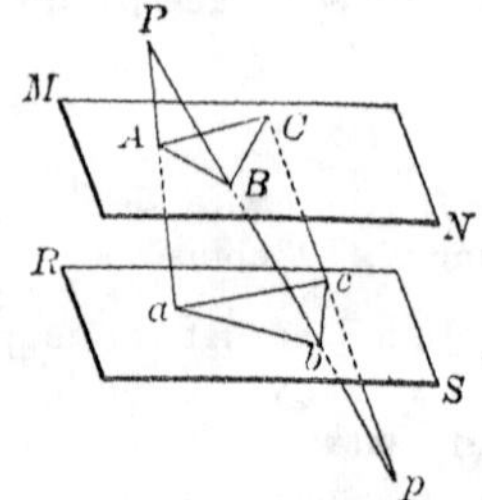

Ex. 27. Given $MN \parallel RS$,
and $PB = pb$;
prove $\angle ABC = \angle abc$,
and $\triangle ABC \backsim \triangle abc$.

Ex. 28. Two isosceles symmetrical trihedral angles are equal.

Ex. 29. Any two symmetrical trihedral angles are equivalent.

[SUG. Take SA, SB, SC, $S'A'$, $S'B'$, $S'C'$, all equal. Pass planes ABC, $A'B'C'$. Draw SO and $S'O'$ ⊥ these planes. Then the trihedral ∠s are divided into three pairs of isosceles symmetrical trihedral ∠s, etc.]

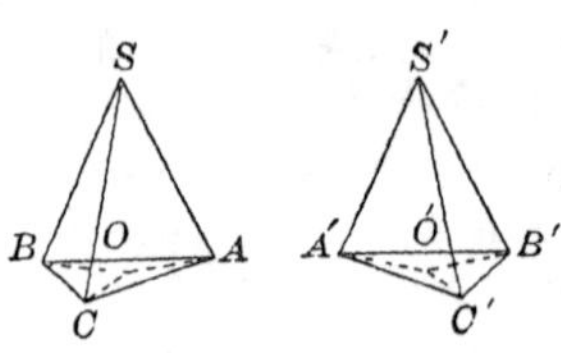

EXERCISES. GROUP 65

LOCI IN SPACE

Find the locus of a point equidistant from

Ex. 1. Two parallel planes.

Ex. 2. Two given points.

Ex. 3. Three given points.

Ex. 4. Two intersecting lines.

Ex. 5. The three faces of a trihedral angle.

Ex. 6. The three edges of a trihedral angle.

Find the locus

Ex. 7. Of all lines passing through a given point and parallel to a given plane.

Ex. 8. Of all lines perpendicular to a given line at a given point in the line.

Ex. 9. Of all points in a given plane equidistant from a given point outside the plane.

Ex. 10. Of all points equidistant from two given points and from two parallel planes.

Ex. 11. Of all points equidistant from two given points and from two intersecting planes.

Ex. 12. Of all points at a given distance from a given plane and equidistant from two intersecting lines.

EXERCISES. GROUP 66

PROBLEMS CONCERNING THE POINT, LINE AND PLANE IN SPACE

Ex. 1. Through a given point pass a plane parallel to a given plane.

Ex. 2. Through a given point pass a plane perpendicular to a given plane.

Ex. 3. Through a given point to construct a plane parallel to two given lines which are not in the same plane.

Prove that only one plane can be constructed fulfilling the given conditions.

Ex. 4. Bisect a given dihedral angle.

Ex. 5. Draw a plane equally inclined to three lines which meet at a point.

Ex. 6. Through a given point draw a line parallel to two given intersecting planes.

Ex. 7. Find a point in a plane such that lines drawn to it from two given points without the plane make equal angles with the plane.

[SUG. See Ex. 23, p. 176.]

Ex. 8. Find a point in a given line equidistant from two given points.

Ex. 9. Find a point in a plane equidistant from three given points.

Ex. 10. Find a point equidistant from four given points not in a plane.

Ex. 11. Through a given point draw a line which shall intersect two given lines.

[SUG. Pass a plane through the given point and one of the given lines, and pass another plane through the given point and the other given line, etc.]

Ex. 12. Through a given point pass a plane cutting the edges of a tetrahedral angle so that the section shall be a parallelogram.

[SUG. Produce each pair of opposite faces to intersect in a straight line, etc.]

Book VII

POLYHEDRONS

585. A **polyhedron** is a solid bounded by **planes.**

586. The **faces** of a polyhedron are its bounding planes; the **edges** of a polyhedron are the lines of intersection of its faces.

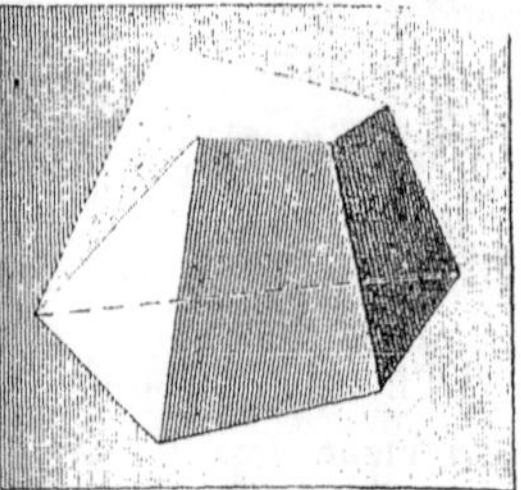
Polyhedron

A **diagonal** of a polyhedron is a straight line joining two of its vertices which are not in the same face. The **vertices** of a polyhedron are the points in which its edges meet or intersect.

587. A **convex polyhedron** is a polyhedron in which a section made by any plane is a convex polygon.

Only convex polyhedrons are to be considered in this book.

588. Classification of polyhedrons. Polyhedrons are sometimes classified according to the number of their faces. Thus, a **tetrahedron** is a polyhedron of four faces; a **hexahedron** is a polyhedron of six faces; an **octahedron** is one of eight, a **dodecahedron** one of twelve, and an **icosahedron** one of twenty faces.

Tetrahedron Cube Octahedron Dodecahedron Icosahedron

The polyhedrons most important in practical life are those determined by their stability, the facility with which they can be made out of common materials, as wood and iron, the readiness with which they can be packed together, etc. Thus, prism means "something sawed off."

PRISMS AND PARALLELOPIPEDS

589. A **prism** is a polyhedron bounded by two parallel planes and a group of planes whose lines of intersection are parallel.

Prism

590. The **bases** of a prism are the faces formed by the two parallel planes; the **lateral faces** are the faces formed by the group of planes whose lines of intersection are parallel.

The **altitude** of a prism is the perpendicular distance between the planes of its bases.

The **lateral area** of a prism is the sum of the areas of the lateral faces.

591. Properties of a prism inferred immediately.

1. *The lateral edges of a prism are equal*, for they are parallel lines included between parallel planes (Art. 589) and are therefore equal (Art. 532).

2. *The lateral faces of a prism are parallelograms* (Art. 160), for their sides formed by the lateral edges are equal and parallel.

3. *The bases of a prism are equal polygons*, for their homologous sides are equal and parallel, each to each, (being opposite sides of a parallelogram), and their homologous angles are equal (Art. 538).

592. A **right section** of a prism is a section made by a plane perpendicular to the lateral edges.

593. A **triangular prism** is a prism whose base is a triangle; a **quadrangular prism** is one whose base is a quadrilateral, etc.

Oblique Prisms Right Prism Regular Prism

594. An **oblique prism** is a prism whose lateral edges are oblique to the bases.

595. A **right prism** is a prism whose lateral edges are perpendicular to the bases.

596. A **regular prism** is a right prism whose bases are regular polygons.

597. A **truncated prism** is that part of a prism included between a base and a section made by a plane oblique to the base and cutting all the lateral edges.

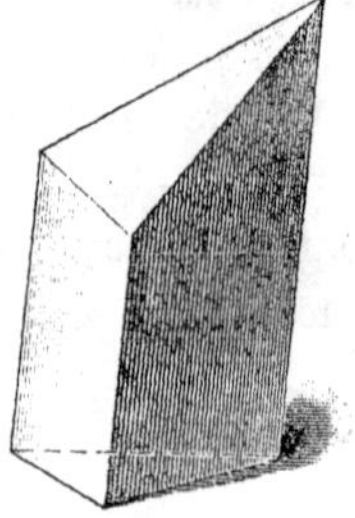

Truncated Prism

598. A **parallelopiped** is a prism whose bases are parallelograms.

Hence, *all the faces of a parallelopiped are parallelograms.*

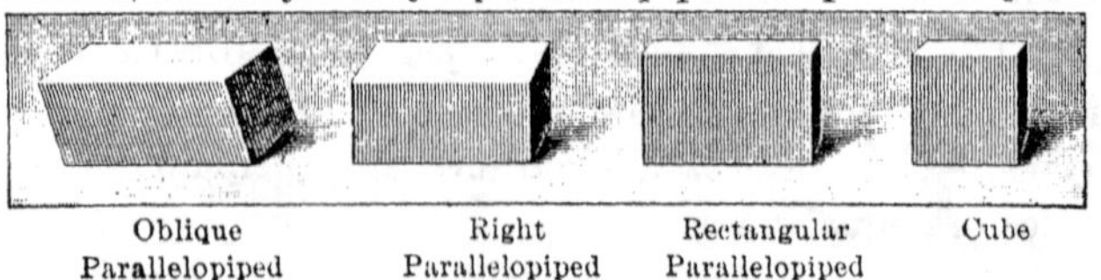

Oblique Parallelopiped Right Parallelopiped Rectangular Parallelopiped Cube

599. A **right parallelopiped** is a parallelopiped whose lateral edges are perpendicular to the bases,

600. A **rectangular parallelopiped** is a right parallelopiped whose bases are rectangles.

Hence, *all the faces of a rectangular parallelopiped are rectangles.*

601. A **cube** is a rectangular parallelopiped whose edges are all equal.

Hence, *all the faces of a cube are squares.*

602. The **unit of volume** is a cube whose edge is equal to some linear unit, as a cubic inch, a cubic foot, etc.

603. The **volume of a solid** is the number of units of volume which the solid contains.

Being a *number*, a volume may often be determined from other numbers in certain expeditious ways, which it is one of the objects of geometry to determine.

604. **Equivalent solids** are solids whose volumes are equal.

Ex. 1. What is the least number of faces which a polyhedron can have?

Ex. 2. A square right prism is what kind of a parallelopiped?

Ex. 3. Are there more right parallelopipeds or rectangular parallelopipeds? That is, which of these includes the other as a special case?

Ex. 4. Prove that if a given straight line is perpendicular to a given plane, and another straight line is perpendicular to another plane, and the two planes are parallel, then the two given lines are parallel.

PROPOSITION I. THEOREM

605. *Sections of a prism made by parallel planes cutting all the lateral edges are equal polygons.*

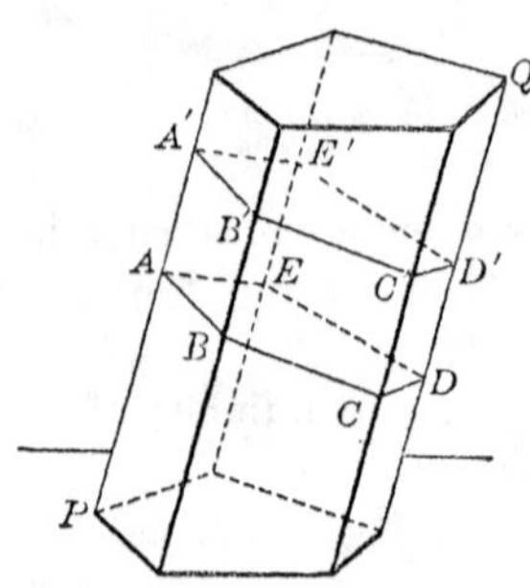

Given the prism PQ cut by || planes forming the sections AD and $A'D'$.

To prove section AD = section $A'D'$.

Proof. AB, BC, CD, etc., are || $A'B'$, $B'C'$, $C'D'$, etc., respectively. Art. 531.

∴ AB, BC, CD, etc., are equal to $A'B'$, $B'C'$, $C'D'$, etc., respectively. Art. 157.

Also ∠s ABC, BCD, etc., are equal to ∠s $A'B'C'$, $B'C'D'$, etc., respectively. Art. 538.

∴ $ABCDE = A'B'C'D'E'$, Art. 47.

(*for the polygons have all their parts equal, each to each, and ∴ can be made to coincide*).

Q. E. D.

606. COR. 1. *Every section of a prism made by a plane parallel to the base is equal to the base.*

607. COR. 2. *All right sections of a prism are equal.*

PROPOSITION II. THEOREM

608. *The lateral area of a prism is equal to the product of the perimeter of a right section by a lateral edge.*

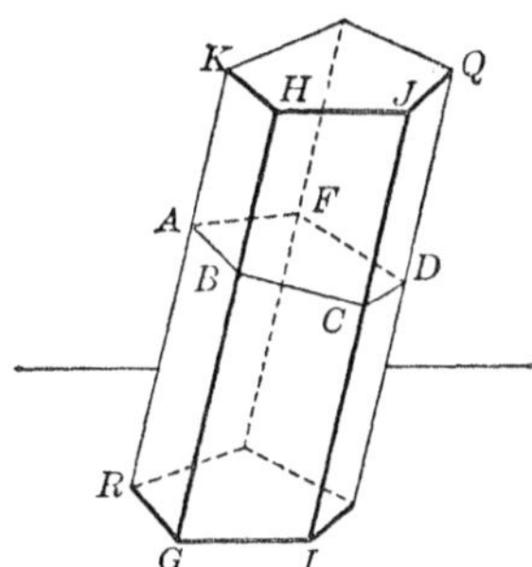

Given the prism RQ, with its lateral area denoted by S and lateral edge by E; and AD a right section of the given prism with its perimeter denoted by P.

To prove $S = P \times E$.

Proof. In the prism RQ, each lateral edge $= E$. Art. 591, 1.

Also $AB \perp GH$, $BC \perp IJ$, etc. Art. 505.

Hence area ▱ $RH = AB \times GH = AB \times E$,
area ▱ $GJ = BC \times E$,
area ▱ $IQ = CD \times E$, etc. } Art. 385.

But S, the lateral area of the prism, equals the sum of the areas of the ▱s forming the lateral surface.

$\therefore$ adding, $S = (AB + BC + CD + \text{etc.}) \times E$. Ax. 2.

Or $S = P \times E$.

Q. E. D.

609. COR. *The lateral area of a right prism equals the product of the perimeter of the base by the altitude.*

Ex. Find the lateral area of a right prism whose altitude is 12 in., and whose base is an equilateral triangle with a side of 6 in. Also find the total area of this figure.

PROPOSITION III. THEOREM

610. *If two prisms have the three faces including a trihedral angle of one equal, respectively, to the three faces including a trihedral angle of the other, and similarly placed, the prisms are equal.*

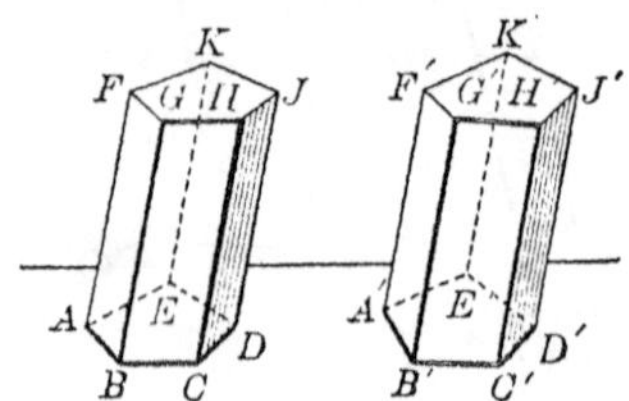

Given the prisms AJ and $A'J'$, having the faces AK, AD, AG equal to the faces $A'K'$, $A'D'$, $A'G'$, respectively, and similarly placed.

To prove $AJ = A'J'$.

Proof. The face ∡ EAF, EAB and BAF are equal, respectively, to the face ∡ $E'A'F'$, $E'A'B'$ and $B'A'F'$. Hyp.

∴ trihedral $\angle A =$ trihedral $\angle A'$ Art. 584.

Apply the prism $A'J'$ to the prism AJ, making each of the faces of the trihedral $\angle A'$ coincide with corresponding equal face of the trihedral $\angle A$. Geom. Ax. 2.

∴ the plane $F'J'$ will coincide in position with the plane FJ, Art. 500.
(*for the points* G', F', K' *coincide with* G, F, K, *respectively*).

Also the point C' will coincide with the point C.

∴ $C'H'$ will take the direction of CH. Geom. Ax. 3.

∴ H' will coincide with H. Art. 508, 1.

In like manner J' will coincide with J.

Hence the prisms AJ and $A'J'$ coincide in all points.

∴ $AJ = A'J'$. Art. 47.

611. COR. 1. *Two truncated prisms are equal if the three faces including a trihedral angle of one are equal to the three faces including a trihedral angle of the other.*

612. Cor. 2. *Two right prisms are equal if they have equal bases and equal altitudes.*

Proposition IV. Theorem

613. *An oblique prism is equivalent to a right prism whose base is a right section of the oblique prism and whose altitude is equal to a lateral edge of the oblique prism.*

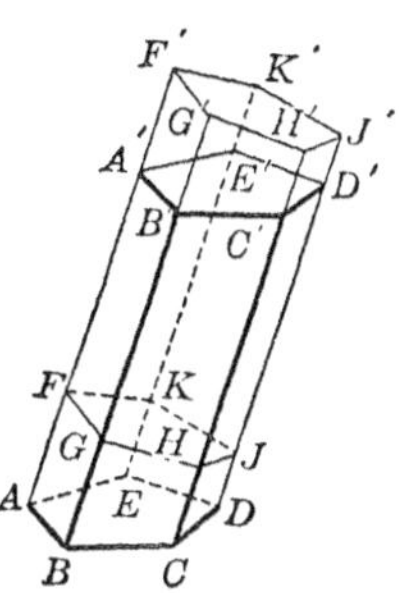

Given the oblique prism AD', with the right section FJ; also the right prism FJ' whose lateral edges are each equal to a lateral edge of AD'.

To prove $AD' \approx FJ'$.

Proof. $AA' = FF'$. Hyp.

Subtracting FA' from each of these, $AF = A'F'$. (Why?)

Similarly $BG = B'G'$.

Also $AB = A'B'$, and $FG = F'G'$. Art. 155.

And ∡ of face AG = homologous ∡ of face $A'G'$. Art. 130.

∴ face AG = face $A'G'$, Art. 47.

(*for they have all their parts equal, each to each, and ∴ can be made to coincide*).

In like manner face AK = face $A'K'$.

But face AD = face $A'D'$. Art. 591, 3.

∴ truncated prism AJ = truncated prism $A'J'$. Art. 611.

To each of these equals add the solid FD'.

∴ $AD' \approx FJ'$. (Why?)

Q. E. D.

PROPOSITION V. THEOREM

614. *The opposite lateral faces of a parallelopiped are equal and parallel.*

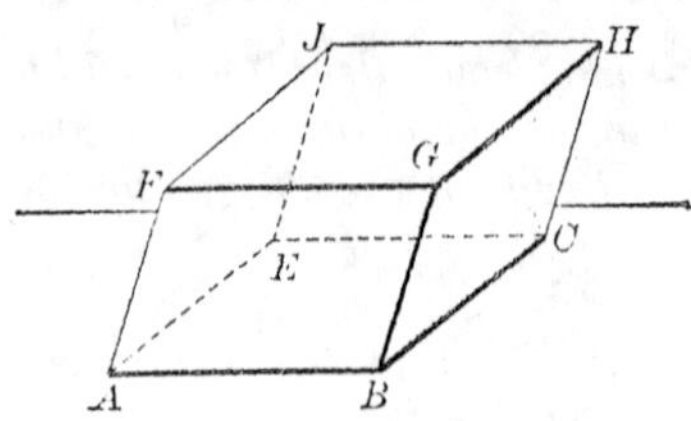

Given the parallelopiped AH with the base AC.

To prove $AG =$ and $\parallel EH$, and $AJ =$ and $\parallel BH$.

Proof. The base AC is a ▱. Art. 598.

$\therefore AB =$ and $\parallel EC$. (Why?)

Also the lateral face AJ is a ▱. Art. 591, 2.

$\therefore AF =$ and $\parallel EJ$. (Why?)

$\therefore \angle BAF = \angle CEJ$. Art. 538.

And ▱ $AG =$ ▱ EH. Art. 162.

Also plane $AG \parallel$ plane EH. Art. 538.

In like manner it may be shown that AJ and BH are equal and parallel.

Q. E. D.

615. COR. *Any two opposite faces of a parallelopiped may be taken as the bases.*

Ex. 1. How many edges has a parallelopiped? How many faces? How many dihedral angles? How many trihedral angles?

Ex. 2. Find the lateral area of a prism whose lateral edge is 10 and whose right section is a triangle whose sides are 6, 7, 8 in.

Ex. 3. Find the lateral area of a right prism whose lateral edge is 16 and whose base is a rhombus with diagonals of 6 and 8 in.

PROPOSITION VI. THEOREM

616. *A plane passed through two diagonally opposite edges of a parallelopiped divides the parallelopiped into two equivalent triangular prisms.*

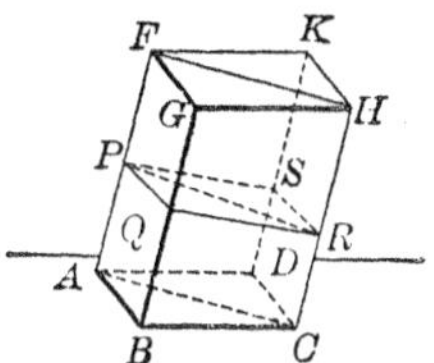

Given the parallelopiped AH with a plane passed through the diagonally opposite edges AF and CH, forming the triangular prisms $ABC\text{-}G$ and $ADC\text{-}K$.

To prove $ABC\text{-}G \approx ADC\text{-}K$.

Proof. Construct a plane $\perp$ to one of the edges of the prism forming the right section $PQRS$, having the diagonal PR formed by the intersection of the plane $FHCA$.

Then $PQ \parallel SR$, and $QR \parallel PS$. (Why ?)

$\therefore PQRS$ is a $\square$. (Why ?)

$\therefore \triangle PQR = \triangle PSR$. (Why ?)

But the triangular prism $ABC\text{-}G \approx$ a prism whose base is the right section PQR and whose altitude is AF. Art. 613.

Also the triangular prism $ADC\text{-}K \approx$ a prism whose base is the right section PSR and whose altitude is AF. (Why ?)

But the prisms having the equal bases, PQR and PSR, and the same altitude, AF, are equal. Art. 612.

$\therefore ABC\text{-}G \approx ADC\text{-}K$. Ax 1.

Q. E. D.

X

Proposition VII. Theorem

617. *If two rectangular parallelopipeds have equal bases, they are to each other as their altitudes.*

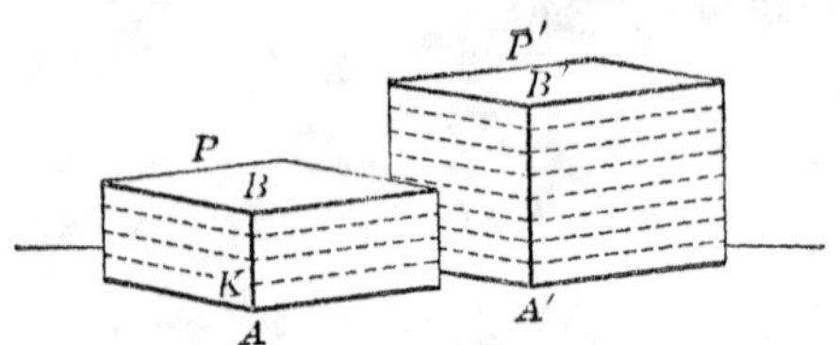

Given the rectangular parallelopipeds P' and P having equal bases and the altitudes $A'B'$ and AB.

To prove $P' : P = A'B' : AB$.

Case I. *When the altitudes $A'B'$ and AB are commensurable.*

Proof. Find a common measure of $A'B'$ and AB, as AK, and let it be contained in $A'B'$ n times and in AB m times.

Then $A'B' : AB = n : m$.

Through the points of division of $A'B'$ and AB pass planes parallel to the bases.

These planes will divide P' into n, and P into m small rectangular parallelopipeds, all equal. Art. 612.

$\therefore P' : P = n : m$.

$\therefore P' : P = A'B' : AB$. (Why?)

Case II. *When the altitudes $A'B'$ and AB are incommensurable.*

Let the pupil supply the proof, using the method of limits. (See Art. 554).

618. Def. The **dimensions of a rectangular parallelopiped** are the three edges which meet at one vertex.

619. COR. *If two rectangular parallelopipeds have two dimensions in common, they are to each other as their third dimensions.*

PROPOSITION VIII. THEOREM

620. *Two rectangular parallelopipeds having equal altitudes are to each other as their bases.*

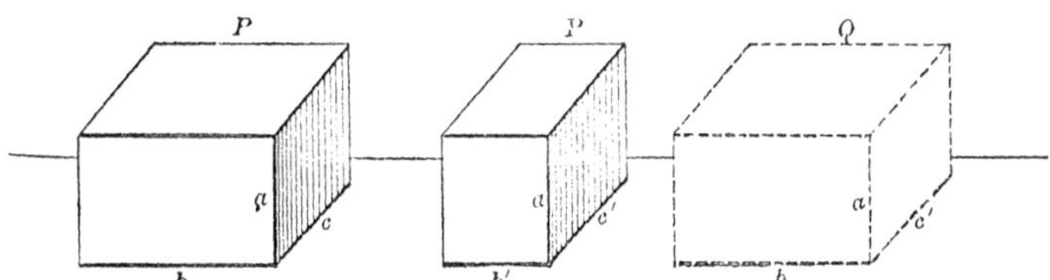

Given the rectangular parallelopipeds P and P' having the common altitude a, and the dimensions of their bases b, c and b', c', respectively.

To prove $$\frac{P}{P'}=\frac{b\times c}{b'\times c'}.$$

Proof. Construct the rectangular parallelopiped, Q, whose altitude is a and the dimensions of whose base are b and c'.

Then $$\frac{P}{Q}=\frac{c}{c'}.$$ Art. 619.

Also $$\frac{Q}{P'}=\frac{b}{b'}.$$ (Why ?)

Multiplying the corresponding members of these equalities,

$$\frac{P}{P'}=\frac{b\times c}{b'\times c'}.$$ Ax. 4.

Q. E. D.

621. COR. *Two rectangular parallelopipeds having one dimension in common are to each other as the products of the other two dimensions.*

PROPOSITION IX. THEOREM

622. *Any two rectangular parallelopipeds are to each other as the products of their three dimensions.*

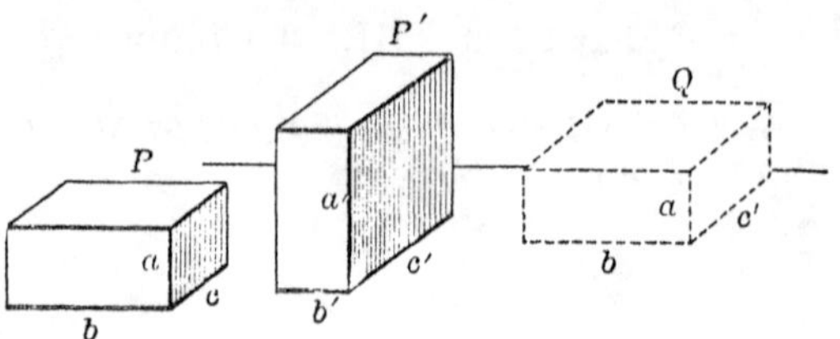

Given the rectangular parallelopipeds P and P' having the dimensions a, b, c and a', b', c', respectively.

To prove $$\frac{P}{P'}=\frac{a\times b\times c}{a'\times b'\times c'}.$$

Proof. Construct the rectangular parallelopiped Q having the dimensions a, b, c'.

Then $$\frac{P}{Q}=\frac{c}{c'}.$$ Art. 619.

Also $$\frac{Q}{P'}=\frac{a\times b}{a'\times b'}.$$ Art. 621.

Multiplying the corresponding members of these equalities,

$$\frac{P}{P'}=\frac{a\times b\times c}{a'\times b'\times c'}.$$ Ax. 4.

Q. E. D.

Ex. 1. Find the ratio of the volumes of two rectangular parallelopipeds whose edges are 5, 6, 7 in. and 7, 8, 9 in.

Ex. 2. Which will hold more, a bin 10 x 2 x 7 ft., or one 8 x 4 x 5 ft.?

Ex. 3. How many bricks 8 x 4 x 2 in. are necessary to build a wall 80 x 6 ft. x 8 in.?

PROPOSITION X. THEOREM

623. *The volume of a rectangular parallelopiped is equal to the product of its three dimensions.*

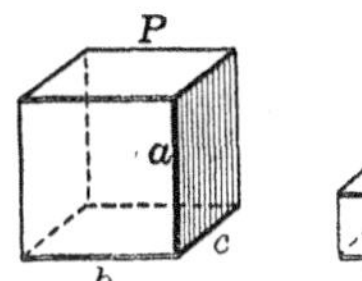

Given the rectangular parallelopiped P having the three dimensions a, b, c.

To prove volume of $P = a \times b \times c$.

Proof. Take as the unit of volume the cube U, whose edge is a linear unit.

Then $$\frac{P}{U} = \frac{a \times b \times c}{1 \times 1 \times 1}.$$ Art. 622.

The volume of P is the number of times P contains the unit of volume U, or $\frac{P}{U}$. Art. 603.

$\therefore$ volume of $P = a \times b \times c$.

Q. E. D.

For significance of this result, see Art. 2.

624. COR. 1. *The volume of a cube is the cube of its edge.*

625. COR. 2. *The volume of a rectangular parallelopiped is equal to the product of its base by its altitude.*

Ex. 1. Find the number of cubic inches in the volume of a cube whose edge is 1 ft. 3 in. How many bushels does this box contain, if 1 bushel = 2150.42 cu. in.?

Ex. 2. The measurement of the volume of a cube reduces to the measurement of the length of what single straight line?

PROPOSITION XI. THEOREM

626. *The volume of any parallelopiped is equal to the product of its base by its altitude.*

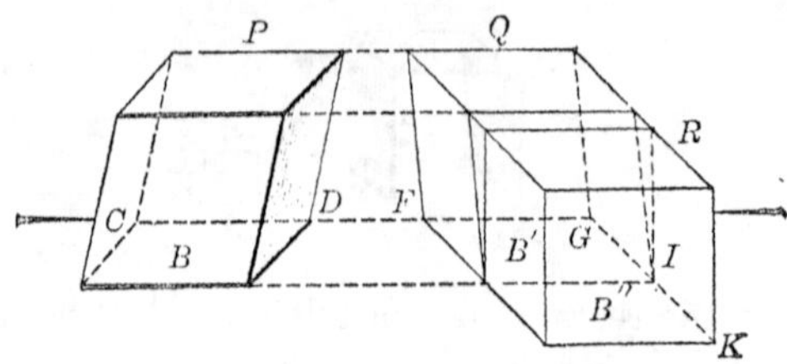

Given the oblique parallelopiped P, with its base denoted by B, and its altitude by H.

To prove volume of $P = B \times H$.

Proof. In P produce the edge CD and all the edges parallel to CD.

On CD produced take $FG = CD$.

Pass planes through F and $G \perp$ the produced edges, forming the parallelopiped Q, with the rectangular base denoted by B'.

Similarly produce the edge GI and all the edges $\parallel GI$.

Take $IK = GI$, and pass planes through I and $K \perp$ the edges last produced, forming the rectangular parallelopiped R, with its base denoted by B''.

Then $P \approx Q \approx R$. Art. 613.

Also $B \approx B' = B''$. Art. 386.

But volume of $R = B'' \times H$. Art. 625.

$\therefore$ volume of $P = B'' \times H$. Ax. 1.

Or volume of $P = B \times H$. Ax. 8.

Q. E. D.

[**Outline Proof.** $P \approx Q \approx R = B'' \times H = B \times H$.]

PROPOSITION XII. THEOREM

627. *The volume of a triangular prism is equal to the product of its base by its altitude.*

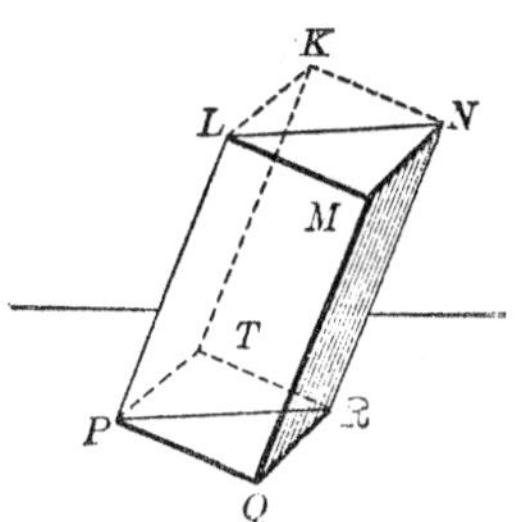

Given the triangular prism $PQR-M$, with its volume denoted by V, area of base by B, and altitude by H.

To prove $V = B \times H$.

Proof. Upon the edges PQ, QR, QM, construct the parallelopiped QK.

Hence $QK =$ twice $PQR-M$. Art. 616.

But volume of $QK =$ area $PQRT \times H$. Art. 626.

$= 2B \times H$. Ax. 8.

$\therefore$ twice volume $PQR-M = 2B \times H$. Ax. 1.

$\therefore$ volume $PQR-M = B \times H$. Ax. 5.

Q. E. D.

Ex. 1. If the altitude of a triangular prism is 18 in., and the base is a right triangle whose legs are 6 and 8 in., find the volume.

Ex. 2 Find the volume of a triangular prism whose altitude is 24, and the edges of whose base are 7, 8, 9. Also find the total surface.

PROPOSITION XIII. THEOREM

628. *The volume of any prism is equal to the product of its base by its altitude.*

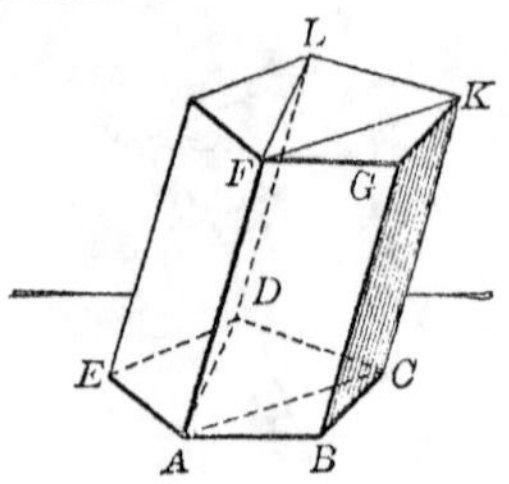

Given the prism AK, with its volume denoted by V, area of base by B, and altitude by H.

To prove $V = B \times H$.

Proof. Through any lateral edge, as AF, and the diagonals of the base, AC and AD, drawn from its foot, pass planes.

These planes will divide the prism into triangular prisms.

Then V, the volume of the prism AK, equals the sum of the volumes of the triangular prisms. Ax. 6.

But the volume of each triangular prism = its base $\times H$.

Hence the sum of the volumes of the $\triangle$ prisms = the sum of the bases of the $\triangle$ prisms $\times H$.

$= B \times H$. Ax. 8.

$\therefore V = B \times H$. Ax. 1.

Q. E. D.

629. COR. 1. *Two prisms are to each other as the products of their bases by their altitudes; prisms having equivalent bases and equal altitudes are equivalent.*

630. COR. 2. *Prisms having equivalent bases are to each other as their altitudes; prisms having equal altitudes are to each other as their bases.*

PYRAMIDS

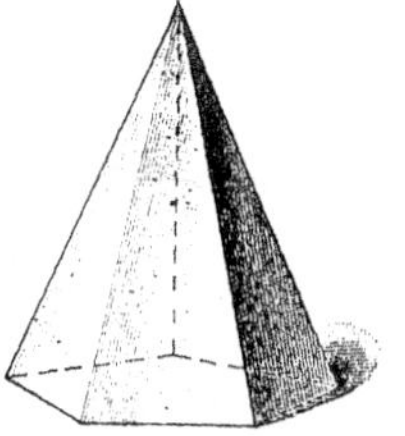
Pyramid

631. A **pyramid** is a polyhedron bounded by a group of planes passing through a common point, and by another plane cutting all the planes of the group.

632. The **base** of a pyramid is the face formed by the cutting plane; the **lateral faces** are the faces formed by the group of planes passing through a common point; the **vertex** is the common point through which the group of planes passes; the **lateral edges** are the intersections of the lateral faces.

The **altitude** of a pyramid is the perpendicular from the vertex to the plane of the base.

The **lateral area** is the sum of the areas of the lateral faces.

633. Properties of pyramids inferred immediately.

1. *The lateral faces of a pyramid are triangles* (Art. 508, 2).

2. *The base of a pyramid is a polygon* (Art. 508, 2).

634. A **triangular pyramid** is a pyramid whose base is a triangle; a **quadrangular pyramid** is a pyramid whose base is a quadrilateral, etc.

A triangular pyramid is also called a tetrahedron, for it has four faces. All these faces are triangles, and any one of them may be taken as the base.

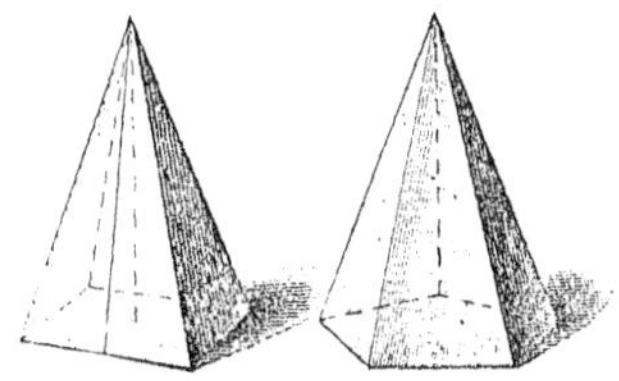

635. A **regular pyramid** is a pyramid whose base is a regular polygon, and the foot of whose altitude coincides with the center of the base.

636. Properties of a regular pyramid inferred immediately.

1. *The lateral edges of a regular pyramid are equal*, for they are oblique lines drawn from a point to a plane cutting off equal distances from the foot of the perpendicular from the point to the plane (Art. 518).

2. *The lateral faces of a regular pyramid are equal isosceles triangles.*

637. The **slant height of a regular pyramid** is the altitude of any one of its lateral faces.

The **axis** of a regular pyramid is its altitude.

638. A **truncated pyramid** is the portion of a pyramid included between the base and a section cutting all the lateral edges.

639. A **frustum of a pyramid** is the part of a pyramid included between the base and a plane parallel to the base.

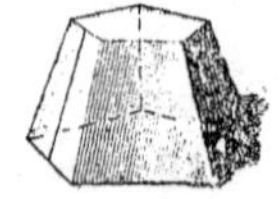

The **altitude of a frustum of a pyramid** is the perpendicular distance between the planes of its bases.

640. Properties of a frustum of a pyramid inferred immediately.

1. *The lateral faces of a frustum of a pyramid are trapezoids.*

2. *The lateral faces of a frustum of a regular pyramid are equal isosceles trapezoids.*

641. The **slant height of the frustum of a regular pyramid** is the altitude of one of its lateral faces.

Ex. 1. Show that the foot of the altitude of a regular pyramid coincides with the center of the circle circumscribed about the base.

Ex. 2. The perimeter of the midsection of the frustum of a pyramid equals one-half the sum of the perimeters of the bases.

Proposition XIV. Theorem

642. *The lateral area of a regular pyramid is equal to half the product of the slant height by the perimeter of the base.*

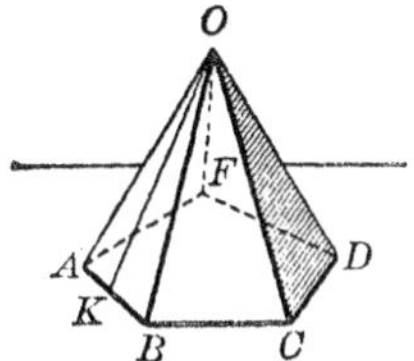

Given $O-ABCDF$ a regular pyramid with its lateral area denoted by S, slant height by L, and perimeter of its base by P.

To prove $S=\frac{1}{2}L\times P$.

Proof. The lateral faces OAB, OBC, etc., are equal isosceles △. Art. 636, 2.

Hence each lateral face has the same slant height, L.

∴ the area of each lateral face $=\frac{1}{2}L\times$ its base.

∴ the sum of all the lateral faces $=\frac{1}{2}L\times$ sum of bases.

$=\frac{1}{2}L\times P$.

∴ $S=\frac{1}{2}L\times P$. Ax. 8.

Q. E. D.

643. Cor. *The lateral area of the frustum of a regular pyramid is equal to one-half the sum of the perimeters of its bases multiplied by its slant height.*

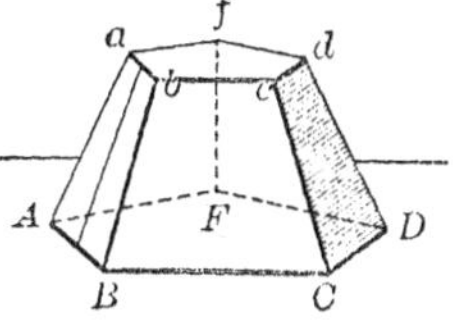

Ex. Find the lateral area of a regular square pyramid whose slant height is 32, and an edge of whose base is 16. Find the total area also.

PROPOSITION XV. THEOREM

644. *If a pyramid is cut by a plane parallel to the base,*

I. *The lateral edges and the altitude are divided proportionally;*

II. *The section is a polygon similar to the base.*

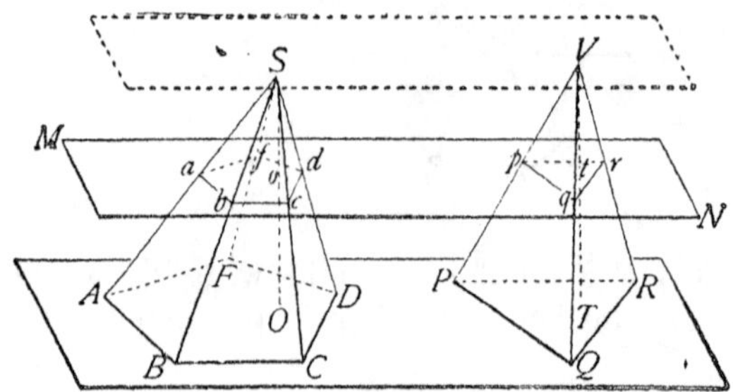

Given the pyramid $S\text{-}ABCDF$, with the altitude SO cut by a plane MN, which is parallel to the base and intersects the lateral edges in a, b, c, d, f and the altitude in o.

To prove I. $\frac{Sa}{SA}=\frac{Sb}{SB}=\frac{Sc}{SC}=\cdots=\frac{So}{SO}$.

II. The section $abcdf$ similar to the base $ABCDF$.

Proof. I. Pass a plane through the vertex $S \parallel MN$.

Then SA, SB, $SC \ldots SO$, are lines intersected by three $\parallel$ planes.

$$\therefore \frac{Sa}{SA}=\frac{Sb}{SB}=\frac{Sc}{SC}=\cdots=\frac{So}{SO}. \quad \text{Art. 539.}$$

II. $ab \parallel AB$. (Why?)

$\therefore$ $\triangle$ Sab and SAB are similar. Art. 328.

In like manner the $\triangle$ Sbc, Scd, etc., are similar to the $\triangle$ SBC, SCD, etc., respectively.

$$\therefore \frac{ab}{AB}=\left(\frac{Sb}{SB}\right)=\frac{bc}{BC}=\left(\frac{Sc}{SC}\right)=\frac{cd}{CD}=\text{ etc.}$$

That is, the homologous sides of *abcdf* and *ABCDF* are proportional.

Also $\angle abc = \angle ABC$, $\angle bcd = \angle BCD$, etc. Art. 538.

$\therefore$ section *abcdf* is similar to the base *ABCDF*. Art. 321.

Q. E. D.

645. Cor. 1. *A section of a pyramid parallel to the base is to the base as the square of its distance from the vertex is to the square of the altitude of the pyramid.*

For $$\frac{abcdf}{ABCDF} = \frac{\overline{ab}^2}{\overline{AB}^2}.$$ (Why?)

But $$\frac{ab}{AB} = \frac{Sa}{SA} = \frac{So}{SO}. \quad \therefore \frac{\overline{ab}^2}{\overline{AB}^2} = \frac{\overline{So}^2}{\overline{SO}^2}.$$ (Why?)

$$\therefore \frac{abcdf}{ABCDF} = \frac{\overline{So}^2}{\overline{SO}^2}.$$ (Why?)

646. Cor. 2. *If two pyramids having equal altitudes are cut by a plane parallel to their bases at equal distances from the vertices, the sections have the same ratio as the bases.*

Let *S–ABCDF* and *V–PQR* be two pyramids cut as described.

Then $\frac{abcdf}{ABCDF} = \frac{\overline{So}^2}{\overline{SO}^2}$ (Art. 645); also $\frac{pqr}{PQR} = \frac{\overline{Vt}^2}{\overline{VT}^2}$. (Why?)

But $VT = SO$, and $Vt = So$. (Why?)

$$\therefore \frac{abcdf}{ABCDF} = \frac{pqr}{PQR}, \text{ or } \frac{abcdf}{pqr} = \frac{ABCDF}{PQR}.$$ (Why?)

647. Cor. 3. *If two pyramids have equal altitudes and equivalent bases, sections made by planes parallel to the bases at equal distances from the vertices are equivalent.*

PROPOSITION XVI. THEOREM

648. *The volume of a triangular pyramid is the limit of the sum of the volumes of a series of inscribed, or of a series of circumscribed prisms of equal altitude, if the number of prisms be indefinitely increased.*

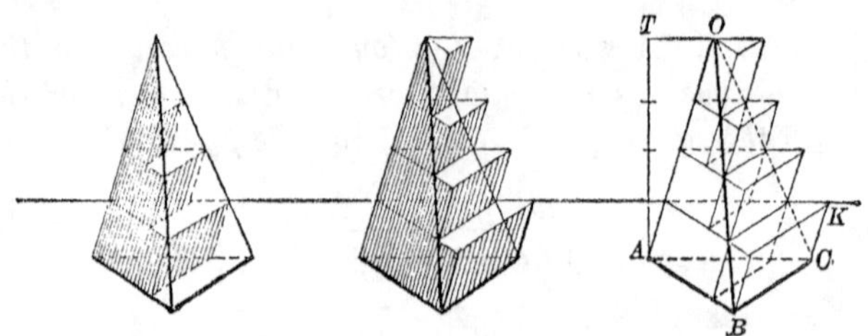

Given the triangular prism $O\text{-}ABC$ with a series of inscribed, and also a series of circumscribed prisms, formed by passing planes which divide the altitude into equal parts, and by making the sections so formed first upper bases, then lower bases, of prisms limited by the next parallel plane.

To prove $O\text{-}ABC$ the limit of the sum of each series, if the number of prisms in each be indefinitely increased.

Proof. Each inscribed prism equals the circumscribed prism immediately above it. Art. 629.

$\therefore$ (sum of circumscribed prisms)—(sum of inscribed prisms) = lowest circumscribed prism, or $ABC\text{-}K$.

If the number of prisms be indefinitely increased, the altitude of each approaches zero as a limit.

Hence volume $ABC\text{-}K \doteq 0$, Art. 253, 2.

(*for its base, ABC, is constant while its altitude* $\doteq 0$).

$\therefore$ (sum of circumscribed prisms)—(sum of inscribed prisms) $\doteq 0$.

$\therefore$ volume $O\text{-}ABC$—(either series of prisms) $\doteq 0$.

(*for this difference* $<$ *difference between the two series, which last difference* $\doteq 0$).

$\therefore$ $O\text{-}ABC$ is the limit of the sum of the volumes of either series of prisms. Q. E. D.

PROPOSITION XVII. THEOREM

649. *If two triangular pyramids have equal altitudes and equivalent bases, they are equivalent.*

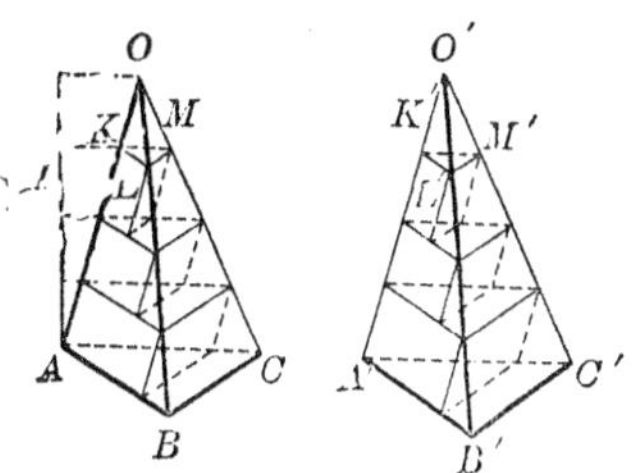

Given the triangular pyramids $O\text{–}ABC$ and $O'\text{–}A'B'C'$ having equivalent bases ABC and $A'B'C'$, and equal altitudes.

To prove $O\text{–}ABC \approx O'\text{–}A'B'C'$.

Proof. Place the pyramids so that they have the common altitude H, and divide H into any convenient number of equal parts.

Through the points of division and parallel to the plane of the bases of the pyramids, pass planes cutting the pyramids.

Using the sections so formed as upper bases, inscribe a series of prisms in each pyramid, and denote the volumes of the two series of prisms by V and V'.

The sections formed by each plane, as KLM and $K'L'M'$, are equivalent. Art. 647.

$\therefore$ each prism in $O\text{–}ABC \approx$ corresponding prism in $O'\text{–}A'B'C'$ (Art. 629). $\therefore V = V'$. Ax. 2.

Let the number of parts into which the altitude is divided be increased indefinitely.

Then V and V' become variables with $O\text{–}ABC$ and $O'\text{–}A'B'C'$ as their respective limits. Art. 648.

But $V \approx V'$ always. (Why?)

$\therefore O\text{–}ABC \approx O'\text{–}A'B'C'$. (Why?)

Q. E. D.

Proposition XVIII. Theorem

650. *The volume of a triangular pyramid is equal to one-third the product of its base by its altitude.*

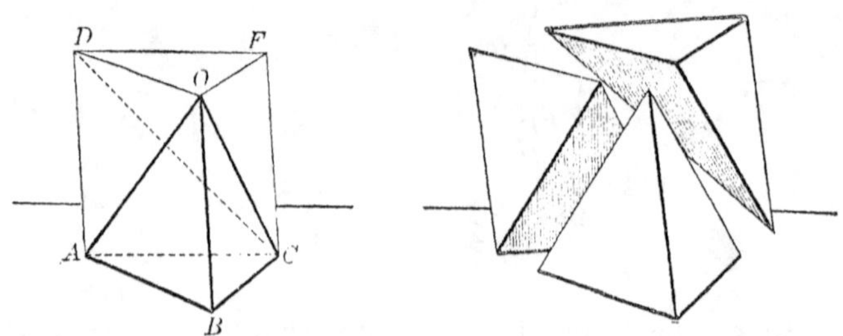

Given the triangular pyramid $O-ABC$, having its volume denoted by V, the area of its base by B, and its altitude by H.

To prove $V = \frac{1}{3} B \times H$.

Proof. On ABC as a base, with OB as a lateral edge, construct the prism $ABC-DOF$.

Then this prism will be composed of the original pyramid $O-ABC$ and the quadrangular pyramid $O-ADFC$.

Through the edges OD and OC pass a plane intersecting the face $ADFC$ in the line DC, and dividing the quadrangular pyramid into the triangular pyramids $O-ADC$ and $O-DFC$.

Then $O-ADC \approx O-DFC$. Art. 649.

(*for they have the common vertex O, and the equal bases ADC and DFC*).

But $O-DFC$ may be regarded as having C as its vertex and DOF as its base. Art. 634.

$\therefore O-DFC \approx O-ABC$. Art. 649.

$\therefore$ the prism is made up of three equivalent pyramids.

$\therefore O-ABC = \frac{1}{3}$ the prism. Ax. 5.

But volume of prism $= B \times H$. Art. 627.

$\therefore O-ABC$, or $V = \frac{1}{3} B \times H$. Ax. 5.

Q. E. D.

Ex. Find the volume of a triangular pyramid whose altitude is 12 ft., and whose base is an equilateral triangle with a side of 15 ft.

PROPOSITION XIX. THEOREM

651. *The volume of any pyramid is equal to one-third the product of its base by its altitude.*

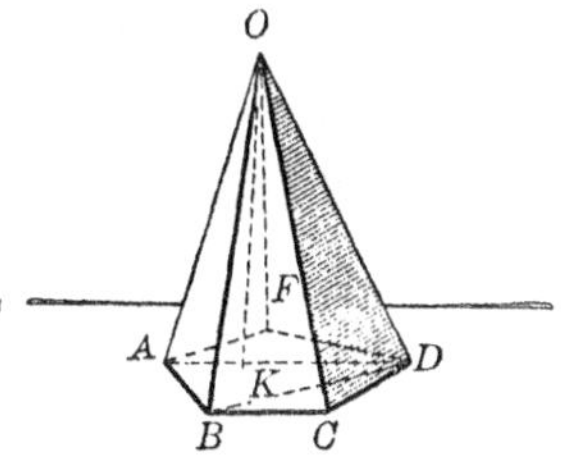

Given the pyramid O-$ABCDF$, having its volume denoted by V, the area of its base by B, and its altitude by H.

To prove $V = \frac{1}{3} B \times H$.

Proof. Through any lateral edge, as OD, and the diagonals of the base drawn from its foot, as AD and BD, pass planes dividing the pyramid into triangular pyramids.

Then V, the volume of the pyramid $O-ABCDF$, will equal the sum of the volumes of the triangular pyramids.

But the volume of each △ pyramid $= \frac{1}{3}$ its base $\times H$. Art. 650.

Hence the sum of the volumes of △ pyramids $= \frac{1}{3}$ sum of their bases $\times H$. Ax. 2.

$= \frac{1}{3} B \times H$. Ax. 8.

$\therefore V = \frac{1}{3} B \times H$. Ax. 1.

Q. E. D.

652. COR. 1. *The volumes of two pyramids are to each other as the products of their bases and altitudes; pyramids having equivalent bases and equal altitudes are equivalent.*

653. COR. 2. *Pyramids having equivalent bases are to each other as their altitudes; pyramids having equal altitudes are to each other as their bases.*

654. SCHOLIUM. *The volume of any polyhedron may be found by dividing the polyhedron into pyramids, finding the volume of each pyramid separately, and taking their sum.*

Y.

PROPOSITION XX. THEOREM

655. *The frustum of a triangular pyramid is equivalent to the sum of three pyramids whose common altitude is the altitude of the frustum, and whose bases are the lower base, the upper base, and a mean proportional between the two bases of the frustum.*

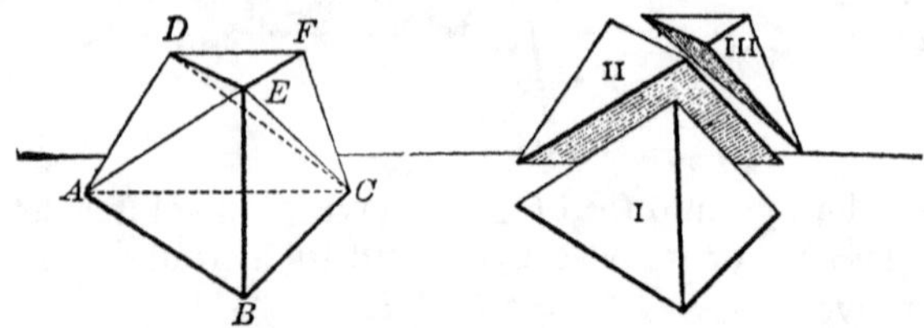

Given $ABC\text{-}DEF$ the frustum of a triangular pyramid, having the area of its lower base denoted by B, the area of its upper base by b, and its altitude by H.

To prove $ABC\text{-}DEF \approx$ three pyramids whose bases are B, b and $\sqrt{Bb}$, and whose common altitude is H.

Proof. Through E and AC, E and DC, pass planes dividing the frustum into three triangular pyramids. Then

1. $E\text{-}ABC$ has the base B and the altitude H.

2. $E\text{-}DFC$, that is, $C\text{-}DEF$, has the base b and the altitude H.

3. It remains to show that $E\text{-}ADC$ is equivalent to a pyramid having an altitude H, and a base that is a mean proportional between B and b.

Denoting the three pyramids by I, II, III,

$$\frac{\text{I}}{\text{II}}=\frac{\triangle ABE}{\triangle ADE}=\frac{AB}{DE}=\frac{AC}{DF}=\frac{\triangle ADC}{\triangle DFC}=\frac{\text{II}}{\text{III}}.$$

(Arts. 653, 391, 644, 321. Let the pupil supply the reason for each step in detail).

$$\therefore \frac{\text{I}}{\text{II}}=\frac{\text{II}}{\text{III}} \text{ (Ax. 1.) or } \text{II}=\sqrt{\text{I}\times\text{III}}. \qquad \text{Art. 303.}$$

$$\therefore E\text{-}ADC=\sqrt{(\tfrac{1}{3}H\times B)(\tfrac{1}{3}H\times b)}=\tfrac{1}{3}H\sqrt{B\times b}.$$

Hence, $ABC\text{-}DEF \approx$ sum of three pyramids, as described.

Q. E. D.

656. Formula for volume of frustum of a triangular pyramid. $V=\frac{1}{3}H(B+b+\sqrt{Bb})$.

PROPOSITION XXI. THEOREM

657. ***The volume of the frustum of any pyramid is equivalent to the sum of the volumes of three pyramids, whose common altitude is the altitude of the frustum, and whose bases are the lower base, the upper base, and a mean proportional between the two bases of the frustum.***

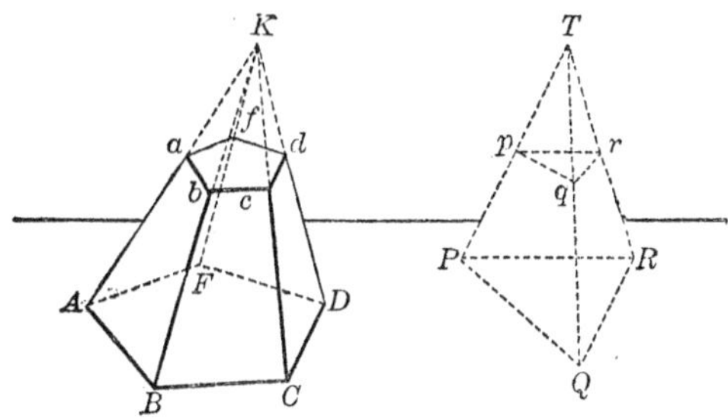

Given the frustum of a pyramid Ad, having its volume denoted by V, the area of its lower base by B, of its upper base by b, and its altitude by H.

To prove $V=\frac{1}{3}H(B+b+\sqrt{Bb})$.

Proof. Produce the lateral faces of Ad to meet in K.

Also construct a triangular pyramid with base PQR equivalent to $ABCDF$, and in the same plane with it, and with an altitude equal to the altitude of K-$ABCDF$. Produce the plane of ad to cut the second pyramid in pqr.

Then $pqr \approx abcdf$. Art. 647.

$\therefore$ pyramid K-$ABCDF \approx$ pyramid T-PQR. Art. 652.

Also pyramid K-$abcdf \approx$ pyramid T-pqr. (Why ?)

Subtracting, frustum $Ad \approx$ frustum Pr. Ax. 3.

But volume $Pr=\frac{1}{3}H(B+b+\sqrt{Bb})$.

$\therefore$ volume $Ad=\frac{1}{3}H(B+b+\sqrt{Bb})$. (Why ?)

Q. E. D.

PROPOSITION XXII. THEOREM

658. *A truncated triangular prism is equivalent to the sum of three pyramids, of which the base of the prism is the common base, and whose vertices are the three vertices of the inclined section.*

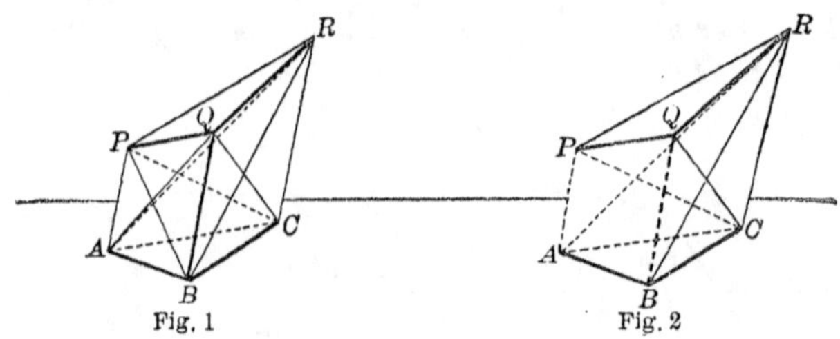

Fig. 1 Fig. 2

Given the truncated triangular prism $ABC-PQR$.

To prove $ABC-PQR \approx$ the sum of the three pyramids $P-ABC$, $Q-ABC$ and $R-ABC$.

Proof. Pass planes through Q and AC, Q and PC, dividing the given figure into the three pyramids $Q-ABC$, $Q-APC$ and $Q-PRC$.

1. $Q-ABC$ has the required base and the required vertex Q.

2. $Q-APC \approx B-APC$, Art. 652.

(*for they have the same base, APC, and the same altitude, their vertices being in a line* $\parallel$ *base APC*).

But $B-APC$ may be regarded as having P for its vertex, and ABC for its base, as desired. Art. 634.

3. $Q-PRC \approx B-ARC$ (see Fig. 2). Art. 652.

(*for the base ARC* $\approx$ *base PRC* (Art. 390); *and the altitudes of the two pyramids are equal, the vertices Q and B being in line* $\parallel$ *plane PACR, in which the bases lie*).

But $B-ARC$ may be regarded as having R for its vertex, and ABC for its base, as desired. Art. 634

$\therefore ABC-PQR \approx$ sum of three pyramids whose common base is ABC, and whose vertices are P, Q, R.

Q. E. D.

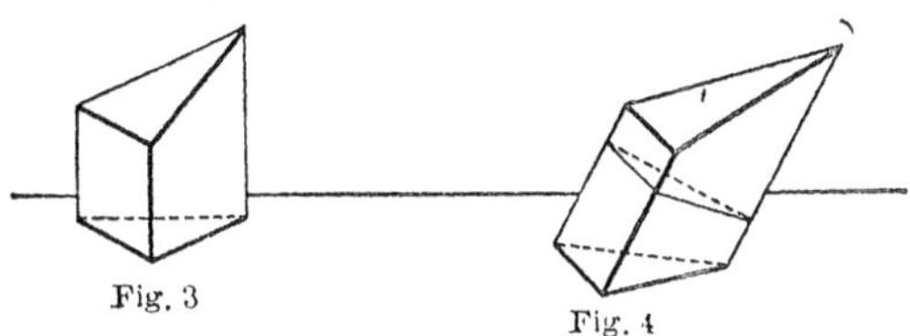

Fig. 3 Fig. 4

659. COR. 1. *The volume of a truncated right triangular prism* (Fig. 3) *is equal to the product of its base by one-third the sum of its lateral edges.*

660. COR. 2. *The volume of any truncated triangular prism* (Fig. 4) *is equal to the product of the area of its right section by one-third the sum of its lateral edges.*

PRISMATOIDS

661. A **prismatoid** is a polyhedron bounded by two polygons in parallel planes, called **bases**, and by lateral faces which are either triangles, trapezoids or parallelograms.

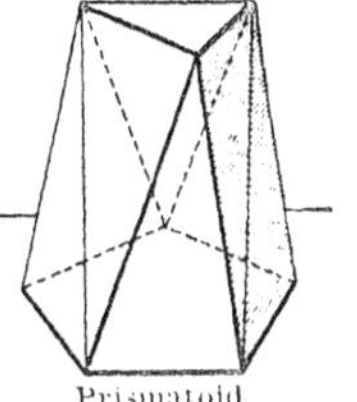

Prismatoid

662. A **prismoid** is a prismatoid in which the bases have the same number of sides and have their corresponding sides parallel.

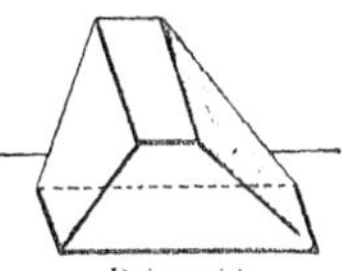

Prismoid

Ex. The volume of a truncated right parallelopiped equals the area of the lower base multiplied by one-fourth the sum of the lateral edges (or by a perpendicular from the center of the upper base to the lower base).

PROPOSITION XXIII. THEOREM

663. *The volume of a prismatoid is equal to one-sixth the product of its altitude by the sum of its bases and of four times the area of its midsection.*

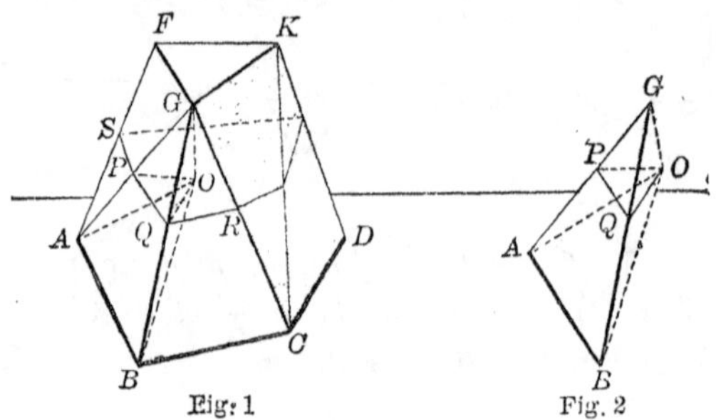

Fig. 1 Fig. 2

Given the prismatoid $ABCD$–FGK, with bases B and b, midsection M, volume V, and altitude H.

To prove $V = \frac{1}{6} H (B + b + 4M)$.

Proof. Take any point O in the midsection, and through it and each edge of the prismatoid let planes be passed.

These planes will divide the figure into parts as follows:

1. A pyramid with vertex O, base $ABCD$ and altitude $\frac{1}{2} H$, and whose volume $\therefore = \frac{1}{6} H \times B$. Art. 651.

2. A pyramid with vertex O, base FGK, and altitude $\frac{1}{2} H$, and whose volume $\therefore = \frac{1}{6} H \times b$. (Why?)

3. Tetrahedrons like O–ABG whose volume may be determined as follows (see Fig. 2):

$AB = 2\ PQ$. (Why?)

$\therefore \triangle AGB = 4 \triangle PGQ$. Art. 398.

$\therefore O\text{–}AGB = 4\ O\text{–}PGQ$. Art. 653.

But O–PGQ (or G–PQO) $= \frac{1}{3} PQO \times \frac{1}{2} H = \frac{1}{6} H \times PQO$.

$\therefore O\text{–}AGB = \frac{1}{6} H \times 4 \triangle PQO$. (Why?)

$\therefore$ the sum of all tetrahedrons like O–$AGB = \frac{1}{6} H \times 4M$.

$\therefore V = \frac{1}{6} H \times B + \frac{1}{6} H \times b + \frac{1}{6} H \times 4M$.

Or $V = \frac{1}{6} H (B + b + 4M)$. Q. E. D.

REGULAR POLYHEDRONS

664. DEF. A **regular polyhedron** is a polyhedron all of whose faces are equal regular polygons, and all of whose polyhedral angles are equal. Thus, the cube is a regular polyhedron.

PROPOSITION XXIV. THEOREM

665. *But five regular polyhedrons are possible.*

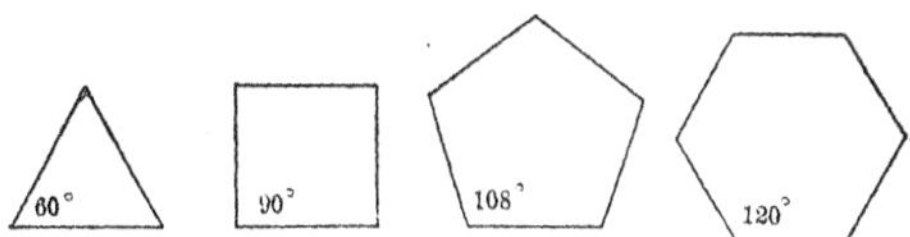

Given regular polygons of 3, 4, 5, etc., sides.

To prove that regular polygons of the same number of sides can be joined to form polyhedral ∡ of a regular polyhedron in but five different ways, and that, consequently, but five regular polyhedrons are possible.

Proof. The sum of the face ∡ of any polyhedral angle $< 360°$. Art. 583.

1. Each ∠ of an *equilateral triangle* is 60°. Art. 134.

$3 \times 60°$, $4 \times 60°$ and $5 \times 60°$ are each less than 360°; but any larger multiple of $60° =$ or $> 360°$.

∴ but *three* regular polyhedrons can be formed with equilateral △ as faces.

2. Each ∠ of a *square* contains 90°. Art. 151.

$3 \times 90°$ is less than 360°, but any larger multiple of $90° =$ or $> 360°$.

∴ but *one* regular polyhedron can be formed with squares as faces.

3. Each ∠ of a regular *pentagon* is 108°. Art. 174.

$3 \times 108°$ is less than 360°, but any larger multiple of $108° > 360°$.

∴ but *one* regular polyhedron can be formed with regular pentagons as faces.

4. Each ∠ of a regular hexagon is 120°, and $3 \times 120° = 360°$.

∴ no regular polyhedron can be formed with hexagons, or with polygons with a greater number of sides as faces.

∴ but five regular polyhedrons are possible.

Q. E. D.

666. The **construction of the regular polyhedrons,** by the use of cardboard, may be effected as follows:

Draw on a piece of cardboard the diagrams given below. Cut the cardboard half through at the dotted lines and entirely through at the full lines. Bring the free edges together and keep them in their respective positions by some means, such as pasting strips of paper over them.

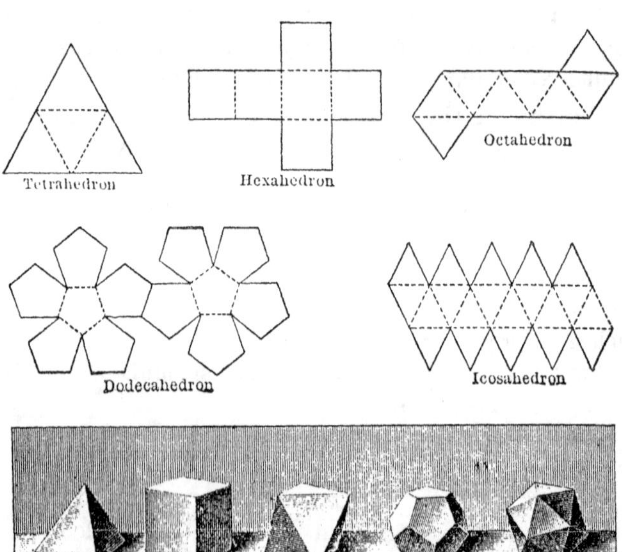

POLYHEDRONS IN GENERAL

PROPOSITION XXV. THEOREM

667. *In any polyhedron, the number of edges increased by two equals the number of vertices increased by the number of faces.*

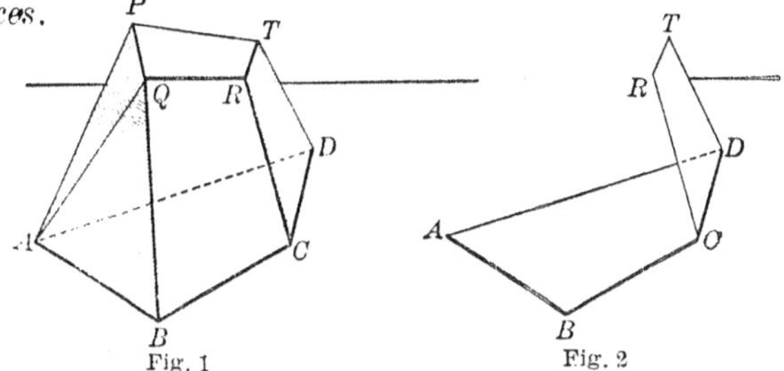

Fig. 1 Fig. 2

Given the polyhedron AT, with the number of its vertices, edges and faces denoted by V, E and F, respectively.

To prove $E + 2 = V + F$.

Proof. Taking the single face $ABCD$, the number of edges equals the number of vertices, or $E = V$.

If another face, $CRTD$, be annexed (Fig. 2), *three* new *edges*, CR, RT, TD, are added and *two* new *vertices*, R and T.

∴ the number of edges gains one on the number of vertices, or $E = V + 1$.

If still another face, $BQRC$, be annexed, two new edges, BQ and QR, are added, and one new vertex, Q. ∴ $E = V + 2$.

With each new face that is annexed, the number of edges gains one on the number of vertices, till but one face is lacking.

The last face increases neither the number of edges nor of vertices.

Hence number of edges gains one on number of vertices, for every face except two, the first and the last, or gains $F - 2$ in all.

∴ for the entire figure, $E = V + F - 2$.

That is $E + 2 = V + F$. Ax. 2.

Q. E. D.

PROPOSITION XXVI. THEOREM

668. *The sum of the face angles of any polyhedron equals four right angles taken as many times, less two, as the polyhedron has vertices.*

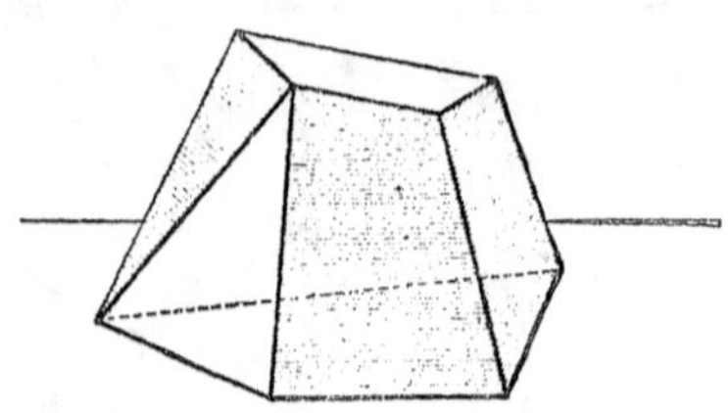

Given any polyhedron, with the sum of its face angles denoted by S, and the number of its vertices, edges and faces denoted by V, E, F, respectively.

To prove $S = (V - 2)\ 4$ rt. ∠s.

Proof. Each edge of the polyhedron is the intersection of two faces, ∴ the number of sides of the faces $= 2\ E$.

∴ the sum of the interior and exterior ∠s of the faces $= 2\ E \times 2$ rt. ∠s, or $E \times 4$ rt. ∠s. Art. 73.

But the sum of the exterior ∠s of each face $= 4$ rt. ∠s. Art. 175.

∴ the sum of exterior ∠s of the F faces $= F \times 4$ rt. ∠s. Ax. 4.

Subtracting the sum of the exterior ∠s from the sum of all the ∠s, the sum of the interior ∠s of the F faces $= (E \times 4$ rt. ∠s$) - (F \times 4$ rt. ∠s$)$.

Or $S = (E - F)\ 4$ rt. ∠s.

But $E + 2 = V + F$. Art. 667.

Hence $E - F = V - 2$. Ax. 3.

Substituting for $E - F$, $S = (V - 2)\ 4$ rt. ∠s. Ax. 8.

Q. E. D.

Ex. Verify the last two theorems in the case of the cube.

COMPARISON OF POLYHEDRONS. SIMILAR POLYHEDRONS

PROPOSITION XXVII. THEOREM

669. *If two tetrahedrons have a trihedral angle of one equal to a trihedral angle of the other, they are to each other as the products of the edges including the equal trihedral angles.*

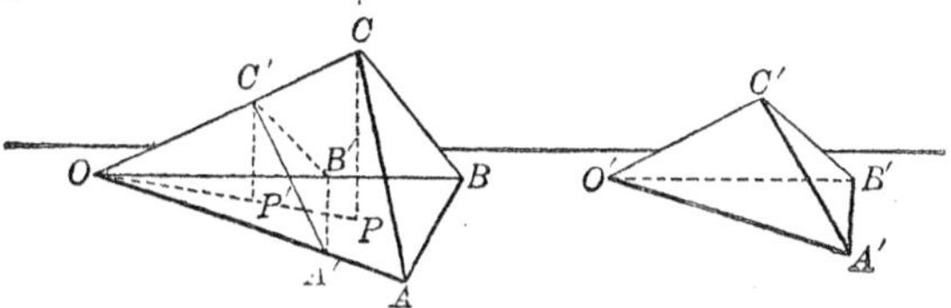

Given the tetrahedrons $O\text{-}ABC$ and $O'\text{-}A'B'C'$, with their volumes denoted by V and V', respectively, and having the trihedral $\angle s$ O and O' equal.

To prove $\dfrac{V}{V'}=\dfrac{OA \times OB \times OC}{O'A' \times O'B' \times O'C'}$.

Proof. Apply the tetrahedron $O'\text{-}A'B'C'$ to $O\text{-}ABC$ so that the trihedral $\angle O'$ shall coincide with its equal, the trihedral $\angle O$.

Draw CP and $C'P' \perp$ plane OAB, and draw OP the projection of OC in the plane OAB.

Taking OAB and $OA'B'$ as the bases, and CP and $C'P'$ as the altitudes of the pyramids $C\text{-}OAB$ and $C'\text{-}OA'B'$, respectively.

$$\frac{V}{V'}=\frac{\triangle OAB \times CP}{\triangle OA'B' \times C'P'}=\frac{\triangle OAB}{\triangle OA'B'}\times\frac{CP}{C'P'}. \quad \text{Art. 652.}$$

But $$\frac{\triangle OAB}{\triangle OA'B'}=\frac{OA \times OB}{OA' \times OB'}. \quad \text{Art. 397.}$$

In the similar rt. $\triangle s$ OCP and $OC'P'$, $\dfrac{CP}{C'P'}=\dfrac{OC}{OC'}$. (Why ?)

$$\therefore \frac{V}{V'}=\frac{OA \times OB \times OC}{OA' \times OB' \times OC'}=\frac{OA \times OB \times OC}{O'A' \times O'B' \times O'C'}. \quad \text{Ax. 8.}$$

Q. E. D.

670. Def. **Similar polyhedrons** are polyhedrons having the same number of faces, similar, each to each, and similarly placed, and having their corresponding polyhedral angles equal.

Proposition XXVIII. Theorem

671. *Any two similar polyhedrons may be decomposed into the same number of tetrahedrons, similar, each to each, and similarly placed.*

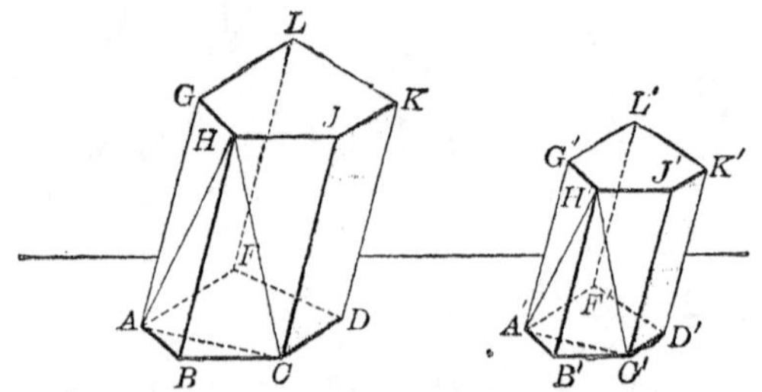

Given P and P', two similar polyhedrons.

To prove that P and P' may be decomposed into the same number of tetrahedrons, similar, each to each.

Proof. Take H and H' any two homologous vertices of P and P'. Draw homologous diagonals in all the faces of P and P' except those faces which meet at H and H', separating the faces into corresponding similar triangles.

Through H and each face diagonal thus formed in P, and through H' and each face diagonal in P', pass planes.

Each corresponding pair of tetrahedrons thus formed may be proved similar.

Thus, in $H\text{-}ABC$ and $H'\text{-}A'B'C'$, the $\triangle$ HBA and $H'B'A'$ are similar. Art. 329.

In like manner $\triangle$ HBC and $H'B'C'$ are similar; and $\triangle$ ABC and $A'B'C'$ are similar.

Also $\frac{HA}{H'A'} = \left(\frac{HB}{H'B'}\right) = \frac{HC}{H'C'} = \left(\frac{BC}{B'C'}\right) = \frac{AC}{A'C'}$. Art. 321.

$\therefore$ ⚠ AHC and $A'H'C'$ are similar. Art. 326.

Hence the corresponding faces of $H-ABC$ and $H'-A'B'C'$ are similar.

Also their homologous trihedral ∡ are equal. Art. 584.

$\therefore$ tetrahedron $H-ABC$ is similar to $H'-A'B'C'$. Art. 670.

After removing $H-ABC$ from P, and $H'-A'B'C'$ from P', the remaining polyhedrons are similar, for their faces are similar, and the remaining polyhedral ∡ are equal. Ax. 3.

By continuing this process, P and P' may be decomposed into the same number of tetrahedrons, similar, each to each, and similarly placed.

Q. E. D.

672. Cor. 1. *The homologous edges of similar polyhedrons are proportional;*

Any two homologous lines in two similar polyhedrons have the same ratio as any other two homologous lines.

673. Cor. 2. *Any two homologous faces of two similar polyhedrons are to each other as the squares of any two homologous edges or lines;*

The total areas of any two similar polyhedrons are to each other as the squares of any two homologous edges.

Ex. 1. In the figure, p. 395, if the edges meeting at O are 8, 9, 12 in., and those meeting at O' are 4, 6, 8 in., find the ratio of the volumes of the tetrahedrons.

Ex. 2. If the linear dimensions of one room are twice as great as the corresponding dimensions of another room, how will their surfaces (and $\therefore$ cost of papering) compare? How will their volumes compare?

Ex. 3. How many 2 in. cubes can be cut from a 10 in. cube?

Ex. 4. If the bases of a prismoid are rectangles whose dimensions are a, b and b, a, and altitude is H, find the formula for the volume.

Proposition XXIX. Theorem

674. *The volumes of two similar tetrahedrons are to each other as the cubes of any pair of homologous edges.*

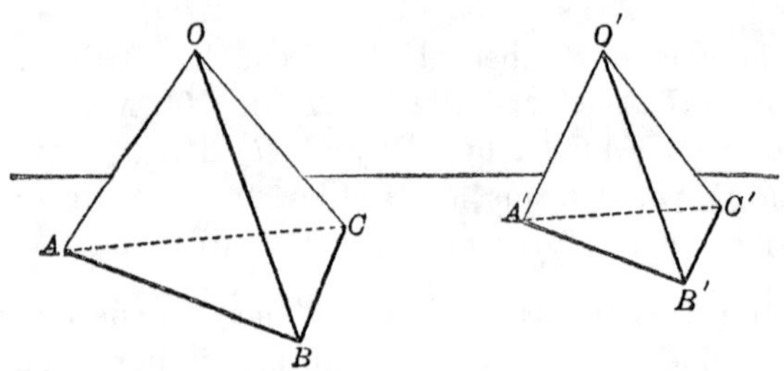

Given the similar tetrahedrons $O\text{–}ABC$ and $O'\text{–}A'B'C'$.

To prove $$\frac{V}{V'}=\frac{\overline{OA}^3}{\overline{O'A'}^3}.$$

Proof. $$\frac{V}{V'}=\frac{OA\times OB\times OC}{O'A'\times O'B'\times O'C'}. \quad \text{(Why?)}$$

$$=\frac{OA}{O'A'}\times\frac{OB}{O'B'}\times\frac{OC}{O'C'}.$$

But $$\frac{OA}{O'A'}=\frac{OB}{O'B'}=\frac{OC}{O'C'}. \quad \text{Art. 672.}$$

$$\therefore \frac{V}{V'}=\frac{OA}{O'A'}\times\frac{OA}{O'A'}\times\frac{OA}{O'A'}=\frac{\overline{OA}^3}{\overline{O'A'}^3}. \quad \text{Ax. 8.}$$

Ex. 1. In the above figures, if $AB=2\ A'B'$, find the ratio of V to V'. Find the same, if $AB=1\frac{1}{2}\ A'B'$.

Ex. 2 The measurement of the volume of a regular triangular prism reduces to the measurement of the lengths of how many straight lines? of a frustum of a regular square pyramid?

Ex. 3. Show how to construct out of pasteboard a regular prism, a parallelopiped, and a truncated square prism.

PROPOSITION XXX. THEOREM

675. ***The volumes of any two similar polyhedrons are to each other as the cubes of any two homologous edges, or of any other two homologous lines.***

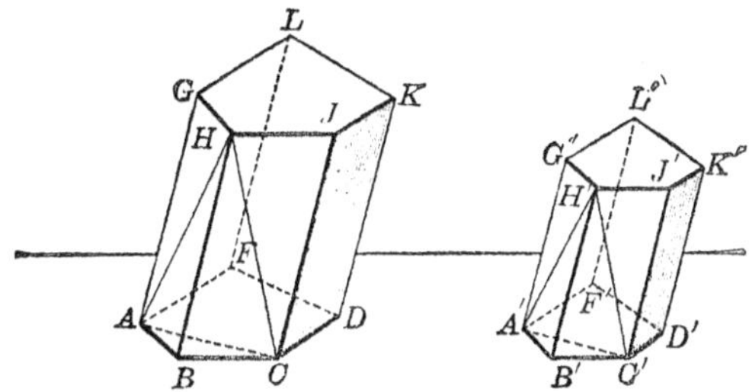

Given the polyhedrons AK and $A'K'$ having their volumes denoted by V and V', and HB and $H'B'$ any pair of homologous edges.

To prove $$\frac{V}{V'}=\frac{\overline{HB}^3}{\overline{H'B'}^3}.$$

Proof. Let the polyhedrons be decomposed into tetrahedrons, similar, each to each, and similarly placed. Art. 671.

Denote the volumes of the tetrahedrons in P by v_1, v_2, v_3 . . . and of those in P' by v'_1, v'_2, v'_3 . . .

Then $$\frac{v_1}{v'_1}=\frac{\overline{HB}^3}{\overline{H'B'}^3}.$$

Also $$\frac{v_1}{v'_1}=\frac{v_2}{v'_2}=\frac{v_3}{v'_3}=\cdots$$ Art. 674, Ax. 1.

(*for each of these ratios* $=\frac{\overline{HB}^3}{\overline{H'B'}^3}$.)

$$\therefore \frac{v_1+v_2+v_3+\text{—}}{v'_1+v'_2+v'_3+\text{—}}=\frac{v_1}{v'_1};\ \text{that is, } \frac{V}{V'}=\frac{v_1}{v'_1}.$$ Art. 312.

$$\therefore \frac{V}{V'}=\frac{\overline{HB}^3}{\overline{H'B'}^3}.$$ Ax. 1.

Q. E. D.

EXERCISES. GROUP 67

THEOREMS CONCERNING POLYHEDRONS

Ex. 1. The lateral faces of a right prism are rectangles.

Ex. 2. A diagonal plane of a prism is parallel to every lateral edge of the prism not contained in the plane.

Ex. 3. The diagonals of a parallelopiped bisect each other.

Ex. 4. The square of a diagonal of a rectangular parallelopiped equals the sum of the squares of the three edges meeting at a vertex.

Ex. 5. Each lateral face of a prism is parallel to every lateral edge not contained in the face.

Ex. 6. Every section of a prism made by a plane parallel to a lateral edge is a parallelogram.

Ex. 7. If any two diagonal planes of a prism which are not parallel to each other are perpendicular to the base of the prism, the prism is a right prism.

Ex. 8. What part of the volume of a cube is the pyramid whose base is a face of the cube and whose vertex is the center of the cube?

Ex. 9. Any section of a regular square pyramid made by a plane through the axis is an isosceles triangle.

Ex. 10. In any regular tetrahedron, an altitude equals three times the perpendicular from its foot to any face.

Ex. 11. In any regular tetrahedron, an altitude equals the sum of the perpendiculars to the faces from any point within the tetrahedron.

Ex. 12. Find the simplest formula for the lateral area of a truncated regular prism of n sides.

Ex. 13. The sum of the squares of the four diagonals of a parallelopiped is equal to the sum of the squares of the twelve edges.

[SUG. Use Art. 352.]

Ex. 14. A parallelopiped is symmetrical with respect to what point?

Ex. 15. A rectanglar parallelopiped is symmetrical with respect to how many planes? (Let the pupil make a definition of a figure symmetrical with respect to a plane. See Arts. 486, 487.)

Ex. 16. The volume of a pyramid whose lateral edges are the three edges of the parallelopiped meeting at a point is what part of the volume of the parallelopiped ?

Ex. 17. If a plane be passed through a vertex of a cube and the diagonal of a face not adjacent to the vertex, what part of the volume of the cube is contained by the pyramid so formed ?

Ex. 18. If the angles at the vertex of a triangular pyramid are right angles and each lateral edge equals a, show that the volume of the pyramid is $\frac{a^3}{6}$.

Ex. 19. How large is a dihedral angle at the base of a regular pyramid, if the apothem of the base equals the altitude of the pyramid ?

Ex. 20. The area of the base of a pyramid is less than the area of the lateral surface.

Ex. 21. The section of a triangular pyramid by a plane parallel to two opposite edges is a parallelogram.

If the pyramid is regular, what kind of a parallelogram does the section become ?

Ex. 22. The altitude of a regular tetrahedron divides an altitude of the base into segments which are as 2 : 1.

Ex. 23. If the edge of a regular tetrahedron is a, show that the slant height is $\frac{a\sqrt{3}}{2}$; and hence that the altitude is $\frac{a\sqrt{6}}{3}$, and the volume is $\frac{a^3\sqrt{2}}{12}$.

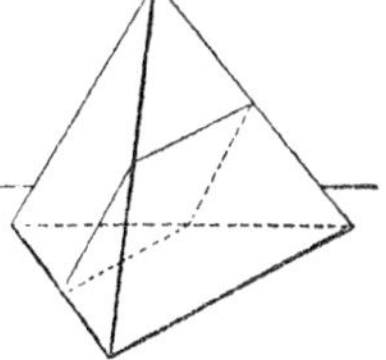

Ex. 24. If the midpoints of all the edges of a tetrahedron except two opposite edges be joined, a parallelogram is formed.

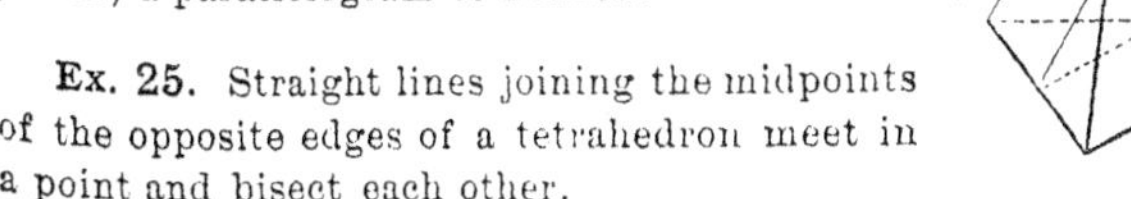

Ex. 25. Straight lines joining the midpoints of the opposite edges of a tetrahedron meet in a point and bisect each other.

Ex. 26. The midpoints of the edges of a regular tetrahedron are the vertices of a regular octahedron.

EXERCISES. GROUP 68

PROBLEMS CONCERNING POLYHEDRONS

Ex. 1. Bisect the volume of a given prism by a plane parallel to the base.

Ex. 2. Bisect the lateral surface of a given pyramid by a plane parallel to the base.

Ex. 3. Through a given point pass a plane which shall bisect the volume of a given parallelopiped.

Ex. 4. Given an edge, construct a regular tetrahedron.

Ex. 5. Given an edge, construct a regular octahedron.

Ex. 6. Pass a plane through the axis of a regular tetrahedron so that the section shall be an isosceles triangle.

Ex. 7. Pass a plane through a cube so that the section shall be a regular hexagon.

Ex. 8. Through three given lines no two of which are parallel pass planes which shall form a parallelopiped.

Ex. 9. From cardboard construct a regular square pyramid each of whose faces is an equilateral triangle.

EXERCISES. GROUP 69

REVIEW EXERCISES

Make a list of the properties of

Ex. 1. Straight lines in space.

Ex. 2. One line and one plane.

Ex. 3. Two or more lines and one plane.

Ex. 4. Two planes and one line.

Ex. 5. Two planes and two lines.

Ex. 6. Polyhedrons in general.

Ex. 7. Similar polyhedrons.

Ex. 8. Prisms in general.

Ex. 9. Right prisms.

Ex. 10. Parallelopipeds in general.

Ex. 11. Rectangular parallelopipeds.

Ex. 12. Pyramids in general.

Ex. 13. Regular pyramids.

Ex. 14. Frusta of pyramids.

Ex. 15. Truncated prisms.

Book VIII

CYLINDERS AND CONES

CYLINDERS

676. A **cylindrical surface** is a curved surface generated by a straight line which moves so as constantly to touch a given fixed curve and constantly be parallel to a given fixed straight line.

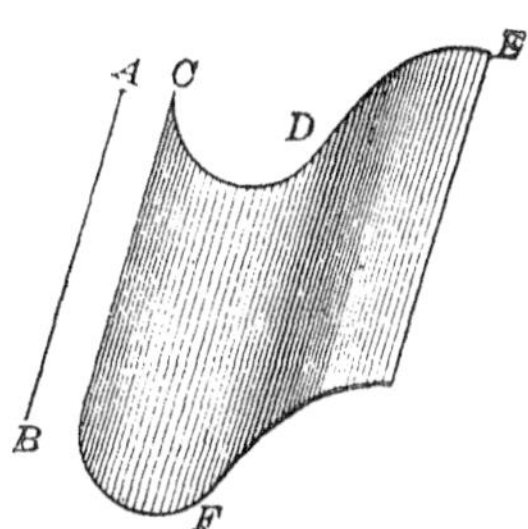

Cylindrical surface

Thus, every shadow cast by a point of light at a great distance, as by a star or the sun, approximates the cylindrical form, that is, is bounded by a cylindrical surface of light. Hence, in all radiations (as of light, heat, magnetism, etc.) from a point at a great distance, we are concerned with cylindrical surfaces and solids.

677. The **generatrix** of a cylindrical surface is the moving straight line; the **directrix** is the given curve, as CDE; an **element** of the cylindrical surface is the moving straight line in any one of its positions, as DF.

678. A **cylinder** is a solid bounded by a cylindrical surface and by two parallel planes.

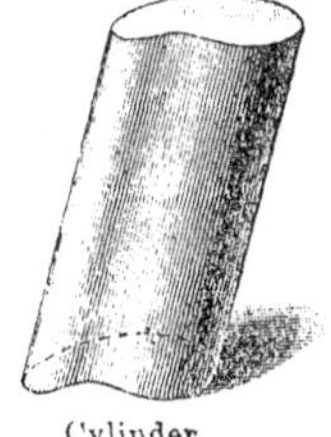
Cylinder

The **bases** of a cylinder are its parallel plane faces; the **lateral surface** is the cylindrical surface included between the parallel planes forming its bases; the **altitude** of a cylinder is the distance between the bases.

The **elements** of a cylinder are the elements of the cylindrical surface bounding it.

679. Property of a cylinder inferred immediately. *All the elements of a cylinder are equal,* for they are parallel lines included between parallel planes (Arts. 532, 676).

The cylinders most important in practical life are those determined by their stability, the ease with which they can be made from common materials, etc.

Oblique circular cylinder

680. A **right cylinder** is a cylinder whose elements are perpendicular to the bases.

681. An **oblique cylinder** is one whose elements are oblique to the bases.

682. A **circular cylinder** is a cylinder whose bases are circles.

683. A **cylinder of revolution** is a cylinder generated by the revolution of a rectangle about one of its sides as an axis.

Hence, a cylinder of revolution is a **right circular cylinder.**

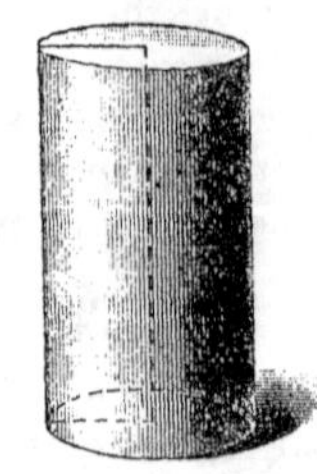

Cylinder of revolution

Some of the properties of this solid are derived most readily by considering it as generated by a revolving rectangle; and others, by regarding it as a particular kind of cylinder derived from the general definition.

684. Similar cylinders of revolution are cylinders generated by similar rectangles revolving about homologous sides.

685. A **tangent plane** to a cylinder is a plane which contains one element of the cylinder, and which does not cut the cylinder on being produced.

Ex. 1. A plane passing through a tangent to the base of a circular cylinder and the element drawn through the point of contact is tangent to the cylinder. (For if it is not, etc.)

Ex. 2. If a plane is tangent to a circular cylinder, its intersection with the plane of the base is tangent to the base.

686. **A prism inscribed in a cylinder** is a prism whose lateral edges are elements of the cylinder, and whose bases are polygons inscribed in the bases of the cylinder.

Inscribed prism

Circumscribed prism

687. **A prism circumscribed about a cylinder** is a prism whose lateral faces are tangent to the cylinder, and whose bases are polygons circumscribed about the bases of the cylinder.

688. **A section of a cylinder** is the figure formed by the intersection of the cylinder by a plane.

A **right section** of a cylinder is a section formed by a plane perpendicular to the elements of the cylinder.

689. **Properties of circular cylinders.** By Art. 441 the area of a circle is the limit of the area of an inscribed or circumscribed polygon, and the circumference is the limit of the perimeters of these polygons; hence

1. *The volume of a circular cylinder is the limit of the volume of an inscribed or circumscribed prism.*

2. *The lateral area of a circular cylinder is the limit of the lateral area of an inscribed or circumscribed prism.*

Also, 3. *By methods too advanced for this book, it may be proved that the perimeter of a right section is the limit of the perimeter of a right section of an inscribed or circumscribed prism.*

Proposition I. Theorem

690. *Every section of a cylinder made by a plane passing through an element is a parallelogram.*

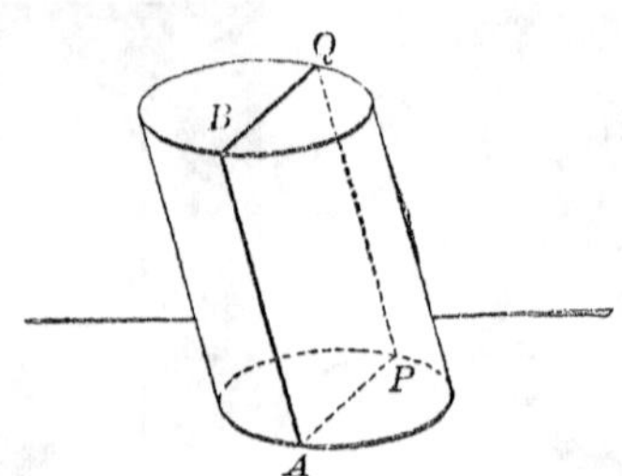

Given the cylinder AQ cut by a plane passing through the element AB and forming the section $ABQP$.

To prove $ABQP$ a ▱.

Proof. $AP \parallel BQ$. Art. 531.

It remains to prove that PQ is a straight line $\parallel AB$.

Through P draw a line in the cutting plane $\parallel AB$.

This line will also lie in the cylindrical surface. Art. 676.

$\therefore$ this line must coincide with PQ,

(*for the line drawn lies in both the cutting plane and the cylindrical surface, hence, it must be their intersection*).

$\therefore$ PQ is a straight line $\parallel AB$.

$\therefore$ $ABQP$ is a ▱. (Why?)

Q. E. D.

691. Cor. *Every section of a right cylinder made by a plane passing through an element is a rectangle.*

Ex. 1. A door swinging on its hinges generates what kind of a solid?

Ex. 2. Every section of a parallelopiped made by a plane intersecting all its lateral edges is a parallelogram.

PROPOSITION II. THEOREM

692. *The bases of a cylinder are equal.*

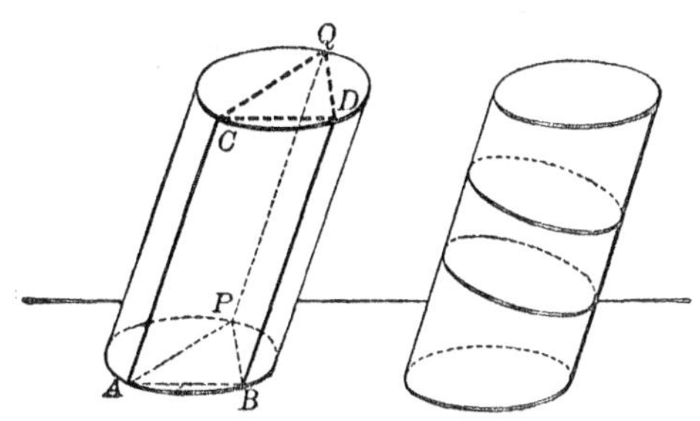

Given the cylinder AQ with the bases APB and CQD.

To prove base APB = base CQD.

Proof. Let AC and BD be any two fixed elements in the surface of the cylinder AQ.

Take P, any point except A and B in the perimeter of the base, and through it draw the element PQ.

Draw AB, AP, PB, CD, CQ, QD.

Then AC and BD are = and $\parallel$. (Why ?)

$\therefore$ AD is a ▱. (Why ?)

Similarly AQ and BQ are ▱s.

$\therefore$ $AB = CD$, $AP = CQ$, and $BP = DQ$. (Why ?)

$\therefore$ $\triangle APB = \triangle CQD$. (Why ?)

Apply the base APB to the base CQD so that AB coincides with CD. Then P will coincide with Q,

(*for* $\triangle APB = \triangle CQD$).

But P is any point in the perimeter of the base APB.

$\therefore$ every point in the perimeter of the lower base will coincide with a corresponding point of the perimeter of the upper base.

$\therefore$ the bases will coincide and are equal. Art. 47.

Q. E. D.

693. Cor. 1. *The sections of a cylinder made by two parallel planes cutting all the elements are equal.*

For the sections thus formed are the bases of the cylinder included between the cutting planes.

694. Cor. 2. *Any section of a cylinder parallel to the base is equal to the base.*

Proposition III. Theorem

695. *The lateral area of a circular cylinder is equal to the product of the perimeter of a right section of the cylinder by an element.*

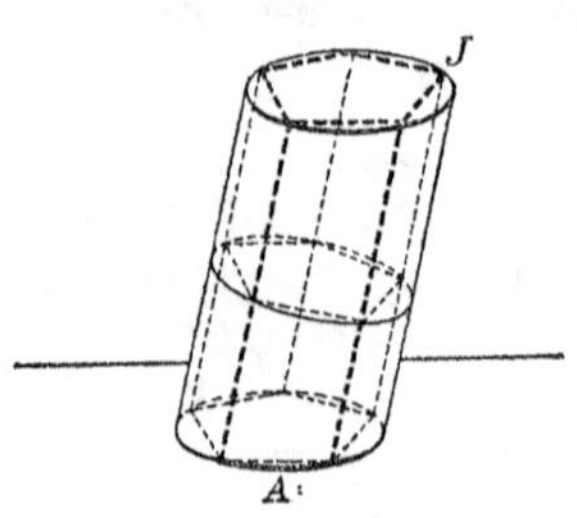

Given the circular cylinder AJ, having its lateral area denoted by S, an element by E, and the perimeter of a right section by P.

To prove $S = P \times E$.

Proof. Let a prism with a regular polygon for its base be inscribed in the cylinder.

Denote the lateral area of the inscribed prism by S', and the perimeter of its right section by P'.

Then the lateral edge of the inscribed prism is an element of the cylinder. Constr

$\therefore\ S' = P' \times E.$ Art. 608.

If the number of lateral faces of the inscribed prism be indefinitely increased,

S' will approach S as a limit. Art. 689, 2.

P' will approach P as a limit. Art. 689, 3.

And $P' \times E$ will approach $P \times E$ as a limit. Art. 253, 2.

But $S' = P' \times E$ always. (Why ?)

$\therefore\ S = P \times E.$ (Why ?)

Q. E. D.

696. Cor. 1. *The lateral area of a cylinder of revolution is equal to the product of the circumference of its base by its altitude.*

697. Formulas for lateral area and total area of a cylinder of revolution. Denoting the lateral area of a cylinder of revolution by S, the total area by T, the radius by R, and the altitude by H.

$$S = 2\ \pi RH.$$

$$T = 2\ \pi RH + 2\ \pi R^2 \ \therefore\ T = 2\ \pi R\ (H + R).$$

Ex. 1. If, in a cylinder of revolution, $H = 10$ in. and $R = 7$ in., find S and T.

Ex. 2. If the altitude of a cylinder of revolution equals the radius of the base ($H = R$), what do the formulas for S and T become in terms of R ? also, in terms of H ?

Ex. 3. What do they become, if the altitude equals the diameter of the base ?

Ex. 4. In a cylinder of revolution, what is the ratio of the lateral area to the area of the base ? to the total area ?

PROPOSITION IV. THEOREM

698. *The volume of a circular cylinder is equal to the product of its base by its altitude.*

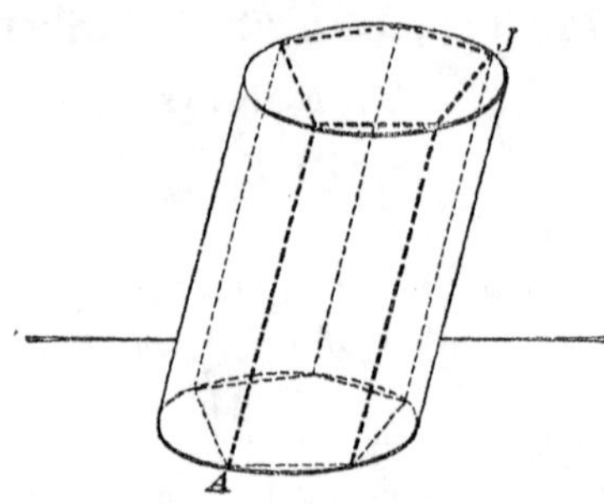

Given the circular cylinder AJ, having its volume denoted by V, its base by B, and its altitude by H.

To prove $V = B \times H$.

Proof. Let a prism having a regular polygon for its base be inscribed in the cylinder, and denote the volume of the inscribed prism by V', and its base by B'.

The prism will have the same altitude, H, as the cylinder.

$\therefore\ V' = B' \times H.$ (Why?)

If the number of lateral faces of the inscribed prism be indefinitely increased,

V' will approach V as a limit. Art. 689, 1.

B' will approach B as a limit. (Why?)

And $B' \times H$ will approach $B \times H$ as a limit (Why?)

But $V' = B' \times H$ always. (Why?)

$\therefore\ V = B \times H.$ (Why?)

Q. E. D.

699. Formula for the volume of a circular cylinder. By use of Art. 450,

$$V = \pi R^2 H.$$

PROPOSITION V. THEOREM

700. *The lateral areas, or the total areas, of two similar cylinders of revolution are to each other as the squares of their radii, or as the squares of their altitudes; and their volumes are to each other as the cubes of their radii, or as the cubes of their altitudes.*

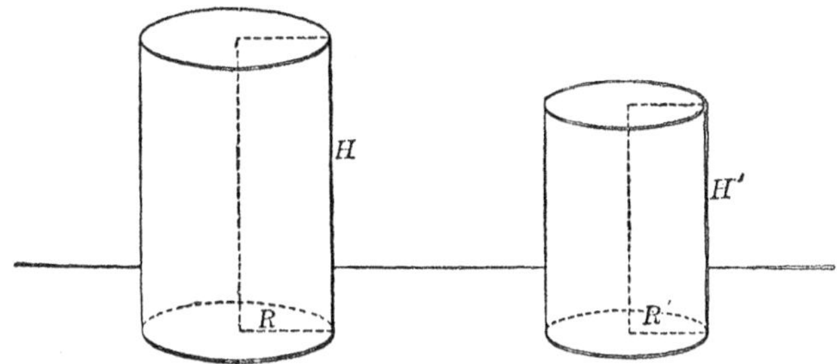

Given two similar cylinders of revolution having their lateral areas denoted by S and S', their total areas by T and T', their volumes by V and V', their radii by R and R', and their altitudes by H and H', respectively.

To prove $S : S' = T : T' = R^2 : R'^2 = H^2 : H'^2$;
and $V : V' = R^3 : R'^3 = H^3 : H'^3$.

Proof. $\frac{H}{H'} = \frac{R}{R'} = \frac{H+R}{H'+R'}$. Arts. 321, 309.

$$\frac{S}{S'} = \frac{2\,\pi RH}{2\,\pi R'H'} = \frac{R \times H}{R' \times H'} = \frac{R}{R'} \times \frac{H}{H'} = \frac{R^2}{R'^2} = \frac{H^2}{H'^2}. \quad \text{(Why?)}$$

Also $$\frac{T}{T'} = \frac{2\,\pi R\,(H+R)}{2\,\pi R'\,(H'+R')} = \frac{R}{R'} \times \frac{H+R}{H'+R'} = \frac{R^2}{R'^2} = \frac{H^2}{H'^2}. \quad \text{(Why?)}$$

Also $$\frac{V}{V'} = \frac{\pi R^2 H}{\pi R'^2 H'} = \frac{R^2}{R'^2} \times \frac{H}{H'} = \frac{R^3}{R'^3} = \frac{H^3}{H'^3}. \quad \text{(Why?)}$$

Q. E. D.

Ex. If a cylindrical cistern is 12 ft. deep, how much more cement is required to line it than to line a similar cistern 6 ft. deep? How much more water will the former cistern hold?

CONES

701. A **conical surface** is a surface generated by a straight line which moves so as constantly to touch a given fixed curve, and constantly pass through a given fixed point.

Thus every shadow cast by a near point of light is conical in form, that is, is bounded by a conical surface of light. Hence, the study of conical surfaces and solids is important from the fact that it concerns all cases of forces radiating from a near point.

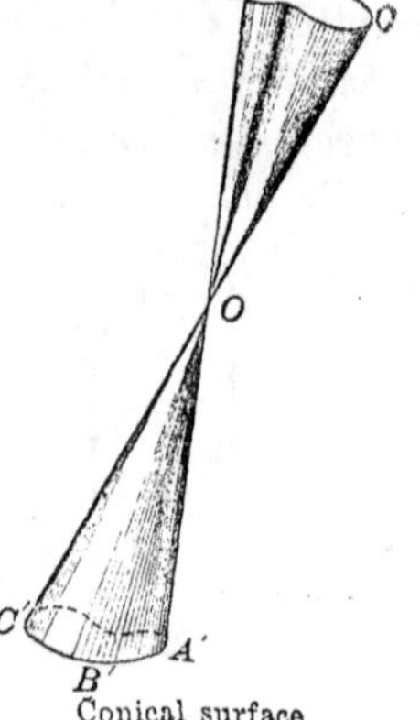

Conical surface

702. The **generatrix** of a conical surface is the moving straight line, as AA'; the **directrix** is the given fixed curve, as ABC; the **vertex** is the fixed point, as O; an **element** is the generatrix in any one of its positions, as BB'.

703. The **upper and lower nappes** of a conical surface are the portions above and below the vertex, respectively, as $O\text{-}ABC$ and $O\text{-}A'B'C'$.

Usually it is convenient to limit a conical surface to a single nappe.

704. A **cone** is a solid bounded by a conical surface and a plane cutting all the elements.

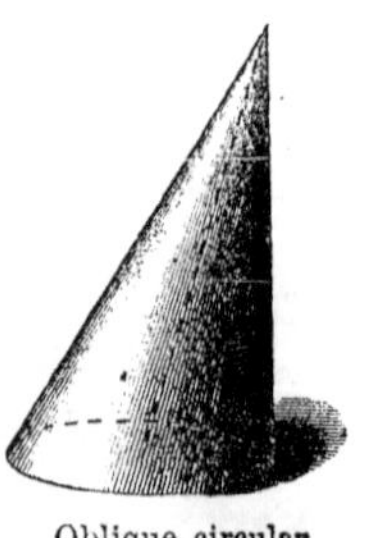
Oblique circular cone

705. The **base** of a cone is the face formed by the cutting plane; the **lateral surface** is the bounding conical surface; the **vertex** of the cone is the vertex of the conical surface; the **elements** of the cone are the elements of the conical surface; the **altitude** of a cone is the perpendicular distance from the vertex to the plane of the base.

706. A **circular cone** is a cone whose base is a circle. **The axis** of a circular cone is the line drawn from the vertex to the center of the base.

707. A **right circular cone** is a circular cone whose axis is perpendicular to the plane of the base.

An **oblique circular cone** is a circular cone whose axis is oblique to the base.

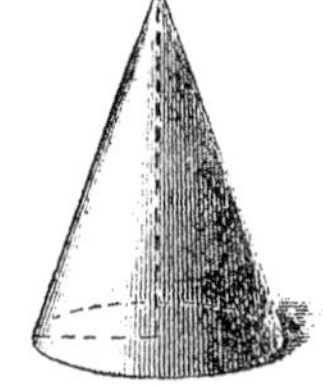
Cone of revolution

708. A **cone of revolution** is a cone generated by the revolution of a right triangle about one of its legs as an axis.

Hence a cone of revolution and a right circular cone are the same solid.

709. Properties of a cone of revolution inferred immediately.

1. *The altitude of a cone of revolution is the axis of the cone.*

2. *All the elements of a cone of revolution are equal.*

710. The **slant height of a cone of revolution** is any one of its elements.

711. Similar cones of revolution are cones generated by similar right triangles revolving about homologous sides.

712. A plane tangent to a cone is a plane which contains one element of the cone, but which does not cut the conical surface on being produced.

Ex. 1. A plane passing through a tangent to the base of a circular cone and the element drawn through the point of contact is tangent to the cone.

Ex. 2. If a plane is tangent to a circular cone, its intersection with the plane of the base is tangent to the cone.

713. A **pyramid inscribed in a cone** is a pyramid whose lateral edges are elements of the cone and whose base is a polygon inscribed in the base of the cone.

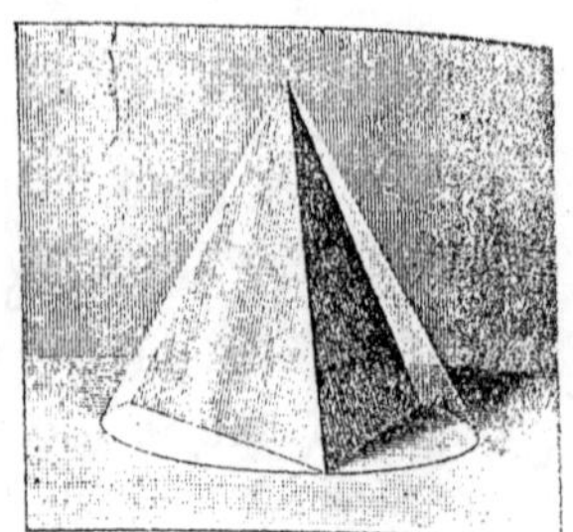

714. A **pyramid circumscribed about a cone** is a pyramid whose lateral faces are tangent to the cone and whose base is a polygon circumscribed about the base of the cone.

715. Properties of circular cones. By Art. 441 the area of a circle is the limit of the area of an inscribed, or of a circumscribed polygon, and the circumference is the limit of the perimeters of these polygons; hence

1. *The volume of a circular cone is the limit of the volume of an inscribed or circumscribed pyramid.*

2. *The lateral area of a circular cone is the limit of the lateral area of an inscribed or circumscribed pyramid.*

716. A **frustum of a cone** is the portion of the cone included between the base of the cone and a plane parallel to the base.

Frustum of a cone.

The **lower base of the frustum** is the base of the cone, and the **upper base of the frustum** is the section made by the plane parallel to the base of the cone.

What must be the **altitude and the lateral surface of a** frustum of a cone; also the **slant height** of the frustum **of a** cone of revolution?

Proposition VI. Theorem

717. *Every section of a cone made by a plane passing through its vertex is a triangle.*

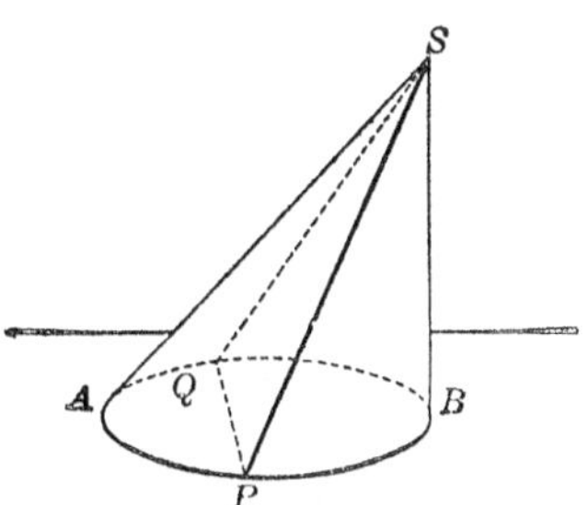

Given the cone S–$APBQ$ with a plane passing through the vertex S, and making the section SPQ.

To prove SPQ a triangle.

Proof. PQ, the intersection of the base and the cutting plane, is a straight line. (Why?)

Draw the straight lines SP and SQ.

Then SP and SQ must be in the cutting plane; Art. 498.

And be elements of the conical surface. Art. 701.

$\therefore$ the straight lines SP and SQ are the intersections of the conical surface and the cutting plane.

$\therefore$ the section SPQ is a triangle, Art. 81.
(*for it is bounded by three straight lines*).

Q. E. D.

Ex. What kind of triangle is a section of a right circular cone made by a plane through the vertex?

PROPOSITION VII. THEOREM

718. *Every section of a circular cone made by a plane parallel to the base is a circle.*

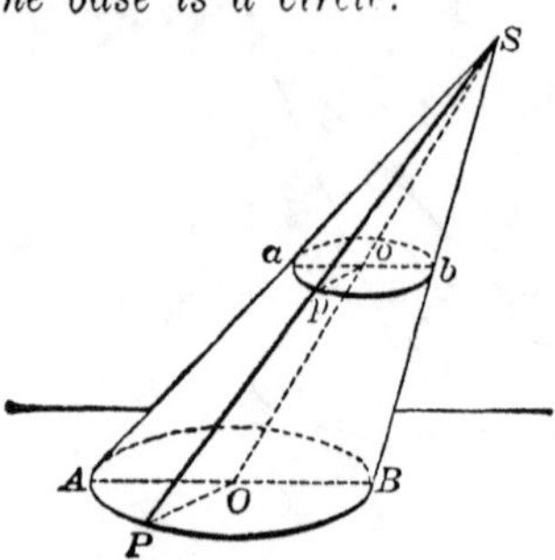

Given the circular cone SAB with apb a section made by a plane parallel to the base.

To prove apb a circle.

Proof. Denote the center of the base by O, and draw the axis, SO, piercing the plane of the section in o.

Through SO and any element, SP, of the conical surface, pass a plane cutting the plane of the base in the radius OP, and the plane of the section in op.

In like manner, pass a plane through SO and SB forming the intersections OB and ob.

$\therefore OP \parallel op$, and $OB \parallel ob$. (Why ?)

$\therefore$ $\triangle$ SPO and SBO are similar to $\triangle$ Spo and Sbo, respectively. Art. 328.

$$\therefore \frac{op}{OP} = \left(\frac{So}{SO}\right) = \frac{ob}{OB}.$$ (Why ?)

But $OP = OB$. (Why ?)

$\therefore op = ob$. (Why ?)

$\therefore opb$ is a circle. (Why ?)

Q. E. D.

719. COR. *The axis of a circular cone passes through the center of every section parallel to the base.*

Proposition VIII. Theorem

720. *The lateral area of a cone of revolution is equal to half the product of the slant height by the circumference of the base.*

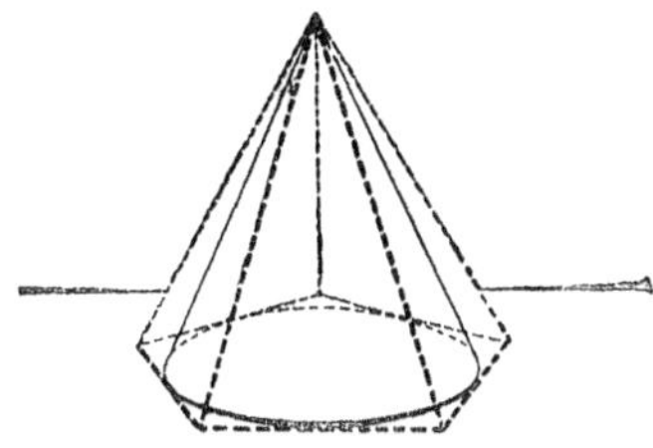

Given a cone of revolution having its lateral area denoted by S, its slant height by L, and the circumference of its base by C.

To prove $S = \frac{1}{2}\, C \times L.$

Proof. Let a regular pyramid be circumscribed about the cone.

Denote the lateral area of the pyramid by S', and the perimeter of its base by P.

Then $S' = \frac{1}{2}\, P \times L.$ Art. 642.

If the number of lateral faces of the circumscribed pyramid be indefinitely increased,

S' will approach S as a limit. Art. 715, 2.

P will approach C as a limit. Art. 441.

And $\frac{1}{2}\, P \times L$ will approach $\frac{1}{2}\, C \times L$ as a limit. Art. 253, 2.

But $S' = \frac{1}{2}\, P \times L$ always. (Why ?)

$\therefore\ S = \frac{1}{2}\, C \times L.$ (Why ?)

Q. E. D.

721. Formulas for lateral area and total area of a cone of revolution. Denoting the radius of the base by R,

$$S = \tfrac{1}{2}\,(2\,\pi R \times L)\ \therefore\ S = \pi RL.$$

Also $T = \pi RL + \pi R^2\ \therefore\ T = \pi R\,(L + R).$

PROPOSITION IX. THEOREM

722. *The volume of a circular cone is equal to one-third of the product of its base by its altitude.*

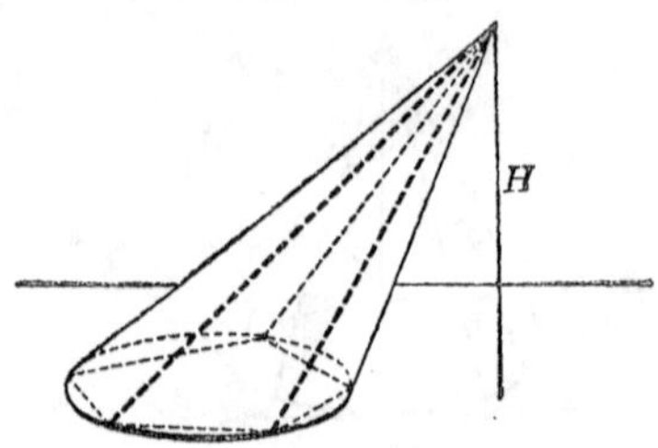

Given a circular cone having its volume denoted by V, its base by B, and its altitude by H.

To prove $V = \frac{1}{3} B \times H$.

Proof. Let a pyramid with a regular polygon for its base be inscribed in the given cone.

Denote the volume of the inscribed pyramid by V', and its base by B'.

Hence $V' = \frac{1}{3} B' \times H$. Art. 651.

If the number of lateral faces of the inscribed pyramid be indefinitely increased,

V' will approach V as a limit. (Why ?)

B' will approach B as a limit. (Why ?)

And $\frac{1}{3} B' \times H$ will approach $\frac{1}{3} B \times H$ as a limit. (Why ?)

But $V' = \frac{1}{3} B' \times H$ always. (Why ?)

$\therefore V = \frac{1}{3} B \times H$. (Why ?)

Q. E. D.

723. Formula for the volume of a circular cone.

$$V = \frac{1}{3} \pi R^2 H.$$

Ex. 1. If, in a cone of revolution, $H=3$ and $R=4$, find S, T and V.

Ex. 2. If the altitude of a cone of revolution equals the radius of the base, what do the formulas for S, T and V become ?

PROPOSITION X. THEOREM

724. *The lateral areas, or the total areas, of two similar cones of revolution are to each other as the squares of their radii, or as the squares of their altitudes, or as the squares of their slant heights; and their volumes are to each other as the cubes of these lines.*

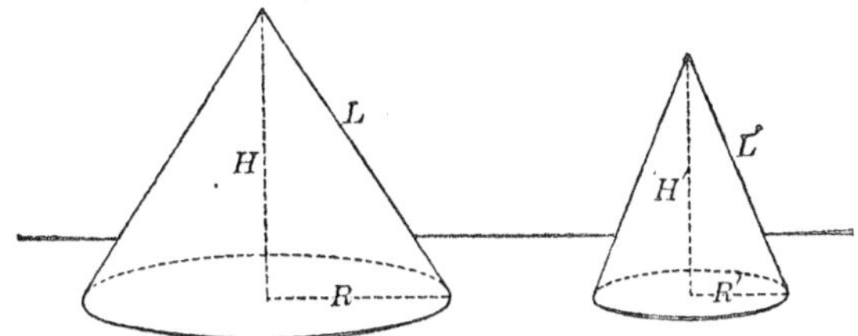

Given two similar cones of revolution having their lateral areas denoted by S and S', their total areas by T and T', their volumes by V and V', their radii by R and R', their altitudes by H and H', and their slant heights by L and L', respectively.

To prove $S : S' = T : T' = R^2 : R'^2 = H^2 : H'^2 = L^2 : L'^2$; and $V : V' = R^3 : R'^3 = H^3 : H'^3 = L^3 : L'^3$.

Proof. $\frac{H}{H'} = \frac{R}{R'} = \frac{L}{L'} = \frac{L+R}{L'+R'}$. (Why ?)

$$\frac{S}{S'} = \frac{\pi RL}{\pi R'L'} = \frac{R}{R'} \times \frac{L}{L'} = \frac{R^2}{R'^2} = \frac{L^2}{L'^2} = \frac{H^2}{H'^2}. \quad \text{(Why ?)}$$

$$\frac{T}{T'} = \frac{\pi R\,(L+R)}{\pi R'\,(L'+R')} = \frac{R}{R'} \times \frac{L+R}{L'+R'} = \frac{R^2}{R'^2} = \frac{L^2}{L'^2} = \frac{H^2}{H'^2}.$$

(Why ?)

$$\frac{V}{V'} = \frac{\frac{1}{3}\pi R^2 H}{\frac{1}{3}\pi R'^2 H'} = \frac{R^2}{R'^2} \times \frac{H}{H'} = \frac{R^3}{R'^3} = \frac{H^3}{H'^3} = \frac{L^3}{L'^3}. \quad \text{(Why ?)}$$

Q. E. D.

725. DEF. An **equilateral cone** is a cone of revolution such that a section through the axis is an equilateral triangle.

PROPOSITION XI. THEOREM

726. *The lateral area of a frustum of a cone of revolution is equal to one-half the sum of the circumferences of its bases multiplied by its slant height.*

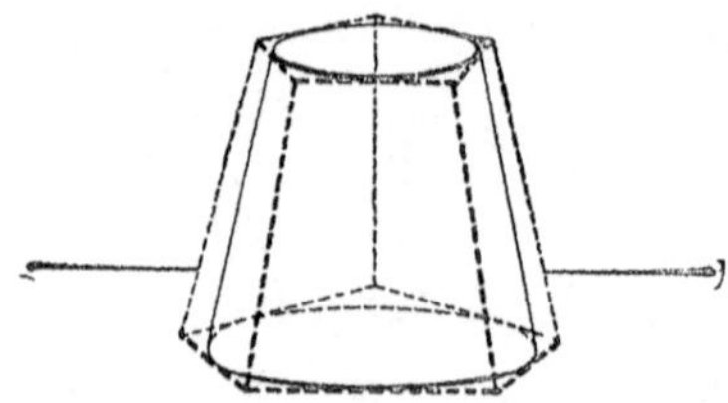

Given a frustum of a cone of revolution having its lateral area denoted by S, its slant height by L, the radii of its bases by R and r, and the circumferences of its bases by C and c.

To prove $S = \frac{1}{2}(C + c) \times L.$

Proof. Let the frustum of a regular pyramid be circumscribed about the given frustum. Denote the lateral area of the circumscribed frustum by S', the perimeter of the lower base by P, and the perimeter of the upper base by p.

The slant height of the circumscribed frustum is L.

Hence $S' = \frac{1}{2}(P + p) \times L.$ Art. 643.

Let the pupil complete the proof.

Q. E. D.

727. Formula for the lateral area of a frustum of a cone of revolution. $S = \frac{1}{2}(2\pi R + 2\pi r) L.$

$$\therefore S = \pi (R + r) L.$$

728. COR. *The lateral area of a frustum of a cone of revolution is equal to the product of the circumference of its midsection by its slant height.*

Proposition XII. Theorem

729. *The volume of the frustum of a circular cone is equivalent to the volume of three cones, whose common altitude is the altitude of the frustum, and whose bases are the lower base, the upper base, and a mean proportional between the two bases.*

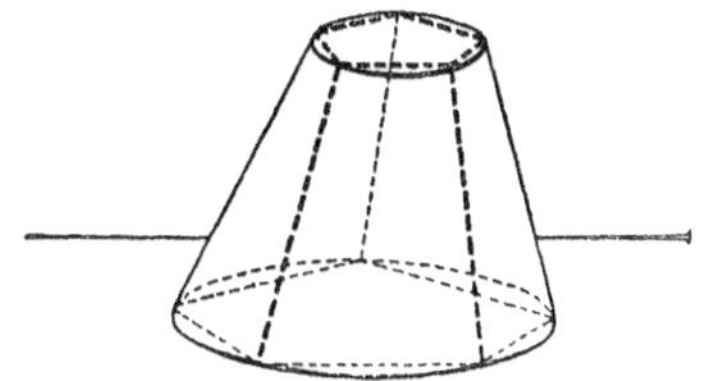

Given a frustum of a circular cone having its volume denoted by V, its altitude by H, the area of its lower base by B, and that of its upper base by b.

To prove $V = \frac{1}{3} H (B + b + \sqrt{B \times b})$.

Proof. Let the frustum of a pyramid with regular polygons for its bases be inscribed in the given frustum. Denote the volume of the inscribed frustum by V', and the areas of its bases by B' and b'.

$$\therefore V' = \frac{1}{3} H (B' + b' + \sqrt{B' \times b'}). \qquad \text{(Why?)}$$

If the number of lateral faces of the inscribed frustum be indefinitely increased, V' will approach V, B' and b' approach B and b respectively, and $B' \times b'$ approach $B \times b$, as limits. **Art. 715.**

Hence, also, $B' + b' + \sqrt{B' \times b'}$ will approach $B + b + \sqrt{B \times b}$ as a limit. **Art. 253.**

But $V' = \frac{1}{3} H (B' + b' + \sqrt{B' \times b'})$ always. (Why?)

$$\therefore V = \frac{1}{3} H (B + b + \sqrt{B \times b}). \qquad \text{(Why?)}$$

Q. E. D.

730. Formula for the volume of the frustum of a circular cone.

$$V = \frac{1}{3} H \left(\pi R^2 + \pi r^2 + \sqrt{\pi R^2 \times \pi r^2}\right).$$

$$\therefore V = \frac{1}{3} \pi H \left(R^2 + r^2 + Rr\right).$$

Ex. 1. The measurement of the volume of a frustum of a cone of revolution reduces to the measurement of the lengths of what straight lines?

Ex. 2. If a conical oil-can is 12 in. high, how much more tin is required to make it than to make a similar oil-can 6 in. high? How much more oil will it hold?

Ex. 3. The linear dimensions of a conical funnel are three times those of a similar funnel. How much more tin is required to make the first? How much more liquid will it hold?

Ex. 4. Make a similar comparison of cylindrical oil-tanks. Of conical canvas tents.

EXERCISES. GROUP 70

THEOREMS CONCERNING CYLINDERS AND CONES

Ex. 1. Any section of a cylinder of revolution through its axis is a rectangle.

Ex. 2. On a cylindrical surface only one straight line can be drawn through a given point.

[SUG. For if two straight lines could be drawn, etc.]

Ex. 3. The intersection of two planes tangent to a cone is a straight line through the vertex.

Ex. 4. If two planes are tangent to a cylinder, their line of intersection is parallel to an element of the cylinder.

[SUG. Pass a plane ⊥ to the elements of the cylinder.]

Ex. 5. If tangent planes be passed through two diametrically opposite elements of a circular cone, these planes intersect in a straight line through the vertex and parallel to the plane of the base.

Ex. 6. In a cylinder of revolution the diameter of whose base equals the altitude, the volume equals one-third the product of the total surface by the radius of the base.

Ex. 7. A cylinder and a cone of revolution have the same base and the same altitude. Find the ratio of their lateral surfaces, and also of their volumes.

Ex. 8. If an equilateral triangle whose side is a be revolved about one of its sides as an axis, find the area generated in terms of a.

Ex. 9. If a rectangle whose sides are a and b be revolved first about the side a as an axis, and then about the side b, find the ratio of the lateral areas generated, and also of the volumes.

Ex. 10. The bases of a cylinder and of a cone of revolution are concentric. The two solids have the same altitude, and the diameter of the base of the cone is twice the diameter of the base of the cylinder. What kind of line is the intersection of their lateral surfaces, and how far is it from the base?

Ex. 11. Determine the same when the radius of the cone is three times the radius of the cylinder. Also when r times.

Ex. 12. Obtain a formula in terms of r for the volume of the frustum of an equilateral cone, in which the radius of the upper base is r and that of the lower base is $3r$.

Ex. 13. A regular hexagon whose side is a revolves about a diagonal through the center as axis. Find, in terms of a, the surface and volume generated.

Ex. 14. Find the locus of a point at a given distance from a given straight line.

Ex. 15. Find the locus of a point whose distance from a given line is in a given ratio to its distance from a fixed plane perpendicular to the line.

Ex. 16. Find the locus of all straight lines which make a given angle with a given line at a given point.

Ex. 17. Find the locus of all straight lines which make a given angle with a given plane at a given point.

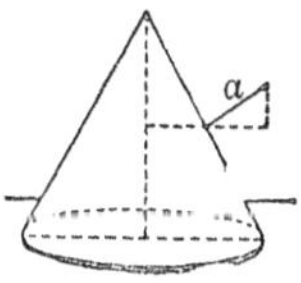

Ex. 18. Find the locus of all points at a given distance from the surface of a given cylinder of revolution.

Ex. 19. Find the locus of all points at a given distance from the surface of a given cone of revolution.

EXERCISES. GROUP 71

PROBLEMS CONCERNING THE CYLINDER AND CONE

Ex. 1. Through a given element of a circular cylinder, pass a plane tangent to the cylinder.

Ex. 2. Through a given element of a circular cone, pass a plane tangent to the cone.

Ex. 3. About a given circular cylinder circumscribe a prism, with a regular polygon for its base.

Ex. 4. Through a given point outside a circular cylinder, pass a plane tangent to the cylinder.

Ex. 5. Through a given point outside a given circular cone, pass a plane tangent to the cone.

[Sug. Through the vertex of the cone and the given point pass a line, and produce it to meet the plane of the base.]

Ex. 6. Into what segments must the altitude of a cone of revolution be divided by a plane parallel to the base, in order that the volume of the cone be bisected?

Ex. 7. Divide the lateral surface of a given cone of revolution into two equivalent parts by a plane parallel to the base.

Ex. 8. If the lateral surface of a cylinder of revolution be cut along one element and unrolled, what sort of a plane figure is formed?

Hence, out of cardboard construct a cylinder of revolution with given altitude and given circumference.

Ex. 9. If the lateral surface of a cone of revolution be cut along one element and unrolled, what sort of a plane figure is formed?

Hence, out of cardboard construct a cone of revolution of given slant height.

Ex. 10. Construct an equilateral cone out of pasteboard.

Ex. 11. Construct a frustum of a cone of revolution out of pasteboard.

Book IX

THE SPHERE

731. A **sphere** is a solid bounded by a surface all points of which are equally distant from a point within called the **center**.

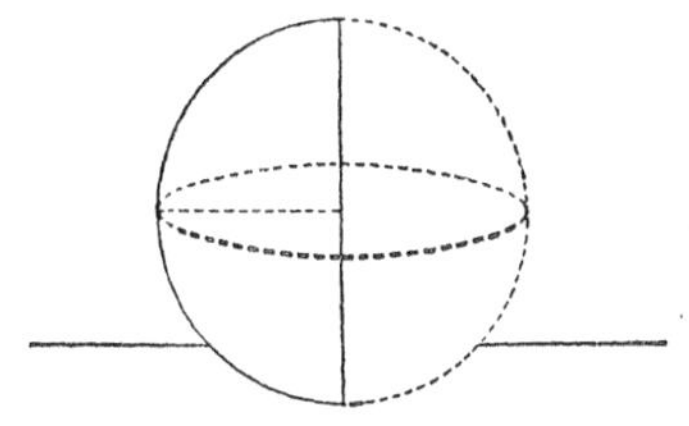

732. A **sphere** may also be defined as a solid generated by the revolution of a semicircle about its diameter as an axis.

Some of the properties of a sphere may be obtained more readily from one of the two definitions given, and some from the other.

A sphere is named by naming the point at its center, or by naming three or more points on its surface.

733. A **radius of a sphere** is a line drawn from the center to any point on the surface.

A **diameter of a sphere** is a line drawn through the center and terminated at each end by the surface of the sphere.

734. A **line tangent to a sphere** is a line having but one point in common with the surface of the sphere, however far the line be produced.

735. **A plane tangent to a sphere** is a plane having but one point in common with the surface of the sphere, however far the plane be produced.

736. **Two spheres tangent to each other** are spheres whose surfaces have one point, and only one, in common.

737. **Properties of a sphere inferred immediately.**

1. *All radii of a sphere, or of equal spheres, are equal.*

2. *All diameters of a sphere, or of equal spheres, are equal.*

3. *Two spheres are equal if their radii or their diameters are equal.*

PROPOSITION I. THEOREM

738. *A section of a sphere made by a plane is a circle.*

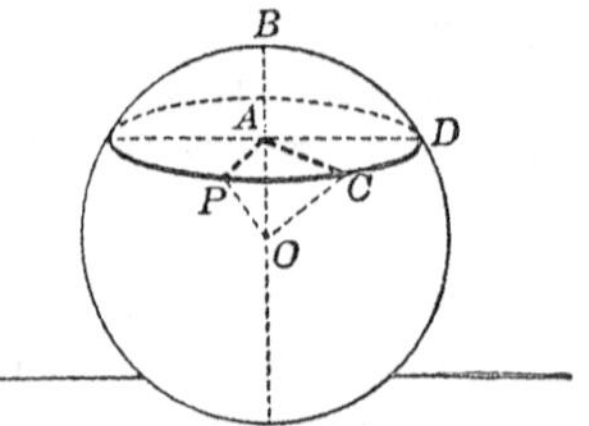

Given the sphere O, and PCD a section made by a plane cutting the sphere.

To prove that PCD is a circle.

Proof. From the center O, draw $OA \perp$ the plane of the section.

Let C be a fixed point on the perimeter of the section, and P any other point on this perimeter.

Draw AC, AP, OC, OP.

Then the $\triangle$s OAP and OAC are rt. $\triangle$s. Art. 505.

$OP = OC$. (Why?)

$OA = OA$. (Why?)

$\triangle OAP = \triangle OAC$. (Why?)

$\therefore AP = AC$. (Why?)

But P is any point on the perimeter of the section PCD.

$\therefore$ every point on this perimeter is at the distance AC from A.

$\therefore PCD$ is a circle with center A. Art. 197.

Q. E. D.

739. Cor. 1. *Circles which are sections of a sphere made by planes equidistant from the center are equal; and conversely.*

740. Cor. 2. *Of two circles on a sphere, the one made by a plane more remote from the center is smaller; and conversely.*

741. Def. A **great circle of a sphere** is a circle whose plane passes through the center of the sphere.

742. Def. A **small circle of a sphere** is a circle whose plane does not pass through the center of the sphere.

743. Def. The **axis of a circle of a sphere** is the diameter of a sphere which is perpendicular to the plane of the circle. Thus, on figure p. 426, BB' is the axis of PCD.

744. Def. The **poles of a circle of a sphere** are the extremities of the axis of the circle. Thus, B and B', of figure p. 426, are poles of the circle PCD.

745. Properties of circles of a sphere inferred imm diately.

1. *The axis of a circle of a sphere passes through th center of the circle; and conversely.*

2. *Parallel circles have the same axis and the same pole*

3. *All great circles of a sphere are equal.*

4. *Every great circle on a sphere bisects the sphere an its surface.*

5. *Two great circles on a sphere bisect each other.*

For the line of intersection of the two planes of th circles passes through the center, and hence is a diamete of each circle.

6. *Through two points (not the extremities of a diameter on the surface of a sphere, one, and only one, great circle ca be passed.*

For the plane of the great circle must also pass throug the center of the sphere (Art. 741), and through thre points not in a straight line only one plane can be passe (Art. 500).

7. *Through any three points on the surface of a spher not in the same plane with the center, one small circle, an only one, can be passed.*

746. Def. **The distance between two points on the su face of a sphere** is the length of the minor arc of a grea circle joining the points.

Ex. 1. If the radius of a sphere is 13 in., find the radius of circle on the sphere made by a plane at a distance of 1 ft. from th center.

Ex. 2. What geographical circles on the earth's surface are great and what small circles?

Ex. 3. What is the largest number of points in which two circle on the surface of a sphere can intersect? Why?

Proposition II. Theorem

747. *All points in the circumference of a circle of a sphere are equally distant from each pole of the circle.*

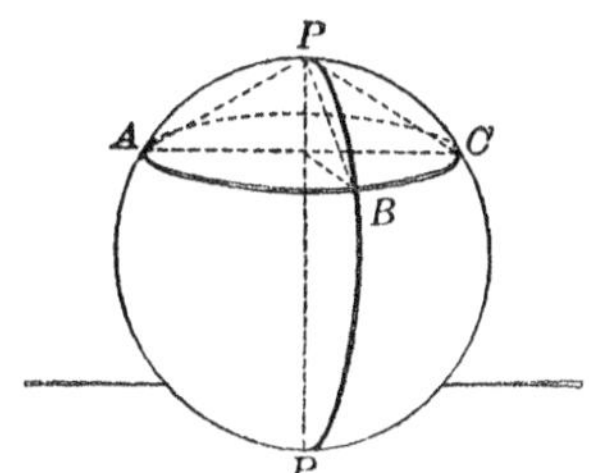

Given ABC a circle of a sphere, and P and P' its poles.

To prove the arcs PA, PB, PC equal, and arcs $P'A$, $P'B$, $P'C$ equal.

Proof. Draw the chords PA, PB, PC.

The chords PA, PB and PC are equal. Art. 518.

$\therefore$ arcs PA, PB and PC are equal. Art. 218.

In like manner, the arcs $P'A$, $P'B$ and $P'C$ may be proved equal.

Q. E. D.

748. Def. The **polar distance of a small circle on a sphere** is the distance of any point on the circumference of the circle from the nearer pole.

The **polar distance of a great circle on a sphere** is the distance of any point on the circumference of the great circle from either pole.

749. Cor. *The polar distance of a great circle is the quadrant of a great circle.*

PROPOSITION III. THEOREM

750. *If a point on the surface of a sphere is at a quadrant's distance from two other points on the surface, it is the pole of the great circle through those points.*

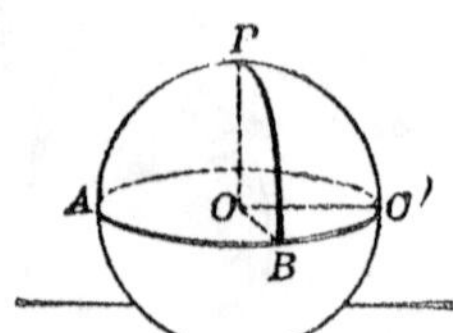

Given PB and PC quadrants on the surface of the sphere O, and ABC a great circle through B and C.

To prove that P is the pole of ABC.

Proof. From the center O draw the radii OB, OC, OP.

The arcs PB and PC are quadrants. (Why ?)

$\therefore$ ∠s POB and POC are rt. ∠s. (Why ?)

$\therefore PO \perp$ plane ABC. (Why ?)

$\therefore P$ is the pole of the great circle ABC. (Why ?)

Q. E. D.

751. COR. *Through two given points on the surface of a sphere to describe a great circle.*

Let A and B be the given points. From A and B as centers, with a quadrant as radius, describe arcs on the surface of the sphere intersecting at P. With P as a center and a quadrant as a radius, describe a great circle.

PROPOSITION IV. THEOREM

752. *A plane perpendicular to a radius at its extremity is tangent to the sphere.*

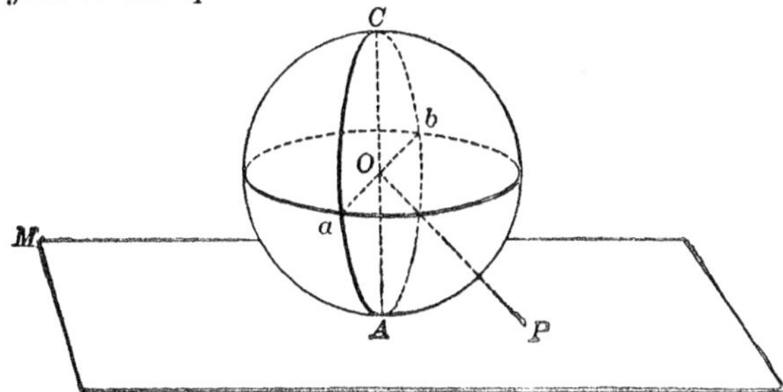

Given the sphere O, and the plane $MN \perp$ the radius OA of the sphere at its extremity A.

To prove MN tangent to the sphere.

Proof. Take P any point in plane MN except A. Draw OP. Then $OP > OA$. (Why?)

$\therefore$ the point P is outside the surface of the sphere.

But P is any point in the plane MN except A.

$\therefore$ plane MN is tangent to the sphere at the point A,

(*for every point in the plane, except A, is outside the surface of the sphere*). Art. 735. Q. E. D.

753. COR. 1. *A plane, or a line, which is tangent to a sphere, is perpendicular to the radius drawn to the point of contact.* Also, *if a plane is tangent to a sphere, a perpendicular to the plane at its point of contact passes through the center of the sphere.*

754. COR. 2. *A straight line perpendicular to a radius of a sphere at its extremity is tangent to the sphere.*

755. COR. 3. *A straight line tangent to a circle of a sphere lies in the plane tangent to the sphere at the point of contact.*

756. COR. 4. *A straight line drawn in a tangent plan and through the point of contact is tangent to the sphere that point.*

757. COR. 5. *Two straight lines tangent to a sphere a given point determine the tangent plane at that point.*

758. DEF. **A sphere circumscribed about a polyhedro** is a sphere in whose surface lie all the vertices of th polyhedron.

759. DEF. **A sphere inscribed in a polyhedron** is sphere to which all the faces of the polyhedron are tangent

PROPOSITION V. PROBLEM

760. *To circumscribe a sphere about a given tetrahedron.*

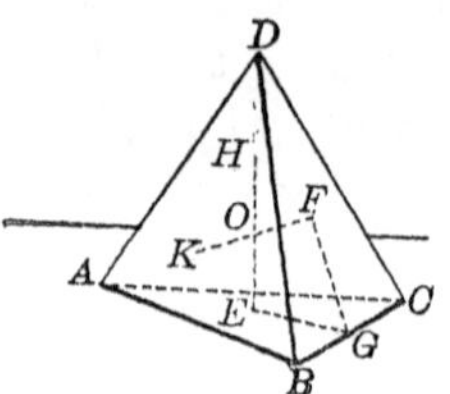

Given the tetrahedron $ABCD$.

To circumscribe a sphere about $ABCD$.

Construction and **Proof.** Construct E and F the centers of circles circumscribed about the $\triangle$ ABC and BCD, respectively. Art. 286.

Draw $EH \perp$ plane ABC and $FK \perp$ plane BCD. Art. 514.

Draw EG and FG to G the midpoint of BC.

Then EG and FG are $\perp$ BC. Art. 113.

$\therefore$ plane $EGF \perp BC$. Art. 509.

$\therefore$ plane $EGF \perp$ plane ABC. Art. 555.

$\therefore$ EH lies in the plane FGE. Art. 558.

In like manner FK lies in the plane FGE.

The lines EG and FG are not $\parallel$,

(*for they meet in the point* G).

$\therefore$ the lines EH and FK are not $\parallel$. Art. 122.

Hence EH must meet FK in some point O.

But EH is the locus of all points equidistant from A, B and C; and FK is the locus of all points equidistant from B, C and D. Art. 520.

$\therefore$ O, which is in both EH and FK, is equidistant from A, B, C and D. (Why?)

Hence a spherical surface constructed with O as a center and OA as a radius will pass through A, B, C and D, and form the sphere required. Q. E. F.

761. Cor. 1. *Four points not in the same plane determine a sphere.*

762. Cor. 2. *The four perpendiculars erected at the centers of the faces of a tetrahedron meet in a point.*

763. Cor. 3. *The six planes perpendicular to the edges of a tetrahedron at their midpoints intersect in a point.*

764. Def. An **angle formed by two curves** is the angle formed by a tangent to each curve at the point of intersection.

765. Def. A **spherical angle** is an angle formed by two intersecting arcs of great circles on a sphere, and hence by tangents to these arcs at the point of intersection.

PROPOSITION VI. PROBLEM

766. *To inscribe a sphere in a given tetrahedron.*

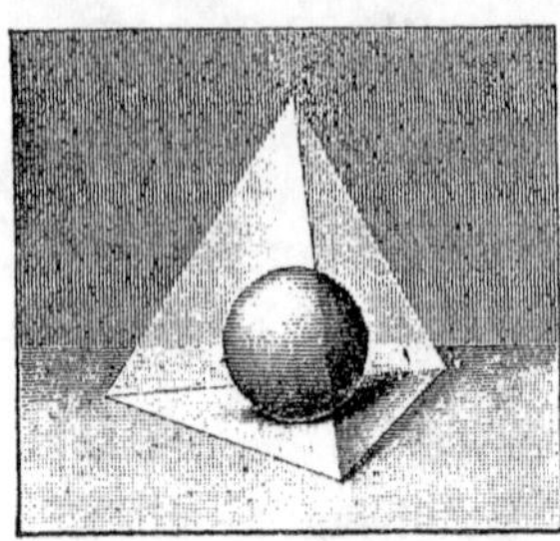

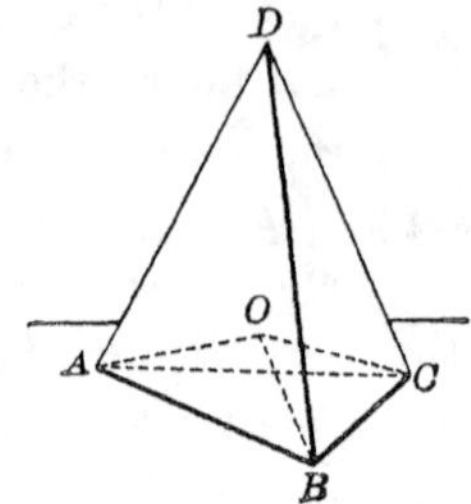

Given the tetrahedron $ABCD$.

To inscribe a sphere in $ABCD$.

Construction and **Proof.** Bisect the dihedral angle D-AB-C by the plane OAB; similarly bisect the dihedral $\angle$s whose edges are BC and AC by the planes OBC and OAC, respectively.

Denote the point in which the three bisecting planes intersect by O.

Every point in the plane OAB is equidistant from the faces DAB and CAB. Art. 562.

Similarly, every point in OBC is equidistant from the two faces intersecting in BC, and every point in OAC is equidistant from the two faces intersecting in AC.

$\therefore$ O is equidistant from all four faces of the tetrahedron. Ax. 1.

Hence, from O as a center, with the $\perp$ from O to any one face as a radius, describe a sphere.

This sphere will be tangent to the four faces of the tetrahedron and $\therefore$ inscribed in the tetrahedron. Art. 759.

Q. E. F.

767. COR. *The planes bisecting the six dihedral angles of a tetrahedron meet in one point.*

Proposition VII. Problem

768. *To find the radius of a given material sphere.*

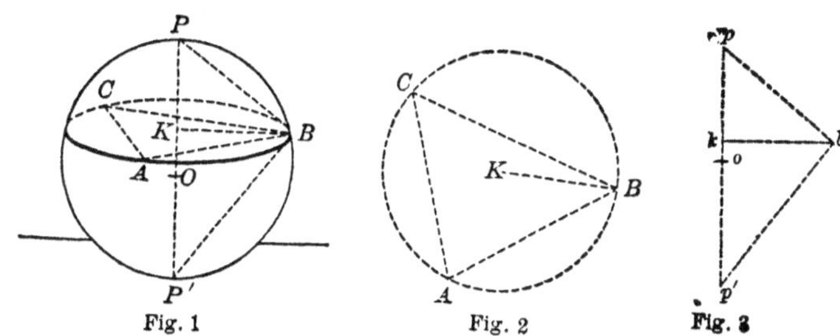

Fig. 1 Fig. 2 Fig. 3

Given the material sphere O.

To construct the radius of the sphere.

Construction. With any point P (Fig. 1) of the surface of the sphere as a pole, describe any convenient circumference on the surface.

On this circumference take any three points A, B and C.

Construct the $\triangle$ ABC (Fig. 2) having as sides the three chords AB, BC, AC, obtained from Fig. 1, by use of the compasses. Art. 283.

Circumscribe a circle about the $\triangle$ ABC. Art. 286.

Let KB be the radius of this circle.

Construct (Fig. 3) the right $\triangle$ kpb, having for hypotenuse the chord pb (Fig. 1) and the base kb. Art. 284.

Draw $bp' \perp bp$ and meeting pk produced at p'.

Bisect pp' at O.

Then op is the radius of the given sphere.

Proof. Let the pupil supply the proof.

Q. E. F.

PROPOSITION VIII. THEOREM

769. *The intersection of two spherical surfaces is the circumference of a circle whose plane is perpendicular to the line joining the centers of the spheres, and whose center is in that line.*

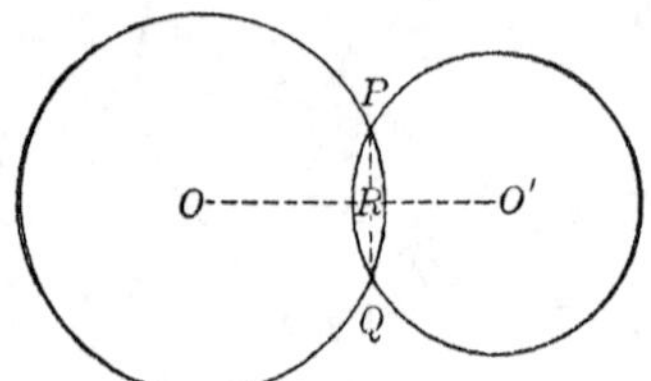

Given two intersecting ⊙ O and O' which, by rotation about the line OO' as an axis, generate two intersecting spherical surfaces.

To prove that the intersection of the spherical surfaces is a ⊙, whose plane ⊥ OO', and whose center lies in OO'.

Proof. Let the two circles intersect in the points P and Q, and draw the common chord PQ.

Then, as the two given ⊙ rotate about OO' as an axis, the point P will generate the line of intersection of the two spherical surfaces that are formed.

But PR is constantly ⊥ OO'. Art. 241.

∴ PR generates a plane ⊥ OO' Art. 510.

Also PR remains constant in length.

∴ P describes a circumference in that plane. Art. 197.

Hence the intersection of two spherical surfaces is a ⊙, whose plane ⊥ the line of centers, and whose center is in the line of centers.

Q. E. D.

The above demonstration is an illustration of the use of the second definition of a sphere (Art. 732).

PROPOSITION IX. THEOREM

770. *A spherical angle is measured by the arc of a great circle described from the vertex of the angle as a pole, and included between its sides, produced, if necessary.*

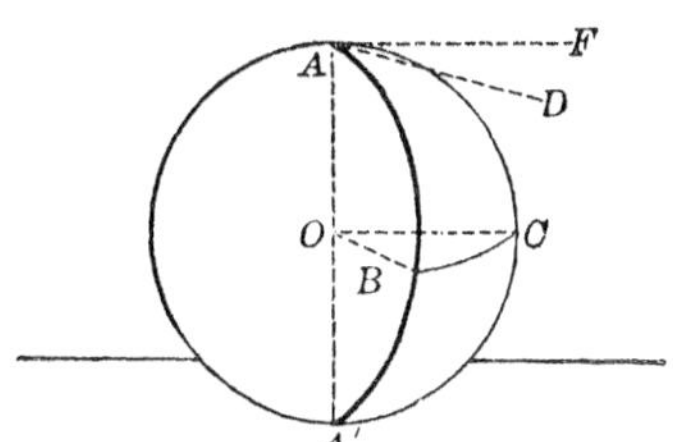

Given $\angle BAC$ a spherical angle formed by the intersection of the arcs of the great circles BA and CA, and BC an arc of a great circle whose pole is A.

To prove $\angle BAC$ measured by arc BC.

Proof. Draw AD tangent to AB, and AF tangent to AC. Also draw the radii OB and OC.

Then $AD \perp AO$. Art. 230.

Also $OB \perp AO$ (*for AB is a quadrant*).

$\therefore OB \parallel AD$. (Why ?)

Similarly $OC \parallel AF$.

$\therefore \angle BOC = \angle DAF$ Art. 538.

But $\angle BOC$ is measured by arc BC. Art. 257.

$\therefore \angle DAF$, that is, $\angle BAC$, is measured by arc BC. (Why ?)

Q. E. D.

771. COR. *A spherical angle is equal to the plane angle of the dihedral angle formed by the planes of its sides.*

SPHERICAL TRIANGLES AND POLYGONS

772. A **spherical polygon** is a portion of the surface of a sphere bounded by three or more arcs of great circles, as $ABCD$.

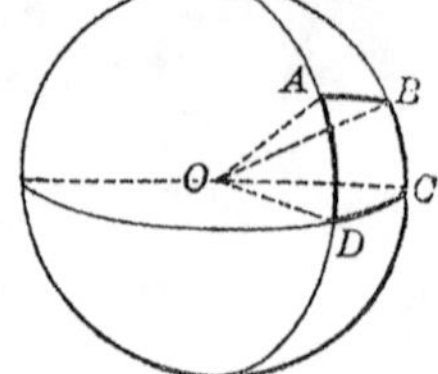

The **sides** of the spherical polygon are the bounding arcs; the **vertices** are the points in which the sides intersect; the **angles** are the spherical angles formed by the sides.

The sides of a spherical polygon are usually limited to arcs less than a semicircumference.

773. A **spherical triangle** is a spherical polygon of three sides.

Spherical triangles are classified in the same way as plane triangles; viz., as isosceles, equilateral, scalene, right, obtuse and acute.

774. Relation of spherical polygons to polyhedral angles. If radii be drawn from the center of a sphere to the vertices of a spherical polygon on its surface (as OA, OB, etc., in the above figure), a polyhedral angle is formed at O, which has an important relation to the spherical polygon $ABCD$

Each **face angle** *of the polyhedral angle equals (in number of degrees contained) the corresponding* **side** *of the spherical polygon;*

Each **dihedral angle** *of the polyhedral angle equals the corresponding* **angle** *of the spherical polygon.*

Hence, *corresponding to each property of a polyhedral angle, there exists a property of a spherical polygon, and conversely.*

Hence, also, a *trihedral angle* and its parts correspond to a *spherical triangle* and its parts.

Of the common properties of a polyhedral angle and a spherical polygon, some are discovered more readily from the one figure and some from the other. In general, the spherical polygon is simpler to deal with than a polyhedral angle. For instance, if a trihedral angle were drawn with the plane angles of its dihedral angles, nine lines would be used, forming a complicated figure in solid space; whereas, the same magnitudes are represented in a spherical triangle by three lines in an approximately plane figure.

On the other hand, the spherical polygon, because of its lack of detailed parts, is often not so suggestive of properties as the polyhedral angle.

Proposition X. Theorem

775. *The sum of two sides of a spherical triangle is greater than the third side.*

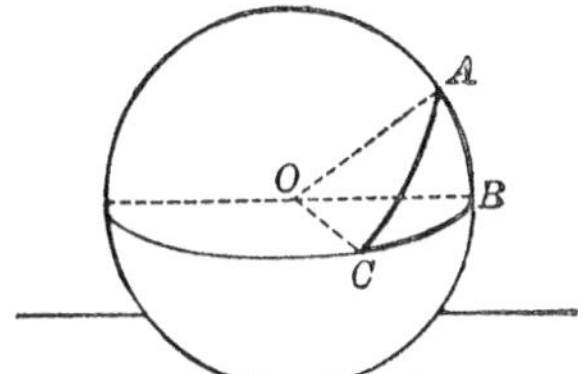

Given the spherical triangle ABC, of which no side is larger than AB.

To prove $AC + BC > AB$.

Proof. From the center of the sphere, O, draw the radii OA, OB, OC.

Then, in the trihedral angle $O\text{–}ABC$,

$$\angle AOC + \angle BOC > AOB. \qquad \text{Art. 582.}$$

$$\therefore AC + BC > AB. \qquad \text{Art. 774.}$$

Q. E. D.

776. Cor. 1. *Any side of a spherical triangle is greater than the difference between the other two sides.*

777. Cor. 2. *The shortest path between two points on the surface of a sphere is the arc of the great circle joining those points.*

For any other path between the two points may be made the limit of a series of arcs of great circles connecting successive points on the path, and the sum of this series of arcs of great circles connecting the two points is greater than the single arc of a great circle connecting them.

Proposition XI. Theorem

778. *The sum of the sides of a spherical polygon is less than* 360°.

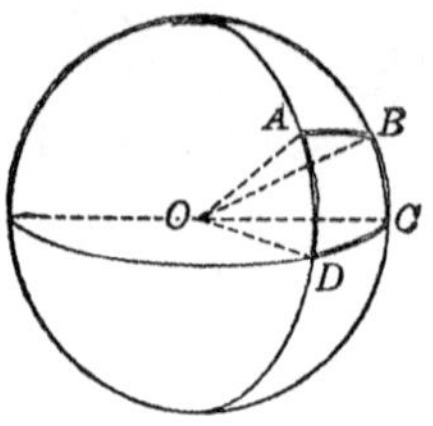

Given the spherical polygon $ABCD$.

To prove the sum of the sides of $ABCD < 360°$.

Proof. From O, the center of the sphere, draw the radii OA, OB, OC, OD.

Then $\angle AOB + \angle BOC + \angle COD + \angle DOA < 360°$. (Why ?)

$\therefore AB + BC + CD + DA < 360°$. Art. 774.

Q. E. D.

779. Def. The **polar triangle** of a given triangle is the triangle formed by taking the vertices of the given triangle as poles, and describing arcs of great circles. (Hence, if each pole be regarded as a center, the radius used in describing each arc is a quadrant.) Thus $A'B'C'$ is the polar triangle of ABC; also $D'E'F'$ is the polar triangle of DEF.

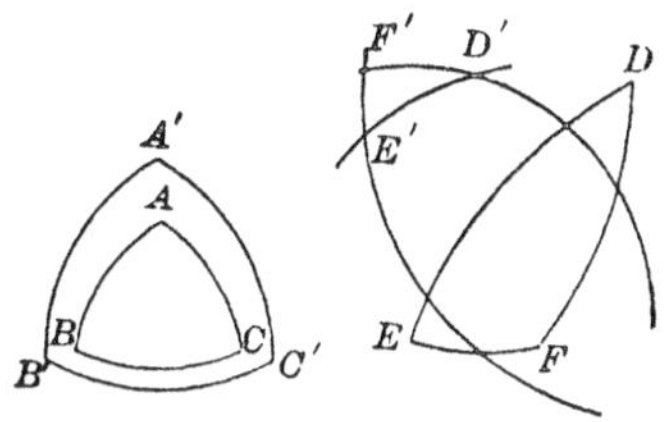

Proposition XII. Theorem

780. *If one spherical triangle is the polar of another, then the second triangle is the polar of the first.*

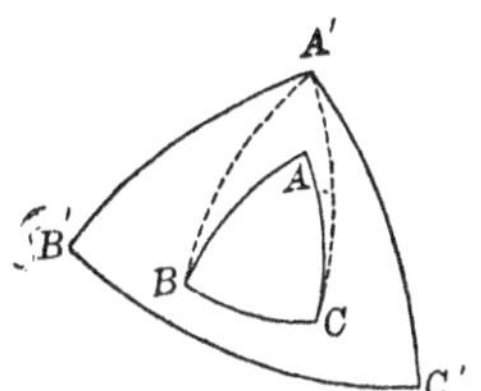

Given $A'B'C'$ the polar triangle of ABC.

To prove ABC the polar triangle of $A'B'C'$.

Proof. B is the pole of the arc $A'C'$. Art. 779.

$\therefore$ arc $A'B$ is a quadrant. (Why ?)

Also C is the pole of the arc $A'B'$. (Why ?)

$\therefore$ arc $A'C$ is a quadrant. (Why ?)

$\therefore$ A' is at a quadrant's distance from both B and C.

$\therefore$ A' is the pole of the arc BC. Art. 750.

In like manner it may be shown that B' is the pole of AC, and C' the pole of AB. Q. E. D.

PROPOSITION XIII. THEOREM

781. *In a spherical triangle and its polar, each angle of one triangle is the supplement of the side opposite in the other triangle.*

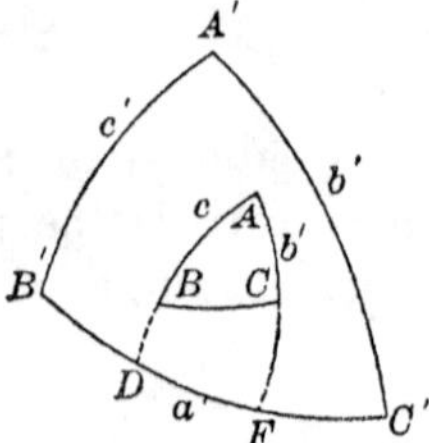

Given the polar $\triangle$ ABC and $A'B'C'$ with the sides of ABC denoted by a, b, c, and the sides of $A'B'C'$ denoted by a', b', c', respectively.

To prove $A+a'=180°$, $B+b'=180°$, $C+c'=180°$,
$A'+a=180°$, $B'+b=180°$, $C'+c=180°$.

Proof. Produce the sides AB and AC till they meet $B'C'$ in the points D and F, respectively.

Then B' is the pole of AF $\therefore$ arc $B'F=90°$. Art. 780.

Also C' is the pole of AD $\therefore$ arc $C'D=90°$. (Why?)

Adding, $B'F+C'D=180°$. (Why?)

Or $B'F+FC'+DF=180°$. Ax. 6.

Or $B'C'+DF=180°$.

But $B'C'=a'$, and DF is the measure of the $\angle A$. Art. 770.

$$\therefore A+a'=180°.$$

In like manner the other supplemental relations may be proved as specified.

Q. E. D.

782. DEF. **Supplemental triangles** are two spherical triangles each of which is the polar triangle of the other.

This new name for two polar triangles is due to the property proved in Art. 781.

Proposition XIV. Theorem

783. *The sum of the angles of a spherical triangle is greater than* 180°, *and less than* 540°.

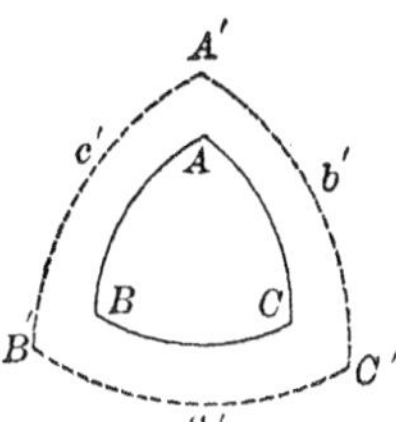

Given the spherical triangle ABC.

To prove $A + B + C > 180°$ and $< 540°$.

Proof. Draw $A'B'C'$, the polar triangle of ABC, and denote its sides by a', b', c'.

Then
$$\left.\begin{aligned} A + a' &= 180° \\ B + b' &= 180° \\ C + c' &= 180° \end{aligned}\right\} \qquad \text{Art. 781.}$$

$$\therefore A + B + C + a' + b' + c' = 540°. \quad . \quad . \quad (1) \qquad \text{Ax. 2.}$$

But
$$\left\{\begin{aligned} a' + b' + c' &< 360° \\ a' + b' + c' &> 0° \end{aligned}\right. \qquad \text{Art. 778.}$$

Subtracting each of these in turn from (1),

$$A + B + C > 180° \text{ and } < 540°. \qquad \text{Ax. 11.}$$

Q. E. D.

784. Cor. *A spherical triangle may have one, two or three right angles; or it may have one, two or three obtuse angles.*

785. Def. A **birectangular spherical triangle** is a spherical triangle containing two right angles.

786. Def. A **trirectangular spherical triangle** is a spherical triangle containing three right angles.

787. COR. *The surface of a sphere may be divided into eight trirectangular spherical triangles.* For let three planes ⊥ to each other be passed through the center of a sphere, etc.

788. DEF. The **spherical excess** of a spherical triangle is the excess of the sum of its angles over 180°.

789. DEF. **Symmetrical spherical triangles** are triangles which have their parts equal, but arranged in reverse order.

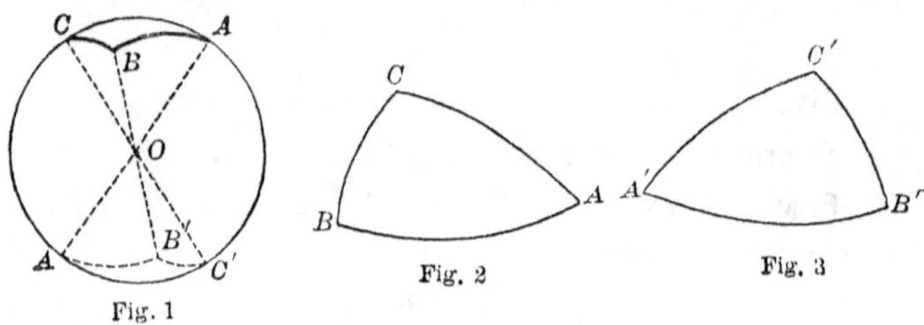

Fig. 1 Fig. 2 Fig. 3

Three planes passing through the center of a sphere form a pair of symmetrical spherical triangles on opposite sides of the sphere (see Art. 580), as ⩓ ABC and $A'B'C'$ of Fig. 1.

790. Equivalence of symmetrical spherical triangles. Two plane triangles which have their parts equal, but arranged in reverse order, may be made to coincide by lifting up one triangle, turning it over in space, and placing it upon the other triangle.

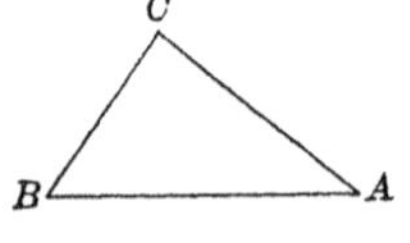

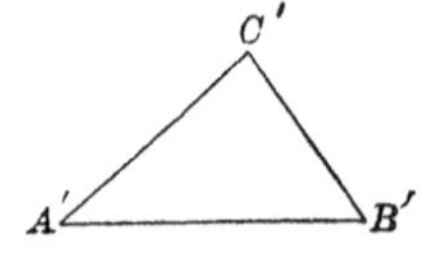

But two symmetrical spherical triangles cannot be made to coincide in this way, because of the curvature of a spherical surface. Hence the equivalence of two symmetrical spherical triangles must be demonstrated in some indirect way.

791. Property of symmetrical spherical triangles. *Two isosceles symmetrical spherical triangles are equal*, for they can be made to coincide.

Proposition XV. Theorem

792. *Two symmetrical spherical triangles are equivalent.*

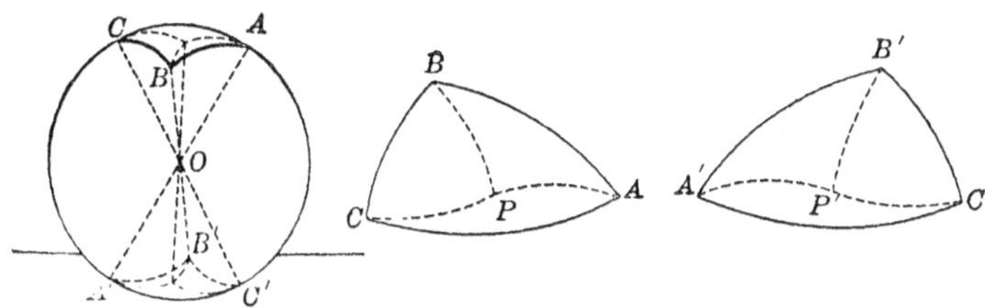

Given the symmetrical spherical $\triangle$ ABC and $A'B'C'$, formed by planes passing through O, the center of a sphere. (See Art. 789.)

To prove $\triangle ABC \approx \triangle A'B'C'$.

Proof. Let P be the pole of a small circle passing through the points A, B, C. Draw the diameter POP'.

Also draw PA, PB, PC, $P'A'$, $P'B'$, $P'C'$, all arcs of great ⊙.

Then $PA = PB = PC$. Art. 747.

Also $P'A' = PA,\ P'B' = PB,\ P'C' = PC$. Arts. 78, 215.

$\therefore P'A' = P'B' = P'C'$. Ax. 1.

Hence PAB and $P'A'B'$ are symmetrical isosceles $\triangle$.

$\therefore \triangle PAB = \triangle P'A'B'$.
Similarly $\triangle PAC = \triangle P'A'C'$. Art. 791.
And $\triangle PBC = \triangle P'B'C'$.

Adding $\triangle PAB + \triangle PAC + \triangle PBC$
$\approx \triangle P'A'B' + \triangle P'A'C' + \triangle P'B'C'$. Ax. 2.

Or $\triangle ABC \approx \triangle A'B'C'$. Ax. 6.

In case the poles P and P' fall outside the $\triangle$ ABC and $A'B'C'$, let the pupil supply the demonstration. Q. E. D.

Proposition XVI. Theorem

793. *On the same sphere, or on equal spheres, two triangles are equal,*

I. *If two sides and the included angle of one are equal to two sides and the included angle of the other; or*

II. *If two angles and the included side of one are equal to two angles and the included side of the other,*

the corresponding equal parts being arranged in the same order in each case.

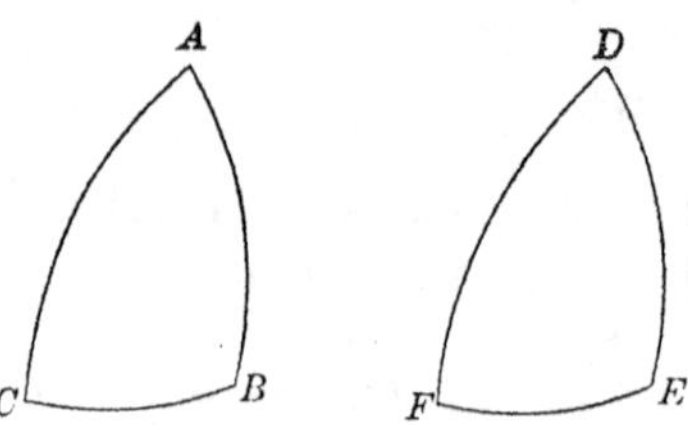

I. **Given** the spherical $\triangle$ ABC and DEF, in which $AC = DF$, $CB = FE$, and $\angle C = \angle F$.

To prove $\triangle ABC = \triangle DEF$.

Proof. Let the pupil supply the proof (see Book I, Prop. VI).

II. **Given** the spherical $\triangle$ ABC and DEF, in which $\angle C = \angle F$, $\angle B = \angle E$, and $CB = FE$.

To prove $\triangle ABC = \triangle DEF$.

Proof. Let the pupil supply the proof (see Book I, Prop. VII).

Ex. 1. If the line of centers of two spheres is 10 in., and the radii are 12 in. and 3 in., how are the spheres situated with reference to each other?

Ex. 2. The tank on a motor car is a cylinder 35 inches long and 15 inches in diameter. How many gallons of gasolene will it hold?

Ex. 3. In an equilateral cone, find the ratio of the lateral area to the area of the base.

PROPOSITION XVII. THEOREM

794. *On the same sphere, or on equal spheres, two triangles are symmetrical and equivalent,*

I. *If two sides and the included angle of one are equal to two sides and the included angle of the other; or*

II. *If two angles and the included side of one are equal to two angles and the included side of the other,*

the corresponding equal parts being arranged in reverse order.

I. **Given** the spherical ⨺ ABC and DEF, in which $AB = DE$, $AC = DF$, and $\angle A = \angle D$, the corresponding parts being arranged in reverse order.

To prove $\triangle ABC$ symmetrical with $\triangle DEF$.

Proof. Construct the $\triangle D'E'F'$ symmetrical with $\triangle DEF$.

Then $\triangle ABC$ may be made to coincide with $\triangle D'E'F'$, Art. 793.

(*having two sides and the included $\angle$ equal and arranged in the same order*).

But $\triangle D'E'F'$ is symmetrical with the $\triangle DEF$.

$\therefore$ $\triangle ABC$, which coincides with $\triangle D'E'F'$, is symmetrical with $\triangle DEF$.

II. The second part of the theorem is proved in the same way.

Q. E. D.

PROPOSITION XVIII. THEOREM

795. *If two triangles on the same sphere, or equal spheres, are mutually equilateral, they are also mutually equiangular, and therefore equal or symmetrical.*

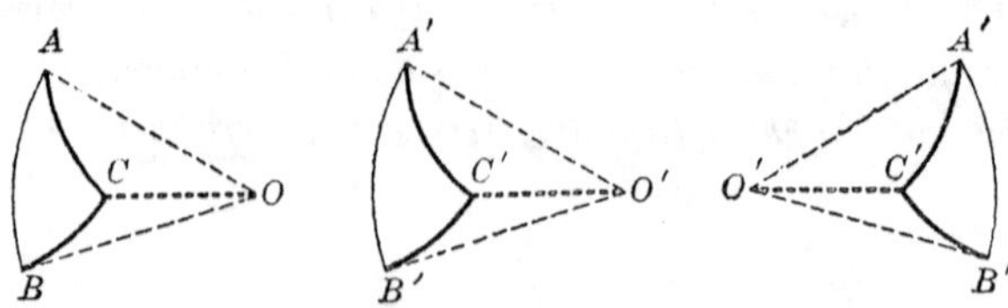

Given two mutually equilateral spherical ≜ ABC and $A'B'C'$ on the same or on equal spheres.

To prove ≜ ABC and $A'B'C'$ equal or symmetrical.

Proof. From O and O', the centers of the spheres to which the given triangles belong, draw the radii OA, OB, OC, $O'A'$, $O'B'$, $O'C'$.

Then the face ∡ at O = corresponding face ∡ at O'. Art. 774.

Hence dihedral ∡ at O = corresponding dihedral ∡ at O'. Art. 584.

∴ ∡ of spherical △ ABC = homologous ∡ of spherical △ $A'B'C'$. Art. 774.

∴ the ≜ ABC and $A'B'C'$ are equal or symmetrical, according as their homologous parts are arranged in the same or in reverse order. Art. 789.

796. NOTE. The conditions in Props. XVI and XVIII which make two spherical triangles equal are the same as those which make two plane triangles equal. Hence many other propositions occur in spherical geometry which are identical with corresponding propositions in plane geometry. Thus, many of the construction problems of spherical geometry are solved in the same way as the corresponding construction problems in plane geometry; as, to bisect a given angle, etc.

PROPOSITION XIX. THEOREM

797. *If two triangles on the same sphere are mutually equiangular, they are also mutually equilateral, and therefore equal or symmetrical.*

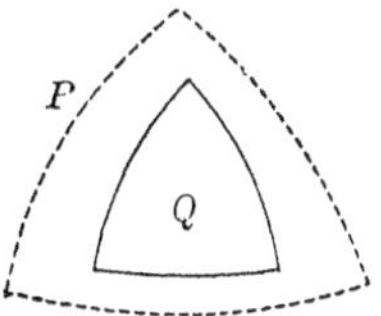

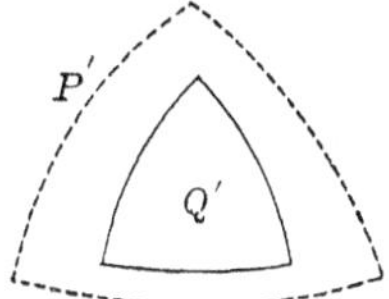

Given the mutually equiangular spherical ⧌ Q and Q' on the same sphere or on equal spheres.

To prove that Q and Q' are mutually equilateral, and therefore equal or symmetrical.

Proof. Construct P and P' the polar ⧌ of Q and Q', respectively.

Then ⧌ P and P' are mutually equilateral. Art. 781.

∴ ⧌ P and P' are mutually equiangular. Art. 795.

But Q is the polar △ of P, and Q' of P'. Art. 780.

∴ ⧌ Q and Q' are mutually equilateral. Art. 781.

Hence Q and Q' are equal or symmetrical, according as their homologous parts are arranged in the same or in reverse order. Art. 789.

Q. E. D.

798. COR. *If two mutually equiangular triangles are on unequal spheres, their corresponding sides have the same ratio as the radii of their respective spheres.*

PROPOSITION XX. THEOREM

799. *In an isosceles spherical triangle the angles opposite the equal sides are equal.*

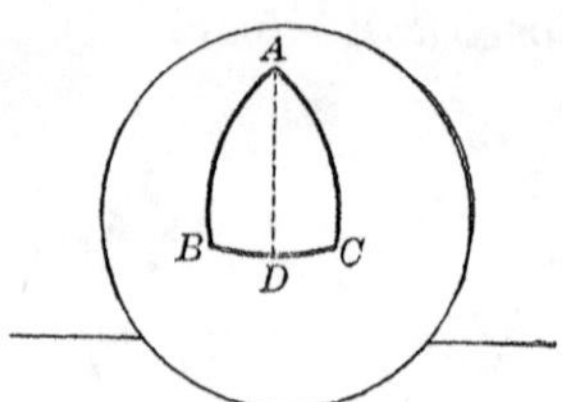

Given the spherical $\triangle$ ABC in which $AB = AC$.

To prove $\angle B = \angle C$.

Proof. Draw an arc from the vertex A to D, the mid-point of the base.

Let the pupil supply the remainder of the proof.

PROPOSITION XXI. THEOREM (CONV. OF PROP. XX)

800. *If two angles of a spherical triangle are equal, the sides opposite these angles are equal, and the triangle is isosceles.*

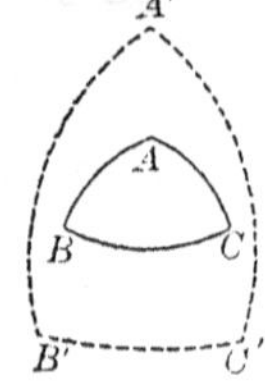

Given the spherical $\triangle$ ABC in which $\angle B = \angle C$.

To prove $AB = AC$.

Proof. Construct $\triangle$ $A'B'C'$ the polar $\triangle$ of ABC.

Then $A'C' = A'B'$. Art. 781.

$\therefore \angle C' = \angle B'$. Art. 799.

$\therefore AB = AC$. Art. 781.

Q. E. D.

Proposition XXII. Theorem

801. *In any spherical triangle, if two angles are unequal, the sides opposite these angles are unequal, and the greater side is opposite the greater angle, and* Conversely.

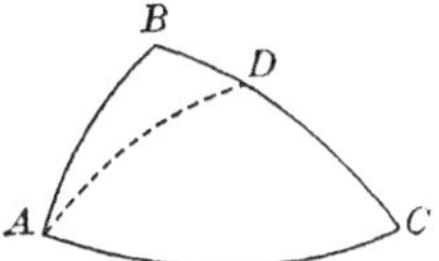

Given the spherical $\triangle ABC$ in which $\angle BAC$ is greater than $\angle C$.

To prove $BC > BA$.

Proof. Draw the arc AD making $\angle DAC$ equal to $\angle C$.

Then $DA = DC$. Art. 800.

To each of these equals add the arc BD.

$\therefore BD + DA = BD + DC$, or BC. (Why?)

But in $\triangle BDA$, $BD + DA > BA$. (Why?)

$\therefore BC > BA$. Ax. 8.

Let the pupil prove the converse by the indirect method (see Art. 106).

Q. E. D.

Ex. 1. Bisect a given spherical angle.

Ex. 2. Bisect a given arc of a great circle on a sphere.

Ex. 3. At a given point in an arc on a sphere, construct an angle equal to a given spherical angle on the same sphere.

Ex. 4. Find the locus of the centers of the circles of a sphere formed by planes perpendicular to a given diameter of the given sphere.

SPHERICAL AREAS

802. Units of spherical surface. A spherical surface may be measured in terms of, either

1. The *customary units of area*, as a square inch, a square foot, etc., or

2. *Spherical degrees*, or *spherids*.

803. A spherical degree, or spherid, is one-ninetieth part of one of the eight trirectangular triangles into which the surface of a sphere may be divided (Art. 787), or $\frac{1}{720}$ part of the surface of the entire sphere.

A **solid degree** is one-ninetieth part of a trirectangular angle (see Art. 774).

804. A **lune** is a portion of the surface of a sphere bounded by two semicircumferences of great circles, as $PBP'C$ of Fig. 1.

The **angle of a lune** is the angle formed by the semicircumferences which bound it, as the angle BPC.

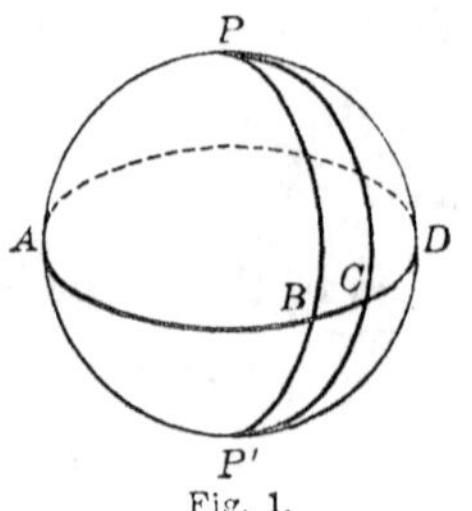

Fig. 1.

805. A **zone** is the portion of the surface of the sphere bounded by two parallel planes.

A zone may also be defined as the surface generated by an arc of a revolving semicircumference. Thus, if QFQ' (Fig. 2) generates a sphere by rotating about QQ', its diameter, any arc of QFQ', as EF, generates a zone.

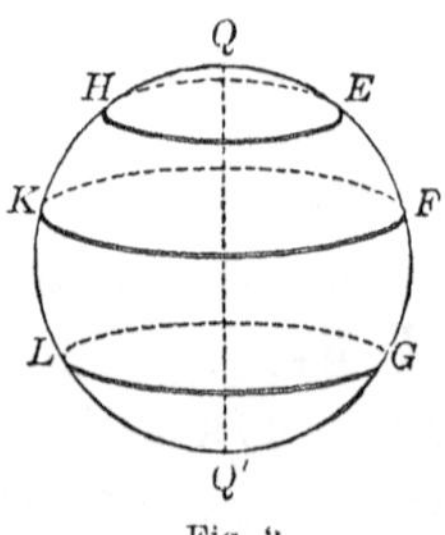

Fig. 2.

806. A zone of one base is a zone one of whose bounding planes is tangent to the sphere, as the zone generated by the arc QE of Fig. 2.

807. The **altitude** of a zone is the perpendicular distance between the bounding planes of the zone.

The **bases** of a zone are the circumferences of the circles of the sphere formed by the bounding planes of the zone.

PROPOSITION XXIII. THEOREM

808. *The area generated by a straight line revolving about an axis in its plane is equal to the projection of the line upon the axis, multiplied by the circumference of a circle whose radius is the perpendicular erected at the midpoint of the line and terminated by the axis.*

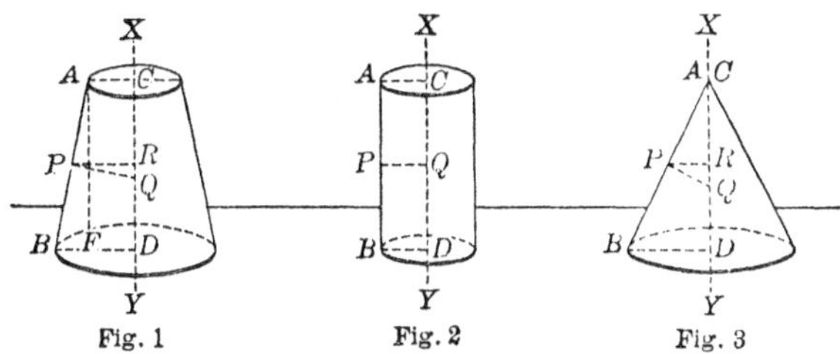

Fig. 1 Fig. 2 Fig. 3

Given AB and XY in the same plane, CD the projection of AB on XY, PQ the $\perp$ bisector of AB; and a surface generated by the revolution of AB about XY, denoted as "area AB."

To prove area $AB = CD \times 2\,\pi PQ$.

Proof. 1. In general, the surface generated by AB is the surface of a frustum of a cone (Fig. 1).

$\therefore$ area $AB = AB \times 2\,\pi PR$. Art. 728.

Draw $AF \perp BD$, then

△ ABF and PQR are similar. Art. 328.

$\therefore AB : AF = PQ : PR$. (Why?)

$\therefore AB \times PR = AF \times PQ$, or $CD \times PQ$. (Why?)

Substituting, area $AB = CD \times 2\,\pi PQ$. Ax. 8.

2. If $AB \parallel XY$ (Fig. 2), the surface generated by AB is the lateral surface of a cylinder.

$\therefore$ area $AB = CD \times 2\,\pi PQ$. Art. 697.

3. If the point A lies in the axis XY (Fig. 3), let the pupil show that the same result is obtained. Q. E. D.

PROPOSITION XXIV. THEOREM

809. *The area of the surface of a sphere is equal to the product of the diameter of the sphere by the circumference of a great circle.*

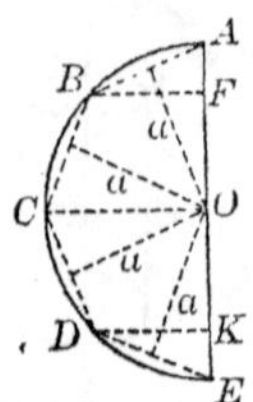

Given a sphere generated by the revolution of the semicircle ACE about the diameter AE, with the surface of the sphere denoted by S, and its radius by R.

To prove $S = AE \times 2\,\pi R$.

Proof. Inscribe in the given semicircle the half of a regular polygon of an even number of sides, as $ABCDE$.

Draw the apothem to each side of the semipolygon, and denote it by a.

From the vertices B, C, D draw ⊥s to AE.

Then
$$\left.\begin{aligned} \text{area } AB &= AF \times 2\,\pi a. \\ \text{area } BC &= FO \times 2\,\pi a. \\ \text{area } CD &= OK \times 2\,\pi a. \\ \text{area } DE &= KE \times 2\,\pi a. \end{aligned}\right\} \quad \text{Art. 808.}$$

Adding, area $ABCDE = AE \times 2\,\pi a$.

If, now, the number of sides of the polygon be indefinitely increased,

area $ABCDE$ approaches S as a limit. Art. 441.

And a approaches R as a limit. (Why?)

$\therefore AE \times 2\,\pi a$ approaches $AE \times 2\,\pi R$ as a limit. (Why?)

But area $ABCDE = AE \times 2\,\pi a$ always.

$\therefore S = AE \times 2\,\pi R$. (Why?)

Q. E. D.

810. Formulas for area of surface of a sphere.

Substituting for AE its equal $2R$, $S = 4\pi R^2$.

Also denoting the diameter of the sphere by D, $R = \frac{1}{2}D$.

$$\therefore\ S = 4\pi\left(\frac{D}{2}\right)^2,\ \text{or}\ S = \pi D^2.$$

811. COR. 1. *The surface of a sphere is equivalent to four times the area of a great circle of the sphere.*

812. COR. 2. *The areas of the surfaces of two spheres are to each other as the squares of their radii, or of their diameters.*

For, if S and S' denote the surfaces, R and R' the radii, and D and D' the diameters of two spheres,

$$\frac{S}{S'} = \frac{4\pi R^2}{4\pi R'^2} = \frac{R^2}{R'^2};\ \text{also}\ \frac{S}{S'} = \frac{\pi D^2}{\pi D'^2} = \frac{D^2}{D'^2}.$$

813. Property of the sphere. The following property of the sphere is used in the proof of Art. 809: *If, in the generating arc of any zone, a broken line be inscribed, whose vertices divide the arc into equal parts, then, as the number of these parts is increased indefinitely, the area generated by the broken line approaches the area of the zone as a limit.* Hence

COR. 3. *The area of a zone is equal to the circumference of a great circle multiplied by the altitude of the zone.*

Thus the area generated by the arc $BC = FO \times 2\pi R$.

814. COR. 4. *On the same sphere, or on equal spheres, the areas of two zones are to each other as the altitudes of the zones.*

Ex. 1. Find the area of a sphere whose diameter is 10 in.

Ex. 2. Find the area of a zone of altitude 3 in., on a sphere whose radius is 10 in.

Proposition XXV. Theorem

815. *The area of a lune is to the area of the surface of the sphere as the angle of the lune is to four right angles.*

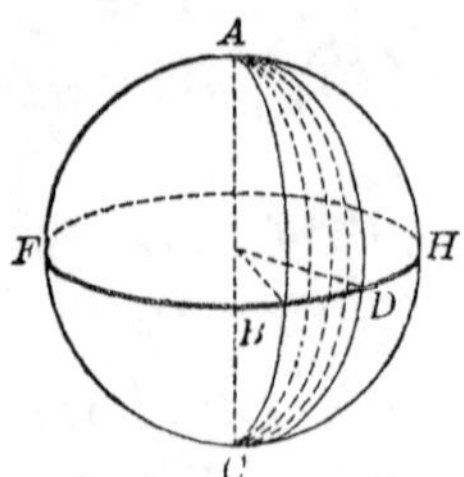

Given a sphere having its area denoted by S, and on the sphere the lune $ABCD$ of $\angle A$ with its area denoted by L.

To prove $L : S = A^\circ : 360^\circ$.

Proof. Draw FBH, the great ⊙ whose pole is A, intersecting the bounding arcs of the lune in B and D.

Case I. *When the arc BD and the circumference FBH are commensurable.*

Find a common measure of BD and FBH, and let it be contained in the arc BD m times, and in the circumference FBH n times.

Then arc BD : circumference $FBH = m : n$.

Through the diameter AC, and the points of division of the circumference FBH pass planes of great ⓢ.

The arcs of these great ⓢ will divide the surface of the sphere in n small equal lunes, m of them being contained in the lune $ABCD$.

$\therefore L : S = m : n$.

$\therefore L : S =$ arc BD : circumference FBH. (Why?)

Or $L : S = A^\circ : 360^\circ$. Art. 257.

Case II. *When the arc BD and the circumference FBH are incommensurable.*

Let the pupil supply the proof. Q. E. D.

816. Formula for the area of a lune in spherical degrees, or spherids. The surface of a sphere contains 720 spherids (Art. 803). Hence, by Art. 815,

$$\therefore \frac{L}{S} \text{ or } \frac{L \text{ spherids}}{720 \text{ spherids}} = \frac{A}{360}, \quad \therefore \frac{L}{2} = A, \text{ or } L = 2A \text{ spherids};$$

that is, *the area of a lune in spherical degrees is equal to twice the number of angular degrees in the angle of the lune.*

817. Formula for area of a lune in square units of area.

$$S = 4\pi R^2 \text{ (Art. 810)} \quad \therefore \frac{L}{4\pi R^2} = \frac{A}{360}, \text{ or } L = \frac{\pi R^2 A}{90}.$$

818. COR. 1. *On the same sphere, or on equal spheres, two lunes are to each other as their angles.*

819. COR. 2. *Two lunes with equal angles, but, on unequal spheres, are to each other as the squares of the radii of their spheres.*

$$\text{For } L : L' = \frac{\pi R^2 A}{90} : \frac{\pi R'^2 A}{90}, \text{ or } L : L' = R^2 : R'^2.$$

Ex. 1. Find the area in spherical degrees of a lune of 27°.

Ex. 2. Find the number of square inches in the area of a lune of 27°, on a sphere whose radius is 10 in.

A solid symmetrical with respect to a plane is a solid in which a line drawn from any point in its surface ⊥ the given plane and produced its own length ends in a point on the surface; hence

Ex. 3. How many planes of symmetry has a circular cylinder? A cylinder of revolution?

Ex. 4. Has either of these solids a center of symmetry?

Ex. 5. Answer the same questions for a circular cone.

Ex. 6. For a cone of revolution. For a sphere.

Ex. 7. For a regular square pyramid. For a regular pentagonal pyramid.

PROPOSITION XXVI. THEOREM

820. *If two great circles intersect on a hemisphere, the sum of two vertical triangles thus formed is equivalent to a lune whose angle is that angle in the triangles which is formed by the intersection of the two great circles.*

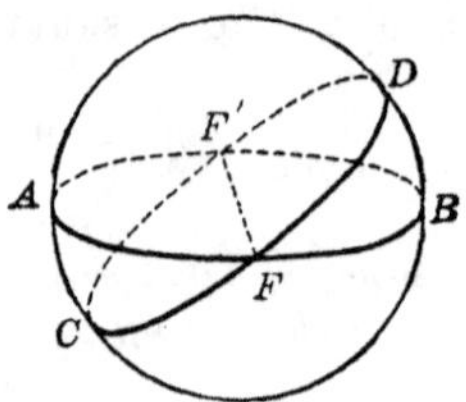

Given the hemisphere $ADBF$, and on it the great circles AFB and DFC, intersecting at F.

To prove $\triangle\ AFC + \triangle\ BFD \approx$ lune whose $\angle$ is BFD.

Proof. Complete the sphere and produce the given arcs of the great circles to intersect at F' on the other hemisphere.

Then, in the $\triangle s$ AFC and $BF'D$,

$$\text{arc } AF = \text{arc } BF',$$

(*each being the supplement of the arc* BF).

In like manner arc CF = arc DF'.

And $\qquad$ arc AC = arc DB.

$$\therefore \triangle\ AFC \approx \triangle\ BF'D. \qquad \text{Art. 795.}$$

Add the $\triangle\ BFD$ to each of these equals;

$$\therefore \triangle\ AFC + \triangle\ BFD \approx \triangle\ BF'D + \triangle\ BFD. \qquad \text{Ax. 3.}$$

Or $\therefore \triangle\ AFC + \triangle\ BFD \approx$ lune $FBF'D$. $\qquad$ Ax. 6.

Q. E. D.

PROPOSITION XXVII. THEOREM

821. *The number of spherical degrees, or spherids, in the area of a spherical triangle is equal to the number of angular degrees in the spherical excess of the triangle.*

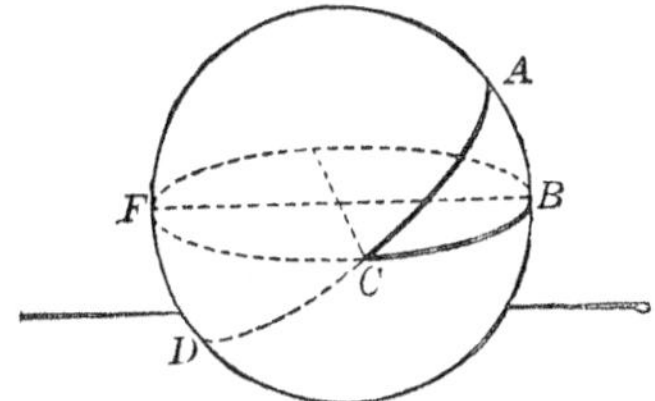

Given the spherical $\triangle ABC$ whose $\angle$s are denoted by A, B, C, and whose spherical excess is denoted by E.

To prove area of $\triangle ABC = E$ spherids.

Proof. Produce the sides AC and BC to meet AB produced in the points D and F, respectively.

$\triangle ABC + \triangle CDB =$ lune $ABDC = 2\,A$ spherids.
$\triangle ABC + \triangle ACF =$ lune $BCFA = 2\,B$ spherids. } Art. 816.

$\triangle ABC + \triangle CFD =$ lune of $\angle BCA = 2\,C$ spherids. Art. 820.

Adding, and observing that $\triangle ABC + \triangle CDB + \triangle ACF + \triangle CFD =$ hemisphere $ABDFC$,

$2\,\triangle ABC +$ hemisphere $= 2\,(A + B + C)$ spherids. Or,
$2\,\triangle ABC + 360$ spherids $= 2\,(A + B + C)$ spherids. Ax. 8.

$\therefore \triangle ABC + 180$ spherids $= (A + B + C)$ spherids. Ax. 5.

$\therefore \triangle ABC = (A + B + C - 180)$ spherids. Ax. 3.

Or area $\triangle ABC = E$ spherids. Art. 788.

Q. E. D.

822. Formula for area of a spherical triangle in square units of area.

Comparing the area of a spherical △ with the area of the entire sphere,

$$\text{area } \triangle : 4\,\pi R^2 = E \text{ spherids} : 720 \text{ spherids}.$$

$$\therefore \text{ area } \triangle = \frac{4\,\pi R^2 \times E}{720}, \text{ or area } \triangle = \frac{\pi R^2 E}{180}.$$

823. The **spherical excess of a spherical polygon** is the sum of the angles of the polygon diminished by $(n-2)\,180°$; that is, it is the sum of the spherical excesses of the triangles into which the polygon may be divided.

Proposition XXVIII. Theorem

824. *The area of a spherical polygon, in spherical degrees or spherids, is equal to the spherical excess of the polygon.*

Given a spherical polygon $ABCDF$ of n sides, with its spherical excess denoted by E.

To prove area of $ABCDF = E$ spherical degrees.

Proof. Draw diagonals from A, any vertex of the polygon, and thus divide the polygon into $n-2$ spherical ⨺.

The area of each △ = (sum of its ∡ − 180) spherids. **Art. 821.**

∴ sum of the areas of the ⨺ = [sum of ∡ of the ⨺ − $(n-2)$ 180] spherids. **Ax. 2.**

∴ area of polygon = E spherids, **Art. 823.**

(*for the sum of ∡ of the ⨺* $-(n-2)\,180° = E°$).

SPHERICAL VOLUMES

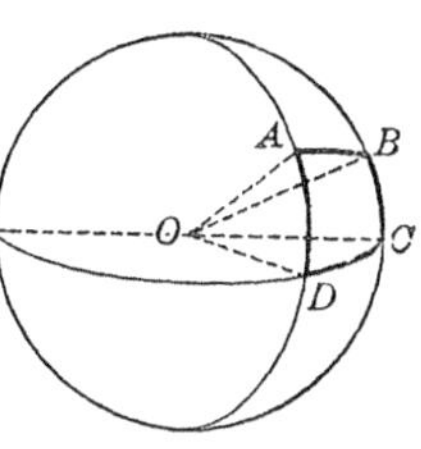

825. A **spherical pyramid** is a portion of a sphere bounded by a spherical polygon and the planes of the great circles forming the sides of the polygon. The **base** of a spherical pyramid is the spherical polygon bounding it, and the **vertex** of the spherical pyramid is the center of the sphere.

Thus, in the spherical pyramid *O–ABCD*, the base is *ABCD* and the vertex is *O*.

826. A **spherical wedge** (or **ungula**) is the portion of a sphere bounded by a lune and the planes of the sides of the lune.

827. A **spherical sector** is the portion of a sphere generated by a sector of that semicircle whose rotation generates the given sphere.

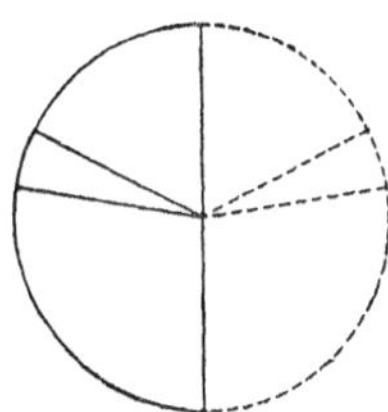

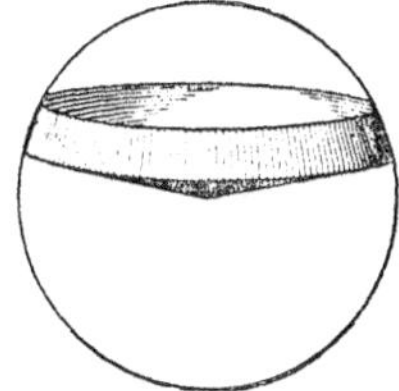

828. The **base** of a spherical sector is the zone generated by the revolution of the arc of the plane sector which generates the spherical sector.

Let the pupil draw a spherical sector in which the base is a zone of one base.

829. A **spherical segment** is a portion of a sphere included between two parallel planes.

The **bases** of a spherical segment are the sections of the sphere made by the parallel planes which bound the given segment; the **altitude** is the perpendicular distance between the bases.

830. A **spherical segment of one base** is a spherical segment one of whose bounding planes is tangent to the sphere.

Proposition XXIX. Theorem

831. *The volume of a sphere is equal to one-third the product of the area of its surface by its radius.*

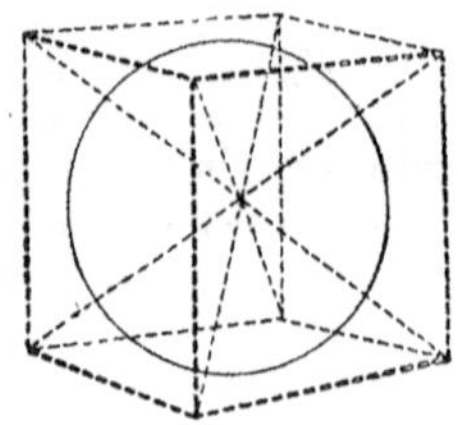

Given a sphere having its volume denoted by V, surface by S, and radius by R.

To prove $V = \frac{1}{3} S \times R$.

Proof. Let any polyhedron be circumscribed about the sphere.

Pass a plane through each edge of the polyhedron and the center of the sphere.

These planes will divide the polyhedron into as many pyramids as the polyhedron has faces, each pyramid having a face of the polyhedron for its base, the center of the sphere for its vertex, and the radius of the sphere for its altitude.

$\therefore$ volume of each pyramid $= \frac{1}{3}$ base $\times R$. (Why?)

$\therefore$ volume of polyhedron $= \frac{1}{3}$ (surface of polyhedron) $\times R$.

If the number of faces of the polyhedron be increased indefinitely, the volume of the polyhedron approaches the volume of the sphere as a limit, and the surface of the polyhedron approaches the surface of the sphere as a limit.

Hence the volume of the polyhedron and $\frac{1}{3}$ (surface of the polyhedron) $\times R$, are two variables always equal.

Hence their limits are equal,

Or $V = \frac{1}{3} S \times R$. (Why?)

Q. E. D.

832. Formulas for volume of a sphere. Substituting $S = 4\pi R^2$, or $S = \pi D^2$ in the result of Art. 831,

$$V = \frac{4\pi R^3}{3}; \text{ also } V = \frac{\pi D^3}{6}.$$

833. Cor. 1. *The volumes of two spheres are to each other as the cubes of their radii, or as the cubes of their diameters.*

For $\frac{V}{V'} = \frac{\frac{4}{3}\pi R^3}{\frac{4}{3}\pi R'^3} = \frac{R^3}{R'^3}$; also $\frac{V}{V'} = \frac{\frac{1}{6}\pi D^3}{\frac{1}{6}\pi D'^3} = \frac{D^3}{D'^3}$.

834. Cor. 2. *The volume of a spherical pyramid is equal to one-third the product of its base by the radius of the sphere.*

835. Cor. 3. *The volume of spherical sector is equal to one-third the product of its base (the bounding zone) by the radius of the sphere.*

836. Formula for the volume of a spherical sector. Denoting the altitude of the sector by H and the volume by V,

$$V = \tfrac{1}{3} \text{ (area of zone)} \times R,$$
$$= \tfrac{1}{3} (2\pi RH) R. \qquad \text{Art. 813.}$$
$$\therefore V = \tfrac{2}{3} \pi R^2 H.$$

PROPOSITION XXX. THEOREM

837. *The volume of a spherical segment is equal to one-half the product of its altitude by the sum of the areas of its bases, plus the volume of a sphere whose diameter is the altitude of the segment.*

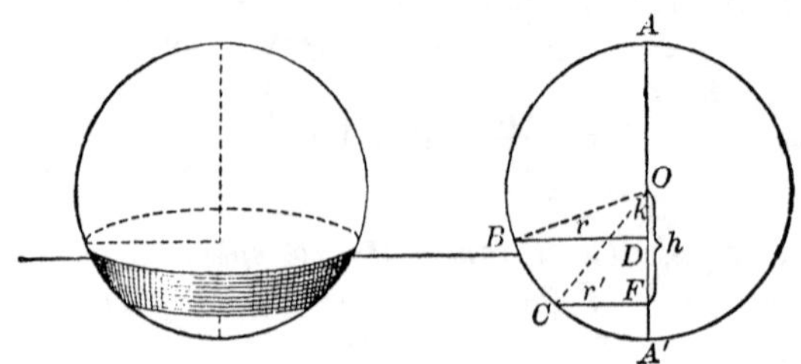

Given the semicircle $ABCA'$ which generates a sphere by its rotation about the diameter AA'; BD and CF semi-chords $\perp AA'$, and denoted by r and r'; and DF denoted by H.

To prove volume of spherical segment generated by $BCFD$, or $V = \frac{1}{2}(\pi r^2 + \pi r'^2) H + \frac{1}{6} \pi H^3$.

Proof. Draw the radii OB and OC.

Denote OF by h, and OD by k.

Then $V = \text{vol. } OBC + \text{vol. } OCF - \text{vol. } OBD$.

$$\therefore V = \tfrac{2}{3} \pi R^2 H + \tfrac{1}{3} \pi r'^2 h - \tfrac{1}{3} \pi r^2 k. \qquad \text{Arts. 836, 723.}$$

But $H = h - k$, $h^2 = R^2 - r'^2$, and $k^2 = R^2 - r^2$. (Why?)

$\therefore V = \frac{1}{3}\pi\ [2\ R^2\,(h-k) + (R^2-h^2)\,h - (R^2-k^2)\,k]$. Ax. 8.

$= \frac{1}{3}\pi\ [2\ R^2\,(h-k) + R^2\,(h-k) - (h^3-k^3)]$.

$= \frac{1}{3}\pi\ H\,[3\ R^2 - (h^2 + hk + k^2)]$.

But $h^2 - 2\ hk + k^2 = H^2$. Ax. 4.

Subtract each member from $3\ h^2 + 3\ k^2$ and divide by 2. Axs. 3, 5.

Then $h^2 + hk + k^2 = \frac{3}{2}\,(h^2 + k^2) - \frac{H^2}{2}$.

$= 3\ R^2 - \frac{3}{2}\,(r^2 + r'^2) - \frac{H^2}{2}$.

$\therefore V = \frac{1}{3}\pi H \left[\frac{3}{2}\,(r^2 + r'^2) + \frac{H^2}{2}\right]$. Ax. 8.

$\therefore V = \frac{1}{2}\,(\pi r^2 + \pi r'^2)\,H + \frac{1}{6}\,\pi H^3$.

Q. E. D.

838. Formula for volume of a spherical segment of one base. In a spherical segment of one base $r' = o$, and $r^2 = (2\ R-H)\ H$ (Art. 343).

Substituting for r and r' these values in the result of Art. 837,

$$V = \pi H^2 \left(R - \frac{H}{3}\right).$$

839. Advantage of measurement formulas. The student should observe carefully that, by the results obtained in Book IX, the measurement of the areas of certain curved surfaces is reduced to the far simpler work of the measurement of the lengths of one or more straight lines; in like manner the measurement of certain volumes bounded by a curved surface is reduced to the simpler work of linear measurements. A similar remark applies to the results of Book VIII.

Ex. 1. Find the volume of a sphere whose radius is 7 in.

Ex. 2. Find the volume of a sphere whose diameter is 7 in.

Ex. 3. In a sphere whose radius is 8 in., find the volume of a spherical segment of one base whose altitude is 3.

EXERCISES. GROUP 72

THEOREMS CONCERNING THE SPHERE

Ex. 1. Of circles of a sphere whose planes pass through a given point within a sphere, the smallest is that circle whose plane is perpendicular to the diameter through the given point.

Ex. 2. If a point on the surface of a given sphere is equidistant from three points on a given small circle of the sphere, it is the pole of the small circle.

Ex. 3. If two sides of a spherical triangle are quadrants, the third side measures the angle opposite that side in the triangle.

Ex. 4. If a spherical triangle has one right angle, the sum of its other two angles is greater than one right angle.

Ex. 5. If a spherical triangle is isosceles, its polar triangle is isosceles.

Ex. 6. The polar triangle of a birectangular triangle is birectangular.

Ex. 7. The polar triangle of a trirectangular triangle is identical with the original triangle.

Ex. 8. Prove that the sum of the angles of a spherical quadrilateral is greater than 4 right angles, and less than 8 right angles. What, also, are the limits of the sum of the angles of a spherical hexagon? Of the sum of the angles of a spherical n-gon?

Ex. 9. On the same sphere, or on equal spheres, two birectangular triangles are equal if their oblique angles are equal.

Ex. 10. Two zones on the same sphere, or on equal spheres, are to each other as their altitudes.

Ex. 11. If one of the legs of a right spherical triangle is greater than a quadrant, another side is also greater than a quadrant.

[Sug. Of the leg which is greater than a quadrant, take the end remote from the right angle as a pole, and describe an arc.]

Ex. 12. If ABC and $A'B'C'$ are polar triangles, the radius OA is perpendicular to the plane $OB'C'$.

Ex. 13. On the same sphere, or on equal spheres, spherical triangles whose polar triangles have equal perimeters are equivalent.

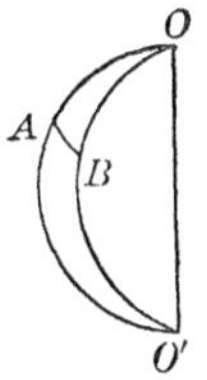

Ex. 14. Given OAO', OBO', and AB arcs of great circles, intersecting so that $\angle OAB = \angle O'BA$; prove that $\triangle OAB = \triangle O'AB$.

[SUG. Show that $\angle OBA = \angle O'AB$.]

Ex. 15. Find the ratio of the volume of a sphere to the volume of a circumscribed cube.

Ex. 16. Find the ratio of the surface of a sphere to the lateral surface of a circumscribed cylinder of revolution; also find the ratio of their volumes.

Ex. 17. If the edge of a regular tetrahedron is denoted by a, find the ratio of the volumes of the inscribed and circumscribed spheres.

Ex. 18. Find the ratio of the two segments into which a hemisphere is divided by a plane parallel to the base of the hemisphere and at the distance $\frac{2}{3}R$ from the base.

EXERCISES. GROUP 73

SPHERICAL LOCI

Ex. 1. Find the locus of a point at a given distance a from the surface of a given sphere.

Ex. 2. Find the locus of a point on the surface of a sphere that is equidistant from two given points on the surface.

Ex. 3. If, through a given point outside a given sphere, tangent planes to the sphere are passed, find the locus of the points of tangency.

Ex. 4. If straight lines be passed through a given fixed point in space, and through another given point other straight lines be passed perpendicular to the first set, find the locus of the feet of the perpendiculars.

EXERCISES. GROUP 74

PROBLEMS CONCERNING THE SPHERE

Ex. 1. At a given point on a sphere, construct a plane tangent to the sphere.

Ex. 2. Through a given point on the surface of a sphere, draw an arc of a great circle perpendicular to a given arc.

Ex. 3. Inscribe a circle in a given spherical triangle.

Ex. 4. Construct a spherical triangle, given its polar triangle.

Given the radius, r, construct a spherical surface which shall pass through

Ex. 5. Three given points.

Ex. 6. Two given points and be tangent to a given plane.

Ex. 7. Two given points and be tangent to a given sphere

Ex. 8. One given point and be tangent to two given planes.

Ex. 9. One given point and be tangent to two given spheres.

Given the radius, r, construct a spherical surface which shall be tangent to

Ex. 10. Three given planes.

Ex. 11. Two given planes and one given sphere.

Ex. 12. Construct a spherical surface which shall pass through three given points and be tangent to a given plane.

Ex. 13. Through a given straight line pass a plane tangent to a given sphere.

[SUG. Through the center of the sphere pass a plane ⊥ given line, etc.]

When is the solution impossible?

Ex. 14. Through a given point on a sphere, construct an arc of a great circle tangent to a given small circle of the sphere.

[SUG. Draw a straight line from the center of the sphere to the given point, and produce it to intersect the plane of the small circle, etc.]

NUMERICAL EXERCISES IN SOLID GEOMETRY

For methods of facilitating numerical computations, see Arts. 493-6.

EXERCISES. GROUP 75

LINES AND SURFACES OF POLYHEDRONS

Find the lateral area and total area of a right prism whose

Ex. 1. Base is an equilateral triangle of edge 4 in., and whose altitude is 15 in.

Ex. 2. Base is a triangle of sides 17, 12, 25, and whose altitude is 20.

Ex. 3. Base is an isosceles trapezoid, the parallel bases being 10 and 15 and leg 8, and whose altitude is 24.

Ex. 4. Base is a rhombus whose diagonals are 12 and 16, and whose altitude is 12.

Ex. 5. Base is a regular hexagon with side 8 ft., and whose altitude is 20 ft.

Ex. 6. Find the entire surface of a rectangular parallelopiped $8 \times 12 \times 16$ in.; of one $p \times q \times r$ ft.

Ex. 7. Of a cube whose edge is 1 ft. 3 in.

Ex. 8. The lateral area of a regular hexagonal prism is 120 sq. ft. and an edge of the base is 10 ft. Find the altitude.

Ex. 9. How many square feet of tin are necessary to line a box $20 \times 6 \times 4$ in.?

Ex. 10. If the surface of a cube is 1 sq. yd., find an edge in inches

Ex. 11. Find the diagonal of a cube whose edge is 5 in.

Ex. 12. If the diagonal of a cube is 12 ft., find the surface.

Ex. 13. If the surface of a rectangular parallelopiped is 208 sq. in., and the edges are as 2 : 3 : 4, find the edges.

In a regular square pyramid

Ex. 14. If an edge of the base is 16 and slant height is 17, find the altitude.

Ex. 15. If the altitude is 15 and a lateral edge is 17, find an edge of the base.

Ex. 16. If a lateral edge is 25 and an edge of the base is 14, find the altitude.

In a regular triangular pyramid

Ex. 17. If an edge of the base is 8 and the altitude is 10, find the slant height.

Ex. 18. Find the altitude of a regular tetrahedron whose edge is 6.

Find the lateral surface and the total surface of

Ex. 19. A regular square pyramid an edge of whose base is 16, and whose altitude is 15.

Ex. 20. A regular triangular pyramid an edge of whose base is 10, and whose altitude is 12.

Ex. 21. A regular hexagonal pyramid an edge of whose base is 4, and whose altitude is 21.

Ex. 22. A regular square pyramid whose slant height is 24, and whose lateral edge is 25.

Ex. 23. A regular tetrahedron whose edge is 4.

Ex. 24. A regular tetrahedron whose altitude is 9.

Ex. 25. A regular hexagonal pyramid each edge of whose base is a, and whose altitude is b.

Ex. 26. In the frustum of a regular square pyramid the edges of the bases are 6 and 18, and the altitude is 8. Find the slant height. Hence find the lateral area.

Ex. 27. In the frustum of a regular triangular pyramid the edges of the bases are 4 and 6, and the altitude is 5. Find the slant height. Hence find the lateral area.

Ex. 28. In the frustum of a regular tetrahedron, if the edge of the lower base is b_1, the edge of the upper base is b_2, and the altitude is a, show that $L = \frac{1}{2}\sqrt{\frac{1}{3}(b_1 - b_2)^2 + 4a^2}$.

Ex. 29. In the frustum of a regular square pyramid the edges of the bases are 20 and 60, and a lateral edge is 101. Find the lateral surface.

EXERCISES. GROUP 76

LINES AND SURFACES OF CONES AND CYLINDERS

Ex. 1. How many square feet of lateral surface has a tunnel 100 yds. long and 7 ft. in diameter.

Ex. 2. The lateral area of a cylinder of revolution is 1 sq. yd., and the altitude is 1 ft. Find the radius of the base.

Ex. 3. The entire surface of a cylinder of revolution is 900 sq. ft. and the radius of the base is 10 ft. Find the altitude.

In a cylinder of revolution

Ex. 4. Find R in terms of S and H.

Ex. 5. Find H in terms of R and T.

Ex. 6. Find T in terms of S and H.

Ex. 7. How many sq. yds. of canvas are required to make a conical tent 20 ft. in diameter and 12 ft. high?

Ex. 8. A man has 400 sq. yds. of canvas and wants to make a conical tent 20 yds. in diameter. What will be its altitude?

Ex. 9. The altitude of a cone of revolution is 10 ft. and the lateral area is 11 times the area of the base. Find the radius of the base.

In a cone of revolution

Ex. 10. Find T in terms of S and L.

Ex. 11. Find R in terms of T and L.

Ex. 12. How many square feet of tin are necessary to make a funnel the diameters of whose ends are 2 in. and 8 in., and whose altitude is 7 in.?

Ex. 13. If the slant height of a frustum of a cone of revolution makes an angle of 45° with the base, show that the lateral area of the frustum is $(r_1^2 - r_2^2)\pi\sqrt{2}$.

EXERCISES. GROUP 77

SPHERICAL LINES AND SURFACES

Ex. 1. Find in square feet the area of the surface of a sphere whose radius is 1 ft. 2 in.

Ex. 2. How many square inches of leather will it take to cover a baseball whose diameter is $3\frac{1}{2}$ in.?

Ex. 3. How many sq. ft. of tin are required to cover a dome in the shape of a hemisphere 6 yds. in diameter?

Ex. 4. What is the radius of a sphere whose surface is 616 sq. in.?

Ex. 5. Find the diameter of a globe whose surface is 1 sq. yd.

Ex. 6. If the circumference of a great circle on a sphere is 1 ft., find the area of the surface of the sphere.

Ex. 7. If a hemispherical dome is to contain 100 sq. yds. of surface, what must its diameter be?

Ex. 8. Find the radius of a sphere in which the area of the surface equals the number of linear units in the circumference of a great circle.

Find the area of a lune in which

Ex. 9. The angle of the lune is 36°, and the radius of the sphere is 14 in.

Ex. 10. The angle of the lune is 18° 20′, and the diameter of the sphere is 20 in.

Ex. 11. The angle of the lune is 24°, and the surface of the sphere is 4 sq. ft.

Find the area of a spherical triangle in which

Ex. 12. The angles are 80°, 90°, 120°, and the diameter of the sphere is 14 ft.

Ex. 13. The angles are 74° 24′, 83° 16′, 92° 20′, and the radius of the sphere is 10.

Ex. 14. The angles are 85°, 95°, 135°, and the surface of the sphere is 10 sq. ft.

Ex. 15. If the sides of a spherical triangle are 100°, 110°, 120° and the radius of the sphere is 16, find the area of the polar triangle.

Ex. 16. If the angles of a spherical triangle are 90°, 100°, 120° and its area is 3900, find the radius of the sphere.

Ex. 17. If the area of an equilateral spherical triangle is one-third the surface of the sphere, find an angle of the triangle.

Ex. 18. In a trihedral angle the plane angles of the dihedral angles are 80°, 90°, 100°; find the number of solid degrees in the trihedral angle.

Ex. 19. Find the area of a spherical hexagon each of whose angles is 150°, on a sphere whose radius is 20 in.

Ex. 20. If each dihedral angle of a given pentahedral angle is 120°, how many solid degrees does the pentahedral angle contain?

Ex. 21. In a sphere whose radius is 14 in., find the area of a zone 3 in. high.

Ex. 22. What is the area of the north temperate zone, if the earth is taken to be a sphere with a radius of 4,000 miles, and the distance between the plane of the arctic circle and that of the tropic of Cancer is 1,800 miles?

Ex. 23. If Cairo, Egypt, is in latitude 30°, show that its parallel of latitude bisects the surface of the northern hemisphere.

Ex. 24. How high must a person be above the earth's surface to see one-third of the surface?

Ex. 25. How much of the earth's surface will a man see who is 2,000 miles above the surface, if the diameter is taken as 8,000 miles?

Ex. 26. If the area of a zone equals the area of a great circle, find the altitude of the zone in terms of the radius of the sphere.

Ex. 27. If sounds from the Krakatoa explosion were heard at a distance of 3,000 miles (taken as a chord) on the surface of the earth, over what fraction of the earth's surface were they heard?

Ex. 28. The radii of two spheres are 5 and 12 in. and their centers are 13 in. apart. Find the area of the circle of intersection and also of that part of the surface of each sphere not included by the other sphere.

EXERCISES. GROUP 78

VOLUMES OF POLYHEDRONS

Find the volume of a prism

Ex. 1. Whose base is an equilateral triangle with side 5 in., and whose altitude is 16 in.

Ex. 2. Whose base is a triangle with sides 12, 13, 15, and whose altitude is 20.

Ex. 3. Whose base is an isosceles right triangle with a leg equal to 2 yds., and whose altitude is 25 ft.

Ex. 4. Whose base is a regular hexagon with a side of 8 ft., and whose altitude is 10 yds.

Ex. 5. Whose base is a rhombus one of whose sides is 25, and one of whose diagonals is 14, and whose altitude is 11.

Ex. 6. Whose base contains 84 sq. yds., and whose lateral faces are three rectangles with areas of 100, 170, 210 sq. yds., respectively.

Ex. 7. How many bushels of wheat are held by a bin 30 x 10 x 6 ft., if a bushel is taken as $1\frac{1}{4}$ cu. ft.?

Ex. 8. How many cart-loads of earth are in a cellar 30 x 20 x 6 ft., if a cart-load is a cubic yard?

Ex. 9. If a cubical block of marble costs $2, what is the cost of a cube whose edge is a diagonal of the first block?

Ex. 10. Find the edge of a cube whose volume equals the sum of the volumes of two cubes whose edges are 3 and 5 ft.

Ex. 11. Find the edge of a cube whose volume equals the area of its surface.

Ex. 12. If the top of a cistern is a rectangle 12 x 8 ft., how deep must the cistern be to hold 10,000 gallons?

Ex. 13. Find the inner edge of a peck measure which is in the shape of a cube.

Ex. 14. A peck measure is to be a rectangular parallelopiped with square base and altitude equal to twice the edge of the base. Find its dimensions.

Ex. 15. Find the volume of a cube whose diagonal is a.

Find the volume of a pyramid

Ex. 16. Whose base is an equilateral triangle with side 8 in., and whose altitude is 12 in.

Ex. 17. Whose base is a right triangle with hypotenuse 29 and one leg 21, and whose altitude is 20.

Ex. 18. Whose base is a square with side 6, and each of whose lateral edges is 5.

Ex. 19. Whose base is a square with side 10, and each of whose lateral faces makes an angle of 45° with the base.

Ex. 20. If the pyramid of Memphis has an altitude of 146 yds. and a square base of side 232 yds., how many cubic yards of stone does it contain? What is this worth at $1 a cu. yd.?

Ex. 21. A church spire 150 ft. high is hexagonal in shape and each side of the base is 10 ft. The spire has a hollow hexagonal interior, each side of whose base is 6 ft., and whose altitude is 45 ft. How many cubic yards of stone does the spire contain?

Ex. 22. If a pyramid contains 4 cu. yds. and its base is a square with one side 2 ft., find the altitude.

Ex. 23. A heap of candy in the shape of a frustum of a regular square pyramid has the edges of its bases 25 and 9 in. and its altitude 12 in. Find the number of pounds in the heap if a pound is a rectangular parallelopiped 4 x 3 x 2 in. in size.

Ex. 24. Find the volume of a frustum of a regular triangular pyramid, the edges of the bases being 2 and 8, and the slant height 12.

Ex. 25. The edges of the bases of the frustum of a regular square pyramid are 24 and 6, and each lateral edge is 15; find the volume.

Ex. 26. If a stick of timber is in the shape of a frustum of a regular square pyramid with the edges of its ends 9 and 15 in., and with a length of 14 ft., find the number of feet of lumber in the stick.

What is the difference between this volume and that of a stick of the same length having the shape of a prism with a base equal to the area of a midsection of the first stick?

Ex. 27. Find the volume of a prismoid whose bases are rectangles 5 x 2 ft. and 7 x 4 ft., and whose altitude is 12 ft.

Ex. 28. How many cart-loads of earth are there in a railroad cut 12 ft. deep, whose base is a rectangle 100 x 8 ft., and whose top is a rectangle 30 x 50 ft.?

Ex. 29. Find the volume of a prismatoid whose base is an equilateral triangle with side 12 ft., and whose top is a line 12 ft. long parallel to one side of the base, and whose altitude is 15 ft.

Ex. 30. If the base of a prismatoid is a rectangle with dimensions a and b, the top is a line c parallel to the side b of the base, and the altitude is h, find the volume.

EXERCISES. GROUP 79

VOLUMES OF CONES AND CYLINDERS

Ex. 1. How many barrels of oil are contained in a cylindrical tank 20 ft. long and 6 ft. in diameter, if a barrel contains 4 cu. ft.?

Ex. 2. How many cu. yds. of earth must be removed in making a tunnel 450 ft. long, if a cross-section of the tunnel is a semicircle of 15 ft. radius?

Ex. 3. A cylindrical glass 3 in. in diameter holds half a pint. Find its height in inches.

Ex. 4. If a cubic foot of brass be drawn out into wire $\frac{1}{30}$ inch in diameter, how long will the wire be?

Ex. 5. A gallon measure is a cylinder whose altitude equals the diameter of the base. Find the altitude.

Ex. 6. Show that the volumes of two cylinders, having the altitude of each equal to the radius of the other, are to each other as $R : R'$.

Ex. 7. In a cylinder, find R in terms of V and H; also V in terms of S and R.

Ex. 8. A conical heap of potatoes is 44 ft. in circumference and 6 ft. high. How many bushels does it contain, if a bushel is $1\frac{1}{4}$ cu. ft.?

Ex. 9. What fraction of a pint will a conical wine-glass hold, if its altitude is 3 in. and the diameter of the top is 2 in.?

Ex. 10. Find the ratio of the volumes of the two cones inscribed in, and circumscribed about, a regular tetrahedron.

Ex. 11. If an equilateral cone contains 1 quart, find its dimensions in inches.

Ex. 12. In a cone of revolution find V in terms of R and L; also find V in terms of R and S.

Ex. 13. Find the volume of a frustum of a cone of revolution, whose radii are 14 and 7 ft., and whose altitude is 2 yds.

Ex. 14. What is the cost, at 50 cts. a cu. ft., of a piece of marble in the shape of a frustum of a cone of revolution, whose radii are 6 and 9 ft., and whose slant height is 5 ft.?

Ex. 15. In a frustum of a cone of revolution, the volume is 88 cu. ft., the altitude is 9 ft., and $R=2r$. Find r.

EXERCISES. GROUP 80

SPHERICAL VOLUMES

Ex. 1. Find the volume of a sphere whose radius is 1 ft. 9 in.

Ex. 2. If the earth is a sphere 7,920 miles in diameter, find its volume in cubic miles.

Ex. 3. Find the diameter of a sphere whose volume is 1 cu. ft.

Ex. 4. What is the volume of a sphere whose surface is 616 sq. in.?

Ex. 5. Find the radius of a sphere equivalent to the sum of two spheres, whose radii are 2 and 4 in.

Ex. 6. Find the radius of a sphere whose volume equals the area of its surface.

Ex. 7. Find the volume of a sphere circumscribed about a cube whose edge is 6.

Ex. 8. Find the volume of a spherical shell whose inner and outer diameters are 14 and 21 in.

Ex. 9. Find the volume of a spherical shell whose inner and outer surfaces are 20π and 12π.

Find the volume of

Ex. 10. A spherical wedge whose angle is 24°, the radius of the sphere being 10 in.

Ex. 11. A spherical sector whose base is a zone 2 in. high, the radius of the sphere being 10 in.

Ex. 12. A spherical segment of two bases whose radii are 4 and 7 and altitude 5 in.

Ex. 13. A wash-basin in the shape of a segment of a sphere is 6 in. deep and 24 in. in diameter. How many quarts of water will the basin hold?

Ex. 14. A plane parallel to the base of a hemisphere and bisecting the altitude divides its volume in what ratio?

Ex. 15. A spherical segment 4 in. high contains 200 cu. in.; find the radius of the sphere.

Ex. 16. If a heavy sphere whose diameter is 4 in. be placed in a conical wine-glass full of water, whose diameter is 5 in. and altitude 6 in., find how much water will run over.

EXERCISES. GROUP 81

EQUIVALENT SOLIDS

Ex. 1. If a cubical block of putty, each edge of which is 8 inches, be molded into a cylinder of revolution whose radius is 3 inches, find the altitude of the cylinder.

Ex. 2. Find the radius of a sphere equivalent to a cube whose edge is 10 in.

Ex. 3. Find the radius of a sphere equivalent to a cone of revolution, whose radius is 3 in. and altitude 6 in.

Ex. 4. Find the edge of a cube equivalent to a frustum of a cone of revolution, whose radii are 4 and 9 ft. and altitude 2 yds.

Ex. 5. Find the altitude of a rectangular parallelopiped, whose base is 3 x 5 in. and whose volume is equivalent to a sphere of radius 7 in.

Ex. 6. Find the base of a square rectangular parallelopiped, whose altitude is 8 in. and whose volume equals the volume of a cone of revolution with a radius of 6 and an altitude of 12 in.

Ex. 7. Find the radius of a cone of revolution, whose altitude is 15 and whose volume is equal to that of a cylinder of revolution with radius 6 and altitude 20.

Ex. 8. Find the altitude of a cone of revolution, whose radius is 15 and whose volume equals the volume of a cone of revolution with radius 9 and altitude 24.

Ex. 9. On a sphere whose diameter is 14 the altitude of a zone of one base is 2. Find the altitude of a cylinder of revolution, whose base equals the base of the zone and whose lateral surface equals the surface of the zone.

EXERCISES. GROUP 82

SIMILAR SOLIDS

Ex. 1. If on two similar solids L, L' and l, l' are pairs of homologous lines; A, A' and a, a' pairs of homologous areas, V, V' and v, v' pairs of homologous volumes,

$$\text{show that } \begin{cases} L : L' = l : l' = \sqrt{A} : \sqrt{A'} = \sqrt[3]{v} : \sqrt[3]{v'}. \\ A : A' = L^2 : L'^2 = a : a' = V^{\frac{2}{3}} : V'^{\frac{2}{3}}. \\ V : V' = L^3 : L'^3 = A^{\frac{3}{2}} : A'^{\frac{3}{2}} = v : v'. \end{cases}$$

Ex. 2. If the edge of a cube is 10 in., find the edge of a cube having 5 times the surface.

Ex. 3. If the radius of a sphere is 10 in., find the radius of a sphere having 5 times the surface.

Ex. 4. If the altitude of a cone of revolution is 10 in., find the altitude of a similar cone of revolution having 5 times the surface.

Ex. 5. In the last three exercises, find the required dimension if the volume is to be 5 times the volume of the original solid.

Ex. 6. The linear dimensions of one trunk are twice as great as those of another trunk. How much greater is the volume ? The surface?

Ex. 7. How far from the vertex is the cross-section which bisects the volume of a cone of revolution? Which bisects the lateral surface?

Ex. 8. If the altitude of a pyramid is bisected by a plane parallel to the base, how does the area of the cross-section compare with the area of the base? How does the volume cut off compare with the volume of the entire pyramid?

Ex. 9. Planes parallel to the base of a cone divide the altitude into three equal parts; compare the lateral surfaces cut off. Also the volumes.

Ex. 10. A sphere 10 in. in diameter is divided into three equivalent parts by concentric spherical surfaces. Find the diameters of these surfaces.

Ex. 11. If the strength of a muscle is as the area of its cross-section, and Goliath of Gath was three times as large in each linear dimension as Tom Thumb, how much greater was his strength? His weight? How, then, does the activity of the one man compare with that of the other?

Ex. 12. If the rate at which heat radiates from a body is in proportion to the amount of surface, and the planet Jupiter has a diameter 11 times that of the earth, how many times longer will Jupiter be in cooling off?

[Sug. How many times greater is the volume, and therefore the original amount of heat in Jupiter? How many times greater is its surface? What will be the combined effect of these factors?]

GROUP 83

MISCELLANEOUS NUMERICAL EXERCISES IN SOLID GEOMETRY

Find S, T and V of

Ex. 1. A right triangular prism whose altitude is 1 ft., and the sides of whose base are 26, 28, 30 in.

Ex. 2. A cone of revolution the radius of whose base is 1 ft. 2 in., and whose altitude is 35 in.

Ex. 3. A frustum of a square pyramid the areas of whose bases are 1 sq. ft. and 36 sq. in., and whose altitude is 9 in.

Ex. 4. A pyramid whose slant height is 10 in., and whose base is an equilateral triangle whose side is 8 in.

Ex. 5. A cube whose diagonal is 1 yd.

Ex. 6. A frustum of a cone of revolution whose radii are 6 and 11 in. and slant height 13 in.

Ex. 7. A rectangular parallelopiped whose diagonal is $2\sqrt{29}$, and whose dimensions are in the ratio 2 : 3 : 4.

Ex. 8. Find the volume of a sphere inscribed in a cube whose edge is 6; also find the area of a triangle on that sphere whose angles are 80°, 90°, 150°.

Ex. 9. Find the volume of the spherical pyramid whose base is the above triangle.

Ex. 10. Find the angle of a lune on the same sphere, equivalent to that triangle.

Ex. 11. On a cube whose edge is 4, planes through the midpoints of the edges cut off the corners. Find the volume of the solid remaining.

Ex. 12. How is V changed if H of a cone of revolution is doubled and R remains unchanged? If R is doubled and H remains unchanged? If both H and R are doubled?

Ex. 13. In an equilateral cone, find S and V in terms of R.

Ex. 14. A piece of lead 20 x 8 x 2 in. will make how many spherical bullets, each $\frac{3}{4}$ in. in diameter?

Ex. 15. How many bricks are necessary to make a chimney in the shape of a frustum of a cone, whose altitude is 90 ft., whose outer diameters are 3 and 8 ft., and whose inner diameters are 2 and 4 ft., counting 12 bricks to the cubic ft.?

Ex. 16. If the area of a zone is 300 and its altitude 6, find the radius of the sphere.

Ex. 17. If the section of a cylinder of revolution through its axis is a square, find S, T, V in terms of R.

Ex. 18. If every edge of a square pyramid is b, find b in terms of T.

Ex. 19. A regular square pyramid has a for its altitude and also for each side of the base. Find the area of a section made by a plane parallel to the base and bisecting the altitude. Find also the volumes of the two parts into which the pyramid is divided.

Ex. 20. If the earth is a sphere of 8,000 miles diameter and its atmosphere extends 50 miles from the earth, find the volume of the atmosphere.

Ex. 21. On a sphere, find the ratio of the area of an equilateral spherical triangle, each of whose angles is 95°, to the area of a lune whose angle is 80°.

Ex. 22. A square right prism has an altitude $6a$ and an edge of the base $2a$. Find the volume of the largest cylinder, sphere, pyramid and cone which can be cut from it.

Ex. 23. Obtain a formula for the area of that part of a sphere illuminated by a point of light at a distance a from the sphere whose radius is R.

Ex. 24. On a sphere whose radius is 6 in., find an angle of an equilateral triangle whose area is 12 sq. in.

Ex. 25. Find the volume of a prismatoid, whose altitude is 24 and whose bases are equilateral triangles, each side 10, so placed that the mid-section of the prismatoid is a regular hexagon.

Ex. 26. On a sphere whose radius is 16, the bases of a zone are equal and are together equal to the area of the zone. Find the altitude of the zone.

Ex. 27. Find the volume of a square pyramid, the edge of whose base is 10 and each of whose lateral edges is inclined 60° to the base.

Ex. 28. An irregular piece of ore, if placed in a cylinder partly filled with water, causes the water to rise 6 in. If the radius of the cylinder is 8 in., what is the volume of the ore?

Ex. 29. Find the volume of a truncated right triangular prism, if the edges of the base are 8, 9, 11, and the lateral edges are 12, 13, 14.

Ex. 30. In a sphere whose radius is 5, a section is taken at the distance 3 from the center. On this section as a base a cone is formed whose lateral elements are tangent to the sphere. Find the lateral surface and volume of the cone.

Ex. 31. The volume of a sphere is 1,437⅓ cu. in. Find the surface.

Ex. 32. A square whose side is 6 is revolved about a diagonal as an axis; find the surface and volume generated.

Ex. 33. Find the edge of a cubical cistern that will hold 10 tons of water, if 1 cu. ft. of water weighs 62.28 lbs.

Ex. 34. A water trough has equilateral triangles, each side 3 ft., for ends, and is 18 ft. long. How many buckets of water will it hold, if a bucket is a cylinder 1 ft. in diameter and 1⅓ ft. high?

Ex. 35. The lateral area of a cylinder of revolution is 440 sq. in., and the volume is 1,540 cu. in. Find the radius and altitude.

Ex. 36. The angles of a spherical quadrilateral are 80°, 100°, 120°, 120°. Find the angle of an equivalent equilateral triangle.

Ex. 37. A cone and a cylinder have equal lateral surfaces, and their axis sections are equilateral. Find the ratio of their volumes.

Ex. 38. A water-pipe ¾ in. in diameter rises 13 ft. from the ground. How many quarts of water must be drawn from it before the water from under the ground comes out? If a quart runs out in 5 seconds, how long must the water run?

Ex. 39. A cube immersed in a cylinder partly filled with water causes the water to rise 4 in. If the radius of the cylinder is 6 in., what is an edge of the cube?

Ex. 40. An auger hole whose diameter is 3 in. is bored through the center of a sphere whose diameter is 8 inches. Find the volume remaining.

Ex. 41. Show that the volumes of a cone, hemisphere, and cylinder of the same base and altitude are as 1 : 2 : 3.

Ex. 42. The volumes of two similar cylinders of revolution are as 8 : 125; find the ratio of their radii. If the radius of the smaller is 10 in., what is the radius of the larger?

Ex. 43. An iron shell is 2 in. thick and the diameter of its outer surface is 28 in. Find its volume.

Ex. 44. The legs of an isosceles spherical triangle each make an angle of 75° with the base. The legs produced form a lune whose area is four times the area of the triangle. Find the angle of the lune.

GROUP 84

EXERCISES INVOLVING THE METRIC SYSTEM

Find S, T, V of

Ex. 1. A right prism the edges of whose base are 6 m., 70 dm., 900 cm., and whose altitude is 90 dm.

Ex. 2. A regular square pyramid an edge of whose base is 30 dm., and whose altitude is 1.7 m.

Ex. 3. A sphere whose radius is 0.02 m.

Ex. 4. A frustum of a cone of revolution whose radii are 10 dm and 6 dm., and whose slant height is 50 cm.

Ex. 5. A cube whose diagonal is 12 cm.

Ex. 6. A cylinder of revolution whose radius equals 2 dm., and whose altitude equals the diameter of the base.

Ex. 7. Find the area of a spherical triangle on a sphere whose radius is 0.02 m., if its angles are 110°, 120°, 130°.

Ex. 8. Find the number of square meters in the surface of a sphere, a great circle of which is 50 dm. long.

Ex. 9. How many liters will a cylindrical vessel hold that is 10 dm. in diameter and 0.25 m. high? How many liquid quarts?

Ex. 10. A liter measure is a cylinder whose diameter is half the altitude. Find its dimensions in centimeters.

Ex. 11. Find the surface of a sphere whose volume is 1 cu. m.

APPENDIX

I. MODERN GEOMETRIC CONCEPTS

840. Modern Geometry. In recent times many new geometric ideas have been invented, and some of them developed into important new branches of geometry. Thus, the idea of symmetry (see Art. 484, etc.) is a modern geometric concept. A few other of these modern concepts and methods will be briefly mentioned, but their thorough consideration lies beyond the scope of this book.

841. Projective Geometry. The idea of projections (see Art. 345) has been developed in comparatively recent times into an important branch of mathematics with many practical applications, as in engineering, architecture, construction of maps, etc.

842. Principle of Continuity. By this principle two or more theorems are made special cases of a single more general theorem. An important aid in obtaining continuity among geometric principles is the application of the concept of negative quantity to geometric magnitudes.

Thus, a negative line is a line opposite in direction to a given line taken as positive.

For example, if OA is $+$, OB is $-$.

Similarly, a negative angle is an angle formed by rotating a line in a plane in a direction opposite from a direction of rotation taken as positive. Thus, if the line OA rotating from the position OA forms the positive angle AOB, the same line rotating in the opposite direction forms the negative angle AOB'. Similarly, positive and negative arcs are formed.

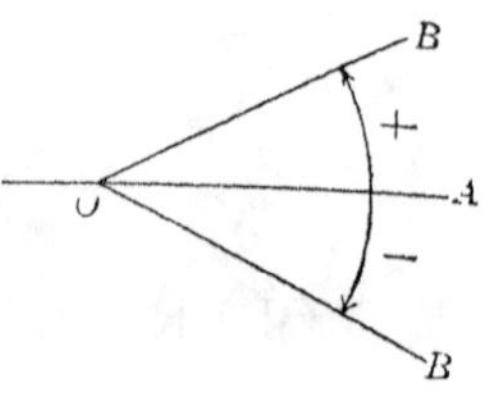

In like manner, if P and P' are on opposite sides of the line AB and the area PAB is taken as positive, the area $P'AB$ will be negative.

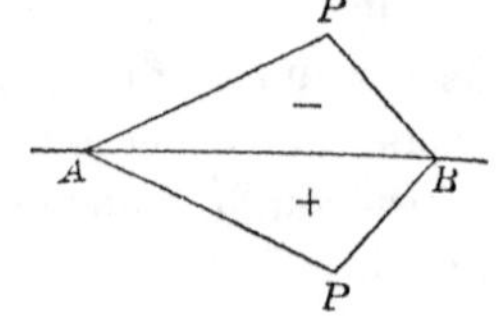

As an illustration of the law of continuity, we may take the theorem that the sum of the triangles formed by drawing lines from a point to the vertices of a polygon equals the area of the polygon.

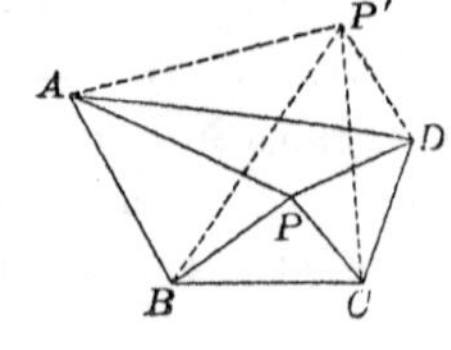

Applying this to the quadrilateral $ABCD$, if the point P falls within the quadrilateral, $\triangle PAB + \triangle PBC + \triangle PCD + \triangle PAD = ABCD$ (Ax. 6).

Also, if the point falls without the quadrilateral at P', $\triangle P'AB + \triangle P'BC + \triangle P'CD + \triangle P'AD = ABCD$, since $\triangle P'AD$ is a negative area, and hence is to be subtracted from the sum of the other three triangles.

843. The Principle of Reciprocity, or Duality, is a principle of relation between two theorems by which each theorem is convertible into the other by causing the words for the same two geometric objects in each theorem to exchange places.

Thus, of theorems VI and VII, Book I, either may be converted into the other by replacing the word "sides" by "angles," and "angles" by "sides." Hence these are termed reciprocal theorems.

The following are other instances of reciprocal geometric properties:

1. *Two points determine a straight line.*	1. *Two lines determine a point.*
2. *Three points not in the same straight line determine a plane.*	2. *Three planes not through the same straight line determine a point.*
3. *A straight line and a point determine a plane.*	3. *A straight line and a plane determine a point.*

The reciprocal of a theorem is not necessarily true.

Thus, two parallel straight lines determine a plane, but two parallel planes do not determine a line.

However, by the use of the principle of reciprocity, geometrical properties, not otherwise obvious, are frequently suggested.

844. Principle of Homology. Just as the law of reciprocity indicates relations between one set of geometric concepts (as lines) and another set of geometric concepts (as points), so the law of homology indicates relations between a set of geometric concepts and a set of concepts outside of geometry: as a set of algebraic concepts, for instance.

Thus, if a and b are numbers, by algebra $(a+b)(a-b) = a^2-b^2$.

Also, if a and b are segments of a line, the rectangle $(a+b)\times(a-b)$ is equivalent to the difference between the squares a^2 and b^2.

By means of this principle, truths which would be overlooked or difficult to prove in one department of thought are made obvious by observing the corresponding truth in another department of thought.

Thus, if a and b are line segments, the theorem $(a+b)^2 + (a-b)^2 = 2(a^2+b^2)$ is not immediately obvious in geometry, but becomes so by observing the like relation between the algebraic numbers a and b.

845. Non-Euclidean Geometry. Hyperspace. By varying the properties of space, as these are ordinarily stated, different kinds of space may be conceived of, each having its own geometric laws and properties. Thus, space, as we ordinarily conceive it, has three dimensions, but it is possible to conceive of space as having four or more dimensions. To mention a single property of four dimensional space, in such a space it would be possible, by simple pressure, to turn a sphere, as an orange, inside out without breaking its surface.

As an aid toward conceiving how this is possible, consider a plane in which one circle lies inside another. No matter how these circles are moved about in the plane, it is impossible to shift the inner circle so as to place it outside the other without breaking the circumference of the outer circle. But, if we are allowed to use the *third* dimension of space, it is a simple matter to lift the inner circle up out of the plane and set it down outside the larger circle.

Similarly if, in space of three dimensions, we have one spherical shell inside a larger shell, it is impossible to place the smaller shell outside the larger without breaking the larger. But if the use of a *fourth* dimension be allowed,—that is, the use of another dimension of freedom of motion,—it is possible to place the inner shell outside the larger without breaking the latter.

846. Curved Spaces. By varying the geometric axioms of space (see Art. 47), different kinds of space may be conceived of. Thus, we may conceive of space such that through a given point *one* line may be drawn parallel to a given line (that is ordinary, or Euclidean space); or such that through a given point *no line* can be drawn parallel to a given line (spherical space); or such that through a given point *more than one* line can be drawn parallel to a given line (pseudo-spherical space).

These different kinds of space differ in many of their properties. For example, in the first of them the sum of the angles of a triangle equals two right angles; in the second, it is greater; in the third, it is less.

These different kinds of space, however, have many properties in common. Thus, in all of them every point in the perpendicular bisector of a line is equidistant from the extremities of the line.

EXERCISES. GROUP 85

Ex. 1. Show by the use of zero and negative arcs that the principles of Arts. 257, 263, 258, 264, 265, are particular cases of the general theorem that the angle included between two lines which cut or touch a circle is measured by one-half the sum of the intercepted arcs.

Ex. 2. Show that the principles of Arts. 354 and 358 are particular cases of the theorem that, if two lines are drawn from or through a point to meet a circumference, the product of the segments of one line equals the product of the segments of the other line.

Ex. 3. Show by the use of negative angles that theorem XXXVIII, Book I, is true for a quadrilateral of the form *ABCD*. [*BCD* is a negative angle; the angle at the vertex *D* is the reflex angle *ADC*.]

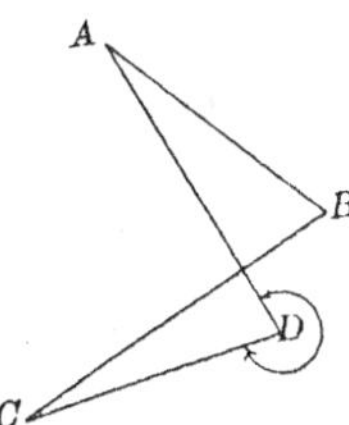

Ex. 4. What is the reciprocal of the statement that two intersecting straight lines determine a plane?

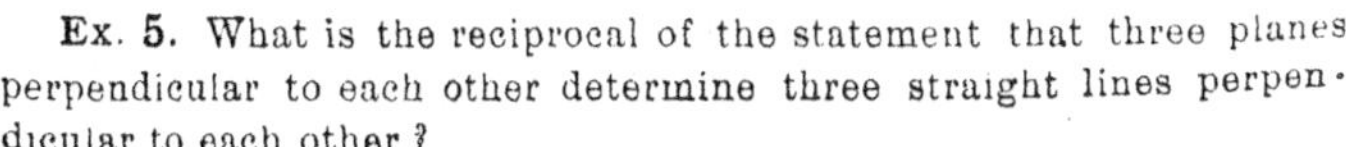

Ex. 5. What is the reciprocal of the statement that three planes perpendicular to each other determine three straight lines perpendicular to each other?

II. HISTORY OF GEOMETRY

847. Origin of Geometry as a Science. The beginnings of geometry as a science are found in Egypt, dating back at least three thousand years before Christ. Herodotus says that geometry, as known in Egypt, grew out of the need of remeasuring pieces of land parts of which had been washed away by the Nile floods, in order to make an equitable readjustment of the taxes on the same.

The substance of the Egyptian geometry is found in an old papyrus roll, now in the British museum. This roll is, in effect, a mathematical treatise written by a scribe named Ahmes at least 1700 B.C., and is, the writer states, a copy of a more ancient work, dating, say, 3000 B. C.

848. Epochs in the Development of Geometry. From Egypt a knowledge of geometry was transferred to Greece, whence it spread to other countries. Hence we have the following principal epochs in the development of geometry:

1. **Egyptian** : 3000 B. C.—1500 B. C.
2. **Greek** : 600 B. C.—100 B. C.
3. **Hindoo** : 500 A. D.—1100 A. D.
4. **Arab** : 800 A. D.—1200 A. D.
5. **European** : 1200 A. D.

In the year 1120 A. D., Athelard, an English monk, visited Cordova, in Spain, in the disguise of a Mohammedan student, and procured a copy of Euclid in the Arabic language. This book he brought back to central Europe, where it was translated into Latin and became the basis of all geometric study in Europe till the year 1533, when, owing to the capture of Constantinople by the Turks, copies of the works of the Greek mathematicians in the original Greek were scattered through Europe.

HISTORY OF GEOMETRICAL METHODS

849. Rhetorical Methods. By rhetorical methods in the presentation of geometric truths, is meant the use of definitions, axioms, theorems, geometric figures, the representation of geometric magnitudes by the use of letters, the arrangement of material in Books, etc. The **Egyptians** had none of these, their geometric knowledge being recorded only in the shape of the solutions of certain numerical examples, from which the rules used must be inferred.

Thales (Greece 600 B.C.) first made an enunciation of an abstract property of a geometric figure. He had a rude idea of the geometric theorem.

Pythagoras (Italy 525, B. C.) introduced formal definitions into geometry, though some of those used by him were not very accurate. For instance, his definition of a point is "unity having position." Pythagoras also arranged the leading propositions known to him in something like logical order.

Hippocrates (Athens, 420 B. C.) was the first systematically to denote a point by a capital letter, and a segment of a line by two capital letters, as the line AB, as is done at present. He also wrote the first text-book on geometry.

Plato (Athens, 380 B. C.) made definitions, axioms and postulates the beginning and basis of geometry.

To **Euclid** (Alexandria, 280 B. C.) is due the division of geometry into Books, the formal enunciation of theorems, the particular enunciation, the formal construction, proof, and conclusion, in presenting a proposition. He also introduced the use of the corollary and scholium.

Using these methods of presenting geometric truths, Euclid wrote a text-book of geometry in thirteen books, which was the standard text-book on this subject for nearly two thousand years.

The use of the symbols △, ▱, ‖, etc., in geometric proofs originated in the United States in recent years.

850. Logical Methods. The **Egyptians** used no formal methods of proof. They probably obtained their few crude geometric processes as the result of experiment.

The **Hindoos** also used no formal proof. One of their writers on geometry merely states a theorem, draws a figure, and says "Behold!"

The use of logical methods of geometric proof is due to the **Greeks.** The early Greek geometricians used *experimental methods* at times, in order to obtain geometric truths. For instance, they determined that the angles at the base of an isosceles triangle are equal, by folding half of the triangle over on the altitude as an axis and observing that the angles mentioned coincided as a fact, but without showing that they *must* coincide.

Pythagoras (525 B. C.) was the first to establish geometric truths by systematic *deduction*, but his methods were sometimes faulty. For instance, he believed that the converse of a proposition is necessarily true.

Hippocrates (420 B. C.) used correct and *rigorous deduction* in geometric proofs. He also introduced specific varieties of such deduction, such as the method of *reducing* one proposition to another (Art. 296), and the *reductio ad absurdum*.

The methods of deduction used by the Greeks, however, were defective in their lack of generality. For instance, it was often thought necessary to have a separate proof of a theorem for each different kind of figure to which the theorem applied. Thus, the theorem that the sum of the angles of a triangle equals two right angles was proved,

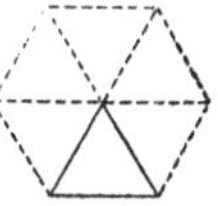

(1) for the equilateral triangle by use of the regular hexagon;

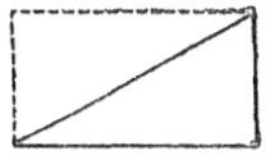

(2) for the right triangle by the use of a rectangle;

(3) for a scalene triangle by dividing the scalene triangle into two right triangles.

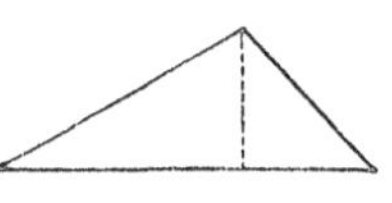

The Greeks appeared to fear that a general proof might be vitiated if it were applied to a figure in any way special or peculiar. In other words, they had no conception of the principle of continuity (Art. 842).

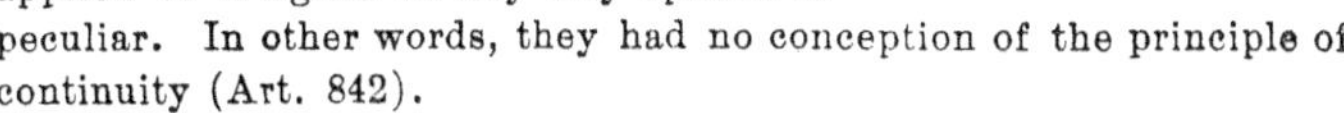

Plato (380 B. C.) introduced the method of proof by *analysis*, that is, by taking a proposition as true and working from it back to known truths (see Art. 196).

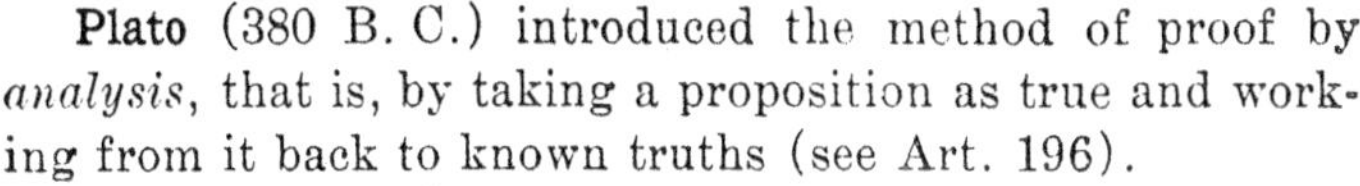

To **Eudoxus** (360 B. C.) is virtually due proof by the method of *limits*; though his method, known as the method of exhaustions, is crude and cumbersome.

Apollonius (Alexandria, 225 B. C.) used *projections*, *transversals*, etc., which, in modern times, have developed into the subject of projective geometry.

851. Mechanical Methods. The **Greeks,** in demonstrating a geometrical theorem, usually drew the figure employed in a bed of sand. This method had certain advantages, but was not adapted to the use of a large audience.

At the time when geometry was being developed in Greece, the interest in the subject was very general. There was scarcely a town but had its lectures on the subject. The news of the discovery of a new theorem spread from town to town, and the theorem was redemonstrated in the sand of each market place.

The Greek treatises, however, were written on vellum or papyrus by the use of the reed, or calamus, and ink.

In **Roman times,** and in the **middle ages,** geometrical figures were drawn in wax smeared on wooden boards, called tablets. They were drawn by the use of the stylus, a metal stick, pointed at one end for making marks, and broad at the other for erasing marks. These wax tablets were still in use in Shakespeare's time (see Hamlet Act I, Sc. 5, l. 107). The blackboard and crayon are **modern** inventions, their use having developed within the last one hundred years.

The **Greeks** invented many kinds of drawing instruments for tracing various curves. It was due to the influence of Plato (380 B. C.) that, in constructing geometric figures, the use of only the ruler and compasses is permitted.

HISTORY OF GEOMETRIC TRUTHS. PLANE GEOMETRY.

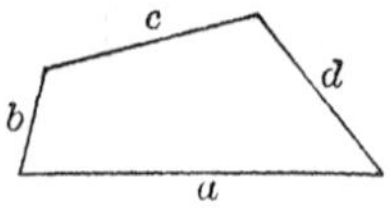

852. Rectilinear Figures. The **Egyptians** measured the area of any four-sided field by multiplying half the sum of one pair of opposite sides by half the sum of the other pair; which was equivalent to using the formula, area $=\frac{a+c}{2}\times\frac{b+d}{2}$.

This, of course, gives a correct result for the rectangle and square, but gives too great a result for other quadrilaterals, as the trapezoid, etc. Hence Joseph, of the Book of Genesis, in buying the fields of the Egyptians for Pharoah in time of famine by the use of this formula, in many cases paid for a larger field than he obtained.

The Egyptians had a special fondness for geometrical constructions, probably growing out of their work as temple builders. A class of workers existed among them called "rope-stretchers," whose business was the marking out of the foundations of buildings. These men knew how to bisect an angle and also to construct a right angle. The latter was probably done by a method essentially the same as forming a right triangle whose sides are three, four and five units of length. Ahmes, in his treatise, has various constructions of the isosceles trapezoid from different data.

Thales (600 B. C.) enunciated the following theorems:

If two straight lines intersect, the opposite or vertical angles are equal;

The angles at the base of an isosceles triangle are equal;

Two triangles are equal if two sides and the included angle of one are equal to two sides and the included angle of the other;

The sum of the angles of a triangle equals two right angles;

Two mutually equiangular triangles are similar.

Thales used the last of these theorems to measure the height of the great pyramid by measuring the length of the shadow cast by the pyramid and also measuring the length of the shadow of a post of known height at the same time and making a proportion between these quantities.

Pythagoras (525 B. C.) and his followers discovered correct formulas for the areas of the principal rectilinear figures, and also discovered the theorems that the areas of similar polygons are as the squares of their homologous sides, and that the square on the hypotenuse of a right triangle equals the sum of the squares on the other two sides. The latter is called the Pythagorean theorem. They also discovered how to construct a square equivalent to a given parallelogram, and to divide a given line in mean and extreme ratio.

To **Eudoxus** (380 B. C.) we owe the general theory of proportion in geometry, and the treatment of incommensurable quantities by the method of Exhaustions. By the use of these he obtained such theorems as that the areas of two circles are to each other as the squares of their radii, or of their diameters.

In the writings of **Hero** (Alexandria, 125 B. C.) we first find the formula for the area of a triangle in terms of its sides, $K=\sqrt{s(s-a)(s-b)(s-c)}$. Hero also was the first to place land-surveying on a scientific basis.

It is a curious fact that Hero at the same time gives an incorrect formula for the area of a triangle, viz., $K=\frac{1}{2}a(b+c)$, this formula being apparently derived from Egyptian sources.

Xenodorus (150 B. C.) investigated isoperemetrical figures.

The **Romans,** though they excelled in engineering, apparently did not appreciate the value of the Greek geometry. Even after they became acquainted with it, they continued to use antiquated and inaccurate formulas for areas, some of them of obscure origin. Thus, they used the Egyptian formula for the area of a quadrilateral, $K=\frac{a+b}{2}\times\frac{c+d}{2}$. They determined the area of an equilateral triangle whose side is a, by different formulas, all incorrect, as $K=\frac{13a^2}{30}$, $K=\frac{1}{2}(a^2+a)$, and $K=\frac{1}{3}a^2$.

853. The **Circle.** **Thales** enunciated the theorem that every diameter bisects a circle, and proved the theorem that an angle inscribed in a semicircle is a right angle.

To **Hippocrates** (420 B. C.) is due the discovery of nearly all the other principal properties of the circle given in this book.

The **Egyptians** regarded the area of the circle as equivalent to $\frac{64}{81}$ of the diameter squared, which would make $\pi = 3.1604$.

The **Jews** and **Babylonians** treated π as equal to 3.

Archimedes, by the use of inscribed and circumscribed regular polygons, showed that the true value of π lies between $3\frac{1}{7}$ and $3\frac{10}{71}$; that is, between 3.14285 and 3.1408.

The **Hindoo** writers assign various values to π, as 3, $3\frac{1}{8}$, $\sqrt{10}$, and **Aryabhatta** (530 A. D.) gives the correct approximation, 3.1416. The Hindoos used the formula $\sqrt{2-\sqrt{4-AB^2}}$ (See Art. 468) in computing the numerical value of π.

Within recent times, the value of π has been computed to 707 decimal places.

The use of the symbol π for the ratio of the circumference of a circle to the diameter was established in mathematics by **Euler** (Germany, 1750).

HISTORY OF GEOMETRIC TRUTHS. SOLID GEOMETRY

854. Polyhedrons. The **Egyptians** computed the volumes of solid figures from the linear dimensions of such figures. Thus, Ahmes computes the contents of an Egyptian barn by methods which are equivalent to the use of the formula $V = a \times b \times \frac{3c}{2}$. As the shape of these barns is not known, it is not possible to say whether this formula is correct or not.

Pythagoras discovered, or knew, all the regular polyhedrons except the dodecahedron. These polyhedrons were supposed to have various magical or mystical properties. Hence the study of them was made very prominent.

Hippasus (470 B. C.) discovered the dodecahedron, but he was drowned by the other Pythagoreans for boasting of the discovery.

Eudoxus (380 B. C.) showed that the volume of a pyramid is equivalent to one-third the product of its base by its altitude.

E. F. August (Germany, 1849) introduced the prismatoid formula into geometry and showed its importance.

855. The Three Round Bodies. **Eudoxus** showed that the volume of a cone is equivalent to one-third the area of its base by its altitude.

Archimedes discovered the formulas for the surface and volume of the sphere.

Menelaus (100 A. D.) treated of the properties of spherical triangles.

Gerard (Holland, 1620) invented polar triangles and found the formulas for the area of a spherical triangle and of a spherical polygon.

856. Non-Euclidean Geometry. The idea that a space might exist having different properties from those which we regard as belonging to the space in which we live, has occurred to different thinkers at different times, but **Lobatchewsky** (Russia, 1793-1856) was the first to make systematic use of this principle. He found that if, instead of taking Geom. Ax. 2 as true, we suppose that through a given point in a plane several straight lines may be drawn parallel to a given line, the result is not a series of absurdities or a general reductio ad absurdum; but, on the contrary, a consistent series of theorems is obtained giving the properties of a space.

III. REVIEW EXERCISES

EXERCISES. GROUP 86

REVIEW EXERCISES IN PLANE GEOMETRY

Ex. 1. If the bisectors of two adjacent angles are perpendicular to each other, the angles are supplementary.

Ex. 2. If a diagonal of a quadrilateral bisects two of its angles, the diagonal bisects the quadrilateral.

Ex. 3. Through a given point draw a secant at a given distance from the center of a given circle.

Ex. 4. The bisector of one angle of a triangle and of an exterior angle at another vertex form an angle which is equal to one-half the third angle of the triangle.

Ex. 5. The side of a square is 18 in. Find the circumference of the inscribed and circumscribed circles.

Ex. 6. The quadrilateral $ADBC$ is inscribed in a circle. The diagonals AB and DC intersect in the point F. Arc $AD = 112°$, arc $AC = 108°$, $\angle AFC = 74°$. Find all the other angles of the figure.

Ex. 7. Find the locus of the center of a circle which touches two given equal circles.

Ex. 8. Find the area of a triangle whose sides are 1 m., 17 dm., 210 cm.

Ex. 9. The line joining the midpoints of two radii is perpendicular to the line bisecting their angle.

Ex. 10. If a quadrilateral be inscribed in a circle and its diagonals drawn, how many pairs of similar triangles are formed?

Ex. 11. Prove that the sum of the exterior angles of a polygon (Art. 172) equals four right angles, by the use of a figure formed by drawing lines from a point within a polygon to the vertices of the polygon.

Ex. 12. In a circle whose radius is 12 cm., find the length of the tangent drawn from a point at a distance 240 mm. from the center.

Ex. 13. If two sides of a regular pentagon be produced, find the angle of their intersection.

Ex. 14. In the parallelogram $ABCD$, points are taken on the diagonals such that $AP=BQ=CR=DS$. Show that $PQRS$ is a parallelogram.

Ex. 15. A chord 6 in. long is at the distance 4 in. from the center of a circle. Find the distance from the center of a chord 8 in. long.

Ex. 16. If B is a point in the circumference of a circle whose center is O, PA a tangent at any point P, meeting OB produced at A, and PD perpendicular to OB, then PB bisects the angle APD.

Ex. 17. Construct a parallelogram, given a side, an angle, and a diagonal.

Ex. 18. Find in inches the sides of an isosceles right triangle whose area is 1 sq. yd.

Ex. 19. Given the line a, construct $\frac{a(\sqrt{2}-1)}{3}$.

Ex. 20. If two lines intersect so that the product of the segments of one line equals the product of the segments of the other, a circumference may be passed through the extremities of the two lines.

Ex. 21. Find the locus of the vertices of all triangles on a given base and having a given area.

Ex. 22. On the figure p. 206, prove that $\overline{BC}^2+\overline{AF}^2=\overline{AB}^2+\overline{FC}^2$.

Ex. 23. The area of a rectangle is 108 and the base is three times the altitude. Find the dimensions.

Ex. 24. If, on the sides AC and BC of the triangle ABC, the squares, AD and BF, are constructed, AF and DB are equal.

Ex. 25. If the angle included between a tangent and a secant is half a right angle, and the tangent equals the radius, the secant passes through the center of the circle.

Ex. 26. The sum of the areas of two circles is 20 sq. yds., and the difference of their areas is 15 sq. yds. Find their radii.

Ex. 27. Construct an isosceles trapezoid, given the bases and a leg.

Ex. 28. Show that, if the alternate sides of a regular pentagon be produced to meet, the points of intersection formed are the vertices of another regular pentagon.

Ex. 29. If a post 2 ft. 6 in. high casts a shadow 1 ft. 9 in. long, how tall is a tree which, at the same time, casts a shadow 66 ft. long?

Ex. 30. If two intersecting chords make equal angles with the diameter through their point of intersection, the chords are equal.

Ex. 31. From a given point draw a secant to a circle so that the external segment is half the secant.

Ex. 32. Find the locus of the center of a circle which touches a given circle at a given point.

Ex. 33. If one diagonal of a quadrilateral bisects the other diagonal, the first diagonal divides the quadrilateral into two equivalent triangles.

Ex. 34. In a given square inscribe a square having a given side.

Ex. 35. A field in the shape of an equilateral triangle contains one acre. How many feet does one side contain?

Ex. 36. If perpendiculars are drawn to a given line from the vertices of a parallelogram, the sum of the perpendiculars from two opposite vertices equals the sum of the other two perpendiculars.

Ex. 37. Any two altitudes of a triangle are reciprocally proportional to the bases on which they stand.

Ex. 38. Construct a triangle equivalent to a given triangle and having two given sides.

Ex. 39. The apothem of a regular hexagon is 20. Find the area of the inscribed and circumscribed circles.

Ex. 40. M is the midpoint of the hypotenuse AB of a right triangle ABC. Prove $8\,\overline{MC}^2=\overline{AB}^2+\overline{BC}^2+\overline{AC}^2$.

Ex. 41. Transform a given triangle into an equivalent right triangle containing a given acute angle.

Ex. 42. The area of a square inscribed in a semicircle is to the area of the square inscribed in the circle as 2 : 5.

Ex. 43. If, on a diameter of the circle O, $OA = OB$ and AC is parallel to BD, the chord CD is perpendicular to AC.

Ex. 44. Find the radius of a circle whose area is equal to one-third the area of the circle whose radius is 7 in.

Ex. 45. State and prove the converse of Prop. XXI, Book III.

Ex. 46. If, in a given trapezoid, one base is three times the other base, the segments of each diagonal are as 1 : 3.

Ex. 47. If two sides of a triangle are 6 and 12 and the angle included by them is 60°, find the length of the other side. Also find this when the included angle is 45°; also, when 120°.

Ex. 48. How many sides has a polygon in which the sum of the interior angles exceeds the sum of the exterior angles by 540°?

Ex. 49. If the four sides of a quadrilateral are the diameter of a circle, the two tangents at its extremities, and a tangent at any other point, the area of the quadrilateral equals one-half the product of the diameter by the side opposite it in the quadrilateral.

Ex. 50. An equilateral triangle and a regular hexagon have the same perimeter; find the ratio of their areas.

Ex. 51. To a circle whose radius is 30 cm. a tangent is drawn from a point 21 dm. from the center. Find the length of the tangent.

Ex. 52. If two opposite sides of a quadrilateral are equal, and the angles which they make with a third side are equal, the quadrilateral is a trapezoid.

Ex. 53. If two circles are tangent externally and two parallel diameters are drawn, one in each circle, a pair of opposite extremities of the two diameters and the point of contact are collinear.

Ex. 54. If, in the triangle ABC, the line AD is perpendicular to BD, the bisector of the angle B, a line through D parallel to BC bisects AC.

Ex. 55. Bisect a given triangle by a line parallel to a given line.

Ex. 56. If two parallelograms have an angle of one equal to the supplement of an angle of the other, their areas are to each other as the products of the sides including the angles.

Ex. 57. The sum of the medians of a triangle is less than the perimeter, and greater than half the perimeter.

Ex. 58. If $PARB$ is a secant to a circle through the center O, PT a tangent, and TR perpendicular to PB, then $PA : PR = PO : PB$.

Ex. 59. Two concentric circles have radii of 17 and 15. Find the length of the chord of the larger which is tangent to the smaller.

Ex. 60. On the figure, p. 244,

(*a*) Find two pairs of similar triangles;

(*b*) Find two dotted lines which are perpendicular to each other;

(*c*) Discover a theorem concerning points, not connected by lines on the figure, which are collinear;

(*d*) Discover a theorem concerning squares on given lines.

Ex. 61. One of the legs, AC, of an isosceles triangle is produced through the vertex, C, to the point F, and F is joined with D, the mid-point of the base AB. DF intersects BC in E. Prove that CF is greater than CE.

Ex. 62. The line of centers of two circles intersects their common external tangent at P. $PABCD$ is a secant intersecting one of the two circles at A and B and the other at C and D. Prove $PA \times PD = PB \times PC$.

Ex. 63. Trisect a given parallelogram by lines drawn through a given vertex.

Ex. 64. Find the area of a triangle the sides of which are the chord of an arc of 120° in a circle whose radius is 1; the chord of an arc of 90° in a circle whose radius is 2; and the chord of an arc of 60° in a circle whose radius is 3.

Ex. 65. Construct a triangle, given the median to one side and a median and altitude on the other side.

Ex. 66. Two circles intersect at P and Q. The chord CQ is tangent to the circle QPB at Q. APB is any chord through P. Prove that AC is parallel to BQ.

Ex. 67. In the triangle ABC, from D, the midpoint of BC, DE and DF are drawn, bisecting the angles ADB and ADC, and meeting AB at E and AC at F. Prove $EF \parallel BC$.

Ex. 68. Produce the side BC of the triangle ABC to a point P, so that $PB \times PC = \overline{PA}^2$.

Ex. 69. In a given circle inscribe a rectangle similar to a given rectangle.

Ex. 70. In a given semicircle inscribe a rectangle similar to a given rectangle.

Ex. 71. The area of an isosceles trapezoid is 140 sq. ft., one base is 26 ft., and the legs make an angle of 45° with the other base. Find the other base.

Ex. 72. Cut off one-third the area of a given triangle by a line perpendicular to one side.

Ex. 73. Find the sides of a triangle whose area is 1 sq. ft., if the sides are in the ratio 2 : 3 : 4.

Ex. 74. Divide a given line into two parts such that the sum of the squares of the two parts shall be a minimum.

Ex. 75. If, from any point in the base of a triangle, lines are drawn parallel to the sides, find the locus of the center of the parallelogram so formed.

Ex. 76. Three sides of a quadrilateral are 845, 613, 810, and the fourth side is perpendicular to the sides 845 and 810. Find the area.

Ex. 77. If BP bisects the angle ABC, and DP bisects the angle CDA, prove that angle $P = \frac{1}{2}$ sum of angles A and C.

Ex. 78. Two circles intersect at P and Q. Through a point A in one circumference lines APC and AQD are drawn, meeting the other in C and D. Prove the tangent at A parallel to CD.

Ex. 79. In a given triangle, draw a line parallel to the base and terminated by the sides so that it shall be a mean proportional between the segments of one side.

Ex. 80. Find the angle inscribed in a semicircle the sum of whose sides is a maximum.

Ex. 81. The bases of a trapezoid are 160 and 120, and the altitude 140. Find the dimensions of two equivalent trapezoids into which the given trapezoid is divided by a line parallel to the base.

Ex. 82. If the diameter of a given circle be divided into any two segments, and a semicircumference be described on the two segments on opposite sides of the diameter, the area of the circle will be divided by the semicircumferences thus drawn into two parts having the same ratio as the segments of the diameter.

Ex. 83. On a given straight line, AB, two segments of circles are drawn, APB and AQB. The angles QAP and QBP are bisected by lines meeting in R. Prove that the angle R is a constant, wherever P and Q may be on their arcs.

Ex. 84. On the side AB of the triangle ABC, as diameter, a circle is described. EF is a diameter parallel to BC. Show that EB bisects the angle ABC.

Ex. 85. Construct a trapezoid, given the bases, one diagonal, and an angle included by the diagonals.

Ex. 86. If, through any point in the common chord of two intersecting circles, two chords be drawn, one in each circle, through the four extremities of the two chords a circumference may be passed.

Ex. 87. From a given point as center describe a circle cutting a given straight line in two points, so that the product of the distances of the points from a given point in the line may equal the square of a given line segment.

Ex. 88. AB is any chord in a given circle, P any point on the circumference, PM is perpendicular to AB and is produced to meet the circle at Q; AN is drawn perpendicular to the tangent at P. Prove the triangles NAM and PAQ similar.

Ex. 89. If two circles $ABCD$ and $EBCF$ intersect in B and C and have common exterior tangents AE and DF cut by BC produced at G and H, then $\overline{GH}^2=\overline{BC}^2+\overline{AE}^2$.

EXERCISES. GROUP 87

REVIEW EXERCISES IN SOLID GEOMETRY

Ex. 1. A segment of a straight line oblique to a plane is greater than its projection on the plane.

Ex. 2. Two tetrahedrons are similar if a dihedral angle of one equals a dihedral angle of the other, and the faces forming these dihedral angles are similar each to each.

Ex. 3. A plane and a straight line, both of which are parallel to the same line, are parallel to each other.

Ex. 4. If the diagonal of one face of a cube is 10 inches, find the volume of the cube.

Ex. 5. Construct a spherical triangle on a given sphere, given the poles of the sides of the triangle.

Ex. 6. Given $AB \perp MN$,
AE and $BF \perp MR$;
prove $EF \perp PM$.

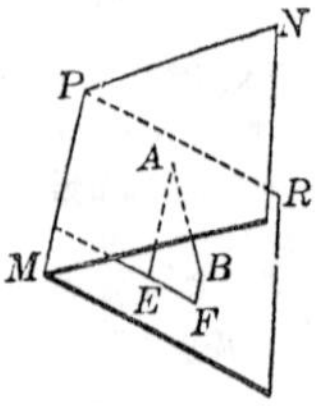

Ex. 7. The diagonals of a rectangular parallelopiped are equal.

Ex. 8. What portion of the surface of a sphere is a triangle each of whose angles is 140°?

Ex. 9. Through a given point pass a plane parallel to two given straight lines.

Ex. 10. Show that the lateral area of a cylinder of revolution is equivalent to a circle whose radius is a mean proportional between the altitude of the cylinder and the diameter of its base.

Ex. 11. The volumes of polyhedrons circumscribed about equal spheres are to each other as the surfaces of the polyhedrons.

Ex. 12. Find S and T of a regular square pyramid an edge of whose base is 14 dm., and whose lateral edge is 250 cm.

Ex. 13. If two lines are parallel and a plane be passed through each line, the intersection of these planes is parallel to the given lines.

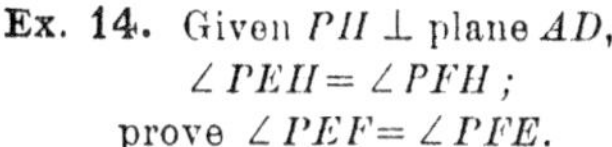

Ex. 14. Given $PH \perp$ plane AD, $\angle PEH = \angle PFH$; prove $\angle PEF = \angle PFE$.

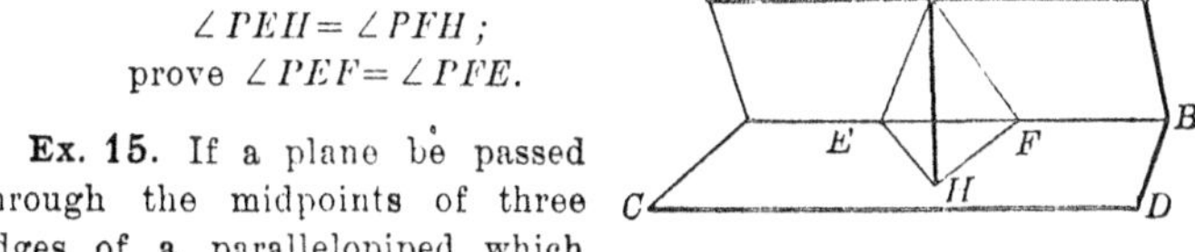

Ex. 15. If a plane be passed through the midpoints of three edges of a parallelopiped which meet at a vertex, the pyramid thus formed is what part of the parallelopiped?

Ex. 16. Find a point in a plane such that the sum of its distances from two given points on the same side of the plane is a minimum.

Ex. 17. Given the points A, B, C, D in a plane and P a point outside the plane, AB perpendicular to the plane PBD, and AC perpendicular to the plane PCD; prove that PD is perpendicular to the plane $ABCD$.

Ex. 18. In a sphere whose radius is 5, find the area of a zone the radii of whose upper and lower bases are 3 and 4.

Ex. 19. Two cylinders of revolution have equal lateral areas. Show that their volumes are as $R : R'$.

Ex. 20. The midpoints of two opposite sides of a quadrilateral in space, and the midpoints of its diagonals, are the vertices of a parallelogram.

Ex. 21. How many feet of two-inch plank are necessary to construct a box twice as wide as deep and twice as long as wide (on the inside), and to contain 216 cu. ft.?

Ex. 22. If two spheres with radii R and r are concentric, find the area of the section of the larger sphere made by a plane tangent to the smaller sphere.

Ex. 23. In the frustum of a regular square pyramid, the edges of the bases are denoted by b_1 and b_2 and the altitude by H; prove that $L = \frac{1}{2}\sqrt{(b_1 - b_2)^2 + 4H^2}$.

Ex. 24. If the opposite sides of a spherical quadrilateral are equal, the opposite angles are equal.

Ex. 25. Obtain the simplest formula for the lateral surface of a truncated triangular right prism, each edge of whose base is a, and whose lateral edges are p, q, and r.

Ex. 26. The area of a zone of one base is a mean proportional between the remaining surface of the sphere and its entire surface. Find the altitude of the zone.

Ex. 27. The lateral edges of two similar frusta are as $1 : a$. How do their areas compare? Their volumes?

Ex. 28. Construct a spherical surface with a given radius, r, which shall be tangent to a given plane, and to a given sphere, and also pass through a given point.

Ex. 29. The volume of a right circular cylinder equals the area of the generating rectangle multiplied by the circumference generated by the point of intersection of its diagonals.

Ex. 30. On a sphere whose radius is $8\frac{1}{3}$ inches, find the area of a zone generated by a pair of compasses whose points are 5 inches apart.

Ex. 31. The perpendicular to a given plane from the point where the altitudes of a regular tetrahedron intersect equals one-fourth the sum of the perpendiculars from the vertices of the tetrahedron to the same plane.

Ex. 32. Two trihedral angles are equal or symmetrical if their corresponding dihedral angles are equal.

Ex. 33. On a sphere whose radius is a, a zone has equal bases and the sum of the bases equals the area of the zone. Find the altitude of the zone.

Ex. 34. A plane which bisects two opposite edges of a tetrahedron bisects the volume of the tetrahedron.

Ex. 35. Find the locus of all points in space which have their distances from two given parallel lines in a given ratio.

Ex. 36. If a, b, c are the sides of a spherical triangle, a', b', c' the sides of its polar triangle, and $a>b>c$, then $a'<b'<c'$.

Ex. 37. A cone of revolution has a lateral area of 4 sq. yd. and an altitude of 2 ft. How much of the altitude must be cut off by a plane parallel to the base, in order to leave a frustum whose lateral area is 2 sq. ft.?

Ex. 38. The total area of an equilateral cone is to the area of the inscribed sphere as 9 : 4.

Ex. 39. Construct a sphere of given radius, r, whose surface shall be tangent to three given spheres.

Ex. 40. The volume of the frustum of an equilateral cone is 300 cu. in. and its altitude is 20 in. Show how to find the radii of the bases.

Ex. 41. On each base of a cylinder of revolution a cone is placed, with its vertex at the center of the opposite base. Find the radius of the circle of intersection of the two conical surfaces.

Ex. 42. The volume of a frustum of a cone of revolution equals the sum of a cylinder and a cone of the same altitude as the frustum, and with radii which are respectively the half sum and the half difference of the radii of the frustum.

Ex. 43. A square whose side is a revolves about a line through one of its vertices and parallel to a diagonal, as axis; find the surface and volume generated.

Ex. 44. If a cone of revolution roll on another fixed cone of revolution so that their vertices coincide, find the kind of surface generated by the axis of the rolling cone.

Ex. 45. An equilateral triangle whose side is a revolves about an altitude as an axis; find the surface and volume generated by the inscribed circle, and also by the circumscribed circle.

Ex. 46. Find the locus of the center of a sphere which is tangent to three given planes.

Ex. 47. If an equilateral triangle whose side is a be rotated about a line through one vertex and parallel to the opposite side, as an axis, find the surface and volume generated.

Ex. 48. What other formulas of solid geometry may be regarded as special cases of the formula for the volume of a prismatoid ?

Ex. 49. Through a given point pass a plane which shall bisect the volume of a given tetrahedron.

Ex. 50. In an equilateral cone and a cone whose opposite elements are perpendicular at the vertex, show that the ratio of the vertical solid angles is as $2-\sqrt{3} : 2-\sqrt{2}$.

PRACTICAL APPLICATIONS OF PLANE GEOMETRY

EXERCISES. GROUP 88

(Book I)

1. Take a piece of paper having a straight edge and fold the paper so as to form a right angle. What geometrical principle or definition have you used?

2. By use of a board with a straight edge, test the accuracy of the outside of a carpenter's square by a method indicated in the diagram. How, then, would you test the accuracy of the inside angle of the square? What geometric principle have you used in each case?

3. In Ex. 2 prove that the error in the outside angle of the carpenter's square, if there be any, equals one-half the angle, x, between the outside lines of the square as shown in the diagram (denote the error by e and show that $e + e = x$).

This principle is important because it is essentially the method used in correcting the axis of a telescope, and hence in correcting instruments of which the telescope is a part, as various surveying and astronomical instruments.

4. By use of a carpenter's square and a given straight edge, lay off a series of parallel lines. What property of parallel lines have you used?

5. Tell how to construct a carpenter's miter box.

6. The diagram shows a drawing instrument called the parallel rulers. The dotted outline shows the rest of the instrument in another position, the part RS remaining fixed. The distance $PQ = RS$, $PR = QS$. Hence show that PQ is parallel to RS.

In like manner show that $P'Q'$ is parallel to RS. Hence show PQ is parallel to $P'Q'$.

If a line be drawn perpendicular to PQ and another perpendicular to $P'Q'$, the perpendiculars thus drawn will be parallel to each other (Art. 122, lines perpendicular to parallel lines are parallel).

The above are cases of linked motion, a kind of mechanism of wide importance. Observe for instance the system of links which connect the driving wheels of a locomotive with the piston in the cylinder, and also the jointed rods connecting the walking beam of a steamboat with the engine. Look up also, in the Century Dictionary, for instance, the words linkage, cell, and parallel motion.

7. The distance between two accessible places separated by an impassable barrier (as between two houses separated by a pond) may be found by the following method when no instrument for measuring angles is at hand.

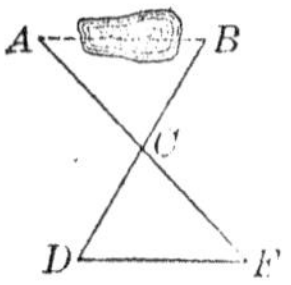

Let A and B be the two places. Take a convenient station C and measure AC and BC. Produce AC to F, making $CF = AC$. Produce BC to D, making $DC = BC$. Measure DF. Prove $AB = DF$.

If $AC = CF = 220$ ft.; $BC = CD = 190$ ft.; and $DF = 210$ ft., how long is AB?

8. In the trusses of steel bridges, why are the beams and rods arranged so as to form a network of triangles as far as possible, and not of quadrilaterals, or pentagons, for instance? (See Ex. 3, p. 76; also Art. 101.)

How is this principle also made use of in forming the frame of a wooden house, or box car, or to strengthen a weak frame or fence of any kind?

9. Draw a map for the following survey notes to the scale of 400 ft. to the inch. In laying off the angles draw a dotted north and south line through each station and then use a protractor.

Stations	Bearings	Distances
A	N. 30° E.	300 ft.
B	N.	200 ft.
C	S. 67° E.	450 ft.
D	S. 60° W.	600 ft.

Keep your drawing for a later use.

10. Obtain or make up a set of survey notes similar to those in Ex. 9 and make a drawing for them.

11. Let DC and FC (p. 512) be two walls perpendicular to the plane of the paper. Let small mirrors be attached to these walls at A and B. Let $\angle C = 45°$. Let a ray of light pass through Q to A, be reflected to B, and thence to P. Prove that $\angle APB$ is a right angle.

[SUG. The law of reflected light is that the angle of incidence equals the angle of reflection, or, in the figure, $x = x$ and $y = y$. Then in the triangle $A\ B\ C$, $x + y + 45° = 180°$, or $x + y = 135°$. About the points A and B, $2\,x + a + 2\,y + b = 360°$ $\therefore a + b = 90°$, etc.]

The preceding is the principle of an instrument called the "optical square" used by foresters in constructing right angles. For a ray of light coming from R through a small hole at B above the mirror will make a right angle with the ray coming from Q through P to A.

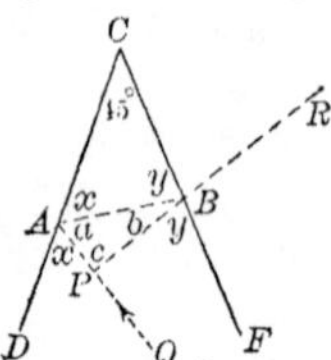

12. The velocity of light is determined by the use of a rotating mirror in the following manner: Let A_1B_1 be a mirror $\perp$ plane of the paper and rotating about O as a pivot; let OP_1 be $\perp$ A_1B_1. Let LO be a ray of light striking the mirror at O and reflected through M to a small stationary mirror some miles distant, whence it is reflected back through M to O. On the return of the ray to O, A_1B_1 will have rotated through a small angle, a, to the position A_2B_2; hence the ray will be reflected in the direction OR. Let the pupil show that $\angle a = \frac{1}{2} \angle LOR$. Since $\angle LOR$ may readily be measured, $\angle a$ is known, and if the rate at which the mirror is rotating is known, the time occupied by the ray in traveling from O to the stationary mirror and back is determined.

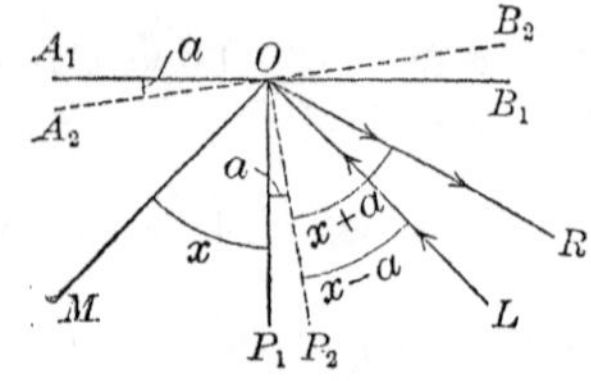

[SUG. $\angle P_1OP_2 = \angle A_1OA_2$ (Art. 132)

$\angle P_1OL = \angle P_1OM = x$ ($\angle$ incidence $= \angle$ reflection)

$\therefore \angle P_2OL = x - a$

$\therefore \angle LOR = x + a - (x - a) = 2a$]

If $\angle LOR = 2° 19'$, $OM = 3$ mi., and the mirror AB rotates 100 times per second, determine the velocity of light per second.

13. To prove that the image of a point in a plane mirror is on a perpendicular from the point to the mirror and as far behind the mirror as the object is in front of it, let MM' be the mirror and P the point, and PAE and $PA'E'$ two reflected rays. Show that

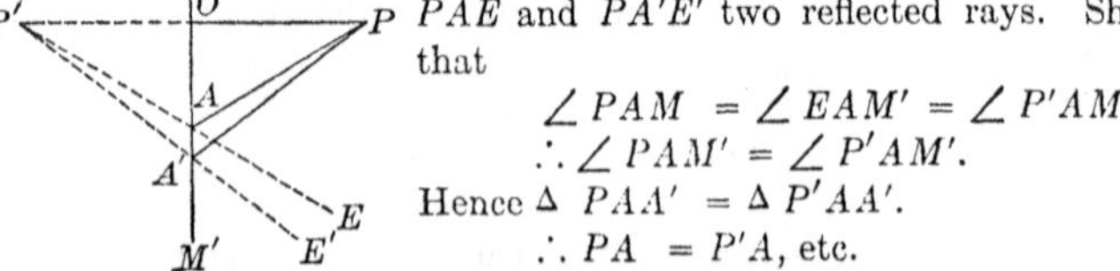

$$\angle PAM = \angle EAM' = \angle P'AM.$$
$$\therefore \angle PAM' = \angle P'AM'.$$

Hence $\triangle PAA' = \triangle P'AA'$.

$$\therefore PA = P'A, \text{ etc.}$$

14. From the above show the direction in which a billiard ball must be sent in order to strike a certain point on the table after striking one side of the table.

15. If two mirrors OM and OM' be perpendicular to each other make a construction to show where the point P will appear to be to an eye at E, after the light from P has been reflected from both mirrors.

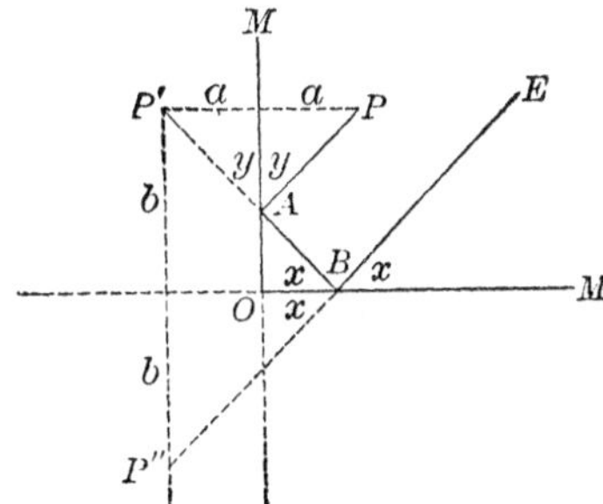

[SUG. Prove $BP'' = BA + AP' = BA + AP$.]

16. What would be the application of Ex. 15 to a ball struck on a billiard table?

17. The path of a ray of light before and after entering a glass prism is given by the lines AB and CD. The entire angle by which a ray of light is deflected on passing through the prism is denoted by x. Prove that $x = i + r - P$. ($\angle$s PBy and PCy are rt. $\angle$s.)

[SUG. By use of a quadrilateral $BPCy$, $y = 180° - P$. Then use the quadrilateral all of whose sides are dotted lines.]

EXERCISES. GROUP 89

(BOOK II)

1. Given a fragment of broken wheel show how to find the radius of the wheel.

2. By use of the carpenter's square find the center of a given circle.

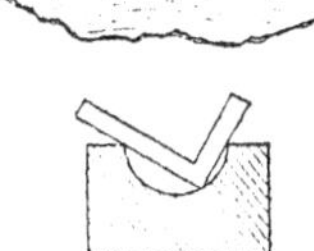

3. Show how a pattern maker by the use of a carpenter's square can determine whether the cavity made in the edge of a board or piece of metal is a semicircle.

4. Bisect a given angle by the use of a carpenter's square.

5. By use of squared paper divide a line $1\frac{1}{2}$ inches long into five equal parts. Into 7 equal parts. Into 3 equal parts. Can you make this division on paper ruled in only one direction?

6. Make up and work a similar example for yourself.

7. Draw a line AB $1\frac{1}{2}$ in. long and let O be its midpoint. Mark off $OC = \frac{1}{4}$ in. and divide AC into four equal parts at D, F, H. With O as a center draw circumferences through C, D, F, H, A. From O draw radii (one of which is OA) dividing the angular space about O into eight equal parts. Beginning at O draw the smooth curve $C123456$ through the points where the radii drawn intersect the circumferences as indicated on the diagram. The result will be the outline of a special form of wheel called a cam, much used in machines, as in the sewing and harvesting machines, printing presses, automobiles, weaving looms, and machinery for making shoes. A cam converts a circular motion into a reciprocating or back and forth motion. The difference of length between OB and OC on the diagram is termed the "throw" of the cam.

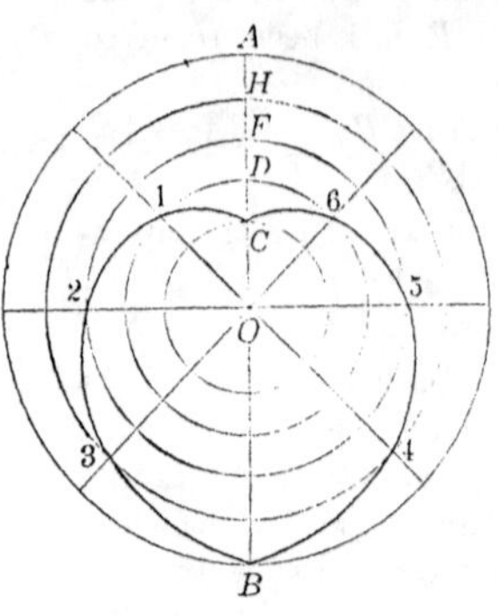

8. Make a drawing of a cam in which the throw shall be 1 in. and the longest radius $1\frac{1}{2}$ in.

9. Stretch a 100 ft. tape or part of it so as to construct an angle of 60° as accurately as possible.

10. To extend a straight line AB beyond an obstacle, as a building, we may proceed as follows:

At B measure off an angle $ABC = 60°$. Produce CB to D, etc. Let the pupil complete the construction and prove that his method is correct.

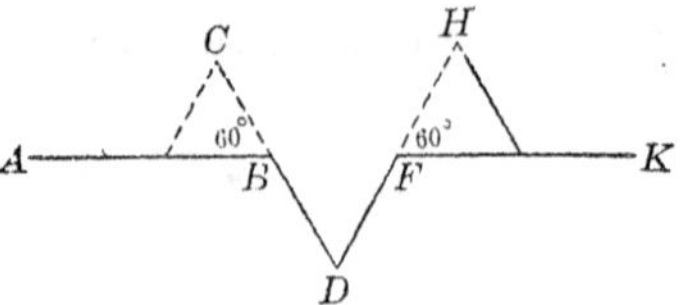

11. Let the pupil solve the problem of Ex. 10 by the construction of right angles instead of angles of 60°.

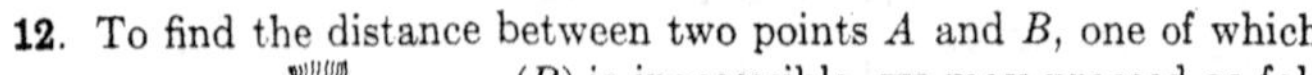

12. To find the distance between two points A and B, one of which (B) is inaccessible, we may proceed as follows: Extend BA to C. Measure a convenient line AD and extend AD to E, making $DE = AD$. From E run a line $\parallel$ AC and meeting BD extended at F. Measure EF. Prove that $EF = AB$.

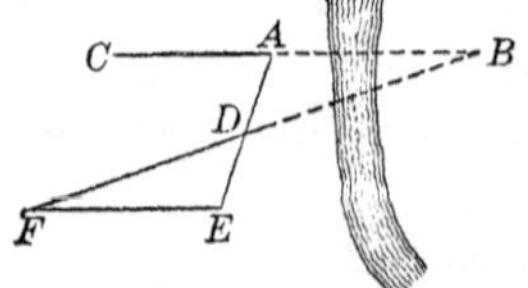

13. Show how to solve the problem of Ex. 12 by constructing a line at right angles to another line instead of one parallel to another.

14. If two streets meet, as in the diagram, show how a curve of given radius r may be made to take the place of the angle A and be tangent with the curb of the two streets.

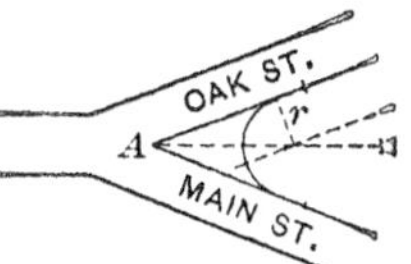

15. The curve in a railroad track is usually an arc of a circle tangent to each straight track which it joins. If two straight tracks AB and CD are joined by a circular curve tangent to both of them, and P is the point where AB and CD would meet if extended, prove $\angle TPD = 2 \angle TAC$.

16. One way of laying off a railroad curve tangent to a given track is as follows: Let ABP be the given straight track. At B construct a small angle PBC (whose size depends on the degree of curvature which the curve is to have). On it mark off $BC = 100$ ft. Construct $\angle CBD = \angle PBC$. Take C as a center and 100 ft. as a radius, and describe an arc cutting BD at D. Construct E from D in like manner. Prove that, B, C, D, E all lie on the arc of a circumference tangent to AB at B.

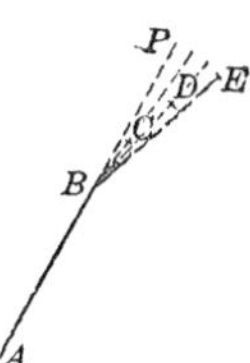

[SUG. Pass a circumference through B, C, D. Prove that E lies on this circumference and that ABP is tangent to it.]

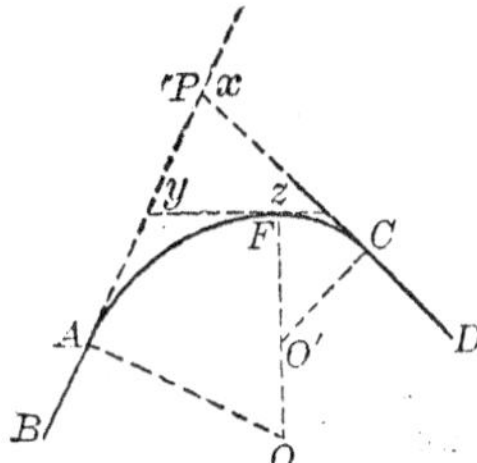

17. Let AB and CD be two straight railroad tracks connected by an arc AF of a circle whose center is O, and an arc FC whose center is O', the arcs having a common tangent at F. Prove that the angle of intersection (x) of the two straight tracks, if produced, equals the sum of the central angles of the two tracks.

[SUG. $x = y + z$. Use Ex. 15.]

A curve like AFC composed of two or more arcs of different radii is called a *compound curve.* Can you suggest why a compound curve should be used in connecting railroad tracks?

18. If the two arcs which compose a compound curve lie on opposite sides of their common tangent, the compound curve is called a *reverse* *curve.* Thus on the diagram, BCD is a reverse curve connecting the straight roads AB and DE.

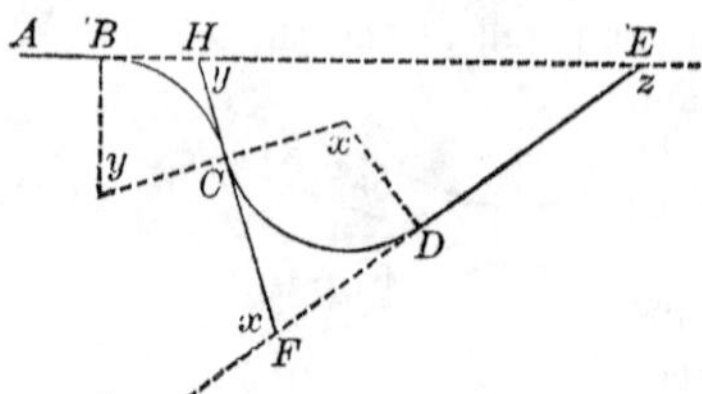

Discover the relation between the angles x, y, z.

[Sug. Use triangle HFE.]

Reverse curves are much used in architecture and ornamental work.

19. What is a railroad frog? If a curved track crosses a straight track, show that the angle of the frog (x) equals the central angle of the curved track (o).

20. If two tracks which curve in the same direction cross each other, find the relation between the angle of the frog and the central angles of the two curved tracks.

21. Find the same when the two curved tracks curve in opposite directions.

22. Show how to locate a gas-generating plant (P) along a given straight road AB so that the length of pipe connecting it with two towns (C and D) shall be a minimum.

(Use Ex. 24, p. 176.)

23. Discover and state an application of Ex. 24, p. 176, similar to one given in Ex. 22.

24. If a treasure has been buried 100 ft. from a certain tree, and equidistant from a given straight road and a path parallel to the road, show how the treasure may be found.

25. Make up and work a similar example for yourself using the principle of Ex. 2, p. 167. Also using that of one of Exs. 3–8, p. 168.

26. Prove that the latitude of a place on the earth's surface equals the elevation of the pole (that is, on the diagram, prove $\angle QEA = \angle PAO$).

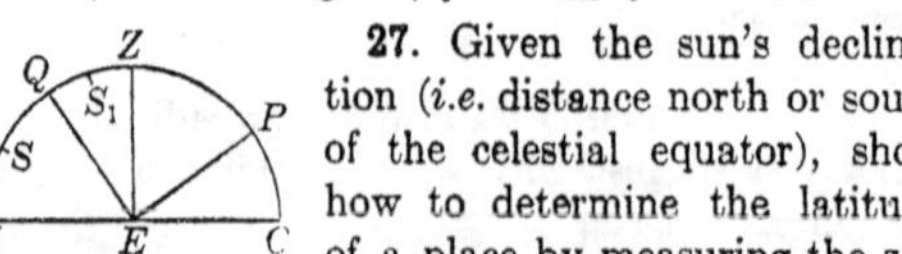

27. Given the sun's declination (*i.e.* distance north or south of the celestial equator), show how to determine the latitude of a place by measuring the zenith distance of the sun. Also by measuring the altitude of the sun above the horizon.

28. How was Peary aided by the principles of Exs. 26 and 27 in determining whether he had arrived at the North Pole?

29. If on April 6 (the day of the year on which Peary was at the North Pole) the sun was 6° 7′ north of the celestial equator, how high above the horizon should the sun have been as observed by Peary? At what hour of the day was this?

30. The sextant is used almost daily by every navigator in determining the altitude of the sun above the horizon, and hence the latitude of the ship. The construction of the sextant is based on the principle that if a ray of light be reflected from two mirrors in succession, the change in the direction of the ray of light equals twice the angle made by the mirrors. Prove this law.

[SUG. Let M and N be the mirrors and the reflected ray be $SMNO$. It is required to prove that $\angle y = 2 \angle x$. It is to be noted that $\angle FNM$, being an exterior angle of triangle $NS'M$, equals $x + i$. Hence angle MNO equals $180° - 2(x + i)$. Hence in triangle MNO, $y + 180° - 2(x + i) + 2i = 180°$, etc.]

In the sextant, angle y is the elevation of the sun above the horizon, and x can be read on the rim of the instrument in a simple manner, since the mirror N is fixed in position while M rotates.

It may be observed that the principle of this example is the same as that proved in Ex. 12, p. 512.

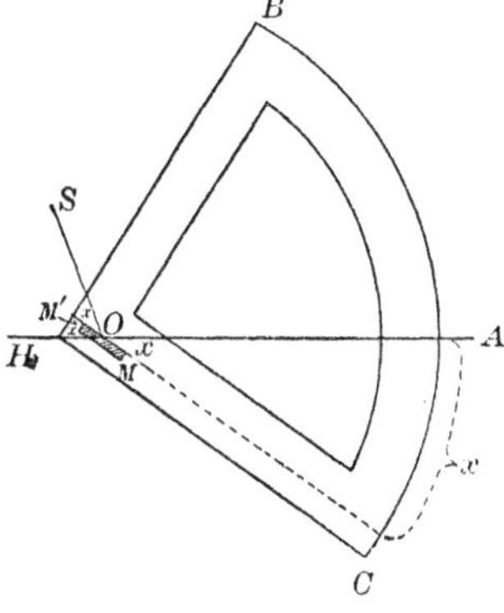

31. The diagram shows a somewhat simple substitute for the sextant called the angle-meter. O is the center of the arc BC and MM' is a fixed mirror. The instrument is held so that a ray SO from the sun when reflected from the mirror passes into the eye (A) in a horizontal direction. Show that the elevation of the sun above the horizon $= 2 \angle x$. The rim BC is graduated so that the angle x can be read on it.

The angle-meter may be used to measure horizontal as well as vertical angles.

What is the advantage in using a mirror on an instrument of this sort? That is, why is not the position of the sun observed directly across a graduated arc?

32. Draw an easement cornice (AB) tangent to the rake cornice BC, and passing through a required point A. (The same construction is used in laying out the easements of stair rails, etc.)

(Use Ex. 7, p. 174.)

33. A segmental arch is a compound curve composed of the arcs of three circles. The method of constructing a segmental arch is as follows: Let AB be the span and CD the altitude of the required arch. Complete the rectangle $GADC$. Draw the diagonal AC. Bisect the angles GAC and GCA. Let the bisectors meet at E. Draw EH perpendicular to AC, meeting AB at N and CD produced at H. Make DK equal to DN. Then show that N is the center and NA the radius, H the center and HE the radius, and K the center and KB the radius for the arcs composing the arch. (See Hanstein's Constructive Drawing.)

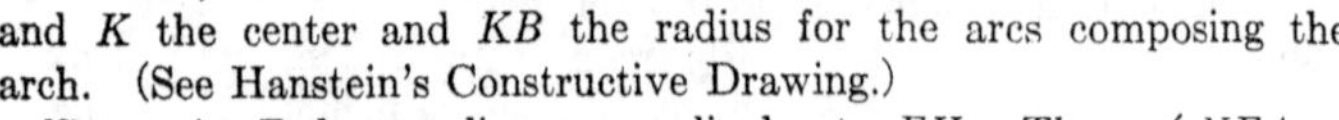

[Sug. At E draw a line perpendicular to EH. Then $\angle NEA = \angle NAE$. (Complements of equal angles are equal) etc.]

34. What is called a Persian arch may be constructed as follows: Let AB be the span and CD the altitude of the required arch. Draw the isosceles triangles ADB. Divide AD into three equal parts at H and G. Construct $HK \perp AG$ at H, and meeting AB produced at K. Produce KG to meet EF which has been drawn through $D \parallel AB$. With K as a center and KA as a radius, and E as a center and EG as a radius, describe arcs meeting at G. Prove that these arcs have a common tangent at G, and therefore form a compound curve. (See Hanstein's Constructive Drawing.)

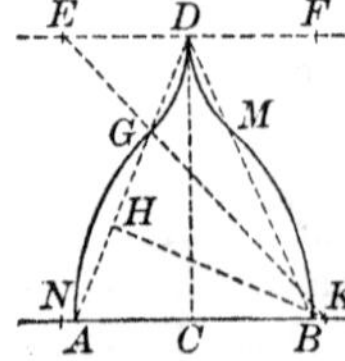

35. Construct a Persian arch in which the arc DG = arc NG.

[Sug. Bisect line AD instead of trisecting it.]

36. Construct Persian arches in which the chords NG and DG have various ratios, and decide which of these arches you think is the most beautiful.

37. Construct segmental arches of various shapes and decide which of these you think is the most beautiful.

38. Construct a trefoil, given the radius of one of its circles.

[SUG. Construct an equilateral triangle, one of whose sides equals twice the given radius. Take each vertex of the triangle as a center, and half of one of the sides as a radius and describe arcs.]

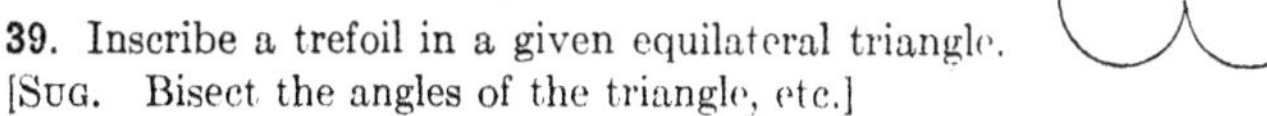

39. Inscribe a trefoil in a given equilateral triangle.

[SUG. Bisect the angles of the triangle, etc.]

40. Discover the method of construction of each of the figures or diagrams on the next page, and then reproduce each of them in a drawing.

Squared paper may be used to advantage as an aid in making some of these constructions. See Fig. 10, p. 520.

41. Construct a diagram similar to Fig. 10, but making a rectangle instead of a square the basis of the drawing.

42. Also one making the rhomboid the basis.

43. Construct a trefoil and develop into an ornamental design by placing small circles and arcs within its parts and larger circles outside.

44. By use of squared paper invent designs similar to Fig. 10 on p. 520, but formed by two intertwining lines.

45. Collect a number of pictures of ornamental designs, tracery, scroll work, etc., such as are used in architecture, wall paper, and similar patterns, whose construction depends on the principles of Book II. Show how these designs are constructed, and reproduce them in drawings.

EXERCISES. GROUP 90

(BOOK III)

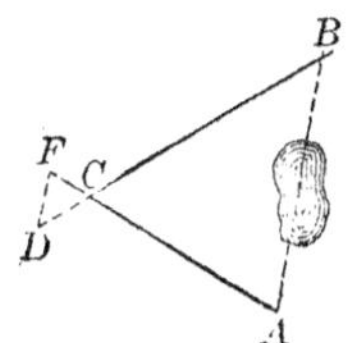

1. To find the distance between two accessible objects A and B separated by a barrier, take a convenient point C, measure AC and BC. Produce BC to D and AC to F so that DC = some fraction, as $\frac{1}{5}$, of CB, and FC = the same fraction of AC, and measure FD. How, then, is AB computed? Prove this.

a
2a
a
1
2
3
[x = 30°]
x
4
5
6
3
3
1
3
1
2
7
8
9
10
11
12

Let $BC = 300$ yd., $DC = 60$ yd., $AC = 240$ yd., $FC = 48$ yd., $FD = 52$ yd., find AB.

Why is this method of finding AB often more convenient than that of Ex. 7, p. 511.

2. What similar method does the adjoining diagram suggest for finding AB?

3. In the diagram of Ex. 2, in case a river ran between A and B, and also between B and F, what measurements would be necessary in order to determine the length of AB? Of BF?

4. Work again Ex. 2, p. 310.

5. In case the sun is not shining and shadows cannot be used, the height of an object like a tree or steeple can often be determined by a method indicated in the drawing. What distance must be measured and why to determine the height of the tree?

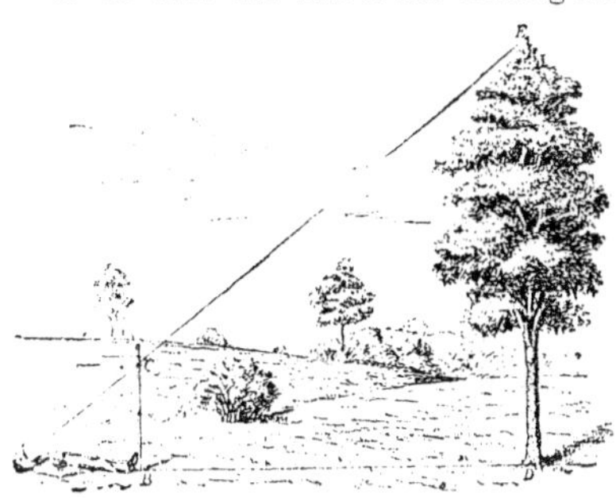

6. Show how to find the height of a tree by placing a mirror in a horizontal position on the ground and standing so as to see the reflection of the top of the tree in the mirror. If the observer's eye is $5\frac{1}{2}$ ft. from the ground, the observer stands 6 ft. from the mirror, and the mirror is 120 ft. from the tree, how high is the tree?

7. Foresters often determine the height of a tree by an instrument called "Faustman's Height Measurer." The principle on which this instrument is constructed is shown in the diagram.

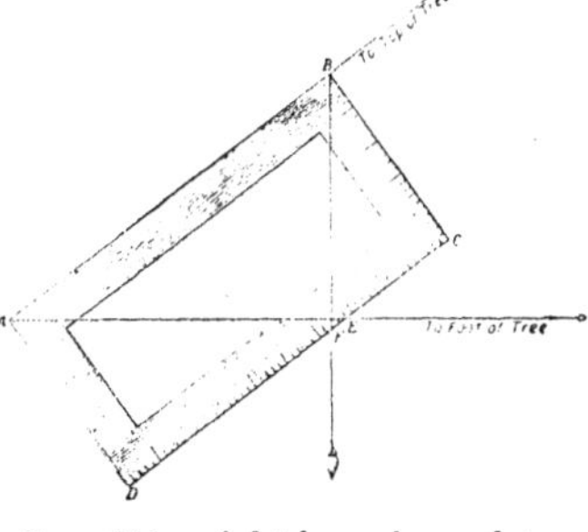

If the distance from A to the foot of the tree is 150 ft., $BC = 6$ inches, and $CF = 4\frac{1}{2}$ inches, find the height of the tree.

8. The distance from A to B in Ex. 7, p. 511, might have been determined by a graphical method as follows: Measure AC, CB, and angle

ACB. On paper make a drawing of the triangle ACB to a convenient scale. On this drawing measure the line which represents AB and hence determine the length of AB.

By use of this method we are saved the labor of marking out and measuring the lines CD, CF, and DF.

Apply this method to the measurement of two objects in your neighborhood which are separated by an impassable barrier.

In like manner show how the distance from a given point to another inaccessible place may be determined.

9. Also the distance between two places both of which are inaccessible.

10. What is the measuring instrument called the diagonal scale? How is the principle of similar triangles used in it? Show how the diagonal scale aids in the accurate measurement of lines and hence in the accurate determination of a distance such as AB in Ex. 7, p. 511.

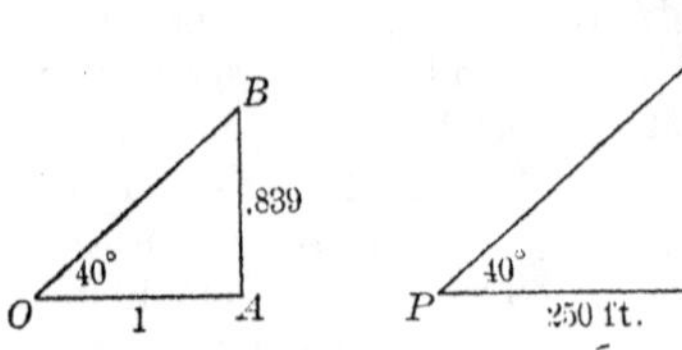

11. In the triangle OAB, A is a right angle, and OA is 1. By a method which is beyond the scope of this book, the length of AB is computed and found to be .839 +. Using this fact find RQ in the second triangle.

12. By use of the accompanying table, find RQ if angle P is 10°. 20°. 70°. 80°.

Tables giving the other sides of all possible right triangles when one side is unity have been computed and when used as in Exs. 11 and 12, form the basis of the subject of Trigonometry. By use of this science, after measuring the length of a single line a few miles long on the earth's surface, we can determine the distances and relative positions of other places thousands of miles away, without measuring any intervening lines. By use of these results as a basis, the distance of the moon is determined as approximately 240,000 miles; of the sun as 92,800,000 miles; and of the nearest fixed star as 20,000,000,000,000 miles.

Angle O	AB	OB
10°	.176	1.015
20°	.364	1.064
30°	.577	1.155
40°	.839	1.305
50°	1.192	1.556
60°	1.732	2.000
70°	2.747	2.924
80°	5.671	5.759

A knowledge of these distances has led to important improvements in methods of navigation and have thus facilitated travel and commerce and increased their benefits for us all.

13. On the diagram given AB is 6,000 ft., and the angles as indicated, compute the length of CD, by use of the table given in Ex. 12.

14. To the diagram in Ex. 13 annex another triangle CFD giving it angles which are multiples of 10°, and compute the length of CF.

15. A *pantograph* is an instrument for drawing a plane figure similar to a given plane figure. It is used for enlarging or reducing maps and drawings. It consists of four bars, parallel in pairs and jointed at c, b, C, and E, as shown in the diagram. $cbEC$ is a parallelogram. The rods may be joined so that any required

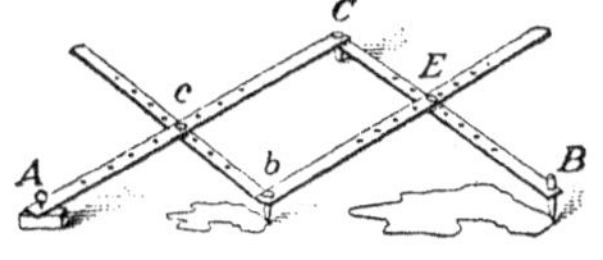

ratio $= \frac{Ac}{AC} = \frac{CE}{CB}$. A turns upon a fixed pivot, and pencils are carried at b and F.

Show that A, b, and B are always in a straight line, and that $\frac{Ab}{AB}$ always equals the given ratio $\frac{Ac}{AC}$.

[Sug. Draw a line from A to b, and a line from A to B. Prove the triangle Acb and ACB similar (Art 193); hence show that Ab and AB coincide, etc.]

16. Using the fact that a triangle whose sides are 3, 4, and 5 units of length is a right triangle, show how, by stretching a 100-ft. tape, to construct a right angle as accurately as possible. (Among the ancient Egyptians a class of workmen existed called rope stretchers, whose business was to construct right angles in this general way.)

17. The strongest beam which can be cut from a given round log is found as follows: Take AB, a diameter of the log, and trisect it at C and D. Draw CE and $DF \perp AB$ and meeting the circumference at E and F respectively. Draw AF, FB, BE, and AE.

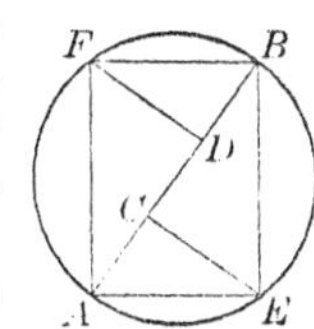

Prove $AFBE$ a rectangle; also $FB : AF = 1 : \sqrt{2}$, or approximately as 5 : 7.

[Sug. FB is a mean proportional between DB and AB. Therefore $\overline{FB}^2 = \frac{1}{3}\overline{AB}^2$ (Art. 343).]

In like manner $\overline{FA}^2 = \frac{2}{3}\overline{AB}^2$, etc.]

18. A sphere S weighing 100 lb. rests on the inclined plane AB. AC contains 8 units of length, BC 6 units. A force which prevents S from rolling down the plane would be equivalent to a lifting force of how many pounds exerted on S and parallel with AB?

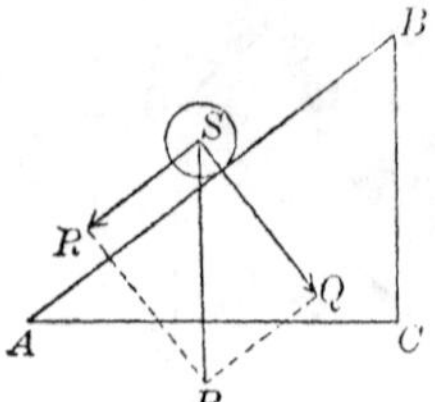

[Sug. Resolve the weight of S (represented by SP) into two forces, one perpendicular to AB and the other parallel to AB. Prove the triangles ACB and SPR similar, and obtain $AB:BC = SP:SR$, or $10:6 = 100$ lb.: SR.]

The principle involved in this example is of great practical importance. Thus in many machines useful results are often obtained by representing a force by the diagonal of a rectangle (or parallelogram) and separating this force into two component forces represented by the sides of the rectangle (or parallelogram), only one of these components being effective. This principle makes possible the action of the propeller of an aeroplane or steamboat, of the best water wheels and wind mills, and indeed of all turbine wheels. It also determines the lifting power of the planes of an aeroplane.

19. A wagon weighing 1800 lb. stands on the side of a hill which has a rise of 18 ft. for every 100 ft. taken horizontally. Hence what force must a horse exert to keep such a wagon from running down hill, friction being neglected?

20. Make up and work a similar example for yourself.

21. Show how the diameter of the earth may be determined by the following method: Drive three stakes in a level piece of ground (or in a shallow piece of water) in line, each two successive stakes being a mile apart, and let each stake project the same distance above the ground (or water). By use of a leveling instrument, determine the amount by which the middle stake projects above a horizontal line connecting the tops of the end stakes. This distance will be found to be 8 in.

[Sug. Use Art. 343. An arc a mile long on the earth's surface may be taken as equal to its chord. Then from the diagram of Art. 343 we obtain the following proportion,

the diameter of the earth: 1 mi. = 1 mi.: 8 in.]

22. Also show that the distance that the middle stake projects above the horizontal line connecting the top of the two end stakes varies as the square of the distance between the end stakes. Thus if the two end stakes were placed three times as far apart as in Ex. 21 (that is, 6 miles apart instead of 2 miles) the bulge of the earth between them would be 3^2 or 9 times what it was originally.

Hence determine the projection of the middle stake (or bulge of the earth) when the end stakes are 4 mi. apart. Also 8 mi. 16 mi. 32 mi.

23. At the seashore an observer whose eye was 10 ft above sea level observed a distant steamboat whose hull was hidden for a height of 12 ft. above water level by the bulge of the earth. About how far off was the steamboat?

24. A seaman in a lookout 42 feet above water level with a glass could barely see the topsail of a distant ship, and estimated this topsail to be 45 ft. above sea level. Estimate the distance of the observed vessel from the seaman.

25. Work again Exs. 7–8, p. 310, Group 56.

26. The moon's distance from the earth's center approximately equals 60 times the earth's radius. A body falling at the earth's surface goes 193 in. in 1 sec. Hence, if the law of gravitation is true, the distance the moon falls in 1 sec. toward the earth will be $\frac{193 \text{ in.}}{60^2}$ In the diagram, let O be the earth's center, ABE the moon's orbit, and CB or AD the distance the moon falls toward the earth in 1 sec. Taking the month as 27 da. 7 hr. 43 min. 11 sec.,

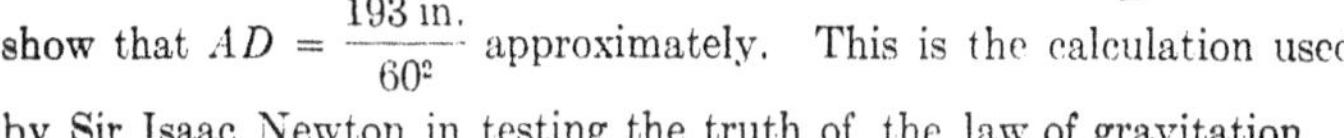

show that $AD = \frac{193 \text{ in.}}{60^2}$ approximately. This is the calculation used by Sir Isaac Newton in testing the truth of the law of gravitation.

27. Any rectangular object, as a book, door, or photograph, is considered to be of the most artistic shape when its length and breadth have the same ratio as the segments of a line divided in mean and extreme ratio (Art. 370). In accordance with this rule, if a window is 6 feet high, how wide should it be?

The division of a line in extreme and mean ratio has been termed the Golden Section. By many the Golden Section is regarded as a fundamental principle of esthetics, having applications in determining the proportions of the ideal human form, in explanation of musical harmonies, etc.

28. Make up and work an example similar to Ex. 27.

EXERCISES. GROUP 91

(Book IV)

1. The supporting power of a wooden beam (of rectangular cross section and of given length) varies as the area of the cross section multiplied by the height of the beam. If the cross section of a given beam is 4″ × 8″, compare the supporting power of the beam when it rests on the narrow edge (4″) with its supporting power when it rests on its wide edge (8″).

2. Two beams of the same length and material have cross sections which are 2″ × 4″ and 3″ × 8″ respectively. Find the ratio of the supporting power of the two beams.

3. In Ex. 17, p. 523, compare the supporting power of a beam cut from a log in the manner indicated, with that of a square beam cut from the same log.

4. Also with that of a beam whose width equals $\frac{1}{4}$ of the diameter.

5. When an irregular area like $ABCD$ is calculated by means of equidistant offsets, the following rule is used: To the half sum of the initial and final offsets add the sum of all the intermediate offsets, and multiply the sum by the common distance between the offsets. Prove this rule.

6. Show how the rule of the preceding problem could be used to calculate an area whose entire boundary is an irregular curved line.

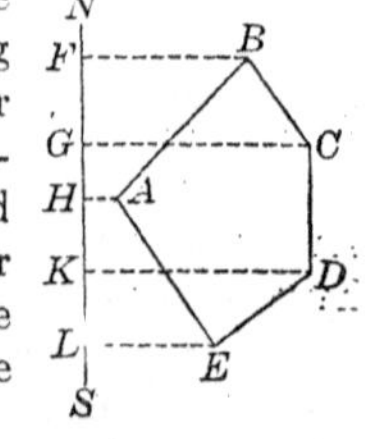

7. Surveyors often determine the area of a piece of land, as of $ABCD$, by taking an auxiliary line as NS, measuring the perpendicular distances from A, B, C, D, E to NS, and the intercepts on NS between these perpendiculars, and combining the areas of the various trapezoids (or triangles) formed. Supply probable numbers for the lengths of lines on the diagram, and compute the area of $ABCDE$.

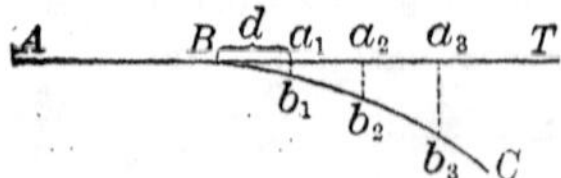

8. Frequently (as when the center of the curve cannot be seen from the curve) a railroad curve is laid out by constructing a series of equidistant offsets perpendicular to the tangent of the curve.

Thus, if ABT is a straight track and BC is a curve to be laid out tangent to AB at B, mark off Ba_1, a_1a_2, a_2a_3, all $= d$, and construct the $\perp$ offsets a_1b_1, a_2b_2, a_3b_3 by using the formula $a_nb_n = r - \sqrt{r^2 - n^2d^2}$ where r is the radius of the curve. Prove that this formula is correct.

9. The diagram represents two straight parallel railroad tracks connected by a compound curve (in this case called a cross-over track) composed of two tracks with common tangents at D and E with centers O_2 and O_1, and having equal radii. Taking the magnitudes as indicated on the figure, show that KPO_2D is a rectangle (use Arts. 122, 160).

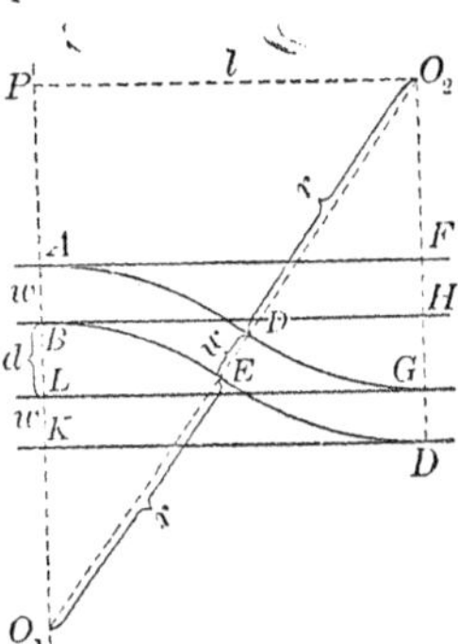

10. Find a formula for r in terms of l, d and w. (Use the right triangle $O_1\,PO_2$ and Art. 400.) Also for l in terms of r, d, and w.

11. If $w = 4'\,8\frac{1}{2}''$, $d = 10$ ft., $r = 150$ ft., find l.

12. If a steamboat is traveling at the rate of 12 miles an hour, and a boy walks across her deck at right angles to her line of motion at the rate of 3 miles an hour, draw a diagram to show the direction of his resultant motion. From this determine the speed at which he is going.

13. Make and work a similar example for yourself concerning a mail bag thrown from a train.

14. Also concerning a breeze blowing into a window of a moving trolley car.

15. Two forces, one of 300 lb., the other of 400 lb., act at right angles on the same body. Find their resultant.

16. Make up and work a similar example for yourself.

17. A river is flowing at a rate of 4.25 mi. per hour, and a man is rowing at right angles with the current at a rate of 3.75 mi. per hour. What is the resultant velocity of the man?

18. If a star has a velocity of 15 mi. a second toward the earth and a velocity of 20 mi. a second at right angles with a line drawn from the star to the earth, find the velocity of the star in its own path.

19. A body is moving in the straight line AD past the point O (p. 528). The distance passed over in a unit of time $= AB = BC = CD$. When the body reaches B it is acted on by a force which impels it toward O.

a distance BF in a unit of time. Prove that the area swept over in the unit of time by a line drawn from the body to O will be unchanged by the action of the new force. (That is, on the diagram prove that $\triangle OCB \approx \triangle OPB$.)

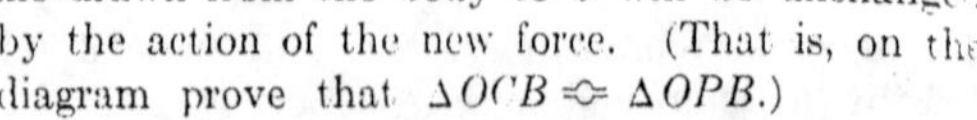

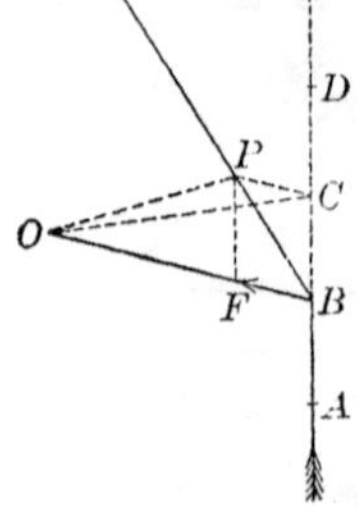

This principle accounts for the fact that the earth (or another planet or a comet) moves faster in its orbit the nearer it is to the sun.

20. By tracing the outline of a map on squared paper, show how to find the area of an irregular figure as of some country or part of a country. By use of this method find the area of some part of the state in which you live.

EXERCISES. GROUP 92

(Book V)

1. A cooper in fitting a head to a barrel takes a pair of compasses and then adjusts them till, when applied six times in succession in the chine, they will exactly complete the circumference. He then takes the distance between the points of the compasses as the radius of the head of the barrel. Why is this?

2. Draw a square and convert it into a regular octagon by cutting off the corners.

[Sug. Draw the diagonals of the square, bisect their angles of intersection, etc.]

3. In heating a house by a hot-air furnace the area of the cross section of the cold-air box should equal the sum of the areas of the pipes conducting hot air from the furnace. If a given furnace has three hot-air pipes, each 6″ in diameter, and one pipe 8″ in diameter, and the width of the cold-air box is 15″, how deep should the box be?

4. What is the most convenient way of determining the diameter of a hot-air pipe if you have no callipers and the ends of the pipe are not accessible?

5. A half-mile running track is to have equal semicircular ends and parallel straight sides. The extreme length of the rectangle together with the semicircular ends is to be 1000 ft. Find the width of the rectangle.

[Sug. Denote the length of the radius of the semicircular ends by x and that of one of the parallel side straight tracks by y, and obtain a pair of simultaneous equations.]

6. A belt runs over two wheels one of which has a diameter of 3 ft. and the other of 6 in. If the first wheel is making 120 revolutions per minute, how many is the second wheel making. How many revolutions must the first wheel make in order that the second may make 300 revolutions per minute?

7. If in laying a track a rail 10 ft. long is bent through an angle of $5° 10'$, what is the radius of the curve?

Assuming (what is not strictly true, owing to friction against the sides of the pipe, etc.) that the rate of flow through a cylindrical pipe is proportional to its area of cross section:

8. Work again Ex. 12, p. 277.

9. If a $1\frac{1}{2}$-in. pipe is replaced by a 1-in. pipe, how much is the flow of water increased?

10. A 3-in. pipe is to be replaced by one which will deliver twice as much water per minute. Find, to the nearest quarter of an inch, the diameter of the new pipe.

11. A 2-in. steam pipe conveying steam from the boiler to the radiators in a school building is found to supply only two thirds the needed amount of steam. What is the smallest even size of pipe that will convey the needed amount?

12. A city of 40,000 people is barely supplied with water by 12-in. mains from the reservoirs. If these mains are torn out, and 18-in. mains substituted, what future population of the city is allowed for?

13. A steel bar 1 in. in diameter will hold up 50,000 lb. What load would a bar $\frac{3}{4}$ in. in diameter carry? What would be the diameter of a round (cylindrical) bar to carry 150,000 lb.?

14. If the center of symmetry of a flat, homogeneous object is the center of mass, find the center of mass of a square; of a rectangle; regular hexagon; circle.

15. It is evident that if the medium of a triangle (BM) be placed on a knife edge the triangle will balance (for if PP' be $\parallel AC$, the pull on P is balanced by the pull on P'). Hence find the center of mass for any triangle. For a regular pentagon.

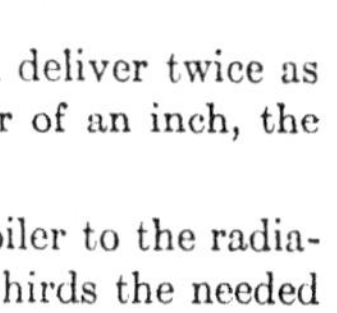

It is useful to be able to determine the center of mass of an object by geometry or by any other means, since a knowledge of the center of mass of a body often enables us to treat the body in a simple way, for example, as if the body were concentrated at a single point.

16. If a box has a triangular end, subject to the same pressure at all points, at what single point on the end must a supporting pressure be applied?

17. The cross section of a cylinder is a circle. The weight-supporting strength of a horizontal cylindrical beam of given material and length varies as the area of the cross section times its radius.

Compare the weight-supporting power of two solid horizontal iron cylindrical beams of the same length and quality of iron, the radii being 3 in. and 6 in. respectively. (Point out and use the short way of getting the desired result.)

18. The cross section of a hollow cylinder (i.e. of a tube) is a circular ring. Denote the outside radius of the tube by R and the inside radius by r. Then it may be shown that the weight-supporting power of a hollow cylindrical tube of given length and material varies as the area of the cross section (i.e. of the ring) times $\dfrac{R^2 + r^2}{R}$.

If $R = 4$ in. and $r = 3$ in., compare the weight-supporting power of the tube, with that of a solid cylindrical beam of the same length and same cross sectional area.

19. Make up and work a similar example.

In general a cylindrical tube is stronger than a solid cylindrical beam of the same length and containing the same amount of material.

Hence in a framework, as in that of an airship where the maximum strength must be obtained from a given amount of material, the metallic rods and posts are all tubular. In like manner, bamboo rods, since they are hollow, are used in an aeroplane instead of solid wooden rods wherever possible. For the same reason, the bones of flying birds, and many bones in men and animals are hollow and not solid.

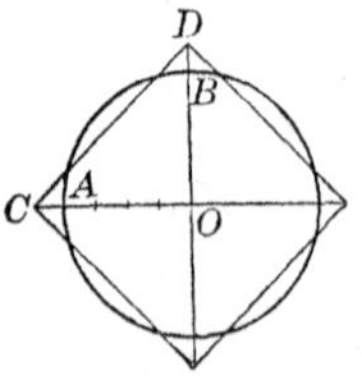

20. To construct a square which shall be approximately equivalent to a given circle O, divide the radius OA into four equal parts and produce each end of two perpendicular diameters a distance equal to one fourth of the radius, and connect the extremities of the lines thus formed. Show that taking the square thus formed as equivalent to the circle is the same as taking $\pi = 3\frac{1}{8}$. Also find the per cent. of error in taking the square as equivalent to the circle.

21. A short way to construct a regular inscribed pentagon and also a five-pointed star (or pentagram) is as follows: Draw a circle O and two diameters AB and CD at right angles. Bisect the radius OB at F, and with F as a center and FC as a radius describe an arc cutting AO at H. Then CH is the length of a side of the regular inscribed pentagon, by joining whose alternate vertices the pentagram may be formed.

As a proof of this, by use of right triangles, prove

$$CH = R\frac{\sqrt{10 - 2\sqrt{5}}}{2} \quad \text{(See Ex. 17, p. 301.)}$$

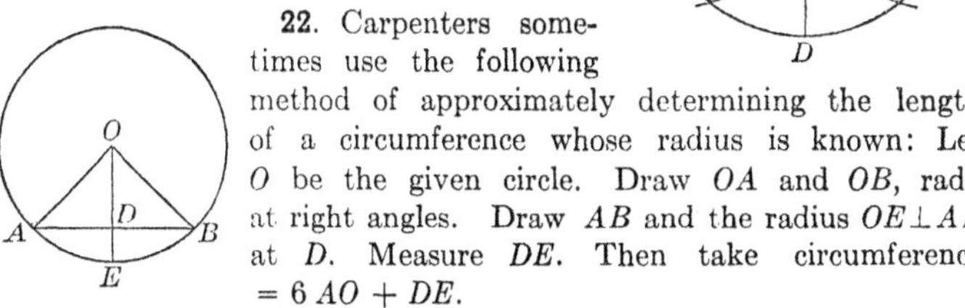

22. Carpenters sometimes use the following method of approximately determining the length of a circumference whose radius is known: Let O be the given circle. Draw OA and OB, radii at right angles. Draw AB and the radius $OE \perp AB$ at D. Measure DE. Then take circumference $= 6\,AO + DE$.

Find the per cent. of error in this method, taking $\pi = 3.14159-$.

23. The following designs are important in architecture or in ornamental work (thus the first is a detail in a stained glass window in an early French cathedral; the second is the plan of the base of a column in an English cathedral). Discover a method of constructing each of the following designs, and reproduce each of them in a drawing:

24. Construct a square. Take each vertex of the square as a center and one half a side of the square as a radius and describe arcs which meet. Erase the square and you have a quatrefoil. By drawing other circles and arcs of circles, elaborate the quatrefoil into an ornamental design (see design 11, p. 520).

25. In like manner construct a cinquefoil by use of a regular pentagon, and develop it into an ornamental design.

26. Treat a regular hexagon in the same way.

27. A pavement or mosaic may be formed out of regular polygons in the following ways. Show how to make each diagram in the simplest way.

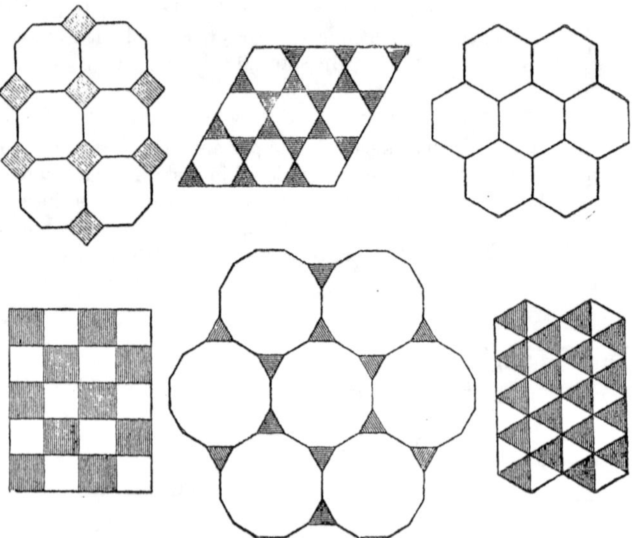

APPLICATIONS OF SOLID GEOMETRY TO MECHANICS AND ENGINEERING

EXERCISES. GROUP 93

(Books VI and VII)

1. A carpenter tests the flatness of a surface by applying a straight edge to the surface in various directions. How does a plasterer test the flatness of a wall surface? What geometric principle is used by these mechanics?

2. Explain why an object with three legs, as a stool or tripod, always rests firmly on the floor while an object with four legs, as a table, does not always rest so. Why do we ever use four-legged pieces of furniture?

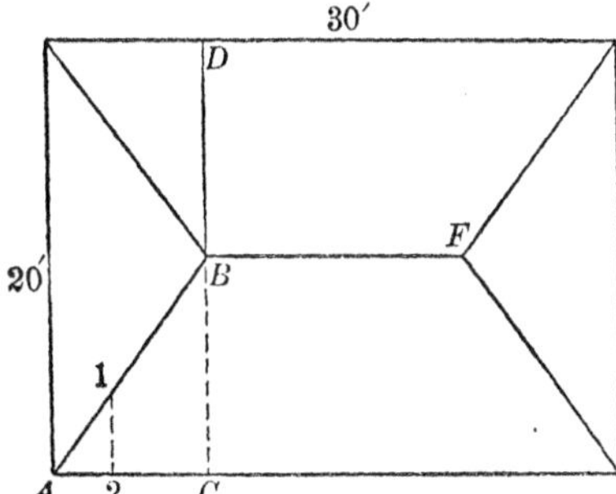

3. How can a carpenter get a corner post of a house in a vertical position by use of a carpenter's square? What geometrical principle does he use?

4. The diagram is the plan of a hip roof. The slope of each face of the roof is 30°. Find the length of a hip rafter as AB.

[Sug. Draw a triangle DBC representing a section of the roof at DBC on the plan. Hence it may be shown that $BC = \frac{20}{3}\sqrt{3}$. In like manner by taking a section through BF, it is found that $AC = 10$. Hence in the triangle ABC, AB may be found.]

5. Find the area of the entire roof represented in Ex. 4.

6. Make drawings showing at what angle the two ends of a rafter like BC in Ex. 4 must be cut.

7. Make drawings showing at what angle a jack rafter like 12 in Ex. 4 must be cut.

[SUG. To determine how the end 1 of the jack rafter must be cut use the principle that two intersecting straight lines determine a plane (Art. 501). The cutting plane at 1 must make an angle at the side of the jack rafter equal to angle CBH, and on the top of the jack rafter equal to angle ABC.]

8. What is a gambrel roof? Make up a set of examples concerning a gambrel roof similar to Exs. 4–7.

9. By use of Art. 645, show that a page of this book held at twice the distance of another page from the same lamp receives one fourth the light the first page receives.

10. The supporting power of a wooden beam varies directly as the area of the cross section times the height of the beam and inversely as the length of the beam. Compare the supporting power of a beam 12 ft. long, 3 in. wide, and 6 in. high with that of a beam 18 ft. long, 4 in. wide, and 10 in. high. Also compare the volumes of the two beams.

EXERCISES. GROUP 94

(BOOKS VIII AND IX)

1. A hollow cylinder whose inside diameter is 6 in. is partly filled with water. An irregularly shaped piece of ore when placed in the water causes the top surface of the water to rise 3.4 in. in the cylinder. Find the volume of the ore.

2. What is a tubular boiler? What is the advantage in using a tubular boiler as compared with a plain cylindrical boiler? If a tubular boiler is 18 ft. long and contains 32 tubes each 3 in. in diameter, how much more heating surface has it than a plain cylindrical boiler of the same length and 36 in. in diameter? (Indicate both the long method and the short method of making this computation and use the short method.)

3. If a bridge is to have its linear dimensions 1000 times as great as those of a given model, the bridge will be how many times as heavy as the model?

Why, then, may a bridge be planned so that in the model it will support relatively heavy weights, yet when constructed according to the model, falls to pieces of its own weight?

Show that this principle applies to other constructions, such as buildings, machines, etc., as well as to bridges.

4. Work again Exs. 23–25, 27, p. 473.

5. Make and work for yourself an example similar to Ex. 24, p. 473.

6. Sound spreads from a center in the form of the surface of an expanding sphere. At the distance of 10 yd. from the source, how will the surface of this sphere compare with its surface as it was at 1 yd.? How, then, does the intensity of sound at 10 yd. from the source compare with its intensity at a distance of 1 yd.?

Does this law apply to all forces which radiate or act from a center as to light, heat, magnetism, and gravitation? Why is it called the law of inverse squares?

7. If a body be placed within a spherical shell, the attractive forces exerted upon the body by different parts of the shell will balance or cancel each other. Hence a body inside the earth, as at the foot of a mine, is attracted effectively only by the sphere of matter whose radius is the distance from the center of the earth to the given body. Hence, prove that the weight of a body below the surface of the earth varies as the distance of the body from the center of the earth.

[SUG. If W denote the weight of the body at the surface, and w its weight when below, R the radius of the earth, and r the distance of the body from the center when below the surface, show that

$$W:w = \frac{R^3}{R^2} : \frac{r^3}{r^2} = R:r$$

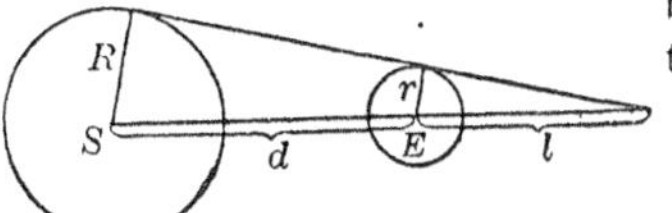

8. The light of the sun falling on a smaller sphere, as on the earth or the moon, causes that body to cast a conical shadow. Denoting the radius of the sun by R, the radius of the smaller sphere by r, the distance between the two spheres by d, and the length of the shadow by l, show that $l = \dfrac{dr}{R - r}$.

Find l when $d = 92{,}800{,}000$ mi., $R = 433{,}000$ mi., $r = 4000$ mi.

9. If in a lunar eclipse the moon's center should pass through the axis of the conical shadow, and the moon is traveling at the rate of 2100 mi. an hour, how long would the total eclipse of the moon last?

How long if the moon's center passed through the earth's conical shadow at a distance of 1000 mi. from the axis of the cone?

10. If the moon's diameter is 2160 mi., find the length of the moon's shadow as caused by sunlight.

11. If the distance of the moon from the earth's center varies from 221,600 mi. to 252,970, show how this explains why some eclipses of the sun are total and others annular.

Why, also, at a given point on the earth's surface is an eclipse of the sun a so much rarer sight than an eclipse of the moon? Why, also, is its duration so much briefer?

FORMULAS OF PLANE GEOMETRY

SYMBOLS

a, b, c = sides of triangle ABC.

$s = \frac{1}{2}(a + b + c)$.

h_c = altitude on side c.

m_c = median on side c.

t_c = bisector of angle opposite side c.

l and m = line segments.

P = perimeter.

S_n = side of a regular polygon of n sides.

R = radius of a circle.

D = diameter of a circle.

C = circumference of a circle.

r = radius of an inscribed circle.

$\pi = \frac{22}{7}$ approx. (or 3.1416—).

K = area.

b = base of a triangle.

h = altitude of a triangle.

b_1 and b_2 = bases of a trapezoid.

LENGTHS OF LINES

1. In a right triangle, C being the right angle,

$$c^2 = a^2 + b^2.$$ Art. 346.

2. In a right triangle, l and m being the projections of a and b on c, and h, the altitude on c,

$$h^2 = l \times m.$$
$$a^2 = l \times c,\ b^2 = m \times c.$$ Art. 342.

3. In an oblique triangle, m being the projection of b on c,
if a is opposite an acute $\angle$, $a^2 = b^2 + c^2 - 2\,c \times m$. Art. 349.
if a is opposite an obtuse $\angle$, $a^2 = b^2 + c^2 + 2\,c \times m$. Art. 350.

4. $h_c = \frac{2}{c}\sqrt{s\,(s-a)\,(s-b)\,(s-c)}$. Art. 393.

5. $m_c = \frac{1}{2}\sqrt{2\,(a^2 + b^2) - c^2}$. Art. 353.

6. $t_c = \dfrac{2}{a+b}\sqrt{a\,b\,s\,(s-c)}$. Art. 363.

7. If l and m are the segments of c made by the bisector of the angle opposite,
$$a : b = l : m.$$ Arts. 332, 336.

8. If l and m are the segments of a line, a, divided in extreme and mean ratio, and $l > m$, $a : l = l : m$. Art. 370.

9. In similar polygons, $\begin{cases} P : P' = a : a'. \\ a : a' = b : b'. \end{cases}$ Art. 341. Art. 321.

10. In circles, $C : C' = R : R'$; also $C : C' = D : D'$. Art. 442.

11. $C = 2\pi R$, or $C = \pi D$. Art. 444.

12. An arc $= \dfrac{\text{central angle}}{180^\circ} \times \pi R$. Art. 445.

13. In inscribed regular polygons,

$$S_{2n} = \sqrt{R\left(2R - \sqrt{4R^2 - S_n^2}\right)}.$$ Art. 467.

AREAS OF PLANE FIGURES

1. In a triangle, $K = \frac{1}{2}bh$. Art. 389.

2. In a triangle, $K = \sqrt{s(s-a)(s-b)(s-c)}$. Art. 393.

3. In an equilateral triangle, $K = \dfrac{a^2\sqrt{3}}{4}$. Ex. 4, p. 257.

4. In a parallelogram, $K = b \times h$. Art. 385.

5. In a trapezoid, $K = \frac{1}{2}h(b_1 + b_2)$. Art. 394.

6. In a regular polygon, $K = \frac{1}{2}r \times P$. Art. 446.

7. In a circle, $K = \pi R^2$ or $K = \frac{1}{4}\pi D^2$. Art. 449.

8. In a sector of a circle, $K = \frac{1}{2}R \times \text{arc}$, Art. 453.

or $K = \dfrac{\text{central } \angle}{360^\circ} \times \pi R^2$. Art. 453.

9. In a segment of a circle, $K = \text{sector} \pm \triangle$ formed by the chord and radii of the segment.

10. In a circular ring, $K = \pi(R^2 - R'^2)$. Art. 449.

11. In any two similar plane figures,

$K : K' = a^2 : a'^2$; Art. 399.

also $a : a' = \sqrt{K} : \sqrt{K'}$. Art. 314.

12. In two circles, $K : K' = R^2 : R'^2 = D^2 : D'^2 = C^2 : C'^2$; Art. 452.

also $R : R' = D : D' = C : C' = \sqrt{K} : \sqrt{K'}$. Art. 314.

FORMULAS OF SOLID GEOMETRY

SYMBOLS

B, b = areas of the lower and upper bases of a frustum.

E = lateral edge (or element); or = spherical excess.

H = altitude.

l, b, h = length, breadth, height.

L = slant height.

M = area of midsection.

P = perimeter of right section.

P, p = perimeters of lower and upper bases of a frustum.

r, r' = radii of bases.

S = area of lateral surface; or = area of surface of sphere, etc.

T = area of total surface.

V = volume.

FORMULAS FOR AREAS

1. In a prism, $S = E \times P$. Art. 608.

2. In a regular pyramid, $S = \frac{1}{2} L \times P$. Art. 641.

3. In a frustum of a regular pyramid, $S = \frac{1}{2}(P + p) L$. Art. 643.

4. In a cylinder of revolution, $S = 2\pi RH$. Art. 697.
   $T = 2\pi R(R + H)$. Art. 697.

5. In a cone of revolution, $S = \pi RL$. Art. 721.
   $T = \pi R(L + R)$. Art. 721.

6. In a frustum of a cone of revolution, $S = \pi L(R + r)$. Art. 727.

7. In a sphere, $S = 4\pi R^2$, or $S = \pi D^2$. Art. 810.

8. In a zone, $S = 2\pi RH$. Art. 813.

9. In a lune, $S = \frac{\pi R^2 A}{90}$. Art. 817.

10. In a spherical triangle, $S = \frac{\pi R^2 E}{180}$. Art. 822.

11. In a spherical polygon, $S = \frac{\pi R^2 E}{180}$ Art. 824.

FORMULAS FOR VOLUMES

1. In a prism, $V = B \times H$. Art. 628.

2. In a parallelopiped, $V = l \times b \times h$. Art. 626.

3. In a pyramid, $V = \frac{1}{3} B \times H$. Art. 651.

4. In a frustum of a pyramid, $V = \frac{1}{3} H (B + b + \sqrt{Bb})$. Art. 656.

5. In a prismatoid, $V = \frac{1}{6} H (B + b + 4M)$. Art. 663.

6. In a cylinder, $V = B \times H$. Art. 698.

7. In a cylinder of revolution, $V = \pi R^2 H$. Art. 699.

8. In a cone, $V = \frac{1}{3} B \times H$. Art. 722.

9. In a circular cone, $V = \frac{1}{3} \pi R^2 H$. Art. 723.

10. In a frustum of a cone, $V = \frac{1}{3} H (B + b + \sqrt{Bb})$. Art. 729.

11. In a frustum of a cone of revolution,
$$V = \frac{1}{3} \pi H (R^2 + r^2 + Rr).$$ Art. 730.

12. In a sphere, $V = \frac{4}{3} \pi R^3$, or $V = \frac{1}{6} \pi D^3$. Art. 832.

13. In a spherical sector, $V = \frac{2}{3} \pi R^2 H$. Art. 836.

14. In a spherical segment of two bases,
$$V = \frac{1}{2} (\pi r^2 + \pi r'^2) H + \frac{1}{6} \pi H^3.$$ Art. 837.

15. In a spherical segment of one base, $V = \pi H^2 (R - \frac{1}{3} H)$. Art. 838.

CONSTANTS

1 acre = 43,560 sq. ft.	$\sqrt[3]{2} = 1.2599 +$
1 bushel = 2150.42 cu. in.	$\sqrt[3]{3} = 1.4422 +$
1 gallon = 231 cu. in.	$\frac{1}{\pi} = .3183 +$
$\sqrt{2} = 1.4142 +$	$\sqrt{\pi} = 1.7725 -$
$\sqrt{3} = 1.7321 -$	$\frac{1}{\sqrt{\pi}} = 0.5642 +$

SUMMARY OF METRIC SYSTEM

TABLE FOR LENGTH

10 millimeters (mm.)	=1 centimeter (cm.)
10 cm.	=1 decimeter (dm.)
10 dm.	=1 meter (m.).
10 m.	=1 Dekameter (Dm.)
10 Dm.	=1 Hektometer (Hm.)
10 Hm.	=1 Kilometer (Km.)
10 Km.	=1 Myriameter (Mm.)

Similar tables are used for the unit of weight, the *gram*; for the unit of capacity, the *liter*; for the unit of land measure, the *are*; and for the unit of wood measure, the *stere*.

TABLE FOR SQUARE MEASURE

100 sq. mm. = 1 sq. cm.
100 sq. cm. = 1 sq. dm., etc.

TABLE FOR CUBIC MEASURE

1000 cu. mm. = 1 cu. cm.
1000 cu. cm. = 1 cu. dm., etc.

A *liter* = 1 cu. dm.
A *gram* = weight of 1 cu. cm. of water at 39.2° Fahrenheit.
An *are* = 100 sq. m.
A *stere* = 1 cu. m.

EQUIVALENTS

1 meter	=39.37 inches.
1 liter	=1.057 liquid quarts, or .9581 dry quarts.
1 kilogram	=2.2046 lbs. av.
1 hektare	=2.471 acres.
1 sq. m.	=1550– sq. in.

INDEX OF DEFINITIONS AND FORMULAS

www.ingramcontent.com/pod-product-compliance
Lightning Source LLC
LaVergne TN
LVHW011252110826
845149LV00001B/111
* 9 7 8 1 4 1 8 1 8 5 3 1 2 *